Crop Improvement with Artificial Intelligence

Crop Improvement with Artificial Intelligence

Methods and Applications

Edited by

Sadhana Singh
School of Sciences, U.P. Rajarshi Tandon Open University, Prayagraj, India

Santosh Kumar Upadhyay
Department of Botany, Panjab University, Chandigarh, India

WILEY

This edition first published 2026

Registered Offices
John Wiley & Sons, Inc., 111 River Street, Hoboken, NJ 07030, USA
John Wiley & Sons Ltd, The Atrium, Southern Gate, Chichester, West Sussex, PO22 8SQ, UK

For details of our global editorial offices, customer services, and more information about Wiley products visit us at www.wiley.com.

The manufacturer' s authorized representative according to the EU General Product Safety Regulation is Wiley-VCH GmbH, Boschstr. 12, 69469 Weinheim, Germany, e-mail: Product_Safety@wiley.com.

Wiley also publishes its books in a variety of electronic formats and by print-on-demand. Some content that appears in standard print versions of this book may not be available in other formats.

Library of Congress Cataloging-in-Publication Data

Names: Upadhyay, Santosh Kumar author | Singh, Sadhana, (Professor of plant biotechnology) author
Title: Crop improvement with artificial intelligence : methods and applications / [editors], Santosh Kumar Upadhyay, Sadhana Singh.
Description: First edition. | Hoboken, NJ : Wiley, 2026. | Includes index.
Identifiers: LCCN 2025031578 (print) | LCCN 2025031579 (ebook) | ISBN 9781394330454 cloth | ISBN 9781394330478 adobe pdf | ISBN 9781394330461 epub
Subjects: LCSH: Crop improvement | Artificial intelligence–Agricultural applications
Classification: LCC SB106.I47 C766 2026 (print) | LCC SB106.I47 (ebook)
LC record available at https://lccn.loc.gov/2025031578
LC ebook record available at https://lccn.loc.gov/2025031579

Cover Design: Wiley
Cover Image: TK

Set in 9.5/12.5pt STIXTwoText by Straive, Chennai, India

Contents

About the Editors

Dr. Sadhana Singh is working as Assistant Professor, School of Sciences, U.P. Rajarshi Tandon Open University (UPRTOU), Prayagraj, India. Dr. Singh has completed her PhD in Plant Biotechnology, from Bharathiar University (DRDO-BU Centre for Life Sciences), Coimbatore, (TN), India. She did her postdoctoral research from the International Crops Research Institute for the Semi-arid Tropics (ICRISAT), Patancheru, India. Her research interests include plant stress tolerance (model and non-model); molecular regulations controlling plant abiotic stresses; role of NAC genes toward abiotic stress tolerance; transcriptional profiling of stress responsive genes (*RAN, NAC, GPAT*); identification and characterization of candidate genes in semiarid legumes against drought; and role of NAC genes toward cold/salt/drought/response, biomass enhancement, carbon trapping and sequestration, etc., using functional genomics approach (genomics, transcriptomics, proteomics, and genome editing). She has published over 20 research/review articles in peer-reviewed international journals in addition to many book chapters (Wiley; Springer) on essential aspects of plant biotechnology and plant stress tolerance. She has been serving as a reviewer of many reputed journals, which include *Functional and Integrative Genomics Genes, 3 Biotech, Physiology and Molecular Biology of Plants, Plant Signaling and Behavior, Molecular Biology Reports, G3: Genes|Genomes|Genetics, Frontiers in Plant Science, Scientific Reports, Forests, Agriculture, Plant Biotechnology Reports, BMC Genomics, BMC Plant Biology*, and *International Journal of Molecular Sciences*. Additionally, she is also an editor of the *Frontiers in Genome Editing* (Review Editor), *Genomics of Plants and the Phytoecosystem* (Guest Associate Editor), and *Biotechnology and Bioinformatics* (Mediterranean Journals). She has been a recipient of several awards and honors that include a National Post Doc Fellowship, DST-SERB, India. She has also been invited for Webinar as a panelist and speaker for [IECPS 2024] SESSION F: Plant Molecular Biology and Plant Genetics, Genomics, and Biotechnology for the Title "CRISPR/Cas-Mediated Genome Editing in Legumes" on January 16, 2024 (MDPI). In short, she works as a vibrant, self-governing expert/scientist to connect her knowledge toward designing stress-smart, better-quality, and high-yielding varieties of crop plants.

Dr. Santosh Kumar Upadhyay has been working as an Assistant Professor at the Department of Botany, Panjab University, Chandigarh, India, for more than 10 years. After completing his doctoral work at the CSIR-National Botanical Research Institute, Lucknow, he served as DST-INSPIRE faculty at the National Agri-Food Biotechnology Institute, Department of Biotechnology, Government of India, Mohali, for one and a half years. Thereafter, he has been serving as an Assistant Professor at Panjab University. He is currently working in the area of functional genomics using various computational approaches. During his doctoral research, he was involved in the characterization of various insect-toxic proteins, such as lectins, chitin-binding proteins, etc., from plant biodiversity. Currently, his research group at Panjab University has characterized numerous important cation transporters including calcium cation antiporters, monovalent cation transporters, calcium ATPases, and important defense-related protein families such as receptor-like kinases, antioxidant enzymes, etc., of bread wheat. They are also characterizing long non-coding RNAs related to the abiotic and biotic stress response. He has also demonstrated the method of genome editing in bread wheat and potato using the CRISPR-Cas system for functional genomics studies and developed a tool, SSinder, for CRISPR target site prediction. He has authored more than 100 publications including research papers in leading journals of international repute, national and international patents, book chapters, and books. In recognition of his strong credentials and contributions, he has been awarded the NAAS Young Scientist Award (2017–2018) and NAAS-Associate (2018) from the National Academy of Agricultural Sciences, India; the INSA Medal for Young Scientist (2013) from the Indian National Science Academy, India; the NASI-Young Scientist Platinum Jubilee Award (2012) from the National Academy of Sciences, India; and the Altech Young Scientist Award (2011). He has also been the recipient of the prestigious DST-INSPIRE Faculty Fellowship (2012) and SERB-Early Career Research Award (2016) from the Ministry of Science and Technology, Government of India, and several research grants from various funding agencies like SERB, DBT, CSIR, etc. He also serves as a member of the editorial board and as a reviewer of many peer-reviewed international journals.

List of Contributors

Tanisha Anand
Amity University
Amity School of Engineering & Technology
Lucknow
Uttar Pradesh
India

Anukriti
Bhagwan Mahavir University
Bhagwan Mahavir College of Basic and Applied Sciences
Department of Biotechnology
Surat
Gujarat
India

Amaan Arif
Amity University Uttar Pradesh
Amity Institute of Biotechnology
Lucknow
India

Vedpriya Arya
Patanjali Herbal Research Division
Patanjali Research Foundation Trust
Uttarakhand
India

Acharya Balkrishna
Patanjali Herbal Research Division
Patanjali Research Foundation Trust
Uttarakhand
India

Akhouri Nishant Bhanu
Rajmata Vijayaraje Scindia Krishi Vishwa Vidyalaya
R.A.K. College of Agriculture
Department of Genetics and Plant Breeding
Sehore
Madhya Pradesh
India

Shalini Bhatt
Patanjali Research Foundation Trust
Patanjali Herbal Research Division
Uttarakhand
India

Arukshita Chandra
University of Delhi, Hindu College
Department of Botany
Delhi
India

Maliheh Eftekhari
University of Tehran
Division of Biotechnology, Department of Agronomy and Plant Breeding, College of Agricultural and Natural Resources
Karaj
Iran

Prekshi Garg
BioinfoCore Solutions (OPC) Pvt. Ltd.
Lucknow
India

Madeeha Gul
Deen Dayal Upadhyaya
Gorakhpur University
Institute of Agriculture and Natural Sciences
Centre of Genomics and Bioinformatics
Gorakhpur
Uttar Pradesh
India

Kapil Gupta
Siddharth University
Department of Biotechnology
Kapilvastu
Uttar Pradesh
India

Kiran Gupta
Siddharth University
Department of Botany
Kapilvastu
Uttar Pradesh
India

Shubhra Gupta
Siddharth University
Department of Biotechnology
Kapilvastu
Uttar Pradesh
India

Nadeem Iqbal
Chengdu University of Technology
State Key Laboratory of Geohazard Prevention and Geoenvironment Protection
Chengdu
Sichuan
PR China

and

Chengdu University of Technology
College of Ecology and Environment
Chengdu
PR China

Hafiz Muhammad Ansab Jamil
COMSATS University Islamabad
Department of Biotechnology
Abbottabad
Pakistan

Diya Kapadia
Gujarat Biotechnology University
Department of Plant Biotechnology
Gandhinagar
Gujarat
India

Ambrina Sardar Khan
Integral University
Department of Environmental Sciences
Lucknow
Uttar Pradesh
India

Dil Khurram
Chengdu University of Technology
State Key Laboratory of Geohazard Prevention and Geoenvironment Protection
Chengdu
Sichuan
PR China

and

Chengdu University of Technology
Key Laboratory of Ministry of Ecology and Environment for Synergetic Control and Joint Remediation of Water & Soil Pollution
Chengdu
PR China

and

Chengdu University of Technology
College of Ecology and Environment
Chengdu
Sichuan
P.R. China

and

Chengdu University of Technology
College of Ecology and Environment
State Environmental Protection Key
Laboratory of Synergetic Control and Joint
Remediation & Water Pollution
Chengdu
P.R. China

Kanti Kiran
Gujarat Biotechnology University
Department of Plant Biotechnology
Gandhinagar
Gujarat
India

Sachin Kumar
Amity University
Amity School of Engineering & Technology
Lucknow
Uttar Pradesh
India

Guo Liu
Chengdu University of Technology
State Key Laboratory of Geohazard
Prevention and Geoenvironment
Protection
Chengdu
Sichuan
PR China

and

Chengdu University of Technology
Key Laboratory of Ministry of Ecology and
Environment for Synergetic Control and
Joint Remediation of Water & Soil Pollution
Chengdu
PR China

and

Chengdu University of Technology
College of Ecology and Environment
Chengdu
PR China

Praveen Kumar Maddheshiya
Siddharth University
Department of Biotechnology
Kapilvastu
Uttar Pradesh
India

Moniba Zahid Mahmood
COMSATS University Islamabad
Department of Environmental Sciences
Abbottabad
Pakistan

Sam A. Masih
Sam Higginbottom University of
Agriculture Technology and Sciences
Department of Molecular and Cellular
Engineering
Prayagraj
Uttar Pradesh
India

Ann Maxton
Sam Higginbottom University of
Agriculture, Technology and Sciences
Genetics and Plant Breeding
Prayagraj
India

Sanjana Mishra
Amity University Uttar Pradesh
Amity Institute of Biotechnology
Lucknow
India

Sumita Mishra
Amity University
Amity School of Engineering & Technology
Lucknow
Uttar Pradesh

Vipin Bihari Mishra
Gujarat Biotechnology University
Department of Plant Biotechnology
Gandhinagar
Gujarat
India

Divya Mohanty
University of Delhi
Hindu College
Department of Botany
Delhi
India

Nupur Mondal
University of Delhi
Atma Ram Sanatan Dharma College
Department of Botany
Delhi
India

Papiya Mukherjee
Department of Botany
Panjab University
Chandigarh
India

Mohammad Reza Naghavi
University of Tehran
College of Agricultural and Natural Resources
Division of Biotechnology, Department of Agronomy and Plant Breeding
Karaj
Iran

Muhammad Nauman
Chengdu University of Technology
State Key Laboratory of Geohazard Prevention and Geoenvironment Protection
Chengdu
Sichuan
PR China

and

Chengdu University of Technology
College of Ecology and Environment
Chengdu
PR China

Rakshit Pathak
Patanjali Herbal Research Division
Patanjali Research Foundation Trust
Haridwar, Uttarakhand
India

Jayshree Pawar
Gujarat Biotechnology University
Department of Plant Biotechnology
Gandhinagar
Ahmedabad, Gujarat
India

Nikhil Raghuwanshi
Deen Dayal Upadhyaya
Gorakhpur University
Institute of Agriculture and Natural
Sciences
Department of Agronomy
Gorakhpur
Uttar Pradesh
India

Saurav Raj
Rajmata Vijayaraje Scindia Krishi Vishwa
Vidyalaya
R.A.K. College of Agriculture
Department of Genetics and Plant Breeding
Sehore
Madhya Pradesh
India

Riyazuddin Riyazuddin
Universidad Miguel Hernández
Instituto de Bioingeniería, Elche
Spain

Satyam Sanodiya
Rajmata Vijayaraje Scindia Krishi Vishwa
Vidyalaya
R.A.K. College of Agriculture
Department of Genetics and Plant Breeding
Sehore
Madhya Pradesh
India

Manan Shah
Pandit Deendayal Energy University
School of Energy Technology
Department of Chemical Engineering
Gandhinagar
Gujarat
India

Satwi Shah
Nirma University
Department of Computer Science and
Engineering
Gujarat
India

Charu Sharma
Panjab University
Department of Botany
Chandigarh
India

Shrishti
Department of Botany
Panjab University
Chandigarh
India

Kalpita Singh
HUN-REN Centre for Agricultural
Research
Agricultural Institute
Martonvásár
Hungary

Noopur Singh
Deen Dayal Upadhyaya
Gorakhpur University
Institute of Agriculture and Natural
Sciences
Centre of Genomics and Bioinformatics
Gorakhpur
Uttar Pradesh
India

and

Deen Dayal Upadhyaya Gorakhpur
University
Institute of Agriculture and Natural
Sciences
Department of Genetics and Plant Breeding
Gorakhpur
Uttar Pradesh
India

Ravindra Pratap Singh
Rajarshi Tandon Open University
School of Sciences (Biochemistry)
Prayagraj
Uttar Pradesh
India

Sadhana Singh
U.P. Rajarshi Tandon Open University
School of Sciences
Prayagraj
Uttar Pradesh
India

Sakshi Singh
University of Delhi
Hindu College
Department of Botany
Delhi
India

Siddharth Singh
Amity University Uttar Pradesh
Amity Institute of Biotechnology
Lucknow
India

Perumalla Srikanth
Mohan Babu University
School of Agriculture
Agricultural Microbiology
Tirupati
Andhra Pradesh
India

Prachi Srivastava
Amity University Uttar Pradesh
Amity Institute of Biotechnology
Lucknow
India

Prateek Srivastava
University of Allahabad
Department of Botany
Prayagraj
Uttar Pradesh
India

Rajesh K. Tiwari
Amity University
Amity Institute of Biotechnology
Lucknow
Uttar Pradesh
India

Mala Trivedi
Amity University
Amity Institute of Biotechnology
Lucknow
Uttar Pradesh
India

Bangar Vaibhav
Rajmata Vijayaraje Scindia Krishi Vishwa Vidyalaya
Gwalior
R.A.K. College of Agriculture
Department of Genetics and Plant Breeding
Sehore
Madhya Pradesh
India

Avani Vasudeva
Panjab University
Department of Botany
Chandigarh
India

Nancy Vora
Nirma University
Department of Computer Science and Engineering
Gujarat
India

Mohammad Yasin
Rajmata Vijayaraje Scindia Krishi Vishwa Vidyalaya Gwalior
R.A.K. College of Agriculture
Department of Genetics and Plant Breeding
Sehore
Madhya Pradesh
India

Preface

Crop Improvement with Artificial Intelligence (AI): Methods and Applications, offers a comprehensive overview of how AI is being integrated into crop development, highlighting the latest advancements and applications in this area. From the expansion of agriculture to the limitations and possibilities of manipulating AI for crop improvement, this book enfolds a broad range of topics important for grasping the innovative prospects of AI toward crop enhancement.

Commencing with an investigation on the documented factors and capacity of innovation in agriculture, readers are presented with the basics or core concepts of AI and its groundbreaking influence on progressive cultivation approaches. The book digs into several practicalities of AI in plant biology and plant breeding, from data retrieval and preprocessing methods to predictive analysis for crop yield and pathogen resistance. Ethical reflections and obstacles in AI-enabled crop improvement are also confronted, offering readers with a comprehensive insight of the field.

Researchers will gain a deeper understanding of AI-assisted pathogen and pest management, which will enhance crop cultivation practices and support the development of climate-resilient crops through AI techniques. Furthermore, the book scrutinizes the amalgamation of AI with the Internet of Things (IoT) along with precision agriculture methods, furnishing pragmatic illustrations and models of effective executions.

Moreover, the book addresses the challenges and opportunities of using AI for crop improvement, highlighting the importance of genetic diversity, resource optimization, and ethical considerations. It explores future directions and emerging trends in AI technologies, emphasizing the potential for interdisciplinary collaboration and the integration of AI with other emerging technologies. In addition to this, to create a balanced view of AI in agriculture, it is important to recognize that many AI approaches are still in the experimental stage and require extensive testing. Data limitations, economic viability, and unproven technologies present significant challenges. Additionally, ethical and social considerations, such as the impact on labor and the need for training, must be addressed for successful implementation.

Delving deeper, this book examines the up-to-date execution of the AI and machine learning technologies ranging from functional genomics of plant and phenotyping, to high-throughput imaging and genomic prediction in plants. It also explores how these models can be utilized to investigate and thus could be incorporated with the multi-omics information to determine complex traits and study plant transcriptomics, providing awareness into the future of AI to transform plant breeding and genetic advancement. This book will also cover large language models (LLMs) and generative AI approaches to understand the application of AI in agriculture. LLMs can enhance the analysis of agricultural data by interpreting complex datasets and generating actionable insights, thereby improving decision-making processes. Generative AI approaches, on the other hand, can facilitate the creation of innovative solutions and predictive models for optimizing crop management, pest control, and resource allocation. These technologies hold significant potential for advancing agricultural practices and achieving greater efficiency in the field.

In the past few years, all these findings have been published separately by various groups in numerous plant species including agro-economically important crop plants like rice, wheat, millet, maize, etc., and thus it is impossible to get all the details on an integrated platform due to the unavailability of any suitable book gathering all the necessary information. Therefore, compiling all the data into a single book will be highly valuable for various groups, ranging from graduate students to scientists and industrialists.

In conclusion, AI has immense potential in boosting crop productivity via quantitative description of several plant phenotypic characters, efficient and rapid disease diagnosis in plants, and efficient utilization of agrochemicals, while supporting growers with location-specific agronomic advice. With its exhaustive and thorough exploration and multidisciplinary approach, *Crop Improvement with Artificial Intelligence: Methods and Applications* is an essential asset for scholars, researchers, academicians, agronomists, policymakers, and anyone interested in capitalizing on AI to address the hurdles of worldwide food security and encourage sustainable agricultural implementation.

Sadhana Singh
School of Sciences,
U.P. Rajarshi Tandon Open University,
Prayagraj, India.

Santosh Kumar Upadhyay
Department of Botany,
Panjab University,
Chandigarh, India.

1

Introduction to Artificial Intelligence for Crop Improvement

Madeeha Gul[1], *Nikhil Raghuwanshi*[2], *and Noopur Singh*[1,3]

[1] *Centre of Genomics and Bioinformatics, Institute of Agriculture and Natural Sciences, Deen Daya lUpadhyaya Gorakhpur University, Civil Lines, Gorakhpur 273009, Uttar Pradesh, India*
[2] *Department of Agronomy, Institute of Agriculture and Natural Sciences, Deen Dayal Upadhyaya Gorakhpur University, Civil Lines, Gorakhpur 273009, Uttar Pradesh, India*
[3] *Department of Genetics and Plant Breeding, Institute of Agriculture and Natural Sciences, Deen Dayal Upadhyaya Gorakhpur University, Civil Lines, Gorakhpur 273009, Uttar Pradesh, India*

1.1 Introduction

The global agricultural sector faces escalating challenges such as food security, climate changes, and sustainable environmental practices. By 2050, the global population is expected to surpass nine billion, demanding a 70% increase in food production (Boora and Ahalawat 2024). Traditional methods of crop improvement are inadequate to address the complexity of these challenges (Ray et al. 2013). These methods tend to lack the accuracy and flexibility needed to satisfy the needs of modern agriculture, especially under environmental circumstances and emerging pests and disease conditions. Artificial intelligence (AI) advancements have become a game-changing tool in agriculture, offering innovative solutions to these persistent challenges through precision agriculture, multi-omics data integration, predictive analytics, and high-throughput phenotyping (HTP) technologies. These AI-driven approaches (Table 1.1) have demonstrated immense potential to accelerate crop breeding, increase resilience, and improve agricultural decision-making (Shakoor et al. 2017). Precision agriculture tracks and operates crop health, soil conditions, and weather patterns in real time using AI-driven tools, including sensors, drones, autonomous tractors, and machine learning (ML) algorithms. This leads to efficient resource management, minimizing environmental risks, and improved crop yields by increasing the accuracy by 30–50% (Hoque and Padhiary 2024). Using AI, farmers can enhance their crop health, maximize resource utilization, and boost crop yields. This helps them optimize crop management, predict environmental risks, and enhance overall productivity.

The AI-powered precision agriculture optimizes resources to reduce water consumption by up to 30% and minimize the need for pesticides and fertilizers, improving environmental sustainability (Gangwani 2024; Jeyalakshmi et al. 2024). Additionally, AI tools enhance crop quality by predicting phenotypic outcomes from large genotypic datasets, achieving high accuracy in yield production tailored to specific environments (Li et al. 2024).

Table 1.1 AI algorithms used in crop improvement.

AI algorithms	Functions	Crop examples	References
Convolutional neural networks (CNNs)	• Image analysis for crop monitoring • Detecting diseases and pests. • Assessing plant growth stages	Wheat, rice, and vegetables	Bhatt et al. (2017)
Random forest (RF)	• Yield prediction • Disease and pest detection and management	Corn, soybeans, and wheat	Dhillon et al. (2023)
Support vector machines (SVMs)	• Classifying types of crops • Disease detection • Pest management	Maize, legumes, and fruits	Kok et al. (2021)
Neural networks	• Detecting early signs of diseases in crops • Pattern recognition • Enhancing precision in disease management	Tomatoes, potatoes, and citrus	Moreno-Revelo et al. (2021)
Linear programming	• Optimizing logistics, storage, and transportation of crops • Enhancing supply chain efficiency • Reducing the wastage of agricultural operations	Fruits and vegetables	Mellaku et al. (2018)
Decision trees	• Providing data-driven farm management • Optimizing water and fertilizer usage • Supporting in decision-making	Rice, sugarcane, and cotton	Motamedi and Villányi (2024)
Best linear unbiased prediction (BLUP)	• Helping in genomic selection for breeding • Predicting genetic value • Enhancing crop resistance to diseases	Corn, soybean, millets, and wheat	Anuradha et al. (2022)
Reinforcement learning (RL)	• Optimizing robotic tasks in agricultural fields • Automating planting, weeding, and harvesting • Improving efficiency in agricultural operations	Tomatoes, potatoes, and lettuce	Chen et al. (2023b)
Autoregressive integrated moving average (ARIMA)	• Predicting weather conditions • Assisting in farm planning • Analyzing historical weather data	All major crops	Praveen and Sharma (2020)
Genetic algorithms (GAs)	• Optimizing water and fertilizer usage • Solving complex problems in farming • Enhancing agricultural sustainability	Rice, sugarcane, and cotton	Bi and Hu (2021)

This integration of AI into plant breeding accelerates the breeding process and leads to more resilient and productive crops. The AI adoption has surged a 50% increase in the agricultural sector over the past five years, expanding modern farming (Aggarwal et al. 2024). The role of AI is to address global challenges like food security, optimize farming practices, and ensure environmental stability through precision agriculture to enhance crop quality (Adewusi et al. 2024; Parashar et al. 2024). Even though AI has the potential to drastically change agriculture, many problems make it worse to use it on a large scale. To completely realize AI's potential in the agricultural sector, these social, economic, ethical, and technical issues must be addressed, underscoring the need for equitable access in the AI-powered agricultural revolution (Rane et al. 2024).

The integration of AI technology into agribusinesses is becoming more widely acknowledged as a critical tactic for addressing issues, such as market volatility, labor shortages, and environmental threats (Choudhary et al. 2024). This chapter explores different aspects of AI-driven agricultural innovations such as genomic selection (GS), digital twins, soil microbiome optimization, and ethical issues. It investigates how AI changes traditional farming systems and addresses socioeconomic matters to produce a resilient and equitable agricultural future.

1.2 AI-Driven Multi-Omics Data Integration

The use of AI in multi-omics data analysis has revolutionized the area of crop improvement. Multi-omics approaches refer to the integration of data from genomics, transcriptomics, proteomics, metabolomics, phenomics, and epigenomics to understand the biological traits in plants (Chaturvedi et al. 2024). AI plays an important role in interpreting large, complex multi-omics datasets, facilitating the identification of desirable traits in crops, and accelerating breeding programs. AI-driven technologies excel in handling complex and large multi-omics datasets. ML algorithms such as support vector machines (SVMs), random forests (RFs), and gradient boosting are commonly used by researchers to interpret complex datasets by identifying correlations and patterns that may not be apparent in traditional farming practices, leading to a holistic and nuanced understanding of plant biology. Hence, it provides complete insights into plant phenotypes and the ability to breed resilient and productive crops (Awasthi 2020; Cembrowska-Lech et al. 2023). For instance, AI can predict specific genetic variations that influence phenotypic traits in crops and interpret the interaction of environmental factors with genetic information that affects plant growth and development. These algorithms are crucial in uncovering genetic, epigenetic, and metabolic markers, addressing the "small n, large p" problems to expedite the breeding process (Al-Mamun et al. 2024).

The research for crop improvement has benefited from the advancements in AI. Deep learning (DL) models such as convolutional neural networks (CNNs) and recurrent neural networks (RNNs) capture nonlinear patterns and complex interactions within large multi-omics datasets. Deep Genotype-to-Phenotype (DeepG2P) is an AI model used in crop improvement to predict phenotypes, thereby enhancing crop production, increasing yield, reducing water and nutrient losses, and minimizing environmental threats. It integrates various types of data, such as genomics, environmental factors, weather, soil, and management fields, by modeling DNA sequences as natural language and predicting phenotypic outcomes based on these inputs provided by scientists (Sharma et al. 2022). Additionally, Soybean Deep Neural Genomic Prediction (SoyDNGP) is a DL AI tool that predicts complex traits in soybeans, including disease resistance and yield. This tool handles complex genetic interactions to provide accurate trait value predictions (Gao et al. 2023). Integrating pan-omics within system biology plays a pivotal role in understanding complex traits and regulatory networks, facilitating decision-making in the breeding process. This integration not only accelerates the development of crop yield and quality but also addresses global food security challenges (James et al. 2021; Zhang et al. 2022; Murmu et al. 2024).

TensorFlow and PyTorch are powerful DL frameworks that play a crucial role in genetic marker discovery linked to heat tolerance in wheat. Genetic markers pinpoint specific regions in the genome that help with heat resistance. They analyze complex multi-omics datasets under heat stress to identify heat-responsive genes in wheat, producing heat-tolerant wheat varieties (Sun et al. 2022). Utilizing a CNN model helps in analyzing complex datasets, including visual and environmental data, improving rice yield production. CNNs extract relevant features, such as leaf color, plant height, and texture, from images that indicate crop health. By combining image data with environmental data such as soil moisture, rainfall, and temperature, it provides accurate yield potential (Chang et al. 2024). AI models excel at extracting meaningful patterns and correlations that may be overlooked by traditional farming practices. These advancements reduce the time required for genetic marker discovery and trait prediction to produce resilient crop varieties capable of withstanding environmental stresses.

1.3 Genomic Selection and Breeding

The advent of AI has revolutionized GS and precision breeding. These advancements analyze complex datasets by identifying genetic markers linked with desirable traits with speed and accuracy. ML algorithms such as RFs, support vector machines (SVMs), gradient boosting machines, and neural networks analyze phenotypic and genotypic data to predict the performance of different types of genotypes based on their genetic makeup (Hamazaki and Iwata 2024). For example, GS models incorporate AI to enhance robust wheat varieties under various environmental conditions. DL algorithms analyze and capture complex genomic datasets, allowing breeders to select wheat varieties suited to specific environmental factors. ML algorithms predict the phenotypic traits, accelerating the development of new climate-resilient wheat varieties supporting food security (Bose et al. 2024). These technologies have gained a great deal of significance for detecting leaf diseases from captured images and preventing major crop losses (Jackulin and Murugavalli 2022).

1.3.1 CRISPR Target Prediction

Clustered Regularly Interspaced Short Palindromic Repeats (CRISPR) target prediction is a technology that pinpoints genes using genome editing with the help of guide RNAs (gRNAs) to direct Cas9 protein to specific regions on DNA, allowing researchers to make changes in crop plants (Kumar et al. 2024; Singh et al. 2024; Upadhyay 2024). To make sure CRISPR technology is working efficiently in predicting the best target sites, AI comes in, which advances the process faster and more accurately in identifying gRNA sequences. DeepFM-Crispr is a DL model that predicts the on-target efficiency and off-target effects of CRISPR-Cas technology (Bao and Liu 2024). These models analyze the genomic context and predict the most effective gRNAs for targeting genes linked with pest resistance in crop plants (Raya González et al. 2024). For instance, in Australia, researchers use the integration of AI and CRISPR technology, benefiting farmers to develop wheat varieties with a 10% increase in yield (Rasheed et al. 2021). In the United States, researchers have edited green mustards to reduce their pungency to make them appealing to consumers, benefiting farmers with new market strategies (Yang et al. 2021). In China, researchers use these technologies to create tomato varieties with increased sugar content to meet consumer preferences, offering high market value for farmers (Liu et al. 2021). These technologies benefit farmers by enhancing taste, yield, and market value, shaping the future of agriculture.

Maize (*Zea mays*), one of the most cultivated crops, faces challenges from pest and disease resistance, such as common rust, southern rust, maydis leaf blight, gray leaf spot, and turcicum leaf blight (Gülmez 2024). To combat these challenges, researchers use hybrid models that combine CNNs and SVMs to enhance CRISPR-based gene editing. The pattern recognition feature-support vector machine (PRF-SVM) model is a hybrid model combining PSPNet, ResNet50, and Fuzzy SVM components to identify diseases in maize leaves. High-quality images of healthy and affected maize leaves are captured and extracted, along with their pattern features such as color, shape, and texture. Moreover, advanced techniques such as the gray-level co-occurrence matrix (GLCM) and histogram of oriented gradients (HOG) are used to capture intricate details of the diseased areas. These features are then fed into an SVM classifier to draw optimal decisions based on different classes. This integration provides the difference between healthy and affected leaves of maize, ensuring sustainable crop improvement (Bachhal et al. 2024). Another example is the use of state-space attention, an advanced mechanism in DL algorithms to detect maize leaf disease patterns with high precision (Liu et al. 2024). DenseNet169, XceptionNet, MobileNet, and VGG16 models are used to detect potato plant leaves that focus on specific parts of an image, improving early detection of diseases (Kasana and Rathore 2024).

1.3.2 DeepVariant in Genome Mapping

DeepVariant is a DL-based variant caller, developed by Google Research that uses CNNs to call genetic variants from high-throughput sequencing data, such as next-generation DNA sequencing (Bose et al. 2024). This tool identifies single-nucleotide polymorphism (SNP) and indels (insertions and deletions) with high accuracy. It surpasses traditional genome mapping methods, making it a valuable tool in GS and breeding practice programs (Gul and Banday 2024). In rice genomics, DeepVariant produces variant detection, facilitating the accurate assembly of genomic sequences associated with desirable traits such as stress resistance and yield (Poplin et al. 2018). In the study of sorghum crops, researchers use DeepVariant to call SNPs and small indels, providing complete race diversity insights and its association with disease resistance and stress tolerance. This leads to the development of improved sorghum cultivars and contributes to efficient breeding practices (Ruperao et al. 2022). The integration of DeepVariant into plant genomics offers a great promise for future innovations in the agricultural sector, leading to improved breeding practices for enhanced crop resilience and productivity.

1.3.3 Advantages of AI in Crop Breeding

AI accelerates breeding programs by reducing the time consumption required for the identification of pattern traits and genetic marker discovery, surpassing traditional methods (Rai 2022). It helps in selecting desirable traits in crop plants by analyzing large, complex genetic and phenotypic data, leading to optimal crop varieties (Wang et al. 2023). It enhances cost efficiency to streamline the breeding process, reducing extensive field trials and manual monitoring, thereby lowering overall expenses for farmers (Mushtaq et al. 2024). AI-assisted breeding programs optimize breeding strategies and enhance decision-making processes, leading to the development of high-yield crop varieties (Khan et al. 2022). For instance, AI models have been used to predict and select the best genotypes to improve maize yields (Liu et al. 2024). Unlike the traditional farming practices that take decades to achieve significant gains, AI can achieve this within a few breeding cycles, benefiting farmers in crop improvement. AI models help in resource management by reducing water usage by 30% and fertilizer consumption by 25% to improve crop yields (Hoque and Padhiary 2024). AI-assisted breeding techniques ensure food security and stable incomes for farmers by providing high-yielding, pest-resistant, stress-tolerant, and reduced usage of pesticide and fertilizer crop varieties, enhancing agricultural sustainability.

1.4 High-Throughput Phenotyping and Imaging

The advent of HTP and imaging technologies using AI has revolutionized crop improvement. It enables accurate, efficient, and large-scale crop monitoring, enhancing the speed of phenotyping processes (Araus et al. 2018). The traditional phenotyping methods were labor-intensive and inconsistent and have been supplanted by AI-powered systems, enabling rapid, precise, and standard data collection, leading to reliable results for sustainable farming practices (Gatkal et al. 2024).

1.4.1 Imaging Tools for Efficient Crop Monitoring

AI imaging tools such as drones, robots, satellites, and machine vision systems are equipped with advanced sensors and cameras, providing real-time monitoring of crops and high-resolution images from large agricultural fields. It detects minute variations in crops and identifies early signs of disease or stress (Kundu et al. 2024). Drones (unmanned aerial vehicles) are equipped with multispectral sensors that capture aerial high-resolution images of the agricultural field, providing a comprehensive view of crop health and their growth patterns. They collect data based on soil moisture, crop health, and pest infestation, which is then analyzed using ML algorithms. For instance, FlyPix AI uses drones and satellites, providing real-time monitoring of crop data, allowing farmers to make decisions

(Das Choudhury et al. 2019). Drones such as the DJI Phantom 4 Multispectral and DJI Mavic 2 Enterprise Advanced with thermal imaging are used for crop monitoring and help detect water stress and irrigation efficiency (Tsvetkov 2023). Similarly, robots collect data at ground level, such as planting, harvesting, and irrigation, by navigating through agricultural fields (Kumari et al. 2024). It carries sensors such as LiDAR, RGB cameras, and fluorescence to measure plant height, biomass, area, and leaf angle, showing early signs of diseases and facilitating real-time monitoring of crops (Suwaied Al Bakri et al. 2024). While satellites monitor large coverage areas and data integration, which is crucial for precision agriculture and plant growth (Johri et al. 2023). For instance, Sentinel-2 and Landsat-8 are satellites that provide high-resolution optical imagery, are useful for detecting weeds, and provide data on water and nitrogen stress (Mancini et al. 2019). Machine vision systems utilize AI algorithms to extract meaningful information on plant phenotypes such as plant height, stem thickness, growth rate, leaf area, and stress levels by analyzing images captured by drones and robots using ML and DL algorithms (Maraveas 2024; Tan et al. 2024). These algorithms identify water stress and nutrient deficiencies, allowing for real-time monitoring of plant health and its growth (Islam et al. 2024; Sharma et al. 2024). For instance, AI-powered tools can detect changes in leaf color and texture, indicating pests, diseases, or environmental stress (Naik et al. 2017). PEAT's Plantix app identifies plant diseases, which benefits farmers by increasing crop yields, reducing losses, and optimizing resource management (Kumar et al. 2024). This automated analysis allows for real-time monitoring of crops and early detection of pests or diseases, enabling timely interventions by reducing labor costs and time spent on manual assessments on crops, helping farmers with sustainable agricultural practices (Thakur et al. 2023).

1.4.2 Transition from Traditional to AI-Driven Phenotyping

Traditional phenotyping methods rely on manual measurements and observations, which are labor-intensive and time-consuming, leading to inconsistencies and errors (Bian et al. 2022; Chang-Brahim et al. 2024). In contrast, AI-driven phenotyping systems have drastically reduced evaluation times from weeks to days, providing accurate measurement at macro- and microscales in minimal time. For instance, the CropQuant-Air system uses drones to capture high-quality images of wheat canopies and utilizes AI algorithms to analyze traits such as spike number and yield performance of wheat crops (Chen et al. 2023a). Portable edge computing systems using various sensors detect and analyze wheat yellow rust infections in real time, improving disease management strategies to enhance food security (Khalid 2024). However, traditional methods require significant labor and time, resulting in incomplete results. On the other hand, AI-driven phenotyping can process vast amounts of data quickly, providing reliable insights into crop improvement performance and enabling more efficient breeding program practices (Mostafa et al. 2023). AI phenotyping involves the collection of data by capturing images and data from agricultural fields using drones and robots equipped with sensors and cameras. AI extracts these collected data on the basis of plant height, leaf area, canopy temperature, and stress indicators such as drought or nutrient deficiency traits. Furthermore, ML models interpret these data, identifying patterns and accurate coefficient correlations ($R^2 > 0.7156$), surpassing traditional phenotyping methods (Ren et al. 2022). Recent studies have reported accuracy rates of over 90% in trait detecting and quantifying plant crops, ensuring reliable data for researchers and farmers, leading to better crop management decisions and improved yields (Maraveas 2024).

1.5 Generative AI in Plant Breeding

Generative AI is a branch of AI designed to create new data from existing data. It comprises generative adversarial networks (GANs), variational autoencoders (VAEs), and generative pre-trained transformer (GPT) (Sengar et al. 2024). Generative models, particularly GANs, are used to enhance various agricultural processes for sustainable farming practices. These technologies analyze complex genetic and phenotypic data, optimizing breeding strategies and reducing resource demands. Traditional breeding methods are time-consuming and labor-intensive; by integrating generative AI models offers innovative approaches to stimulate genetic outcomes, predict phenotypic traits, and reduce breeding cycles (Rai 2022). GenoDrawing tool leverages generative AI to predict complex plant traits based on their genetic information by analyzing genetic markers. This tool can predict complex traits that are difficult to analyze with traditional breeding methods, allowing researchers to quickly predict complex traits, leading to improved crop varieties (Jurado-Ruiz et al. 2023). Conditional generative adversarial networks (CGANs) generate additional synthetic data conditioned on specific traits, predicting and improving desirable traits in crop plants (Drees et al. 2021). For instance, researchers analyze vast datasets to identify and select wheat varieties with improved nutritional quality, particularly higher protein (essential amino acids) content in wheat. It not only accelerates the breeding process but also enhances the nutritional value of wheat (Mushtaq et al. 2024).

GAN were developed by Ian Goodfellow in 2014, consists of a generator that creates new synthetic data that mimics the real data and a discriminator that evaluates them by predicting outcomes (Goodfellow et al. 2014). The synthetic data improves crop recommendation (CR) models by selecting the best crop varieties for specific climatic conditions, which enhances the efficiency and accuracy of crop breeding strategies (Dos Santos and Vasconcelos 2024). GAN stimulates the outcomes of genetic crosses by analyzing the datasets of genotypes and phenotypes. It predicts the best parent plants for crossing, reducing the time needed for traditional breeding methods that would result in obtaining disease-resistant crop varieties (Patil 2023). For instance, the SoyDNGP tool utilizes generative AI to stimulate genetic crosses by predicting disease-resistant genes in soybean crops. It identifies the genetic markers linked with disease resistance and predicts combinations of genes that confer resistance to specific pathogens, such as soybean cyst nematode (SCN) (Feng et al. 2024). GAN helps in designing disease-resistant crop varieties by identifying disease symptoms that can be missed by the human eye. It also generates the genetic blueprints for crops with enhanced resistance against diseases, accelerating the breeding methods (Farooq et al. 2024). For instance, GAN helps in detecting cassava leaf diseases by generating synthetic images from diseased leaves, which balances the datasets by achieving 99.51% accuracy in classifying healthy and diseased leaves of cassava crops (Lokesh et al. 2024). Moreover, the marker-assisted breeding and GS using ML models identify genetic markers that are associated with late blight and potato cyst nematodes in potatoes to develop blight-resistant potato varieties (Sood et al. 2023).

Hybrid vigor, or heterosis, is an important aspect of crop improvement in which two different parent genes develop offspring with enhanced performance. GAN analyzes the genetic combinations and predicts hybrid performance, helping breeders in selecting higher-yielding crop traits, such as plant height, fruit weight, and number of fruits per cluster (Rasheed et al. 2018). For instance, GAN helps in detecting maize diseases with high accuracy using attention mechanisms that focus on significant parts in the captured images, improving the accuracy of disease detection. It generates training samples that address the challenges by achieving high accuracy and mean average precision (mAP) in detecting maize diseases to predict hybrid vigor and reduce trial-and-error methods (Song et al. 2023). Generative AI stimulates and predicts outcomes, reducing breeding cycle duration by helping plant breeders to produce efficient, precise, and resource-effective crops. It reduces the need for trial-and-error approaches to expedite trait selection and the resulting disease-resistant, high-yielding crop varieties, contributing to global food security (Rai 2022).

1.6 Advanced Pest and Disease Management

AI utilizes ML and DL methods, making significant strides in pest and disease management. It enables precise identification of pests or diseases across different crop varieties and timely intervention, which is important for mitigating crop damage (Rahman and Ravi 2022). Plant diseases and pests cause significant agricultural losses to farmers, threatening global food security. Earlier, traditional crop inspection was carried out by experts in the field, resulting in incorrect decisions. The advent of AI has brought transformative solutions to address complex problems in agriculture. These technologies accurately detect early signs of diseases, crop stress, and pest infestations by leveraging sensors, cloud computing, and high-quality datasets, enhancing the productivity of agricultural farming practices (Yang et al. 2024). PlantVillage is a powerful tool introduced by Penn State University to detect diseases in cassava plants. Farmers use their smartphones to capture cassava leaf images, boasting an accuracy of 95% in detecting diseases in cassava leaves. DL models analyze and detect symptoms of cassava mosaic disease (CMD) and cassava brown streak disease (CBSD), empowering farmers to mitigate crop losses (Mrisho et al. 2020). However, AI enhances pest and disease management in agriculture through image-based detection, predictive analytics, and Internet of Things (IoT) integration, enabling early detection of crop diseases, forecasting of pest outbreaks, and real-time monitoring, promoting sustainable farming practices and crop health.

1.6.1 Image-Based Detection

AI-driven image-based detection has emerged as a transformative tool in pest and disease management by enabling early detection of crop stress, diseases, and pest infestations in modern agriculture. DL, particularly CNNs, plays a pivotal role in analyzing high-quality crop images captured through drones or robots to detect early signs of infections or infestations in crops. CNNs can achieve accuracy rates of up to 95.71% in identifying crop diseases, as they process images of stems, leaves, fruits, and soil to identify the presence of pests or diseases, allowing for timely interventions (Dhyani et al. 2024). For instance, AI-driven image-based detection helps in detecting diseases in cotton leaves from captured images using drones or robots for further analysis using AI models. CNN recognizes patterns associated with diseases and classifies the images into healthy and infected cotton leaves (Shahid et al. 2024). The traditional approaches to pest and disease detection are labor-intensive and prone to human error. In contrast, AI-driven image-based detection analyzes thousands of images in a matter of seconds, allowing for early detection of diseases, preventing the spread of infestations, and mitigating crop damages (Paul et al. 2024). Despite the promising potential of AI, the variability of environmental conditions affects the accuracy of image-based detection approaches and requires a sophisticated AI algorithm to account for these variations, such as shadows, lighting, and background noises (Liu and Wang 2021). Additionally, the availability of labeled datasets is crucial for improving the quality of training datasets, ensuring crop health and global food security.

1.6.2 Predictive Analytics for Pest Outbreaks

Predictive analytics utilizes AI to forecast pest outbreaks by analyzing weather patterns, soil conditions, and historical data. It analyzes vast datasets and predicts pest outbreaks before they occur, allowing farmers to take proactive measures and minimize crop losses (Gul and Banday 2024). For instance, DL models use environmental data such as air temperature, carbon dioxide concentration, and humidity to predict the risks of crop pests and diseases (Lee and Yun 2023). Additionally, integrating RNNs with weather data such as temperature and humidity has significantly enhanced the prediction accuracy by recognizing patterns in sequence data. AI models analyze large datasets to forecast when and where pests might strike, allowing farmers to take proactive measures rather than reactive ones (Maheswari et al. 2024). Researchers use predictive analytics along with ML algorithms such as linear regression, RFs, SVMs, and artificial neural networks to forecast rice yields and enable early disease detection in Malaysia by analyzing weather data such as temperature, rainfall, soil, and historical data, providing accurate predictions and decision-making by farmers (Marong et al. 2024). These practices lead to better economic outcomes and contribute to sustainable agriculture.

1.6.3 IoT Integration for Real-Time Monitoring

The integration of AI with IoT provides real-time monitoring of crop health. IoT uses interconnected devices that collect and transmit data through the Internet. In agriculture, IoT devices such as drones, robots, and soil moisture sensors are used to gather critical data, providing real-time monitoring of nutrients in soil and optimizing fertilizer usage (Zhang et al. 2020). By continuously monitoring the crops in real time, it detects early diseases, minimizing the risk of widespread infestations and providing precise recommendations for pest control and nutrient management strategies. The data collected by IoT devices utilizes a CNN architecture, specifically GoogleNet, to analyze thousands of images, enabling the early detection of pest outbreaks by identifying changes in plant leaves using mobile applications (Yulita et al. 2023). AI models analyze weather and soil data generated by IoT devices to forecast the risk of fungal infections, allowing farmers to take proactive measures in crops before a disease spreads (Liu and Wang 2021). The integration of IoT devices and

AI algorithms facilitates irrigation schedules based on real-time soil moisture data from IoT devices, reducing wastage of water and ensuring adequate amounts of water in crops (Patel et al. 2023). For instance, smart farming leverages IoT sensors for data collection and AI algorithms for decision-making, resulting in a 25% increase in crop yield and a 30% reduction in pesticide usage (Chen et al. 2022). Additionally, in vineyards, IoT devices equipped with AI algorithms detect the risk of fungal infections such as powdery mildew. Sensors collect data on the basis of humidity, temperature, and leaf wetness. These AI algorithms analyze these data to predict mildew development. Farmers receive alerts on their smartphones to take proactive measures such as adjusting irrigations and applying fungicides only when required (Delfani et al. 2024; Wang et al. 2024b). By leveraging IoT technology, farmers achieve more efficient and sustainable pest and disease management, enhancing crop productivity, resource optimization, reducing chemical use, and reducing environmental impact.

1.7 Climate-Smart Agriculture Using AI

Climate-smart agriculture is revolutionizingagriculture by providing tools amidst challenges posed by environmental change. By integrating IoT and AI technologies, researchers can analyze vast datasets such as soil conditions, weather patterns, and crop health, enhance resource management, and mitigate the impacts caused by climate variations. AI technologies enhance the efficient use of water, fertilizers, and soil, crucial for adapting to climate change. ML algorithms such as RF and SVMs estimate evopotranspiration (E_{To}) using IoT sensor data, ensuring accurate water delivery and conserving freshwater resources. The real-time monitoring of data from IoT sensors with AI-driven models provides early warnings about droughts, floods, and heat waves, enabling farmers to take preemptive measures to safeguard their crops (Lewis et al. 2024; Nawaz and Babar 2024). Climate-smart farming with AI-driven algorithms plays a pivotal role in enhancing crop productivity and reducing greenhouse gas emissions. By adopting climate-smart farming practices, farmers can ensure global food security goals. For instance, in Ethiopia, the heat-resistant wheat varieties such as Hidase, Digelu, Dandeha, and Kubsa contribute to improved global food security and crop resilience (Hailu 2022).

AI models help in predicting how crops respond to different climate conditions by analyzing historical data along with current data to stimulate crop growth, yields, and stress tolerance under different climate scenarios (Hadeed et al. 2024). For instance, predictive models such as XGBoost for legume crops have shown potential in forecasting yields under various climatic conditions, anticipating agricultural outcomes (Na and Na 2024). AI recommends climate-resilient crop varieties tailored to specific areas by analyzing nutrient composition in soil and weather patterns that can withstand climate challenges. For instance, the drought-tolerant maize varieties and heat-tolerant wheat varieties are identified to enhance resilience against climate change (Hadeed et al. 2024). CropX is an advanced farm management system platform combining ML algorithms with soil sensors, offering disease control, precise irrigation systems, water efficiency, and nutrient management, providing real-time insights to help farmers adapt to climate challenges. Additionally, Plantix is a diagnostic tool that utilizes AI algorithms to identify pests, plant diseases, and nutrient deficiencies, helping farmers reduce crop losses and manual monitoring (Golubin and Kopnin 2024). AI-driven climate adaptation has shown success in Africa but faces challenges such as inadequate water distribution systems, ethical concerns, data inadequacy, and a lack of localized AI technologies. To address these challenges, UN Global Pulse and the African Regional Data Cube have taken initiatives by leveraging satellites and geospatial technologies, developing AI solutions tailored to Africa's unique climate conditions, and fostering agricultural resilience and sustainability (Rutenberg et al. 2021).

1.8 AI-Powered Digital Twins for Agriculture

In agriculture, digital twins are virtual representations of physical systems such as farming fields, crops, and livestock. They stimulate crop-yielding processes, enhancing crop production and optimizing farming practices through real-time data collection from various sources such as weather, soil conditions, and crop health status, benefiting farmers by continuous monitoring and improving decision-making in irrigation, fertilization, and planting selection (Skobelev et al. 2024). Digital twins help farmers with detailed insights into their operations by providing real-time monitoring of crops, reducing costs and risks, predictive analysis, and decision-making, which enhance efficiency, productivity, and sustainability (Peladarinos et al. 2023). Data is collected through sensors, satellite imagery, farm records, weather stations, and IoT devices, which monitor environmental conditions and crop health status. This data is processed using AI and ML algorithms that simulate yield forecasts and growth patterns. Farmers receive real-time alerts on corrective actions and recommendations (Skobelev et al. 2024). For instance, digital twins are used to predict wheat yields using a multi-agent system that provides detailed information on wheat crops. By integrating data from sensors, drones, robots, weather forecasts, and historical data into digital twin models, employing AI and ML algorithms along with Unmanned Aerial Vehicle RGB imagery for real-time wheat yield predictions, accuracy rates of up to 91.69% can be achieved, helping farmers optimize yield outcomes (Ma et al. 2023; Rajeswari et al. 2024). By leveraging digital twins in large-scale farming, farmers can achieve by enabling precision farming techniques with higher efficiency, productivity, and sustainability (Basa 2024; Mohammed et al. 2024).

1.9 AI for Soil Microbiome Optimization

Soil health is important for crop productivity and sustainability as it directly influences the availability of nutrients, pollutant degradation, rhizosphere immunity, water retention, and disease resistance (Zhang et al. 2021). Healthy soil facilitates nutrient cycling, organic matter decomposition, and disease suppression and improves soil structure. The soil microbiome is a key player in maintaining soil health, plant growth, and stress tolerance (Tiwari et al. 2021). AI technologies that utilize ML and DL techniques have revolutionized by identifying beneficial microbes in the soil (Zhang et al. 2021). Traditional methods are time-consuming and limited in scope for soil analysis. In contrast, AI algorithms can process vast amounts of data from soil samples by identifying patterns and correlations in less time. AI algorithms facilitate DNA sequencing of data to identify different microbial species, pathogenic or nonpathogenic strains, in less time. This rapid identification allows a comprehensive understanding of the soil microbiome and interaction with the host, supporting plant health (Rangwala et al. 2014; Wang et al. 2024a). AI-based microbiome analysis enhances nutrient uptake and plant growth as soil samples are collected from various locations and undergo high-throughput DNA sequencing, generating genetic data that is analyzed by AI and ML algorithms to detect patterns and correlations between microbial communities and soil health. AI models predict relationships between microbes and nutrient uptake, helping in understanding which microbial community is beneficial for plant growth and resilience (Awais et al. 2023). For instance, the use of plant growth-promoting rhizobacteria (PGPR) as biofertilizers promotes sustainable crop production by improving nutrient uptake such as nitrogen, phosphorus, and potassium, reducing the need for chemical fertilizers (Alzate Zuluaga et al. 2024). The integration of AI with omics technologies facilitates a comprehensive analysis of plant microbiomes and identifies beneficial microbial strains and their functions, leading to improved crop yields (Beattie et al. 2024). Predictive models optimize the soil microbiome for rice crops, as rice crops suffer from yield plateaus due to poor soil health. These models utilize AI to analyze data from soil samples, weather patterns, and crop performance, promoting improved rice yield. Rice grains require beneficial bacteria during specific growth stages to improve rice productivity. It helps farmers implement targeted microbes that enhance soil health and boost rice productivity (Chen et al. 2024). The integration of AI into rice farming can increase productivity by 15% through soil microbiome management (Ritambara et al. 2024). The use of mycorrhizal fungi and nitrogen-fixing bacteria augments soil fertility and plant growth, promoting sustainable crop production (Douglas 2022). The integration of AI into soil microbiome analysis marks a significant advancement in crop improvement by identifying beneficial microbes and fostering agricultural practices.

1.10 Ethical and Socioeconomic Aspects of AI in Agriculture

The integration of AI into the agriculture sector holds significant promise for enhancing crop productivity and sustainability. However, it possesses ethical and socioeconomic challenges that need to be addressed for equitable benefits:

1) **Cost of AI Tools:** AI technologies can be too expensive initially, particularly for smallholder farmers. The AI adoption may be hampered by costs associated with hardware, software, equipment, and maintenance. This high cost can be a significant barrier to AI adoption in agriculture (Dara et al. 2022).
2) **Data Privacy Concerns:** Privacy and ownership concerns are brought up by AI systems' acquisition and use of farm data. Farmers are concerned about the privacy and security of their data. They may distrust AI solutions because they are concerned about who might access their data and how it will be used. Gaining farmers' trust requires addressing ownership rights and ensuring data security (Dara et al. 2022).
3) **Inclusivity for Smallholder Farmers:** Smallholder farmers face a multitude of interconnected challenges, including inadequate infrastructure, limited access to technical

expertise, restricted Internet connectivity, and financial constraints. Smallholders may be disadvantaged as a result of this digital divide, which can worsen the already-existing disparities. In order to prevent the gap between large- and small-scale farming practices from growing, AI applications must take into account the difficulties faced by smallholder farmers (Cafer and Rikoon 2018).

Governments and organizations subsidize AI programs, encouraging wider AI adoption by alleviating the financial burden on smallholder farmers. Incorporating AI into government-funded insurance plans encourages the adoption of new technologies (Osorio et al. 2024). The India AI mission aims to democratize AI and provide funding and resources to AI startups and innovators. Developing open-source AI platforms reduces the cost of AI tools by providing free and customizable solutions, helping farmers to access advanced tools (Talaviya et al. 2020). Providing workshops and hands-on training programs helps farmers' bridge the knowledge gap. Governments offer financial assistance to farmers so that they can invest in AI technologies. By addressing these challenges, AI tools become more accessible for all farmers, leading to more sustainable and productive agricultural practices.

1.11 Challenges, Limitations, and Future Directions

The integration of AI into crop improvement faces several challenges and limitations while paving the way for future advancements in the agricultural sector.

1.11.1 Challenges

1) **High Initial Costs:** The AI adoption in agriculture involves expenses for hardware, software, maintenance, and infrastructure. These technologies can deter farmers, especially in developing countries. This can be a financial barrier for smallholder and resource-poor farmers (Dara et al. 2022).
2) **Data Privacy and Ownership:** Farmers hesitate to adopt AI solutions due to privacy and ownership concerns of their data. They worry about how their data is stored, shared, and utilized (Dara et al. 2022).
3) **Digital Divide:** The gap between those who can access digital technologies and those who cannot is known as Digital Divide. In rural areas, smallholder farmers lack technical knowledge, Internet connection, and infrastructure (Cafer and Rikoon 2018).
4) **Technical and Data Quality Issues:** In agriculture, AI applications rely on high-quality datasets. However, a lack of labeled datasets and environmental data reduces the accuracy and reliability of AI models, impacting decision-making (Bose et al. 2024).
5) **Environmental and Ethical Concerns:** In agriculture, overuse of some resources and ecological imbalances result in overreliance on AI for automation. Ethical issues such as skewed algorithms that can penalize particular groups could have a negative impact on the environment (Rutenberg et al. 2021).

1.11.2 Limitations

1) **Algorithmic Bias and Generalization:** AI algorithms struggle to generalize agricultural systems, climates, and crop types. Bias in data collection methods can limit their applicability, making them less effective, and reduce their accuracy and reliability (Liu and Wang 2021).
2) **Limited Integration of AI into Traditional Practices:** The lack of integration between traditional farming practices and AI-based tools creates resistance among farmers, who often rely on indigenous knowledge. It hinders the widespread AI adoption in agriculture (Bose et al. 2024).
3) **Dependency on Interconnected Systems:** AI algorithms rely on IoT devices and cloud computing, making them vulnerable to connectivity issues, cyber threats, and system outages, disrupting operations and reliability (Patel et al. 2023).
4) **Scalability Issues:** AI models developed in controlled environments encounter challenges in real-world farming practices due to pests, diseases, and climate variability that affect the performance of AI (Rutenberg et al. 2021).

1.11.3 Future Directions

1) **Equitable Access and Affordability:** Governments and private sectors must collaborate and subsidize AI technologies and promote open-source platforms to make these tools accessible to smallholder farmers (Osorio et al. 2024).
2) **Improved Data Collection and Sharing:** Developing a standardized data-sharing architecture can address concerns about privacy, ensuring AI models are trained and rely on high-quality datasets (Dara et al. 2022).
3) **Integrating AI with Traditional Knowledge:** Future AI systems combine with traditional farming practices, creating hybrid models that are culturally acceptable and sustainable, thereby encouraging farmers to adopt AI tools (Bose et al. 2024).
4) **Sustainability and Environmental Focus:** AI tools should improve resource efficiency, reduce greenhouse gas emissions, and enhance soil health, providing eco-friendly climate change (Nawaz and Babar 2024).
5) **Ethical AI Development:** Addressing algorithmic bias and transparency and implementing regulations for AI ethics will be important to build trust and maximize the societal benefits of AI in the agricultural sector (Dara et al. 2022).

1.12 Conclusion

The integration of AI into crop improvement has marked a significant leap forward in agricultural practices, offering multifaceted challenges such as global food security, environmental change, and sustainability. AI technologies such as GS, multi-omics data analysis, HTP, generative AI models, soil microbiome optimization, and predictive modeling have revolutionized the way crops are bred, monitored, and managed. These advancements enable the development of resilient, high-yielding, and sustainable crop varieties. However, challenges such as high costs, data privacy concerns, and technology disparity exist for smallholder farmers in developing regions. To address these challenges, governments subsidize AI technologies and promote open-source platforms to make these AI tools accessible to smallholder farmers. The AI technology evolves, AI becomes the cornerstone of a future thriving in sustainability, ensuring global food security.

References

Adewusi, A.O., Asuzu, O.F., Olorunsogo, T. et al. (2024). AI in precision agriculture: a review of technologies for sustainable farming practices. *World Journal of Advanced Research and Reviews* 1: 2276–2285. https://doi.org/10.30574/wjarr.2024.21.1.0314.

Aggarwal, S., Bansal, S., and Goel, R. (2024). AI in agriculture:a looming challenge, a gleaming opportunity. *International Journal of Engineering Science and Humanities* 14 (Special Issue 1): 43–52. https://doi.org/10.62904/3jnm5k59.

Al-Mamun, H.A., Danilevicz, M.F., Marsh, J.I. et al. (2024). Exploring genomic feature selection: a comparative analysis of GWAS and machine learning algorithms in a large-scale soybean dataset. *The Plant Genome* 18 (1): e20503. https://doi.org/10.1002/tpg2.20503.

AlzateZuluaga, M.Y., Fattorini, R., Cesco, S., and Pii, Y. (2024). Plant-microbe interactions in the rhizosphere for smarter and more sustainable crop fertilization: the case of PGPR-based biofertilizers. *Frontiers in Microbiology* 15: 1440978. https://doi.org/10.3389/fmicb.2024.1440978.

Anuradha, N., Patro, T.S.S.K., Singamsetti, A. et al. (2022). Comparative study of AMMI- and BLUP-based simultaneous selection for grain yield and stability of finger millet [*Eleusinecoracana* (L.) Gaertn.] genotypes. *Frontiers in Plant Science* 12: 786839. https://doi.org/10.3389/fpls.2021.786839.

Araus, J.L., Kefauver, S.C., Zaman-Allah, M. et al. (2018). Translating high-throughput phenotyping into genetic gain. *Trends in Plant Science* 23 (5): 451–466. https://doi.org/10.1016/j.tplants.2018.02.001.

Awais, M., Naqvi, S.M.Z.A., Zhang, H. et al. (2023). AI and machine learning for soil analysis: an assessment of sustainable agricultural practices. *Bioresources and Bioprocessing* 10 (1): 90. https://doi.org/10.1186/s40643-023-00710-y.

Awasthi, Y. (2020). Press "A" for artificial intelligence in agriculture: areview. *International Journal on Informatics Visualization* 4 (3): 112–116. https://doi.org/10.30630/joiv.4.3.387.

Bachhal, P., Kukreja, V., Ahuja, S. et al. (2024). Maize leaf disease recognition using PRF-SVM integration: a breakthrough technique. *Scientific Reports* 14 (1): 10219. https://doi.org/10.1038/s41598-024-60506-8.

Bao, C., and Liu, F. (2024). Prediction of CRISPR on-target effects via deep learning. *Proceedings of the 1st Workshop on NLP for Science (NLP4Science)*, Miami, FL,USA. Association for Computational Linguistics. 9–15. November 16, 2024. https://doi.org/10.18653/v1/2024.nlp4science-1.2.

Basa, R. (2024). AI in agriculture: revolutionizing precision farming and sustainable crop management. *International Journal of Scientific Research in Computer Science, Engineering and Information Technology* 10 (5): 535–543. https://doi.org/10.32628/CSEIT241051040.

Beattie, G.A., Bayliss, K.L., Jacobson, D.A. et al. (2024). From microbes to microbiomes: applications for plant health and sustainable agriculture. *Phytopathology®* 114 (8): 1742–1752. https://doi.org/10.1094/PHYTO-02-24-0054-KC.

Bhatt, P., Sarangi, S., and Pappula, S. (2017). Comparison of CNN models for application in crop health assessment with participatory sensing. *IEEE Global Humanitarian Technology Conference (GHTC)* 2017: 1–7. https://doi.org/10.1109/GHTC.2017.8239295.

Bi, L. and Hu, G. (2021). A genetic algorithm-assisted deep learning approach for crop yield prediction. *Soft Computing* 25 (16): 10617–10628. https://doi.org/10.1007/s00500-021-05995-9.

Bian, L., Zhang, H., Ge, Y. et al. (2022). Closing the gap between phenotyping and genotyping: review of advanced, image-based phenotyping technologies in forestry. *Annals of Forest Science* 79 (1): 22. https://doi.org/10.1186/s13595-022-01143-x.

Boora, N. and Ahalawat, N. (2024). The transformative role of AI in agriculture: revolutionizing farming practices and sustainability. In: *Futuristic Trends in Renewable & Sustainable Energy Volume 3 Book 5*, 1ste (ed. D. Neog, U. Nadeem, S. Muthubalaji, et al.), 1–18. Iterative International Publishers, Selfypage Developers Pvt Ltd. https://doi.org/10.58532/V3BARS5P1CH1.

Bose, S., Banerjee, S., Kumar, S. et al. (2024). Review of applications of artificial intelligence (AI) methods in crop research. *Journal of Applied Genetics* 65 (2): 225–240. https://doi.org/10.1007/s13353-023-00826-z.

Cafer, A.M. and Rikoon, J.S. (2018). Adoption of new technologies by smallholder farmers: the contributions of extension, research institutes, cooperatives, and access to cash for improving tef production in Ethiopia. *Agriculture and Human Values* 35 (3): 685–699. https://doi.org/10.1007/s10460-018-9865-5.

Cembrowska-Lech, D., Krzemińska, A., Miller, T. et al. (2023). An integrated multi-omics and artificial intelligence framework for advance plant phenotyping in horticulture. *Biology* 12 (10): 1298. https://doi.org/10.3390/biology12101298.

Chang, C.-H., Lin, J., Chang, J.-W. et al. (2024). Hybrid deep neural networks with multi-tasking for rice yield prediction using remote sensing data. *Agriculture* 14 (4): 513. https://doi.org/10.3390/agriculture14040513.

Chang-Brahim, I., Koppensteiner, L.J., Beltrame, L. et al. (2024). Reviewing the essential roles of remote phenotyping, GWAS and explainable AI in practical marker-assisted selection for drought-tolerant winter wheat breeding. *Frontiers in Plant Science* 15: 1319938. https://doi.org/10.3389/fpls.2024.1319938.

Chaturvedi, P., Pierides, I., Zhang, S. et al. (2024). Multiomics for crop improvement. In: *Frontier Technologies for Crop Improvement* (ed. M.K. Pandey, A. Bentley, H. Desmae, et al.), 107–141. Springer Nature Singapore. https://doi.org/10.1007/978-981-99-4673-0_6.

Chen, J., Zhou, J., Li, Q. et al. (2023a). CropQuant-Air: an AI-powered system to enable phenotypic analysis of yield- and performance-related traits using wheat canopy imagery collected by low-cost drones. *Frontiers in Plant Science* 14: 1219983. https://doi.org/10.3389/fpls.2023.1219983.

Chen, L., Bao, H., Yang, J. et al. (2024). Dynamics of rice seed-borne bacteria from acquisition to seedling colonization. *Microbiome* 12 (1): 253. https://doi.org/10.1186/s40168-024-01978-8.

Chen, Q., Li, L., Chong, C., and Wang, X. (2022). AI-enhanced soil management and smart farming. *Soil Use and Management* 38 (1): 7–13. https://doi.org/10.1111/sum.12771.

Chen, Y., Yu, Z., Han, Z. et al. (2023b). A decision-making system for cotton irrigation based on reinforcement learning strategy. *Agronomy* 14 (1): 11. https://doi.org/10.3390/agronomy14010011.

Choudhary, R., Mantri, S.P., Barse, M.V., and Chitnis, S. (2024). Leveraging AI in smart agro-informatics: areview of data science applications. *International Research Journal on Advanced Engineering and Management (IRJAEM)* 2 (6): 1964–1975. https://doi.org/10.47392/IRJAEM.2024.0291.

Dara, R., HazratiFard, S.M., and Kaur, J. (2022). Recommendations for ethical and responsible use of artificial intelligence in digital agriculture. *Frontiers in Artificial Intelligence* 5: 884192. https://doi.org/10.3389/frai.2022.884192.

Das Choudhury, S., Samal, A., and Awada, T. (2019). Leveraging image analysis for high-throughput plant phenotyping. *Frontiers in Plant Science* 10: 508. https://doi.org/10.3389/fpls.2019.00508.

Delfani, P., Thuraga, V., Banerjee, B., and Chawade, A. (2024). Integrative approaches in modern agriculture: IoT, ML and AI for disease forecasting amidst climate change. *Precision Agriculture* 25 (5): 2589–2613. https://doi.org/10.1007/s11119-024-10164-7.

Dhillon, M.S., Dahms, T., Kuebert-Flock, C. et al. (2023). Integrating random forest and crop modeling improves the crop yield prediction of winter wheat and oil seed rape. *Frontiers in Remote Sensing* 3: 1010978. https://doi.org/10.3389/frsen.2022.1010978.

Dhyani, A., Singh, H., Rawat, P. et al. (2024). Deep learning solution for crop health monitoring using CNN in smart agriculture. *International Conference on Electrical Electronics and Computing Technologies (ICEECT)* 2024: 1–6. https://doi.org/10.1109/ICEECT61758.2024.10739188.

Dos Santos, D. P., and Vasconcelos, J. C. S. (2024). Using Gaussian copulas and generative adversarial networks for generating synthetic data in beet productivity analysis. *Sugar Tech*. https://doi.org/10.1007/s12355-024-01506-w.

Douglas, A.E. (2022). Plant microbiomes in agriculture and food production. In: *Microbiomes: A Very Short Introduction*, 1ste (ed. I.A.E. Douglas), 88–106. Oxford University PressOxford. https://doi.org/10.1093/actrade/9780198870852.003.0006.

Drees, L., Junker-Frohn, L.V., Kierdorf, J., and Roscher, R. (2021). Temporal prediction and evaluation of Brassica growth in the field using conditional generative adversarial networks. *Computers and Electronics in Agriculture* 190: 106415. https://doi.org/10.1016/j.compag.2021.106415.

Farooq, M.A., Gao, S., Hassan, M.A. et al. (2024). Artificial intelligence in plant breeding. *Trends in Genetics* 40 (10): 891–908. https://doi.org/10.1016/j.tig.2024.07.001.

Feng, W., Gao, P., and Wang, X. (2024). AI breeder: genomic predictions for crop breeding. *New Crops* 1: 100010. https://doi.org/10.1016/j.ncrops.2023.12.005.

Gangwani, N. (2024). AI-driven precision agriculture: optimizing crop yield and resource efficiency. *International Journal for Multidisciplinary Research* 6 (6): 29913. https://doi.org/10.36948/ijfmr.2024.v06i06.29913.

Gao, P., Zhao, H., Luo, Z. et al. (2023). SoyDNGP: a web-accessible deep learning framework for genomic prediction in soybean breeding. *Briefings in Bioinformatics* 24 (6): bbad349. https://doi.org/10.1093/bib/bbad349.

Gatkal, N., Dhar, T., Prasad, A. et al. (2024). Development of a user-friendly automatic ground-based imaging platform for precise estimation of plant phenotypes in field crops. *Journal of Field Robotics* 41 (7): 2355–2372. https://doi.org/10.1002/rob.22254.

Golubin, A.V. and Kopnin, A.A. (2024). Prospects and opportunities for using AI technologies in the agricultural sector. *E3S Web of Conferences* 537: 08012. https://doi.org/10.1051/e3sconf/202453708012.

Goodfellow, I.J., Pouget-Abadie, J., Mirza, M. et al. (2014). Generative adversarial networks (*arXiv:1406.2661*). *arXiv*. https://doi.org/10.48550/arXiv.1406.2661.

Gul, D. and Banday, R.U.Z. (2024). Transforming crop management through advanced AI and machine learning: insights into innovative strategies for sustainable agriculture. *AI, Computer Science and Robotics Technology* 3: https://doi.org/10.5772/acrt.20240030.

Gülmez, B. (2024). Advancements in maize disease detection: a comprehensive review of convolutional neural networks. *Computers in Biology and Medicine* 183: 109222. https://doi.org/10.1016/j.compbiomed.2024.109222.

Hadeed, M.Z., Ali, A., Malik, A. et al. (2024). Harnessing AI and GIS technologies for climate-resilient agriculture and environmental sustainability. In: *Practice, Progress, and Proficiency in Sustainability* (ed. B. Singh, C. Kaunert, K. Vig, and S. Dutta), 333–364. IGI Global. https://doi.org/10.4018/979-8-3693-6336-2.ch013.

Hailu, H. (2022). Impacts of adopting improved wheat varieties on household food security in GirarJarsodistrict, Ethiopia. *Review of Agricultural and Applied Economics* 25 (1): https://doi.org/10.15414/raae.2022.25.01.73-86.

Hamazaki, K. and Iwata, H. (2024). AI-assisted selection of mating pairs through simulation-based optimized progeny allocation strategies in plant breeding. *Frontiers in Plant Science* 15: 1361894. https://doi.org/10.3389/fpls.2024.1361894.

Hoque, A. and Padhiary, M. (2024). Automation and AI in precision agriculture: innovations for enhanced crop management and sustainability. *Asian Journal of Research in Computer Science* 17 (10): 95–109. https://doi.org/10.9734/ajrcos/2024/v17i10512.

Islam, S., Reza, M.N., Samsuzzaman, A.S. et al. (2024). Machine vision and artificial intelligence for plant growth stress detection and monitoring: a review. *Precision Agriculture Science and Technology* 6 (1): 33–57. https://doi.org/10.22765/PASTJ.20240003.

Jackulin, C. and Murugavalli, S. (2022). A comprehensive review on detection of plant disease using machine learning and deep learning approaches. *Measurement: Sensors* 24: 100441. https://doi.org/10.1016/j.measen.2022.100441.

James, D., Rennya, P.R., Mallavarapu, M.D. et al. (2021). Integrating pan-omics data in a systems approach for crop improvement: opportunities and challenges. In: *Omics Technologies for Sustainable Agriculture and Global Food Security*, vol. II (ed. A. Kumar, R. Kumar, P. Shukla, and H.K. Patel), 215–246. Springer Singapore. https://doi.org/10.1007/978-981-16-2956-3_8.

Jeyalakshmi, J., Vijay, K., Eugene Berna, I., and Sowmia, K.R. (2024). AI in optimisation of agricultural output in precision agriculture. In: *Artificial Intelligence for Precision Agriculture*, 1ste (ed. P. Raj, N. Gayathri, and G.J.W. Kathrine), 221–237. Auerbach Publications. https://doi.org/10.1201/9781003504900-11.

Johri, P., Sharma, K.P., Chauhan, A., and Sunilkkhatri. (2023). Role of satellites in agriculture. In: *Evolving Networking Technologies*, 1ste (ed. K.P. Sharma, S. Gupta, A. Sharma, and D. Le), 165–175. Wiley. https://doi.org/10.1002/9781119836667.ch10.

Jurado-Ruiz, F., Rousseau, D., Botía, J.A., and Aranzana, M.J. (2023). GenoDrawing: an autoencoderframework for image prediction from SNP markers. *Plant Phenomics* 5: 0113. https://doi.org/10.34133/plantphenomics.0113.

Kasana, S.S. and Rathore, A.S. (2024). Attention-based light weight deep learning models for early potato disease detection. *Applied Sciences* 14 (17): 8038. https://doi.org/10.3390/app14178038.

Khalid, H. (2024). Portable edge computing system analysis for real-time detection and classification of wheat yellow rust infection types using embedded AI. *2024 International Conference on Engineering & Computing Technologies (ICECT)*, 1–4. https://doi.org/10.1109/ICECT61618.2024.10581008.

Khan, M.H.U., Wang, S., Wang, J. et al. (2022). Applications of artificial intelligence in climate-resilient smart-crop breeding. *International Journal of Molecular Sciences* 23 (19): 11156. https://doi.org/10.3390/ijms231911156.

Kok, Z.H., Mohamed Shariff, A.R., Alfatni, M.S.M., and Khairunniza-Bejo, S. (2021). Support vector machine in precision agriculture: a review. *Computers and Electronics in Agriculture* 191: 106546. https://doi.org/10.1016/j.compag.2021.106546.

Kumar, A., Rajput, R., Bihari, C. et al. (2024). Role of artificial intelligence in vegetable production: areview. *Journal of Scientific Research and Reports* 30 (9): 950–963. https://doi.org/10.9734/jsrr/2024/v30i92423.

Kumari, A., Khan, M.N., and Sinha, A.K. (2024). Harvesting intelligence. In: *Data-Driven Farming*, 1ste (ed. S.N.H. Bukhari), 126–141. Auerbach Publications. https://doi.org/10.1201/9781003485179-7.

Kundu, S., Saini, D.K., Meena, R.K. et al. (2024). High-throughput phenotyping and AI technologies for deciphering crop resilience to heat stress. *Plant Physiology Reports* 29 (4): 699–715. https://doi.org/10.1007/s40502-024-00821-4.

Lee, S. and Yun, C.M. (2023). A deep learning model for predicting risks of crop pests and diseases from sequential environmental data. *Plant Methods* 19 (1): 145. https://doi.org/10.1186/s13007-023-01122-x.

Lewis, J.I., Toney, A., and Shi, X. (2024). Climate change and artificial intelligence: assessing the global research landscape. *Discover Artificial Intelligence* 4 (1): 64. https://doi.org/10.1007/s44163-024-00170-z.

Li, J., He, Z., Zhou, G. et al. (2024). DeepAT: adeep learning wheat phenotype prediction model based on genotype data. *Agronomy* 14 (12): 2756. https://doi.org/10.3390/agronomy14122756.

Liu, B., Jiang, H., and Li, L. (2024). The innovative applications and prospects of artificial intelligence in agricultural research: gene editing, crop improvement, and intelligent data analysis. *Resources Economics Research Board* 4: https://doi.org/10.50908/arr.4.4_624.

Liu, J. and Wang, X. (2021). Plant diseases and pests detection based on deep learning: a review. *Plant Methods* 17 (1): 22. https://doi.org/10.1186/s13007-021-00722-9.

Liu, Q., Yang, F., Zhang, J. et al. (2021). Application of CRISPR/Cas9 in crop quality improvement. *International Journal of Molecular Sciences* 22 (8): 4206. https://doi.org/10.3390/ijms22084206.

Lokesh, G.H., Chandregowda, S.B., Vishwanath, J. et al. (2024). Intelligent plant leaf disease detection using generative adversarial networks: acase-study of cassava leaves. *The Open*

Agriculture Journal 18 (1): e18743315288623. https://doi.org/10.2174/0118743315288623240223072349.

Ma, J., Wu, Y., Liu, B. et al. (2023). Wheat yield prediction using unmanned aerial vehicle RGB-imagery-based convolutional neural network and limited training samples. *Remote Sensing* 15 (23): 5444. https://doi.org/10.3390/rs15235444.

Maheswari, S., Dhilip Kumar, V., Geman, O., and Chiuchisan, I. (2024). Intelligent IoTenvironment for early crop disease prediction and management. *2024 International Symposium on Sensing and Instrumentation in 5G and IoT Era (ISSI)*, 1–6. https://doi.org/10.1109/ISSI63632.2024.10720484.

Mancini, A., Frontoni, E., and Zingaretti, P. (2019). Satellite and UAV data for precision agriculture applications. *2019 International Conference on Unmanned Aircraft Systems (ICUAS)*, 491–497. https://doi.org/10.1109/ICUAS.2019.8797930.

Maraveas, C. (2024). Image analysis artificial intelligence technologies for plant phenotyping: current state of the art. *AgriEngineering* 6 (3): 3375–3407. https://doi.org/10.3390/agriengineering6030193.

Marong, M., Husin, N.A., Zolkepli, M., and Affendey, L.S. (2024). Data-driven rice yield predictions and prescriptive analytics for sustainable agriculture in Malaysia. *International Journal of Advanced Computer Science and Applications* 15 (3): https://doi.org/10.14569/IJACSA.2024.0150337.

Mellaku, M.T., Reynolds, T.W., and Woldeamanuel, T. (2018). Linear programming-based cropland allocation to enhance performance of smallholder crop production: apilot study in AbaroKebele, Ethiopia. *Resources* 7 (4): 76. https://doi.org/10.3390/resources7040076.

Mohammed, M.A, Jameel, S.H., Hussein, A.J. et al. (2024). The transformation of agriculture by artificial intelligence in smart farming. *2024 36th Conference of Open Innovations Association (FRUCT)*, 515–526. https://doi.org/10.23919/FRUCT64283.2024.10749901.

Moreno-Revelo, M.Y., Guachi-Guachi, L., Gómez-Mendoza, J.B. et al. (2021). Enhanced convolutional-neural-network architecture for crop classification. *Applied Sciences* 11 (9): 4292. https://doi.org/10.3390/app11094292.

Mostafa, S., Mondal, D., Panjvani, K. et al. (2023). Explainable deep learning in plant phenotyping. *Frontiers in Artificial Intelligence* 6: 1203546. https://doi.org/10.3389/frai.2023.1203546.

Motamedi, B. and Villányi, B. (2024). A predictive analytics model with Bayesian-optimized ensemble decision trees for enhanced crop recommendation. *Decision Analytics Journal* 12: 100516. https://doi.org/10.1016/j.dajour.2024.100516.

Mrisho, L.M., Mbilinyi, N.A., Ndalahwa, M. et al. (2020). Accuracy of a smartphone-based object detection model, PlantVillageNuru, in identifying the foliar symptoms of the viral diseases of cassava–CMD and CBSD. *Frontiers in Plant Science* 11: 590889. https://doi.org/10.3389/fpls.2020.590889.

Murmu, S., Sinha, D., Chaurasia, H. et al. (2024). A review of artificial intelligence-assisted omics techniques in plant defense: current trends and future directions. *Frontiers in Plant Science* 15: 1292054. https://doi.org/10.3389/fpls.2024.1292054.

Mushtaq, M.A., Ahmed, H.G.M.-D., and Zeng, Y. (2024). Applications of artificial intelligence in wheat breeding for sustainable food security. *Sustainability* 16 (13): 5688. https://doi.org/10.3390/su16135688.

Na, M.H. and Na, I.S. (2024). AI-powered predictive modelling of legume crop yields in a changing climate. *LegumeResearch—An International Journal* 47: https://doi.org/10.18805/LRF-790.

Naik, H.S., Zhang, J., Lofquist, A. et al. (2017). A real-time phenotyping framework using machine learning for plant stress severity rating in soybean. *Plant Methods* 13 (1): 23. https://doi.org/10.1186/s13007-017-0173-7.

Nawaz, M. and Babar, M.I.K. (2024). IoT and AI: a panacea for climate change-resilient smart agriculture. *Discover Applied Sciences* 6 (10): 517. https://doi.org/10.1007/s42452-024-06228-y.

Osorio, C.P., Leucci, F., and Porrini, D. (2024). Analyzing the relationship between agricultural AI adoption and government-subsidized insurance. *Agriculture* 14 (10): 1804. https://doi.org/10.3390/agriculture14101804.

Parashar, A., Mabrouki, J., and Sharma, J. (2024). AI and smart technologies for smart agriculture environment. In: *Advanced Systems for Environmental Monitoring, IoT and the application of Artificial Intelligence*, vol. 143 (ed. J. Mabrouki and M. Azrour), 95–107. Springer Nature Switzerland. https://doi.org/10.1007/978-3-031-50860-8_6.

Patel, A., Kethavath, A., Kushwaha, N.L. et al. (2023). Review of artificial intelligence and internet of things technologies in land and water management research during 1991–2021: a bibliometric analysis. *Engineering Applications of Artificial Intelligence* 123: 106335. https://doi.org/10.1016/j.engappai.2023.106335.

Patil, A. (2023). Use of artificial intelligence to hasten progress in plant genetics. https://doi.org/10.20944/preprints202309.1067.v1.

Paul, B., Paul, M., and Rub, A. (2024). Advancements in AI-based pest and disease detection in agriculture: acomprehensive review of image recognition and disease modelling. *SSRN Electronic Journal*. https://doi.org/10.2139/ssrn.4843850.

Peladarinos, N., Piromalis, D., Cheimaras, V. et al. (2023). Enhancing smart agriculture by implementing digital twins: acomprehensive review. *Sensors* 23 (16): 7128. https://doi.org/10.3390/s23167128.

Poplin, R., Chang, P.-C., Alexander, D. et al. (2018). A universal SNP and small-indel variant caller using deep neural networks. *Nature Biotechnology* 36 (10): 983–987. https://doi.org/10.1038/nbt.4235.

Praveen, B. and Sharma, P. (2020). Climate variability and its impacts on agriculture production and future prediction using autoregressive integrated moving average method (ARIMA). *Journal of Public Affairs* 20 (2): e2016. https://doi.org/10.1002/pa.2016.

Rahman, S.M. and Ravi, G. (2022). Role of artificial intelligence in pest management. In: *Current Topics in Agricultural Sciences Vol. 7* (ed. M. Tripathi), 64–81. Book Publisher International (a part of ScienceDomain International). https://doi.org/10.9734/bpi/ctas/v7/2141B.

Rai, K.K. (2022). Integrating speed breeding with artificial intelligence for developing climate-smart crops. *Molecular Biology Reports* 49 (12): 11385–11402. https://doi.org/10.1007/s11033-022-07769-4.

Rajeswari, D., Venkatachalam Parthiban, A., and Ponnusamy, S. (2024). Digital twin-based crop yield prediction in agriculture. In: *Advances in Business Information Systems and Analytics* (ed. S. Ponnusamy, M. Assaf, J. Antari, et al.), 99–110. IGI Global. https://doi.org/10.4018/979-8-3693-3234-4.ch008.

Rane, N.L., Desai, P., and Choudhary, S. (2024). Challenges of implementing artificial intelligence for smart and sustainable industry: technological, economic, and regulatory barriers. In: *Artificial Intelligence and Industry in Society 5.0* (ed. N.L. Rane). Deep Science Publishing. https://doi.org/10.70593/978-81-981271-1-2_5.

Rangwala, H., Charuvaka, A., and Rasheed, Z. (2014). Machine learning approaches for metagenomics. In: *Machine Learning and Knowledge Discovery in Databases*, vol. 8726 (ed. T. Calders, F. Esposito, E. Hüllermeier, and R. Meo), 512–515. Berlin Heidelberg: Springer. https://doi.org/10.1007/978-3-662-44845-8_47.

Rasheed, A., Ahmed, S., and Wassan. (2018). Estimation of hybrid vigor for yield and yield related traits in tomato (*Solanum lycopersicon* MI II). *International Journal of Biosciences (IJB)* 12 (1): 160–167. https://doi.org/10.12692/ijb/12.1.160-167.

Rasheed, A., Gill, R.A., Hassan, M.U. et al. (2021). A critical review: recent advancements in the use of CRISPR/Cas9 technology to enhance crops and alleviate global food crises. *Current Issues in Molecular Biology* 43 (3): 1950–1976. https://doi.org/10.3390/cimb43030135.

Ray, D.K., Mueller, N.D., West, P.C., and Foley, J.A. (2013). Yield trends are insufficient to double global crop production by 2050. *PLoS One* 8 (6): e66428. https://doi.org/10.1371/journal.pone.0066428.

Raya González, L. E., Saldaña-Robles, N., Camarena, V. A. A.et al. (2024). Artificial intelligence model for disease and pest detection in maize using a hybrid convolutional neural networks and support vector machines approach. https://doi.org/10.2139/ssrn.4937609.

Ren, A., Jiang, D., Kang, M. et al. (2022). Evaluation of an intelligent artificial climate chamber for high-throughput crop phenotyping in wheat. *Plant Methods* 18 (1): 77. https://doi.org/10.1186/s13007-022-00916-9.

Ritambara, Kaushal, S., and Shubham (2024). Frontiers of artificial intelligence in agricultural sector: trends and transformations. *Journal of Scientific Research and Reports* 30 (10): 970–980. https://doi.org/10.9734/jsrr/2024/v30i102518.

Ruperao, P., Gandham, P., Odeny, D. A. et al. (2022). DeepVariant calling provides insights into race diversity and its implication for sorghum breeding. https://doi.org/10.1101/2022.09.06.505536.

Rutenberg, I., Gwagwa, A., and Omino, M. (2021). Use and impact of artificial intelligence on climate change adaptation in Africa. In: *African Handbook of Climate Change Adaptation* (ed. N. Oguge, D. Ayal, L. Adeleke, and I. Da Silva), 1107–1126. Springer International Publishing. https://doi.org/10.1007/978-3-030-45106-6_80.

Sengar, S. S., Hasan, A. B., Kumar, S., and Carroll, F. (2024). Generative artificial intelligence: asystematic review and applications (*arXiv:2405.11029*). *arXiv*. https://doi.org/10.48550/arXiv.2405.11029.

Shahid, M.F., Khanzada, T.J.S., Aslam, M.A. et al. (2024). An ensemble deep learning models approach using image analysis for cotton crop classification in AI-enabled smart agriculture. *Plant Methods* 20 (1): 104. https://doi.org/10.1186/s13007-024-01228-w.

Shakoor, N., Lee, S., and Mockler, T.C. (2017). High throughput phenotyping to accelerate crop breeding and monitoring of diseases in the field. *Current Opinion in Plant Biology* 38: 184–192. https://doi.org/10.1016/j.pbi.2017.05.006.

Sharma, A., Patel, R.K., Pranjal, P. et al. (2024). Computer vision-based smart monitoring and control system for crop. In: *Applications of Computer Vision and Drone Technology in*

Agriculture 4.0 (ed. S.S. Chouhan, U.P. Singh, and S. Jain), 65–82. Springer Nature Singapore. https://doi.org/10.1007/978-981-99-8684-2_5.

Sharma, S., Partap, A., Balaguer, M. A.deL., Malvar, S., and Chandra, R. (2022). DeepG2P: fusing multi-modal data to improve crop production (version 1). *arXiv*. https://doi.org/10.48550/ARXIV.2211.05986.

Singh, S., Mishra, V., Riyazuddin, R. et al. (2024). CRISPR/Cas9-mediated genome editing for trait improvement and stress tolerance in leguminosae (Legume family). *Plant Breeding* 2–19. https://doi.org/10.1111/pbr.13234.

Skobelev, P., Tabachinskiy, A., Strizhakov, A. et al. (2024). Experimental studies of the digital twin of plant based on ontologies and multi-agent technologies. *WSEAS Transactions onEnvironment and Development* 20: 624–632. https://doi.org/10.37394/232015.2024.20.60.

Song, Y., Zhang, H., Li, J. et al. (2023). High-accuracy maize disease detection based on attention generative adversarial network and few-shot learning. *Plants* 12 (17): 3105. https://doi.org/10.3390/plants12173105.

Sood, S., Bhardwaj, V., Bairwa, A. et al. (2023). Genome-wide association mapping and genomic prediction for late blight and potato cyst nematode resistance in potato (*Solanum tuberosum* L.). *Frontiers inPlant Science* 14: 1211472. https://doi.org/10.3389/fpls.2023.1211472.

Sun, L., Wen, J., Peng, H. et al. (2022). The genetic and molecular basis for improving heat stress tolerance in wheat. *aBIOTECH* 3 (1): 25–39. https://doi.org/10.1007/s42994-021-00064-z.

Suwaied Al Bakri, K.Z., Khalfan Saif Al Flaiti, M., Saleh Al-Balushi, S.A., and Poorngalingam, J. (2024). Revolutionizing agriculture: harnessing the power of artificial intelligence for sustainable farming practices. *International Journal of Advanced IT Research and Development* 1 (1): https://doi.org/10.69942/1920184/20240101/01.

Talaviya, T., Shah, D., Patel, N. et al. (2020). Implementation of artificial intelligence in agriculture for optimisation of irrigation and application of pesticides and herbicides. *Artificial Intelligence in Agriculture* 4: 58–73. https://doi.org/10.1016/j.aiia.2020.04.002.

Tan, G.D., Chaudhuri, U., Varela, S. et al. (2024). Machine learning-enabled computer vision for plant phenotyping: a primer on AI/ML and a case study on stomatal patterning. *Journal of Experimental Botany* 75 (21): 6683–6703. https://doi.org/10.1093/jxb/erae395.

Thakur, A., Venu, S., and Gurusamy, M. (2023). An extensive review on agricultural robots with a focus on their perception systems. *Computers and Electronics in Agriculture* 212: 108146. https://doi.org/10.1016/j.compag.2023.108146.

Tiwari, P., Bose, S.K., and Bae, H. (2021). Plant growth-promoting soil microbiomes: beneficial attributes and potential applications. In: *Soil Microbiomes for Sustainable Agriculture*, vol. 27 (ed. A.N. Yadav), 1–30. Springer International Publishing. https://doi.org/10.1007/978-3-030-73507-4_1.

Tsvetkov, M.Y. (2023). Satellite and drone multispectral and thermal images data fusion for intelligent agriculture monitoring and decision making support. In: *Remote Sensing for Agriculture, Ecosystems, and Hydrology XXV* (ed. C.M. Neale and A. Maltese), 65. SPIE. https://doi.org/10.1117/12.2679922.

Upadhyay, S.K. (2024). *Applications of Genome Engineering in Plants*. Wiley. https://doi.org/10.1002/9781394183913.

Wang, X., Zeng, H., Lin, L. et al. (2023). Deep learning-empowered crop breeding: intelligent, efficient and promising. *Frontiers in Plant Science* 14: 1260089. https://doi.org/10.3389/fpls.2023.1260089.

Wang, X.-W., Wang, T., and Liu, Y.-Y. (2024a). Artificial intelligence for microbiology and microbiome research (version 1). *arXiv*. https://doi.org/10.48550/ARXIV.2411.01098.

Wang, Z., Qiao, X., Wang, Y. et al. (2024b). IoT-based system of prevention and control for crop diseases and insect pests. *Frontiers in Plant Science* 15: 1323074. https://doi.org/10.3389/fpls.2024.1323074.

Yang, J., Liu, T., Yoon, S. et al. (2024). Editorial: advanced AI methods for plant disease and pest recognition. *Frontiers in Plant Science* 15: 1434320. https://doi.org/10.3389/fpls.2024.1434320.

Yang, Y., Saand, M.A., Huang, L. et al. (2021). Applications of multi-omics technologies for crop improvement. *Frontiers in Plant Science* 12: 563953. https://doi.org/10.3389/fpls.2021.563953.

Yulita, I.N., Rambe, M.F.R., Sholahuddin, A., and Prabuwono, A.S. (2023). A convolutional neural network algorithm for pest detection using GoogleNet. *AgriEngineering* 5 (4): 2366–2380. https://doi.org/10.3390/agriengineering5040145.

Zhang, F., Zhu, Y., Peng, J. et al. (2021). Linking the soil microbiome to soil health. *ScientiaSinica Vitae* 51 (1): 1–11. https://doi.org/10.1360/SSV-2020-0320.

Zhang, R., Zhang, C., Yu, C. et al. (2022). Integration of multi-omics technologies for crop improvement: status and prospects. *Frontiers in Bioinformatics* 2: 1027457. https://doi.org/10.3389/fbinf.2022.1027457.

Zhang, X., Cao, Z., and Dong, W. (2020). Overview of edge computing in the agricultural internet of things: key technologies, applications, challenges. *IEEE Access* 8: 141748–141761. https://doi.org/10.1109/ACCESS.2020.3013005.

2

Advances in Artificial Intelligence for Plant Biology and Crop Breeding: An Overview

Diya Kapadia, Jayshree Pawar, and Kanti Kiran

Gujarat Biotechnology University, Department of Plant Biotechnology, GIFT City-382355 Gandhinagar, Gujarat, India

2.1 Introduction

This chapter is intended to introduce the implementation of artificial intelligence (AI) in plant breeding and agriculture, including various other factors and strategies implemented. By minimizing the gap between phenotypes and genotypes, these approaches seem to be promising in revolutionizing agriculture and food security as per modern needs (Yoosefzadeh Najafabadi et al. 2023).

The rapidly increasing population around the globe has led to increased food demands. The global population is expected to reach about 9.5 billion by 2050 (He and Li 2020), which would exert tremendous pressure to meet the requirements. Alongside the changes in food patterns, rapid and unpredictable climate change has also posed a challenge to agriculture, hence demanding more resilient and productive crop varieties. While conventional crop breeding is helpful in achieving increased grain production and economic development, such traditional methods are often time- and resource-consuming. These challenges are well addressed by high-throughput computing, bioinformatics, AI, and machine learning (ML), along with other integrated technologies that have emerged to accelerate crop improvement by facilitating informed decision-making in management strategies and resource utilization (Sheikh et al. 2024).

Large datasets are available and are also continuously generated by high-throughput technologies, such as genomics, phenomics, and enviromics. AI-powered tools can effectively utilize these datasets to understand the genetic architecture of desired traits, predict performance, optimize strategies, and develop resilient varieties (Figure 2.1).

2.2 Big Data: Handling and Analytics

Big data refers to "massive amounts of structured and unstructured data that are difficult to process via traditional approaches." Big data is often described by the three V's—volume, variety, and velocity, and is expressed in terms of bytes, kilobytes, etc.

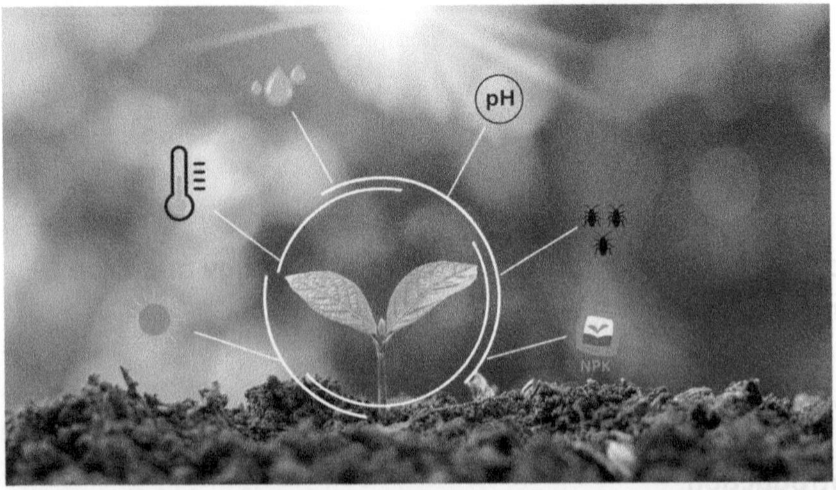

Figure 2.1 Important components determining plant growth. Plant metabolism, enzyme activity, and growth are all regulated by temperature; different species have different ideal ranges. Water is required for photosynthesis, the exchange of nutrients, and the maintenance of plant turgidity; both excessive and insufficient moisture can be stressful. The pH of the soil affects the availability of nutrients. Most plants prefer a pH range of 6.0–7.0 to grow, which is slightly acidic to neutral. Sunlight impacts the cycles of plant growth, flowering, and fruiting by supplying energy for photosynthesis. Insects and pests may damage plant tissues, spread illnesses, and lower yields; therefore, pest management techniques are crucial. The growth of plants is supported by essential macronutrients (nitrogen, phosphorus, and potassium [NPK]) and micronutrients; deficits result in poor health (Figure 2.1 Generated with AI using Canva).

Here, velocity determines how fast the data is produced and analyzed, influencing the efficiency of data processing; volume is the amount of data; and variety is the diversity of data (structured—numerical, unstructured—text and images) that needs to be processed for decision-making.

Big data in plant biology is generated using high-throughput technologies, such as omics. This data can be used for the identification and improvement of desirable traits, e.g., plant yield. By tracking this data, one can identify the best experiment type, an area that requires improvement, the effects of environmental conditions, germplasms, breeding materials, and more. It can be used to check whether the rate of data generation can keep up with the rate of the breeding process so that strategies can be planned accordingly (Figure 2.3).

The two main types of datasets include:

a) Structured data (e.g., genomics data that are organized in rows and columns)
b) Unstructured data (e.g., field notes, images, photos, and videos)

Unstructured datasets are difficult to analyze but play a vital role in determining plant behavior and characteristics.

Big data in plant biology can also identify new genetic markers linked with particular traits, the production of different crops under various conditions. The advancement of big data has made it possible to investigate in-depth interactions between DNA and RNA, which are important in plant breeding and crop improvement. Data collection is a crucial step in plant breeding, which involves collecting data for desired traits from a variety of settings, trials, and plant populations. Since the data is huge, it needs to be efficiently analyzed using

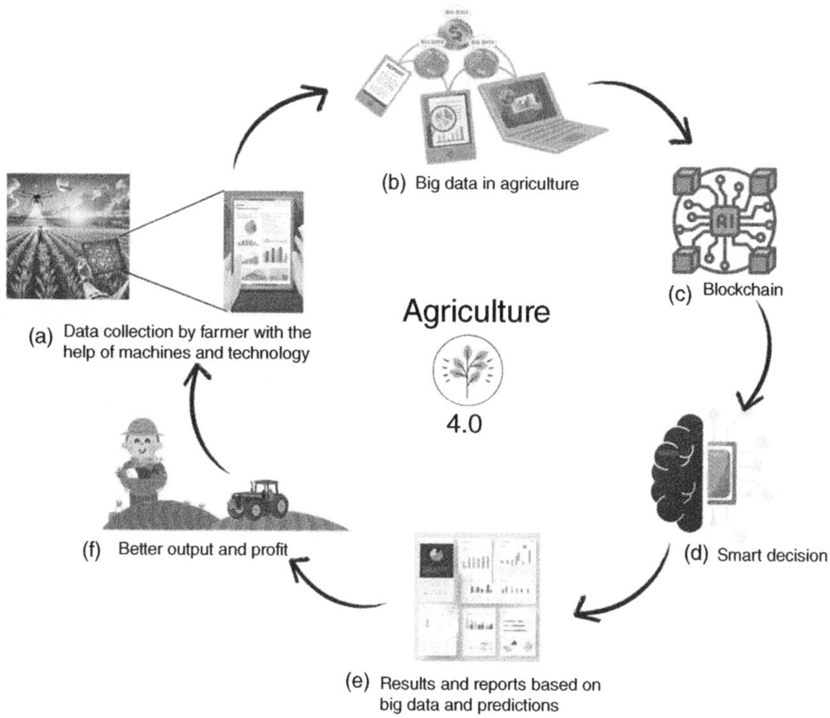

Figure 2.2 AGRICULTURE 4.0; this figure represents the integration of machines, technology, and AI with agriculture to provide the best benefits in yield, breeding, variety improvement, and farmers' economic profit. *Source:* Generated with AI using Canva version 2.298. (a) The farmer can collect all the data directly into a device, capture images with the help of a camera, with regular monitoring of drones, and also connect to satellites for data related to weather conditions and farms. (b) This big data of agriculture and breeding can be stored in BC for data integrity and security. (c) BC data can be facilitated with the help of AI. (d) AI and predictive machine learning models, using big data, can analyze the data in real time and also store it for future reference to make informed decisions. (e) All of this can generate reports based on which better decisions could be made, giving (f) better agricultural output to farmers with minimum losses and controlled resource utilization.

appropriate statistical approaches. AI can also help with decision-making when it comes to creating idiotype types for breeding and cultivation. In the end, this results in more effective and prosperous crop production systems, enhancing food security and averting world poverty (Yoosefzadeh Najafabadi et al. 2023) (Figure 2.2).

2.3 Blockchain Technology to Trace and Store Huge Breeding Data

Nearly 10% of people worldwide suffer from foodborne illnesses each year, which result in more than a million deaths (WHO, 2015, https://www.who.int/news-room/fact-sheets/detail/food-safety; https://reliefweb.int/node/1287036). Saving lives apart from nurturing is one of the main reasons for increasing efficiency in the food supply chain. Real-time tracking technologies make a huge difference by reducing food waste and getting it to those who need it on time, thus improving global food security. The demand–supply network of the agriculture supply chain starts from production, processing, transportation, storage, and finally distribution to consumers. This is accomplished by a complex process of acquiring the right information and followed further by proper management skills to deal with the logistics and capital involved. Blockchain (BC) plays a role here which includes a technology that creates decentralized networks available to each person, enhancing the connectivity, availability, and transparency of total available data (reference). Unlike traditional methods, BC allows the records to be maintained by a global network, which ensures the authenticity of data (Figure 2.2).

Satoshi (2008) developed the BC as a community of peers based on the concept of cryptography. Cryptography is used to protect information and communication, ensuring access to only authorized people; however, any person connected to the network can view all entries within the BC. A transaction is the exchange of a token between two participants, and a process of mining is used to group the transactions into a block. Whenever a transaction is done, it is verified by computer algorithms to determine its authenticity, after which the new one is cryptographically linked with the previous one, forming a chain of transactions called blockchain. Once data is added to a BC, it cannot be modified (Sarmah 2018), and new data is added through a process called distributed consensus, which is an agreement between various nodes of the network (NITI Aayog). Additionally, BC has self-executing smart contracts, which are agreements between buyers and sellers that are programmed into codes. The data is incorruptible and transparent (Figure 2.3). For better security and modification prevention, the BC system has many nodes and each node consists of the same BC, thereby creating a public account book. Hence, decentralization, safety, immutability, and transparency are therefore the four main features of the BC.

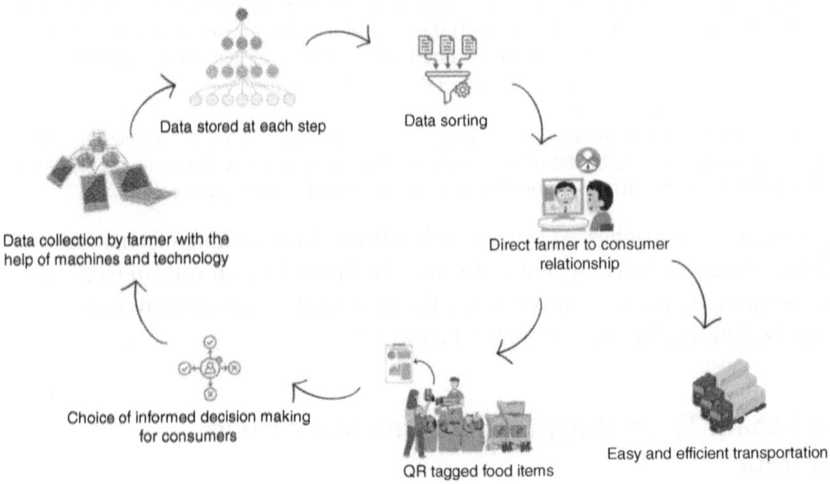

Figure 2.3 Farmer–consumer benefits—The data collected by farmers at each step of production is stored as huge datasets on BC, where it remains unchanged, maintaining transparency. This data is then sorted as per the requirements of markets and consumers' demand. Based on these data, including market demands, farmers can decide whether and where to sell their produce without third-party interference, helping them earn higher profits. The food can be transported to places where it is actually needed, reducing the wastage. While purchasing, consumers can also check the quality of the product as they are QR-coded, which includes every small detail of its journey from farm to market, which would help make wise choices. *Source:* Created using Canva version 2.298.

2.4 3D Printing in Plant Science

Using multidisciplinary strategies from systems biology, material science, and computational design, bioprinting is a modern technique for creating multiple biological objects with intrinsic functionality. The use of plant-based materials as bioinks in three-dimensional (3D) printing has grown in popularity recently because of their non-toxic, sustainable, cost-effective, biodegradable, and biocompatible qualities (Mehrotra et al. 2020). Bioink is a mixture of living cells propped by scaffolds that provide structural support and essential nutrients (Van Den Broeck et al. 2022). Gelinsky et al. in 2015 pioneered this technique of green bioprinting, based on the totipotency of plant cells to engineer tissues with precise architectures and desired functionalities.

Extrusion-based printing is used for plant-based bioprinting (Sundarsingh et al. 2024). Alginate, a non-Newtonian fluid and a biocompatible hydrogel, is a suitable material for 3D bioprinting due to its shear-thinning behavior, which enables the smooth extrusion of cell-laden inks through fine nozzles. It is also used for storing seeds for longer periods and increasing their growth (Landerneau et al. 2022). The process begins with the design of a digital 3D model in CAD software, sliced into layers that will then create instructions for the 3D printer. It is an automated process, where one can precisely control printing parameters, such as layer thickness, infill density, and speed (Landerneau et al. 2022).

Traditional methods for cultivating plant cells are unable to replicate the complex 3D environment found in plants, whereas 3D printing allows for the creation of precisely engineered 3D structures that mimic these natural conditions and create life-sized models of plants that offer a standardized reference for phenotype. Bioprinting allows us to investigate crucial cell behavior and patterns, including:

- **Cell Viability:** Cell survival in a 3D environment
- **Cell Division:** Multiplication and growth, along with neighboring components
- **Cell Identity:** Special features of cells
- **Cell-to-Cell Communication:** How a cell interacts and exchanges signals with each other (Van Den Broeck et al. 2022).

Dynamic cellular components, like mitochondria and the cell wall (CW), can be studied because they play a crucial role in understanding how cells respond morphophysiologically to various growth environments, thereby helping in the improvement of crop varieties.

Root architecture and geometry can be better understood through the manufacture of bioprinted root models/biorobots and plant surfaces, which provides the scope to expand our understanding and implement them to improve crop production. 3D bioprinting has made it possible to combine electric signals with the chemical processes of the plants, making an e-plant which is an advancement in 3D bioprinting in plant science (Mehrotra et al. 2020). 3D bioprinting is also useful in the production of nutritionally packed food items and customized meals for people with special diets. It promotes future crops that can be produced in massive amounts due to technological advancements (Sundarsingh et al. 2024).

By enabling direct control, green bioprinting is promising for addressing critical challenges in agriculture, such as the need to increase crop yields while minimizing environmental impact, reducing the reliance on traditional agricultural practices, and mitigating the competition for arable land (Landerneau et al. 2022).

Four-dimensional (4D) printing technology is an enhanced version of 3D printing technology that allows a structure to dynamically change its shape over time. As a result, the structure can perform self-folding/unfolding under certain stimuli. It is useful in the field of plant bionics, as it allows for targeted nanoparticle delivery to increase bioavailability. It is also particularly useful in developing slow-release fertilizers and long-lasting pesticides by creating such structures that can control the release (Huang et al. 2021).

2.5 Machine Learning-Based Approaches in Modernized Plant Biology and Biotechnology

ML is a branch of AI that develops algorithms capable of identifying patterns and making decisions from large amounts of data. These algorithms are utilized in predictive models, image/object recognition, robot control, and other applications to anticipate crop yields, monitor growth, practice precision agriculture, and automate irrigation.

Three categories of ML algorithms exist: supervised learning, unsupervised learning, and reinforcement learning. The most popular applications for supervised learning algorithms are in plant biology and breeding that are trained on labeled datasets. They can, at an early stage, predict complicated traits, discover and identify genomic regions associated with certain traits, and enable the selection of superior genotypes by genomic selection. Unsupervised learning algorithms, on the other hand, run on unlabeled data and identify patterns and similarities within datasets. Such algorithms are vital for determining genotype correlations, constructing kinship matrices, and organizing unstructured data in plant breeding applications. Reinforcement learning is a relatively new field in plants and involves training models through interactions with an environment where actions are rewarded or penalized.

Plant breeding programs prioritize repeatability and reproducibility using automated tuning methods like grid search and Bayesian optimization. These methods optimize parameters by splitting data into training, validation, and testing datasets.

A number of pre-trained model components are employed to minimize errors, including transfer learning, which uses less data and compute, and feature engineering, which improves ML model accuracy by generating features from raw data. Multiple models are combined in ensemble learning to produce more potent and precise models that are used to increase the precision and resilience of ML models. Another popular supervised ML approach for regression and classification is called random forest (RF), which mixes several decision trees to get a prediction that is accurate and reliable. Additionally, the idea of "Breeding 4.0" was recently introduced and exploits advances in ML to create sustainable systems around the world (Yoosefzadeh Najafabadi et al. 2023).

2.6 Supervised and Unsupervised Learning Algorithms

Modern crop breeding utilizes both supervised and unsupervised ML techniques to increase productivity with accuracy and sustainability. These algorithms are able to predict a variety of characteristics and breeding innovations, enabling the development of improved crop varieties using huge amounts of data (Gabur et al. 2022).

Algorithms for supervised learning function with training sets containing initial data, producing the expected result. The former makes predictions for new, unidentified datasets using the labeled examples. Supervised learning is crucial for phenotypic prediction, where the decision tree, support vector machine (SVM), and neural network models harness environmental and genotypic data to forecast features such as disease resistance, yield, and drought tolerance (Xavier 2021; Guo and Li 2023).

This makes the breeding process much more efficient by assisting breeders in selecting genotypes exactly dependent on particular environmental conditions. In addition to transformations such as convolutional neural networks (CNNs), supervised learning is used to accurately describe growth patterns and plant phenotypes from photos and illnesses using high-throughput imaging systems. It also helps increase the rate of selection of superior parental stocks. For this purpose, sources such as direct genomic values are utilized to forecast breeding values from genotypes, employing models like RFs and gradient boosting machines (GBMs) in genomic selection.

On the other hand, unsupervised learning algorithms utilize unstructured data and try to identify patterns within it without any reference outputs. These techniques were designed for data exploration and, therefore, are even more effective at crop breeding. For a better understanding or choosing different types for breeding material, the use of data analysis techniques such as k-means and hierarchical clustering provides germplasm identification based on the genetic similarity coefficient or characteristics. Principal component analysis (PCA) and other dimensionality reduction techniques simplify massive genomic datasets to visualize and relate genotype to phenotype that are useful in genetic architecture investigations. The use of unsupervised learning in aberration detection aims to discover variations in plant health or performance that indicate special characteristics relevant to breeding experiments or the possible problems with a particular variety. Furthermore, these algorithms enable the simultaneous analysis of multi-omics data by combining genomic, transcriptomic, and protein sequence datasets that identify latent relationships between genes, phenotypes, and their environments (Yan and Wang 2023).

Crop breeding and selection strategies can immensely benefit from the combination of supervised and unsupervised learning, where supervised learning could perform better in prediction and decision-making. While unsupervised learning could offer more curious suggestions and uncover unknown information within datasets.

The effective use of ML techniques in crop breeding represents an advancement in the struggle against global agricultural problems. For the benefit of breeders in fulfilling the demands of climate-smart, sustainable agriculture and food security globally, the combination of supervised and unsupervised learning techniques will continue to be important (Yan and Wang 2023).

2.7 Artificial Neural Network and Genetic Algorithm Usage in Crops

Two quite essential AI technologies that have been incorporated into crop cultivation are artificial neural networks (ANNs) and genetic algorithms (GAs). These tools provide unique solutions for the generation of superior quality varieties from conventional crops with enhanced yields, better quality, and improved stress adaptation. They address major problems, including data handling, resource utilization, and the identification of climate-resilient crops.

In crop breeding, ANNs are applied in the assessment of phenotypic characteristics from genotypic and environmental information. Incorporating multiple omics data into the model, ANNs allow breeders to accurately define the genotype–phenotype mapping of plants. For example, ANNs used to identify and forecast plant yield potentialities depending on environmental conditions provide the choice of appropriate parental genotypes. Also, high-throughput phenotyping involves the use of ANNs to correlate with imaging techniques to assess plant height, biomass, or even disease symptoms without much human interface with precise results. They also have immense potential in studying plant responses to abiotic stresses to strategize the development of genetically improved plants with improved stress tolerance against factors such as drought and salinity (Yan and Wang 2023).

GAs, on the other hand, are optimization algorithms based on the ideas of natural evolution. To identify the most effective solution to challenging problems, it mimics a number of bio-genetic processes, including crossover, mutation, and selection. GAs are often used to simultaneously improve yield, insect resistance, and drought tolerance. GAs help breeders select out alleles with interesting characteristics to provide optimal solutions. Also, they are employed in setting up breeding schedules and patterns, helping to regulate the design of trials for increasing the usage of available resources. GAs are able to tune the crops directly by adjusting the parameters and matching with the field data for improving the simulation model's predictive capability.

Combining the techniques of ANNs and GAs could bring up solutions to be utilized in crop breeding. Although ANNs are considered as best at data analysis and pattern recognition, GAs have good optimization techniques on their own. For instance, GAs may apply the type of breeding strategy resulting in great genetic gains, and ANNs, on the other hand, can use breeding values and the results of separate breeding operations (Olakulehin and Omidiora 2014).

As these technologies develop more and get integrated with other unconventional platforms such as precision agricultural systems and high-throughput phenotyping, they will become even more powerful. All in all, both ANNs and GAs have offered direction to address important issues relevant to resource inefficiency, genetic variability and environmental adaptability in the search for a sustainable and resilient agriculture.

2.8 Predictive Analytics for Plant Biology

Food consumption for the growing population, accompanied by rapidly shifting climate conditions for plants to thrive, is posing major problems for the agriculture sector (Elufioye et al. 2024). The advent of technology and intelligent machinery is causing a global shift in the established methods and approaches used in the agriculture sector (Kumar and Garg 2018). Predictive analysis has emerged as a powerful tool to empower farmers to navigate these complexities by predicting the future. Predictive analytics is an advanced technique that enables analyzing both current and historical data to predict future events, utilizing statistics, data mining, and AI-ML-based approaches. This can benefit growers in several ways by predicting seasonal changes, evaluating market demands, and analyzing product supply, soil and crop nutrient requirements, as well as comparing the performance of different breeding and cultivation methods employed in the breeding and cultivation of crops. The transition from traditional agriculture to data-driven approaches has transformed crop production by facilitating informed decision-making in various cultivation practices.

There are three types of predictive analytics, namely "predictive models predictive models," "descriptive models, descriptive models" and "decision models decision models." Predictive models forecast future outcomes based on data, descriptive models identify patterns and relationships within the data, and decision models use this revised data to support decision-making. Working with predictive models is essential nowadays to improve the growing competition of more tasks in less time (Kumar and Garg 2018). In agriculture, predictive analysis is mainly based on a regression model to forecast crop yields. Regression-based statistical technique identifies key patterns within large datasets and estimates the relationships between variables. However, the use of linear regression models and geographic information systems further refines these analyses by considering both environmental and geographical conditions (Kumar and Garg 2018; Elufioye et al. 2024).

Several predictive models utilized in plant biotechnology and agriculture are ML-based models that learns from a vast dataset, improving prediction accuracy over time. Similarly, statistical models are used to identify relationships between different factors influencing crop yield, and the AI-based models process complex data and learn patterns to improve prediction accuracy. Collectively, these models provide an early and accurate prediction for all problems related to crop performance, disease resistance, pest outbreaks, nutrient deficiencies, weather patterns, temperature, rainfall, soil health, and transportation logistics based on market trends and other relevant data.

Real-time data on crop and environmental conditions by remote sensing technologies like satellite imaging and drones have significantly contributed to improving the advanced agricultural system. Similarly, the role of big data is indispensable in storing the huge amount of data collected from numerous genotype, phenotype, and envirotype analyses. Envirotyping refers to the collection and analysis of environmental factors that influence the phenotype of organisms, mainly in agriculture. This data is used to train the models for precise and reliable predictions, thereby optimizing agricultural practices even more specifically and precisely (Asimiyu 2024).

Predictive analytics empowers farmers to use precision agriculture practices, helping them to determine irrigation schedules and optimize fertilizer/pesticide applications according to the needs of individual plants. AI-based predictive analysis has enabled farmers to get higher returns on investment through proper utilization of resources and crop management strategies (Elufioye et al. 2024).

2.9 Usage of Agents and Robotics in Breeding Programs

Plant breeding is aimed at agricultural productivity and growth to create new and improved crop varieties from crop genetic resource (CGR) recombination and modification. Sustainable Development Goal 2.5 is the emphasis of CGRs, or "genetic diversity of seeds, cultivated plants, and their related wild species" (Gerullis and Schulz 2023). The majority of the agricultural work was done by people, either by hand or with the assistance of machinery or other human-operated devices. The term "agriculture 4.0" has recently emerged, referring to the automation of skilled manual agricultural work through the use of cooperative robots, such as unmanned aerial aircraft and ground robot vehicles, for agricultural applications (Lytridis et al. 2021).

Agricultural robots, harnessing recent advances in robotics, sensing technologies, and data science, offer a potential solution by automating plant phenotyping and genotyping. They contribute to environmental stability and enhance breeding programs with vast information (Gerullis and Schulz 2023) (Figure 2.2).

By automated soil preparation, seeding, weeding, pest control, and harvesting, agricultural robots reduce reliance on manual labor, enhance efficiency, and contribute to a more resilient and sustainable agricultural sector while also enabling precise seed mapping, weed mapping, and micro-spraying. They also play a crucial role in mitigating greenhouse gas emissions, leading to more sustainable and resource-efficient agricultural practices (Ørum et al. 2023).

For example, weeds are generally being controlled using synthetic chemicals and manual weeding. These methods pose environmental and health hazards, require extensive labor with low efficiency, and have contributed to the emergence of herbicide-resistant weed varieties. All of this have led to the development of robotic weeders for cost-effective and efficient weeding purposes, reducing reliance on herbicides. These robots can selectively remove weeds based on the program that we provide to them while preserving crops. There are also specific robots for detecting and eliminating insects and pests, detecting and removing diseased leaves, plucking fruits and vegetables, and spraying fertilizers in the field based on AI recognition (Fennimore and Cutulle 2019). Some of the robotics models in agriculture include: (a) "human–robot," (b) cooperative unmanned aerial vehicles (UAVs), (c) cooperative unmanned ground vehicles (UGVs), (d) a hybrid of UAV and UGV, and (e) cooperative manipulation by multi-arm systems.

Nevertheless, a significant challenge in regulating highly automated and autonomous agricultural equipment (HAAAE) is the constant human supervision to be functional, though this reduces agricultural production costs (Maritan et al. 2023). Environmental factors such as rain, wind, and dust can also interfere with sensor measurements and damage delicate equipment. Furthermore, it is crucial for phenotype plants under field conditions to accurately assess their responses to real-world environmental stresses such as drought and pests. To deal with this, robust and adaptable sensor carriers, like autonomous ground robots and drones, are always being developed and upgraded over time to generate accurate data and record plant responses to real-world environmental stresses (Gerullis and Schulz 2023).

In breeding programs, the term "agent" implies self- or at least semi-autonomous systems able to detect processes and respond in a manner that achieves predetermined objectives. Marker-assisted selection (MAS) is extremely useful in breeding projects for being able to coordinate multiple tasks in different scenarios. They are used for breeding, genotyping, phenotyping, environmental simulation, experiment control and improvement, and the integration of several types of data. Based on big data and patterns found in the datasets, all of these systems assist breeders in making decisions. Additionally, they maximize resource allocation, eliminate unnecessary and repetitive cycles, and maximize breeding program efficacy (Cobb et al. 2019).

Robotics, on the other hand, uses complex computer operations for field and lab breeding techniques. High-throughput phenotyping involves using robots fitted with cameras, sensors, and AI to accurately determine elements such as height and size of the plant's leaves and their rates of growth. This automation reduces the time and the cost and increases accuracy. Generally, field robots, including the UGVs and drones, are employed for supervising large-scale breeding trials. They can easily operate on fields on their own and collect data on the health status of plants, state of soils, and climatic conditions (Virlet et al. 2016).

Subordinate processes such as growing, watering, and collecting specimens are also automated and carried out in fully contained spaces like greenhouses. Robotic arms equipped with computer vision technology allow precise experimental setups by allowing the genotyping of individual plants or plant groups. In addition, precise breeding techniques like genome editing and MAS enhance robotics. They enable accurate biological specimen manipulation and successful molecular procedure completion in order to minimize variability mediated by skill or expertise (Atefi et al. 2021).

Agents and robotics are improving the breeding programs through minimizing labor activities, increasing the accuracy of data collected, and reducing the waste of resources. Thus, their incorporation into crop breeding work plans allows breeders to mitigate the demands of food security and environmental degradation globally. As the expertise in the use of AI, sensors, and robotics grows, these technologies in the near future will become even more critical in defining the further development of agriculture (Ersoz et al. 2020).

2.10 Sensors and Interpretations of Networking

Breeders are benefiting from the data generated through networking and sensing systems, which are applied in breeding programs to develop improved crops. In high-throughput phenotyping, sensors play an important role in evaluating plant traits such as height, leaf area, growth rate, and response to environmental factors. Plant health sensors involving multi- and hyper-spectral cameras, humidifiers, pH meters, and nutrient meters search for indications of diseases or stresses, while weather sensors provide real-time data on atmospheric parameters like temperature, humidity, and precipitation to analyze genotype–environment interactions.

The huge amount of data that these sensors capture is easily interfaced and analyzed due to networking technologies. In order to transmit data in real time to centralized servers, sensors in breeding fields are connected to wireless sensor networks (WSNs) and Internet of Things (IoT) systems. Big data can be processed and stored in cloud computing with the help of distant servers, while edge computing generates the data locally to enable quick responses and take appropriate action based on the results (Alahmad et al. 2023).

The interpreted data has diverse applications in breeding programs. High-metabolism phenotyping with the support of sensors enables rapid assessment of germplasm, which allows breeders to determine high-performing genotypes. Multidimensional molecular profiling technologies enable sufficient connectivity of multi-omics data, including phenotyping for understanding the multiple genotype–phenotype–environment interfaces. It improves the eligibility criteria processes, leading to the development of new plant varieties with better characteristics such as higher yields, disease tolerance, and better ability to withstand climate change (Mahmood et al. 2022).

2.11 Object Image Capture and Analysis of Data Used in Plant Biology

Morphological characterization using object images and data analysis is important and has gained popularity when performing research within plant systems. The modern technologies of imaging growth phenotyping, with the help of analyzing relevant data, enable accelerating disease diagnosis and trait analysis, which is extremely important for increasing crop yield and sustainable agriculture.

A technique for visually depicting all characteristics related to the morphology, growth, and physiological traits of plants is imaging. A broad-spectrum investigation of plant health and efficiency can be obtained using conventional methods utilizing multispectral, hyperspectral, thermal, and/or fluorescent imaging as well as RGB cameras. Multi- and hyperspectral imaging captures the content and vitality of plants by additionally examining their thermal structure, water content, and stress response, while RGB captures characteristics like color and texture. It detects variations in plant temperature and reports on heat stress and water use efficiency. Quantification of chlorophyll fluorescence is facilitated by fluorescence images, which enables the evaluation of photosynthetic intensity and efficiency (Ge et al. 2016).

ANNs involving complex algorithms are used to assess image-derived data and gather relevant details. High-throughput phenotyping platforms achieve this by analyzing thousands of plants. They quantify plant density, plant spread, canopy cover, and root structure using image analysis. Due to advancements in ML, it is now possible to examine attributes with high accuracy by using patterns and relationships found in datasets. CNNs are used to detect diseases, pests, or nutrient deficiencies by observing recognizable patterns within plant photos (Mohanty et al. 2016).

Image-based data enables the examination of genotype–phenotype relationships in plant biology research. Using an integrated analytical strategy that combines genetics, environmetrics, and imaging data, one can explain how specific genotypes influence phenotypes in a particular environment. This understanding is helpful when implementing new crop varieties, such as high-yielding crops, crops resistant to pests, and drought-tolerant crops. Furthermore, by identifying phenotypic variation at the genetic level, image analysis may enhance selection in breeding programs, even for characteristics that are hard for the human eye to recognize (Minervini et al. 2015).

Another important area is disease identification and stress control. Both biotic and abiotic stresses in plants can be detected early enough through imaging systems to make appropriate interventions. Hyperspectral imaging is capable of identifying the difference in the reflection spectrum of the leaves due to pathogens or nutrient-locked infection.

Another benefit of thermal images is that they track heat stress or water deficiency, thereby guiding toward efficient management strategies for water consumption and irrigation.

Consequently, the key to future advancements that have a major effect on plant biology and its evaluation of plant characteristics, health, and interactions with the environment is object picture capture and data processing. They have proven to be very helpful in enhancing crop yields, controlling diseases, and ensuring sustainable agriculture that are essential resolutions to the challenges of food security and climate change.

2.12 Application of ANNs in Plant Science

Neural networks have emerged as critical tools for understanding plants by performing cutting-edge analytical and predictive tasks in place of traditional biological and agronomic concerns. ANNs are inspired by the human brain, which learns and identifies a large number of specific inputs to offer patterns, predictions, and conclusions, making them crucial for breed, phenotype, and stress detection.

One of the most common applications of ANNs in plant research is to model the relationship between genotype and phenotype. Using whole-genome or phenotypic information, genomics-enabled ANNs may identify how specific genes affect plant performance and quality, such as yield, drought tolerance, or disease resistance. This predictive capacity raises the pace of reproduction by utilizing diverse crop breeds that do not require intensive cultivation. ANNs are also employed in high-throughput phenotyping for traits such as canopy cover, leaf area, and chlorophyll content, which are measured by RGB, hyperspectral, or thermal camera imaging (Dijk et al. 2021).

ANNs process changes in plant appearance or spectral signatures in stress detection and disease diagnosis to identify the presence of factors such as nutrition deficiency, pest attack, or abiotic stress such as heat waves or droughts. As a result, crop losses are reduced in the event of a disease outbreak or almost ideal weather. ANNs are also used in environmental modeling to study how plants respond to climate change factors such as temperature, humidity, and soil moisture content to develop climate-adaptive crops (Gago et al. 2010).

ANNs also analyze large omics datasets such as genomics, proteomics, and metabolomics to explain the molecular basis of plant development and acclimation. This integration aids in ranking relevant genes for enhancement in the absence of genetic improvement. Hence, ANN applications are now widely used in plant science to enhance plant breeding, precision agriculture, and efficient farming in the context of climate change and food security.

2.13 Future Perspectives of AI for Crop Improvements

AI is predicted to emerge as a leading tool for crop enhancement in the future, meeting the world's food demands while addressing climate change. AI can give unmatched opportunities for optimizing crop improvement, boosting resource efficiency, and increasing yields through methodologies such as data analysis, ML, and automated technologies. Once AI merges with high-throughput phenotyping and genotyping techniques, another exponential growth is expected. Furthermore, massive datasets from hyperspectral, thermographic, 3D, and other imaging technologies can be filtered out using specialized AI-based tools to identify minor phenotypic changes associated with beneficial features. As a result, breeders can develop a high-yielding crop variety that can resist stress in a shorter period of time. Additionally, AI may use genomic, transcriptomic, proteomic, and metabolomic data to demonstrate the true genotype–phenotype interaction, resulting in a precision breeding model (Sheikh et al. 2024).

The use of predictive modeling by AI will add a progressive dimension to breeding. Breeders can use ML to predict plant performance under different environmental conditions and develop plants that can withstand climate change factors such as drought, salinity, and diseases. These prediction models improve scenario analysis, allowing farmers to use adaptive management to cut losses and increase yields.

Precision agriculture is another area where AI can be applied. AI systems can track soil health, pest presence, and other parameters in real time using IoT sensors and drones, allowing farmers to make efficient use of their resources. We continue to use water, fertilizers, and pesticides efficiently, thereby reducing environmental impact while increasing productivity. Though it might be a point of argument, the use of AI in crop improvement is quite promising, as it can transform traditional breeding, improve precision in farming, and contribute to sustainable production systems. With full advancement of the potential of AI in agriculture, it is conceivable to make this sector more productive, responsive to the changing needs of the planet, and offer solutions for the future (Naresh et al. 2020).

2.14 Conclusion

This chapter explores the incorporation of advanced AI-based technologies into agriculture to transform farming and breeding practices, with the goal of enhancing the development of resilient and high-yielding crop varieties. Utilizing the big data generated by high-throughput technologies, AI can further analyze the datasets to understand the patterns, predict performances, and assist in optimizing strategies. All these advanced processes can be maintained, ensuring complete transparency and security, while the innovations in 3D printing and automated robotics can further prove to be assets to enable the achievement of the specific goals in the area of crop biology.

Conventional plant biology and crop development strategies are being referred by AI, which offers novel and creative solutions to problems in agriculture. AI is assisting in the analysis of large data and the production of accurate findings, ranging from the most basic forms of ML, such as supervised and unsupervised learning, to more complex structures like ANNs or GAs. They provide these technologies as instruments for assisting researchers in understanding the relationships between genotypes and phenotypes, enhancing both genotyping and phenotyping, and improving agriculture.

The extension of sensors and networking technologies has now magnified the role of AI in monitoring breeding data for increased breeding success. This chapter highlights the following imaging tools in quantifying plant traits, plant health, and their interactions with the environment in terms of high-throughput phenotyping and stress detection, with the help of AI data analysis. New digital technologies, such as robotics and autonomous agents, are improving field management, which involves multi-omics integration in breeding strategies for handling biotic and abiotic stresses.

The future applications of AI for crop improvement should involve further use of this technology for predictive modeling and precision farming, including other applications for direct delivery to smallholder farmers. AI solutions are expected to improve the utilization, optimal management, and better resilience of crops to climate change with cost-effective measures for the benefit of growers and consumers at the global level.

Full-Forms

AI:	Artificial Intelligence,
ANNs:	Artificial Neural Networks,
BC:	Blockchain,
CAD:	Computer-Aided Design,
CGRs:	Crop Genetic Resources,
CNNs:	Convolutional Neural Networks,
GA:	Genetic Algorithms,
HAAAE:	Highly Automated and Autonomous Agriculture Equipment,
IoT:	Internet of Things,
MAS:	Marker-Assisted Selection,
ML:	Machine Learning.

References

Alahmad, T., Neményi, M., and Nyéki, A. (2023). Applying IoT sensors and big data to improve precision crop production: a review. *Agronomy* 13 (10): 2603.

Asimiyu, Z. (2024). Advancing crop resilience and yield through predictive analytics in plant biotechnology. https://www.researchgate.net/publication/385629116.

Atefi, A., Ge, Y., Pitla, S., and Schnable, J. (2021). Robotic technologies for high-throughput plant phenotyping: contemporary reviews and future perspectives. *Frontiers in Plant Science* 12: 611940.

Cobb, J.N., Biswas, P.S., and Platten, J.D. (2019). Back to the future: revisiting MAS as a tool for modern plant breeding. *Theoretical and Applied Genetics* 132: 647–667.

van Dijk, A.D.J., Kootstra, G., Kruijer, W., and de Ridder, D. (2021). Machine learning in plant science and plant breeding. *Iscience* 24 (1).

Elufioye, O.A., Ike, C.U., Odeyemi, O. et al. (2024). AI-driven predictive analytics in agricultural supply chains: a review: assessing the benefits and challenges of AI in forecasting demand and optimizing supply in agriculture. *Computer Science & IT Research Journal* 5 (2): Article 2. https://doi.org/10.51594/csitrj.v5i2.817.

Ersoz, E.S., Martin, N.F., and Stapleton, A.E. (2020). On to the next chapter for crop breeding: convergence with data science. *Crop Science* 60 (2): 639–655.

Fennimore, S.A. and Cutulle, M. (2019). Robotic weeders can improve weed control options for specialty crops. *Pest Management Science* 75 (7): 1767–1774. https://doi.org/10.1002/ps.5337.

Gabur, I., Simioniuc, D.P., Snowdon, R.J., and Cristea, D. (2022). Machine learning applied to the search for nonlinear features in breeding populations. *Frontiers in Artificial Intelligence* 5: 876578.

Gago, J., Landín, M., and Gallego, P. (2010). Strengths of artificial neural networks in modeling complex plant processes. *Plant Signaling & Behavior* 5 (6): 743–745.

Ge, Y., Bai, G., Stoerger, V., and Schnable, J.C. (2016). Temporal dynamics of maize plant growth, water use, and leaf water content using automated high throughput RGB and hyperspectral imaging. *Computers and Electronics in Agriculture* 127: 625–632.

Gerullis, M.K. and Schulz, W. (2023). Robustness of plant breeding systems under automated phenotyping. *Smart Agricultural Technology* 5: 100225. https://doi.org/10.1016/j.atech.2023.100225.

Guo, T. and Li, X. (2023). Machine learning for predicting phenotype from genotype and environment. *Current Opinion in Biotechnology* 79: 102853.

He, T. and Li, C. (2020). Harness the power of genomic selection and the potential of germplasm in crop breeding for global food security in the era with rapid climate change. *The Crop Journal* 8 (5): 688–700.

Huang, J., Xia, S., Li, Z. et al. (2021). Applications of four-dimensional printing in emerging directions: review and prospects. *Journal of Materials Science and Technology* 91: 105–120. https://doi.org/10.1016/j.jmst.2021.02.040.

Kumar, V. and Garg, M.L. (2018). Predictive analytics: a review of trends and techniques. *International Journal of Computer Applications* 182: 31–37. https://doi.org/10.5120/ijca2018917434.

Landerneau, S., Lemarié, L., Marquette, C., and Petiot, E. (2022). Green 3D bioprinting of plant cells: a new scope for 3D bioprinting. *Bioprinting* 27: e00216. https://doi.org/10.1016/j.bprint.2022.e00216.

Lytridis, C., Kaburlasos, V.G., Pachidis, T. et al. (2021). An overview of cooperative robotics in agriculture. *Agronomy* 11 (9): Article 9. https://doi.org/10.3390/agronomy11091818.

Mahmood, U., Li, X., Fan, Y. et al. (2022). Multi-omics revolution to promote plant breeding efficiency. *Frontiers in Plant Science* 13: 1062952.

Maritan, E., Lowenberg-DeBoer, J., Behrendt, K., and Franklin, K. (2023). Economically optimal farmer supervision of crop robots. *Smart Agricultural Technology* 3: 100110. https://doi.org/10.1016/j.atech.2022.100110.

Mehrotra, S., Kumar, S., Srivastava, V. et al. (2020). 3D bioprinting in plant science: an interdisciplinary approach. *Trends in Plant Science* 25 (1): 9–13. https://doi.org/10.1016/j.tplants.2019.10.014.

Minervini, M., Scharr, H., and Tsaftaris, S.A. (2015). Image analysis: the new bottleneck in plant phenotyping [applications corner]. *IEEE Signal Processing Magazine* 32 (4): 126–131.

Mohanty, S.P., Hughes, D.P., and Salathé, M. (2016). Using deep learning for image-based plant disease detection. *Frontiers in Plant Science* 7: 1419.

Naresh, R.K., Chandra, M.S., Charankumar, G.R. et al. (2020). The prospect of artificial intelligence (AI) in precision agriculture for farming systems productivity in sub-tropical India: a review. *Current Journal of Applied Science and Technology* 39 (48): 96–110.

Olakulehin, O.J. and Omidiora, E.O. (2014). A genetic algorithm approach to maximize crop yields and sustain soil fertility. *Netherlands Journal of Agricultural Science* 2: 94–103.

Ørum, J.E., Tamirat, T.W., Pedersen, S.M. et al. (2023). Optimal use of agricultural robot in arable crop rotation: a case study from the Netherlands. *Smart Agricultural Technology* 5: 100261. https://doi.org/10.1016/j.atech.2023.100261.

Sarmah, S.S. (2018). Understanding blockchain technology. *Computer Science and Engineering* 8 (2): 23–29.

Satoshi N. (2008). *Bitcoin: A Peer-to-Peer Electronic Cash System*. https://bitcoin.org/bitcoin.pdf

Sheikh, M., Iqra, F., Ambreen, H. et al. (2024). Integrating artificial intelligence and high-throughput phenotyping for crop improvement. *Journal of Integrative Agriculture* 23 (6): 1787–1802.

Sundarsingh, A., Zhang, M., Mujumdar, A.S., and Li, J. (2024). Research progress in printing formulation for 3D printing of healthy futurefoods. *Food and Bioprocess Technology* 17 (11): 3408–3439. https://doi.org/10.1007/s11947-023-03265-0.

Van Den Broeck, L., Schwartz, M.F., Krishnamoorthy, S. et al. (2022). Establishing a reproducible approach to study cellular functions of plant cells with 3D bioprinting. *Science Advances* 8 (41): eabp9906. https://doi.org/10.1126/sciadv.abp9906.

Virlet, N., Sabermanesh, K., Sadeghi-Tehran, P., and Hawkesford, M.J. (2016). Field Scanalyzer: an automated robotic field phenotyping platform for detailed crop monitoring. *Functional Plant Biology* 44 (1): 143–153.

World Health Organization. (2015). WHO estimates of the global burden of foodborne diseases: foodborne disease burden epidemiology reference group 2007–2015. ISBN 978 92 4 156516 5. https://reliefweb.int/node/1287036.

Xavier, A. (2021). Technical nuances of machine learning: implementation and validation of supervised methods for genomic prediction in plant breeding. *Crop Breeding and Applied Biotechnology* 21: e381421S2.

Yan, J. and Wang, X. (2023). Machine learning bridges omics sciences and plant breeding. *Trends in Plant Science* 28 (2): 199–210.

Yoosefzadeh Najafabadi, M., Hesami, M., and Eskandari, M. (2023). Machine learning-assisted approaches in modernized plant breeding programs. *Genes* 14 (4): Article 4. https://doi.org/10.3390/genes14040777.

3

The Role of Artificial Intelligence in Modern Agriculture and Crop Innovation

Dil Khurram[1,2,3], Nadeem Iqbal[1,3], Muhammad Nauman[1,3], Riyazuddin Riyazuddin[4], and Guo Liu[1,2,3]

[1] *State Key Laboratory of Geohazard Prevention and Geoenvironment Protection, Chengdu University of Technology, Dongsanlu, Erxianqiao, Chengdu, Sichuan, 610059, PR China*
[2] *Key Laboratory of Ministry of Ecology and Environment for Synergetic Control and Joint Remediation of Water & Soil Pollution, Chengdu University of Technology, Dongsanlu, Erxianqiao, Chengdu 610059, PR China*
[3] *College of Ecology and Environment, Chengdu University of Technology, Dongsanlu, Erxianqiao, Chengdu 610059, PR China*
[4] *Instituto de Bioingeniería, Universidad Miguel Hernández, Elche 03202, Spain*

3.1 Introduction

Agriculture has always been the backbone for the evolution and sustainability of civilizations throughout human history (Fuller and Stevens 2019). The advancement of society is reciprocal to the increase in population corresponding to additional food resources (Ben Ayed and Hanana 2021). The world population is increasing at alarming levels, but the arable land and other resources are shrinking in the face of climate change (Rao et al. 2018). Crop improvement is one of the basic concepts to tackle the growing demand for food, aiming to develop varieties with enhanced yield, stress resistance, adaptability, and nutrient quality, dating back thousands of years (Rao et al. 2018; Upadhyay 2021). Selective breeding and hybridization have been the primary tools for crop improvement since ancient times, along with the more recent innovative molecular breeding and genomic approaches (Bailey-Serres et al. 2019; Upadhyay 2024). There is a catch for all these approaches: these techniques are time-consuming and require additional efforts and resources without clear chances of success (Farooq et al. 2024). Moreover, modern agricultural challenges are growing and becoming increasingly complex with the advancement of technology, requiring modern solutions that are efficient, data-driven, precise, and offer more chances of success than mere anticipation.

Artificial intelligence (AI) is an innovative and transformative tool miming human intelligence using machines, revolutionizing numerous industries and human concepts (Rattan et al. 2022). It has immense potential to transform the agriculture sector in various ways (Saadane et al. 2022), from predictive analytics in crop management (Su et al. 2023) to capitalizing on multi-omic datasets for plant breeding (Farooq et al. 2024). AI is changing the concepts of how we approach crop improvement

(Ben Ayed and Hanana 2021), including the interpretations of vast and high-dimensional datasets, i.e., genomic, phenomic, and enviromic datasets (Jubair and Domaratzki 2023), which are key components of modern crop improvement strategies, with unprecedented speed and accuracy. Traditional statistical methods and expert-driven models fall short behind modern AI techniques like machine learning (ML) (Gill et al. 2022), deep learning (DL) (Razavi 2021), and computer vision (CV) (Ghazal et al. 2024) because these techniques provide advanced capabilities for the identification of patterns, prediction of outcomes, and optimization of various processes and parameters.

The global food security challenges are increasing due to anthropogenic factors (Ben Ayed and Hanana 2021). Increasing agricultural productivity and sustainability of resources is critical, and this can be safeguarded by bridging the gap between modern technologies, agricultural sciences, and AI (Javaid et al. 2023). The simplification and accessibility of AI tools and techniques in agriculture are critical, requiring further consideration and allocation of resources (Saadane et al. 2022). This chapter provides a detailed introduction to the imperative role of AI in agriculture in general and crop improvement in specific. This chapter starts with an overview of key definitions and concepts involving AI and crop improvement, followed by various AI techniques and tools employed in agricultural research and crop improvement, including crop growth models, ML and DL-based algorithms, CV, object detection, remote sensing, natural language processing, multi-omics analysis, and symbolic AI. An overview of the pros and cons, as well as future directions for these techniques, is also provided in each section (Figure 3.1).

3.2 Crop Improvement

Crop improvement is the systematic and scientific process of developing crop varieties with enhanced agronomic, physiological, and genetic traits to increase crop productivity, adaptability, quality, and overall value (Rao et al. 2018). The desirable list of plant traits is very long, including increasing yield potential (Bailey-Serres et al. 2019), plant resistance to various biotic (pests, diseases, and weeds) (González-Camacho et al. 2018) and abiotic (salinity, drought, and climate) factors (Ben Ayed and Hanana 2021), nutrient quality, and adaptability to a diverse range of environmental conditions (Bailey-Serres et al. 2019). The concept of crop improvement is ancient and dates back to the origin of civilization and the domestication of different crops for food and animals based on their desirable traits compared to modern times, where the primary objective is to ensure food security, increase crop productivity, and meet the food demands of the ever-increasing population (Rao et al. 2018).

Ancient crop improvement primarily consisted of various natural selection and traditional breeding methods involving selecting plants based on desirable traits and cross-breeding them repeatedly through successive generations until desired traits are achieved (Rao et al. 2018). These traditional methods laid the foundation for developing various high-yielding varieties of crops and hybrids. However, these techniques were highly time-consuming, and the efficiency of these approaches is primarily limited by multiple factors, including genetic diversity within closely related plant species (Fuller and Stevens 2019). Recent advancements in crop improvement techniques like

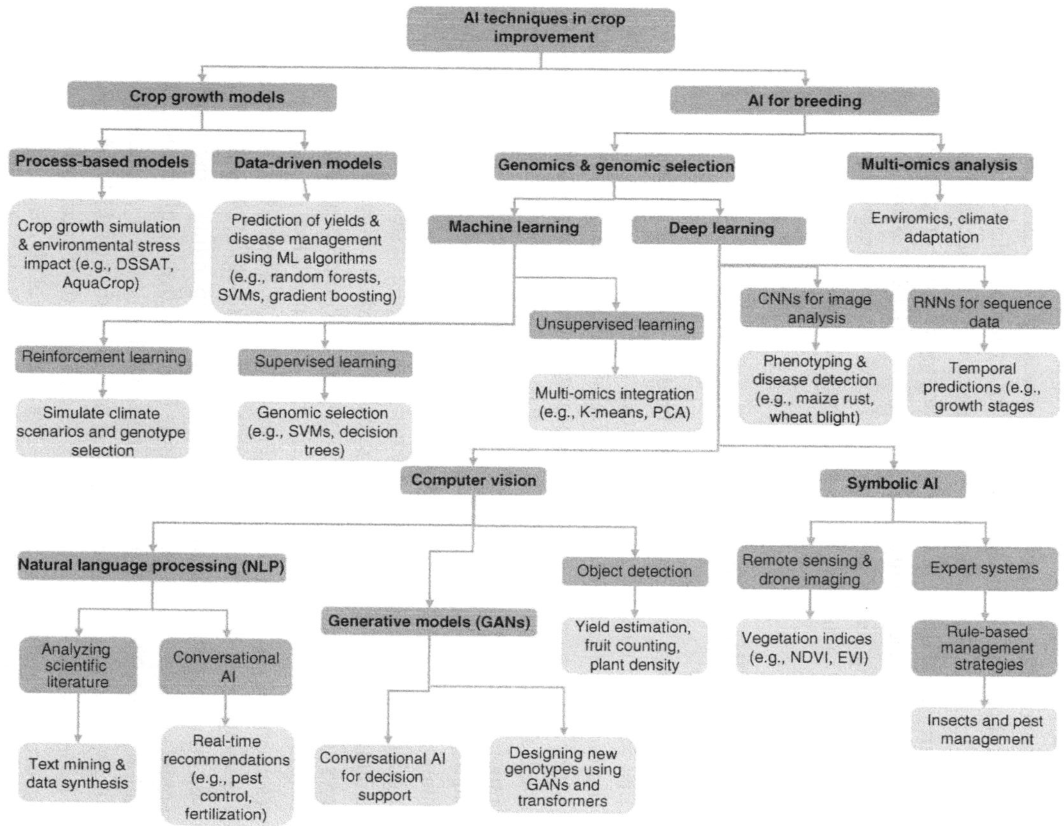

Figure 3.1 AI techniques in crop improvement.

molecular breeding, genomics, and Marker-Assisted Selection (MAS) techniques employ genetic markers to identify selected traits and offer greater precision (Rao et al. 2018). The most recent methods that have revolutionized the field of crop improvement include high-throughput phenotyping (Sheikh et al. 2024), gene editing (e.g., CRISPR/Cas9) (Liu et al. 2022; Kumar et al., 2024), and Genome-Wide Association Studies (GWAS) (Tibbs Cortes et al. 2021).

These advanced methods and techniques have several advantages over traditional approaches but generate large datasets, making it increasingly difficult to process using conventional statistical or other methods and requiring advanced calculations for predicting trends and traits (Rao et al. 2018). Moreover, climate change, the decline in arable land and agricultural resources, and the increasing demand for food require crop improvement technologies to be fast and efficient (Fuller and Stevens 2019; GNAFC 2024). These challenges can be met by employing recent advancements in AI for crop improvement, making it data-driven and efficient in meeting the increasing global food challenges.

3.3 Artificial Intelligence

Artificial intelligence (AI) is a much broader field in computer science and machine computing that deals with a variety of simulations similar to the human brain using computers and similar technologies, and it contains cognitive functions of learning, reasoning, problem-solving, and perception, and sometimes contains decision-making capabilities (Ben Ayed and Hanana 2021; Farooq et al. 2024). It can consist of several layers and a broad range of computational systems and techniques capable of processing data, recognizing patterns, adapting to the induction of new information, and decision-making capabilities either autonomously or semi-autonomously (Saadane et al. 2022). The baseline of AI lies in developing algorithms and models that enable computing systems to mimic human behavior, facilitate decision-making, recognize patterns and outcomes, make predictions, and sometimes surpass human capabilities (Sharma 2021; Ghazal et al. 2024).

There are various classifications and categories of AI defined based on different computing processes, mechanisms, and evolution (Rattan et al. 2022). The earliest forms of AI date back to the development of rule-based systems from 1950 to 1970 (Davis and King 1984), which evolved into expert systems from 1970 to 1980 by incorporating knowledge bases and reasoning engines (Schoech et al. 1985). Learning from data instead of predefined rules led to the formation of early ML in 1980–1990 (Tong and Nikoloski 2021). Handling imprecise data led to fuzzy logic that consisted of logical reasoning that could handle uncertainty and partial information (Dubois et al. 1999). The capability of recognizing patterns led to the emergence of neural networks in 1990–2000 (Tulbure et al. 2022). Advancements in computational power, ML, and neural networks with multiple layers laid the foundation of DL in 2010, and research is still ongoing (Attri et al. 2023). Another similar aspect is reinforcement learning, which focuses on training different systems like precision drones and irrigation systems (Gautron et al. 2022). Advancements in connectivity, sensors, and devices led to the creation of the Internet of Things (IoT), which integrates input and output functions for monitoring, automation, and data analysis (Saadane et al. 2022). Generative AI is another similar field focusing on creating new content and finding innovative solutions based on available data and information (Feuerriegel et al. 2024). The most advanced form of AI is autonomous AI systems (Javaid et al. 2023) that integrate reinforcement learning, robotics, IoT, and other tools to function independently without human intervention.

Various AI techniques and tools have diverse applicability for agriculture and crop improvement (Javaid et al. 2023). AI can transform traditional agricultural practices by analyzing and interpreting large and high-dimensional datasets from various sources like remote sensing, genomic sequencing, IoT, and historical datasets (Saadane et al. 2022; Javaid et al. 2023). Identifying nonlinear relationships and making subsequent real-time predictions are invaluable tools that can transform agriculture and crop improvement practices.

3.4 The Role of AI in Agriculture

The global population is expected to reach 9.7 billion people by 2050 (Falcon et al. 2022), in addition to other multilateral challenges of climate change, soil degradation, and resource scarcity (Hamed et al. 2024). An increase of 24 million people suffering from high levels of acute food insecurity was recorded in 2023, reaching 281.6 million overall (GNAFC 2024). These multifaceted challenges require additional efforts, resources, and innovative research frontiers to increase agriculture and food productivity for an ever-increasing population that is already suffering from food insecurity (Sheikh et al. 2024). These modern challenges have become insurmountable using traditional agricultural and crop improvement practices, and AI has the potential to surmount these challenges (Javaid et al. 2023). AI has singular capabilities among the currently available platforms and technologies, including predictive analyses, precision, efficiency, and decision support systems to transform the agriculture sector (Hamed et al. 2024), i.e., weather and climate patterns, local soil conditions, crop health, and similar data can be analyzed and processed in real-time to guide plant breeders to schedule planting time and associated resource utilization efficiently (Javaid et al. 2023). CV and similar approaches can assist in accurately and efficiently automating the phenotyping processes, disease detection, yield estimation, etc.

The optimization of crop improvement strategies using AI is also a gateway to sustainability through efficient allocation of resources (Hamed et al. 2024). It can help to conserve resources in various ways, ranging from predictive and enviromic data analyses, thereby saving additional experimental costs, to optimizing input parameters such as water, fertilizers, pesticides, and management practices (Ben Ayed and Hanana 2021). Concomitantly, AI can also help improve the food supply chain by improving logistics, thereby reducing food waste and enhancing traceability (Dora et al. 2022). Accordingly, the availability of AI tools and similar efficient mobile apps for ordinary farmers and crop growers can also help increase crop productivity by providing data-driven decisions for crop cultivation.

3.5 Overview of AI Integration in Crop Improvement

AI is considered among those groundbreaking inventions that have changed the course of history, i.e., gunpowder, firearms, printing, electricity, computers, digitalization, and mobile phones (Dora et al. 2022; Rattan et al. 2022). These inventions paved the way for the creation of new types of industries while simultaneously causing obsolescence, a series of sectors and industries that decline or vanish with time due to the induction of new technologies. Therefore, like the survival of species, the survival of industries or practices also depends on adaption and adaptability according to new and emerging technologies to avoid obsolescence, which is even applicable to people and the workforce (Maillard 2021). Therefore, traditional agricultural and crop improvement methods also need to be improvised according to new and emerging technologies; otherwise, like any other obsolete industry, crop production can also undergo serious challenges, which are critical for the survival of humans on the planet (Javaid et al. 2023). These contemplations seem slightly exaggerated, but the chances of becoming a reality cannot be neglected.

Considering the true potential and applicability of AI in agriculture and crop improvement, it can be stated that AI can be a game-changer (Hamed et al. 2024) by complementing traditional breeding methods through enhancing efficiency (Ben Ayed and Hanana 2021), reducing timeframes (Dora et al. 2022), and enabling the integration of multi-omics data (genomics, phenomics, and enviromics) (Xu et al. 2022). ML and DL algorithms, for instance, can be used for genomic selection to identify best-performing genotypes based on the genetic markers instead of conducting a series of breeding experiments, thereby saving a considerable amount of time for crop breeding and improvement (Jubair and Domaratzki 2023). The precision and accuracy of breeders using these techniques can never be met with those utilizing traditional or semi-traditional approaches, thereby influencing the obsolescence of conventional crop improvement techniques. Moreover, these techniques can reach new levels of advancements with time and with the induction and optimization of new and currently available techniques like CV (Divyanth et al. 2022), which have the capacity to streamline phenotyping and enable crop breeders to analyze plant traits such as biomass, disease resistance, and stress tolerance to particular environmental conditions with unprecedented speed and accuracy (Elango et al. 2024; Ghazal et al. 2024). Furthermore, AI-driven models can be used to predict crop performance under diverse environmental conditions (Attri et al. 2023), which can facilitate the development of crop varieties that are resilient to various biotic and abiotic factors like pests, disease, and climate.

AI is a relatively recent innovation, and it is evolving with time. There are possibilities that this technology will further develop, and its integration into crop improvement will expand over time. This evolution in AI for crop improvement can be diverse, ranging from developing virtual genotypes to AI-augmented CRISPR techniques for precise gene editing to developing autonomous platforms for plant breeding and genetics. Agriculture and food production can reach new heights and keep pace with global food security challenges if these recent advancements are appropriately investigated and integrated into crop breeding and genetics techniques.

3.6 Role of Crop Growth Models

3.6.1 Process-Based Models

Process-based models do not directly fall under the category of AI; instead, they are considered essential and fundamental tools that simulate plant growth and development, thereby being basic components in agricultural research and crop improvement (Hajjarpoor et al. 2022). The fundamentals of these models integrate complex simulations and mathematical equations that replicate and describe the interactions of various biological, soil, and environmental processes that shape the growth, development, and yield of crops, thereby replicating individual physiological mechanisms, such as photosynthesis, transpiration, nutrient uptake, and biomass accumulation for simulations (Hajjarpoor et al. 2022). These models can predict crop performance under a diverse range of scenarios depending on soil properties, weather conditions, and management practices (Ellis and Paustian 2024). These traditional models and simulation techniques have been parallelly used in combination with data-driven approaches to predict various crop performance parameters to expand the space of predictors in data-driven models.

There is a range of various commonly used process-based models, and these models have evolved over time (Ellis and Paustian 2024). The Decision Support System for Agrotechnology Transfer (DSSAT) is a widely used example, consisting of a suite of crop models designed to evaluate the effects of various agronomic strategies, including soil, weather, and crop management and based on observational and experimental datasets, including the support of the crop genetic data and modeling (Hoogenboom et al. 2024). The DSSAT has been used for a long term in agricultural crop research and has been instrumental in guiding resource-efficient agricultural practices with improved yields, especially in areas that are prone to environmental stress (Ellis and Paustian 2024). Another similar model is the Agricultural Production Systems Simulator (APSIM), consisting of a suite of models that are used for simulating and modeling crops considering a diverse range of management interactions, including climate, soil, plant, and management practices (Chapagain et al. 2023).

There are various similar process-based models with different functionalities; however, despite their diverse utility and application prospects, there are several limitations to using process-based models (Ellis and Paustian 2024). These models are very complex, require a large number of parameters to function correctly, and are difficult to set up and run (Hajjarpoor et al. 2022). Extensive field and experimental data, such as plant physiological parameters and soil nutrient levels, are required for proper model calibration, and such data may not be readily available (Ellis and Paustian 2024). These models contain inherent complexities due to the oversimplification of assumptions and real-world parameters, i.e., various environmental factors and associated crop responses; therefore, these models may not accurately reflect real-world conditions (Chapagain et al. 2023). AI can provide solutions for these limitations and is increasingly being used to enhance these process-based models (Javaid et al. 2023). ML and similar approaches can be used to effectively and tirelessly calibrate and validate these models, which can increase the efficiency and reliability of these models, i.e., refining parameter estimates to reduce uncertainties in model predictions (Tong and Nikoloski 2021). There are greater prospects for hybrid models consisting of process-based simulations coupled with AI-driven technologies, which can incur revolutions in precision agriculture and crop breeding and genetics and help ensure sustainable crop production in the future.

3.6.2 Data-Driven Predictive Models

Data-driven predictive models are considered an alternative to process-based models, which rely on ML algorithms to analyze historical and real-time data to predict crop performance and similar future outcomes of various processes and mechanisms (Tong and Nikoloski 2021; Bwambale et al. 2023). These models leverage various ML algorithms, including time series (Hamilton 2020), decision trees (De Ville 2013), outlier models (Boukerche et al. 2020), random forests (Montesinos López et al. 2022), gradient boosting and predictive models (Bentéjac et al. 2021), linear regression (Hajjarpoor et al. 2022), support vector machines (SVMs) (Pisner and Schnyer 2020), neural networks (Khaki et al. 2020; Tulbure et al. 2022), etc., to identify patterns and trends in large datasets. These predictive models are helpful and effective in scenarios where abundant data are available, but mechanistic understanding is limited (Bwambale et al. 2023). These models can analyze different historical data variables, i.e., weather patterns, soil composition, crop performance, etc., to accurately forecast crop yield or similar parameters (Tong and Nikoloski 2021). These models can be trained on multi-year datasets and enable farmers and plant breeders to determine parameters for optimum planting times, resource allocation, management practices, etc., thereby saving valuable time and resources, improving profitability, and minimizing waste (González-Camacho et al. 2018). A similar use case scenario is pest and disease prediction by integrating data from remote sensing, weather forecasts, and pest monitoring systems, thereby identifying areas and regions with higher risks of infection and infestation and allowing farmers and plant breeders to make timely decisions (Elango et al. 2024). These approaches can aid in significantly reducing crop losses due to pest and disease outbreaks, and identifying high-risk areas, like locust infestation and rice blasts (Ben Ayed and Hanana 2021), etc.

There are several limitations to using these data-driven predictive models because these models rely on historical data, which is not always reliable, especially when future scenarios change and result in inaccuracies (Bwambale et al. 2023), skewed training data can also lead to bias in model prediction (Kazerouni et al. 2020), complex models also sometimes become difficult to interpret, overfitting or poor generalization of training data due to missing data (Xuan et al. 2023), the requirement of the high-quality and large amount of data for the model to function correctly (Tong and Nikoloski 2021), and changes in the dynamic conditions of the system being modeled (Rattan et al. 2022). Robust data preprocessing techniques, including imputation (Attri et al. 2023), feature engineering (Razavi 2021), and explainable AI (Feuerriegel et al. 2024), can be applied to address or eliminate these challenges. Further advancements and research in data-driven models, data analytics, and AI will lead to enhancing the capacity and capabilities of these models (Razavi 2021). The real-time data streams from various sources, i.e., satellite imagery, drone footage, IoT devices, and similar sources, can be integrated for the models to enhance the prediction accuracy under dynamic conditions (Divyanth et al. 2022). Moreover, explainable AI is also an emerging field that can provide higher transparency for the capabilities of the model predictions.

3.7 Machine Learning (ML) Models and Algorithms

3.7.1 Supervised Learning

It is an ML technique that depends on training models on labeled datasets where inputs, i.e., genetic markers or environmental conditions, are paired with known outputs, i.e., crop yield or stress resistance, to teach neural networks to identify patterns (Jubair and Domaratzki 2019; Tulbure et al. 2022; Li et al. 2023). These models learn to map the inputs to outputs, which enables them to make predictions for new data (Tong and Nikoloski 2021). This technique is among the most widely used ML techniques in agriculture, particularly in crop improvement (Jubair and Domaratzki 2019). These techniques are particularly suitable for genomic selection (González-Camacho et al. 2018), which is a method used to predict the performance of genotypes based on genetic makeup (Xu et al. 2022). Different algorithms like SVMs (Pisner and Schnyer 2020), random forests (Montesinos López et al. 2022), neural networks (Khaki et al. 2020; Tulbure et al. 2022), Naïve Bayes (Sambo et al. 2012), and K-Nearest neighbor (Pfeifer et al. 2020) have been employed and tested to identify different genetic markers associated with various desirable traits in plants including increased yield, disease resistance, drought tolerance, etc. (Jubair and Domaratzki 2019), and predict yield potential of newly developed varieties, which accelerates the selection process of high-performing varieties.

Supervised learning can also aid plant breeders in detecting and identifying desirable plant traits, i.e., traits responsible for diseases and pests (González-Camacho et al. 2018). It can also be used to process high-resolution crop imagery to identify various disease or pest attack symptoms (Meivel and Maheswari 2021; Su et al. 2023), such as leaf rust, late blight, etc., which can help to reduce dependency on manual scouting, allowing early detection and intervention. Although supervised learning is a relatively modern technique and it is evolving with time, it also has shortcomings that may need to be considered while using this approach for crop improvement or agriculture in general. Accurate training of models for these techniques requires large and high-quality datasets (Jubair and Domaratzki 2019). Data collection and generation is a labor-intensive process that can be broadly expensive (Tae and Whang, 2021). The costs can increase further when considering the agriculture sector, where intensive labor is required for data (Xu et al. 2022), and phenotyping is a step further, which requires detailed field investigations (Gill et al. 2022), thereby increasing the costs substantially. Moreover, although data collection and other expenses are met and models are sufficiently trained, there are chances of generalization (Li et al. 2023), where trained models applied to new regions or crops with different initial and environmental conditions generate different results (Attri et al. 2023). New techniques and approaches are emerging to address these challenges, i.e., transfer learning (Hosna et al. 2022), where a model trained on one dataset is fine-tuned for a related task using minimal new data. These techniques can help to make supervised learning more scalable and accessible across a diverse range of agricultural systems, as well as crop breeding and genetics programs.

3.7.2 Unsupervised Learning

Unsupervised learning is an advanced ML technique that analyzes unlabeled data using algorithms and subsequently identifies patterns in the data (Tyagi et al. 2022). The ability to identify patterns from unlabeled data makes this technique particularly suitable for the integration and analyses of complex data (Naeem et al. 2023), which is usually high-dimensional from various multi-omics sources (Rao et al. 2018), including genomics (Yan and Wang 2022), phenomics (Xu et al. 2022), and enviromics (Resende et al. 2024). This method does not rely on predefined outputs like supervised learning (Rattan et al. 2022); instead, this technique focuses on identifying clusters, associations, or anomalies that may provide novel biological insights into the data (Naeem et al. 2023).

These techniques can be invaluable in crop improvement, which allows researchers and plant breeders to identify hidden insights and patterns about plant variations and traits, which can be beneficial to develop enhanced crop varieties without prior knowledge of identified and desirable traits (Tong and Nikoloski 2021). The most commonly used algorithms in crop improvement programs are k-means clustering (Badapanda et al. 2022), hierarchical clustering (Das et al. 2021), and principal component analysis (PCA) (Tong and Nikoloski 2021). These techniques can be used to group various genotypes based on genetic similarities and differences or the performance of multiple traits (Yan and Wang 2022), i.e., clustering can be used to identify genetic diversity in crop populations, which can guide plant breeders to select parents for cross-breeding and maximize heterosis or hybrid vigor (Gill et al. 2022). Phenotype–genotype-environment (P-G-E) integration is another application prospect of unsupervised learning, which is used to analyze how genes, phenotypes, and the environment interact using diverse datasets from different sources (Genaev et al. 2012). This technique can help discover and identify the genetic basis for the novel interactions between genes, environmental stressors, diseases, etc., for developing climate-resilient or disease-resistant crop varieties (Tong and Nikoloski 2021; Farooq et al. 2024). An emerging field of research using this technique is to identify biomarkers for complex traits (Hamany Djande et al. 2021), i.e., unsupervised learning algorithms can be used to analyze metabolomic data to detect compounds that are associated with disease resistance or stress tolerance (Liu et al. 2023). These biomarkers can help accelerate crop breeding or improve crop management practices.

The primary challenge in using unsupervised learning is that interpreting the results can be challenging, primarily due to the lack of labeled outputs, making validation less straightforward (Tyagi et al. 2022). It can be resolved by combining unsupervised learning with supervised learning approaches or domain expertise to enhance the reliability and utility of findings (Naeem et al. 2023). Furthermore, recent advancements in DL and the induction of newer techniques, i.e., autoencoders (Berahmand et al. 2024) and t-distributed stochastic neighbor embedding (t-SNE) (Balamurali 2023), can provide powerful tools for the visualization and understanding of multi-omics data. Further advancements in these methods are expected to provide deeper insights into the genetic and environmental factors related to crop performance.

3.7.3 Reinforcement Learning

The reinforcement learning (RL) is a powerful AI tool in ML where an agent learns to make decisions based on rewards and penalties to achieve the best results by interacting with the environment (Shakya et al. 2023). The ML tools are increasingly used in agriculture and crop improvement for dynamic decision-making tasks requiring adaptive strategies (Gautron et al. 2022). The RL models have advantages over traditional ML algorithms as RL can continuously learn and adapt based on real-world interactions rather than static datasets used in ML (Shakya et al. 2023), making RL suitable for precision farming, automated crop management, and devising climate adaptation strategies. It can be used to simulate future climate scenarios for optimizing genotype selection according to various climate change scenarios for the future (Gautron et al. 2022). Different RL models can be trained on historical climate and genomic datasets to identify crop varieties with the highest probability of resilience in shifting climatic patterns (Tong and Nikoloski 2021). The RL models can also simulate and mimic natural experimentation processes, make decisions, simulate outcomes, and adjust subsequent scenarios based on previous feedback received (Gautron et al. 2022).

The challenges of using RL models involve the requirement of complex datasets in large amounts to effectively train the models (Attri et al. 2023). Training these models also requires higher consumption of computational resources, hardware, and software, and these trained models may not perform well according to changing input conditions, i.e., climate and soil (Gautron et al. 2022). The efficiency of RL models can be enhanced by integrating these models with real-time data support from cameras, drones, field robots, edge computing, and IoT sensors.

3.8 Deep Learning

3.8.1 Convolutional Neural Networks

The convolutional neural networks (CNNs) are among the latest advancements and are at the forefront of deep learning (DL) applications in agriculture and crop production systems, particularly useful for high-throughput phenotyping (Khaki et al. 2020; Tulbure et al. 2022). The CNNs are DL models that use a series of layers for data analysis and pattern identification, i.e., audio, times series, signal analysis, recognition of objects in images, segmentation, etc. (Ketkar et al. 2021), making them invaluable for automating the evaluation of plant traits (Gill et al. 2022). CNNs consist of a convolutional layer that filters and scans input images for various features like textures, edges, and shapes (Khaki et al. 2020), where the activation layer applies nonlinear functions for the pooling layers output that allow the model to identify complex data representations (Kiliçarslan and Celik 2021), followed by a pooling layer (Zafar et al. 2022). The final layer is the fully connected layer, which connects every neuron from previous layers to this layer and identifies relationships and global patterns in data (Ketkar et al. 2021).

Various CNNs can be handy in crop improvement programs (Roy et al. 2021), where they are widely used for morphological traits of plants, including plant height, leaf area (Jiang et al. 2024), and root architecture (Peters et al. 2023). The field application scenario for crop improvement can be the application of CNN-based systems to analyze images of fields captured from drones, identify patterns, and quantify biomass with high precision and accuracy (Yu et al. 2023). These capabilities can be helpful for plant breeding programs to screen large plant populations, thereby accelerating the selection process in crop improvement (Farooq et al. 2024). Disease diagnosis can be another application prospect for CNNs to detect subtle symptoms of various plant diseases (Tulbure et al. 2022) that may have been skipped during human observation; for instance, CNNs have been used previously to identify early-stage symptoms of wheat rust (Schirrmann et al. 2021) and maize leaf blight (Wenxia et al. 2021) which enables timely interventions and avoids losses due to early detection. The accuracy rates of these models can reach up to 90% (Haider and Lavanya 2024), demonstrating their potential to replace or augment traditional scouting methods.

The shortcomings of using this technique, like previously mentioned techniques, involve the prerequisite of large datasets required to train these models, which can be millions of labeled data, which can be an expensive and time-consuming process (Tulbure et al. 2022). Furthermore, the performance of a trained model on a specific dataset may not perform well when it is applied to new crops or changed environmental conditions, requiring additional resources and time to train the model according to newer conditions (Khaki et al. 2020). These shortcomings can be overcome using various techniques, i.e., data augmentation by rotating or cropping images (Divyanth et al. 2022) and transfer learning (Hosna et al. 2022), which can improve model performance, adaptability, and robustness. There are further prospects for seamless integration of CNNs with IoT devices and robotics, which can further enable and enhance real-time automated crop monitoring and phenotyping, etc., in a diverse range of agricultural and crop improvement systems.

3.8.2 Recurrent Neural Networks

The recurrent neural networks (RNNs) are specialized DL models designed to handle and are trained on time series or sequential data to create ML models that are ideal for making sequential conclusions or predictions based on sequential inputs (Khaki et al. 2020; Patel et al. 2024). The CNNs primarily focus on spatial data, but RNNs can adapt and capture temporal patterns, i.e., crop growth changes and environmental conditions over time (Mienye et al. 2024). The RNNs can be particularly applicable in crop improvement programs to model crop growth and development dynamics by using sequential data on rainfall, temperature, soil moisture, and crop performance to predict various growth stages, flowering and harvest time, and crop yield (Khaki et al. 2020; Patel et al. 2024). This capability is also considered useful in precision farming (Divyanth et al. 2022; Elango et al. 2024). RNNs are also regarded as efficient for gene expression analysis (Patel et al. 2024), in which temporal data processing can identify gene activation responses to various stressors, i.e., heat and drought (Tulbure et al. 2022). It can help plant breeders identify genetic pathways that contribute to stress tolerance, thereby contributing to the development of climate-resilient crop varieties.

The use of RNNs also poses some shortcomings, including the requirement for relevant, sequential, and high-quality data (Mienye et al. 2024), requiring higher computational power to train complex RNNs (Tulbure et al. 2022), model interpretation can also be challenging, requiring additional analyses (Khaki et al. 2020), and vanishing gradients (Mienye et al. 2024). The performance and accuracy of RNNs can be further enhanced by utilizing more advanced architectures, i.e., Long Short-Term Memory (LSTM) networks (Gnanasaravanan et al. 2021) and Gated Recurrent Units (GRUs) (Wang et al. 2023). These models have already been applied to assess pest outbreaks and disease progression (Patel et al. 2024), offering valuable decision-support tools for farmers and researchers.

There are prospects of combining RNNs with other DL techniques, i.e., CNNs, which can be used to create hybrid models capable of simultaneously analyzing spatial and temporal data (Khaki et al. 2020). Further research and development on these integrations can revolutionize crop improvement practices like phenotyping, where plant traits evolve over time and across environments.

3.9 Computer Vision

3.9.1 Object Detection

Object detection is an advanced technique in computer vision (CV) that depends on ML or DL algorithms to identify, quantify, and locate specific objects, i.e., fruits, seeds, or plants, in images and videos captured from cameras or drones, replicating human abilities to recognize objects (Zou et al. 2023). It is among the vital applications of CV in agriculture, enabling precise estimations of various plant characteristics (Wu et al. 2021). For example, automation of yield estimation using object detection techniques can significantly reduce labor costs, save counting time, and improve accuracy (Ghazal et al. 2024), i.e., CV can be used in orchards to analyze images to count the number of fruits on trees. These tasks can be more efficiently accomplished using models such as YOLO (You Only Look Once) (Mirhaji et al. 2021) and Faster R-CNN (Region-Based Convolutional Neural Networks) (Mienye et al. 2024), which offer real-time processing capabilities. These models have been successfully applied to various crops to estimate yields, harvesting schedules, and effective resource allocation. i.e., apples, oranges, grapes, etc. (Khaki et al. 2020; Mirhaji et al. 2021; Tulbure et al. 2022; Su et al. 2023). The object detection algorithms can be applied in row crops to identify individual plants, which can provide insights into plant density and help estimate potential yields by counting ears of corn or wheat spikes (Divyanth et al. 2022; Ghazal et al. 2024), etc., which can be critical for the allocation of management practices, including fertilization and irrigation to maximize productivity.

There are some disadvantages associated with using this technique, including viewpoint variation, variability in lighting (Wu et al. 2021), deformation, occlusion (e.g., leaves blocking fruits), cluttered or textured background (Ghazal et al. 2024), and diverse growth patterns in crops (Wu et al. 2021). These issues can be addressed using various techniques, i.e., data augmentation (Divyanth et al. 2022) and domain adaptation (Li et al. 2023). The prospects of object detection include integrating it with IoT, AI, and robotic systems for robust and integrated identification, monitoring, and management of crops or crop breeding trials.

3.9.2 Remote Sensing and Drone Imaging

Remote sensing is an advanced and emerging technology that contributes to images and other remote sensing data through drones or satellites that can be integrated with various AI models and used to monitor multiple field crop parameters and conditions (Meivel and Maheswari 2021). The ability to capture high-resolution images and data to provide a bird's-eye view of the crop enables early detection of crop issues, i.e., nutrient deficiency, water stress, pest infestations, etc. (Wu et al. 2021; Su et al. 2023), and helps to improve crop management. Advancements in drone cameras, i.e., multispectral or hyperspectral cameras, can measure vegetation indices, such as the Normalized Difference Vegetation Index (NDVI) (Meivel and Maheswari 2021), which can provide information about plant vigor based on chlorophyll content (Parida et al. 2024). The Enhanced Vegetation Index (EVI) is an improved version for agricultural fields with dense vegetation (Su et al. 2023). Stress mapping is another practical application of drone imaging (Su et al. 2023), which enables the detection of subtle changes in crop indices, i.e., leaf color or canopy structure, which then allows diseases or drought detection before even becoming visible to the naked eye (Meivel and Maheswari 2021). These indices are primarily based on different CV algorithms capable of accessing crop and field conditions, thereby identifying required patterns and information that can be used for further intervention and minimizing losses.

Besides drones, satellites like Sentinel-2 and Landsat can provide imagery at larger scales (Jiang et al., 2022), enabling regional assessments of crop productivity and land-use patterns (Su et al. 2023). These datasets are particularly valuable for large-scale crop improvement programs or governmental organizations focusing on monitoring food production and optimizing resource allocation. Various drone and satellite imagery can be combined with different ML and DL algorithms for enhanced processing and analysis of data and field conditions (Attri et al. 2023; Resende et al. 2024). The limitations of using drone and satellite imagery include cloud cover, sensor limitations, and high costs of drone deployment and operations (Meivel and Maheswari 2021). Recent advancements in AI-driven image analysis and open-access satellite data can be used to overcome these challenges (Elango et al. 2024). The field drones and satellite data can be seamlessly integrated in real-time with field drones, IoT, and AI systems and algorithms (Su et al. 2023), enabling autonomous functions, i.e., field scouting, resource allocation, and even harvesting, thereby saving resources and labor costs for farmers and crop breeders.

3.10 AI for Multi-Omics Data Analysis

3.11 Natural Language Processing

3.11.1 Analyzing Scientific Literature and Data Reports

The natural language processing (NLP) is an advanced tool emerging in every field of life, including agriculture and crop improvement. It relies on extracting insights by analyzing and interpreting textural data of agricultural research and reports in large volumes using various computer algorithms (Domínguez et al. 2024). The NLP can accelerate the discovery of relevant trends and findings by automating the analysis of unstructured text data, thereby reducing the time required for literature reviews and decision-making (Domínguez et al. 2024). One of the peculiar application case scenarios is text mining, where algorithms are used to analyze scientific articles, patents, and field reports to identify patterns, correlations, and knowledge gaps (D'Souza 2024). In the case of crop improvement, NLP can be used to extract information on gene-trait associations from various genomic studies and guide the breeders in selecting target genes required for crop improvement (Tibbs Cortes et al. 2021). For example, SciBERT is a transformer-based language model that can be trained explicitly on scientific texts and extract the required information, thereby saving a considerable amount of time (D'Souza 2024). Similarly, data synthesis using NLP is another approach that can be used to compile data insights from diverse sources, identify patterns and relationships in data, and make summaries (Domínguez et al. 2024). For example, field trial reports can be analyzed to determine the best and most effective crop management practices for a specific crop or variety under different environmental conditions (Javaid et al. 2023; Farooq et al. 2024).

Presently, NLP is in the initial innovation stage and is facing some challenges, including the need for domain-specific training data and the complexity of jargon used in agriculture and plant breeding programs (Saravanan and Bhagavathiappan 2024). The development of several agriculture-specific ontologies is underway, which can help standardize terminologies and improve the accuracy of NLP models, i.e., Agronomy Ontology (AgrO) (Jonquet et al. 2018), and agriculture domain ontology-based QAS (AOQAS) (Saravanan and Bhagavathiappan 2024). Further research and development can be conducted on developing NLP models capable of integrating text-based insights from other data, i.e., images, videos, and genomic data, thereby creating a unified platform for agricultural research.

3.11.2 Conversational AI for Decision Support

Conversational AI is an advanced and evolving technology revolutionizing every life sector. It can simulate human conversation using NLP models to understand and process human language (Biswas and Goel 2023; Lee and Baek 2023). It is available in various forms and shapes, like chatbots and virtual assistants (Jonquet et al. 2018; Hijazi et al. 2023; Saravanan and Bhagavathiappan 2024), and it is increasingly becoming a valuable resource for farmers and agricultural professionals (Hijazi et al. 2023; Lee and Baek 2023; Samal et al. 2023). These tools can provide handy and real-time recommendations on crop management by leveraging advanced NLP algorithms to interpret user queries and provide actionable and required information and content (Xu et al. 2022). These chatbots can be integrated with various systems, like cameras, robots, weather stations, etc., to advise farmers and breeders on timely and efficient crop management and resource allocation (Biswas and Goel 2023). Various tools are evolving into conversational interfaces that can provide tailored responses based on user descriptions or provided images to answer queries on various crop parameters like plant health and diseases, i.e., Plantix (Samal et al. 2023) and FarmBot (Hijazi et al. 2023).

The accessibility and user-friendliness of conversational AI are predominately high compared to other AI techniques, as they have straightforward and user-friendly interfaces (Biswas and Goel 2023; Lee and Baek 2023). These tools also provide multiple language and voice-based supports, easily accessible from small to large stakeholders, thereby bridging the gap in regions and areas with limited resources (Elango et al. 2024). The primary challenge of using these tools is uncertainties associated with the responses generated by these chatbots, which sometimes make mistakes (Biswas and Goel 2023). Moreover, variations in input data or questions to these chatbots can also lead to non-uniform recommendations, leading to misunderstandings (Dwivedi et al. 2023). The digital literacy barrier may also affect the use and output responses of these chatbots (Lee and Baek 2023). Newer techniques are emerging that can supposedly help to make chatbot recommendations trustworthy and transparent, which include providing the source of information (Biswas and Goel 2023) or techniques like explainable AI (XAI) (Dwivedi et al. 2023). The accuracy and relevance of chatbots improve with training, especially using region-specific data (Lee and Baek 2023). The prospects of conversational AI include its integration with different cameras, drones, robots, IoT devices, and similar configurations (Su et al. 2023) to enable seamless and automated decision-support systems for precision agriculture and crop improvement programs.

3.11.3 Genomics and Genomic Selection

Genomics is the study of the entire genome of organisms, consisting of all genes and genetic information. It has revolutionized crop improvement through insights into the genetic basis of complex plant traits, i.e., yield, stress tolerance, and disease resistance (Jubair and Domaratzki 2023). Genomic selection is the breeding technique used to predict the genetic value and performance of genotypes of an individual using high-density genetic markers from the entire genome, which allows breeders to select desirable traits in a population accurately (Xu et al. 2022). The primary challenge in genomic selection lies in translating massive genomic datasets into usable and actionable information, taking a considerable amount of time and effort (Jubair and Domaratzki 2023). Here, various AI-assisting technologies, i.e., ML (González-Camacho et al. 2018) and DL (Jubair and Domaratzki 2023), can play a crucial role compared to traditional approaches. Various AI models and algorithms can identify optimal combinations of alleles for the desirable traits by analyzing markers across the entire genome, thereby significantly reducing the time, cost, and effort of traditional breeding techniques and cycles (Xu et al. 2022). There are various algorithms available for this purpose, including random forests (Montesinos López et al. 2022), gradient boosting (Bentéjac et al. 2021), and SVMs (Pisner and Schnyer 2020), etc. DL models like CNNs have taken genomic selection to advanced levels by capturing various intricate patterns in genotype–phenotype relationships (Tulbure et al. 2022). The application use case scenario is the prediction of maize yield under diverse environmental conditions by analyzing genomic data alongside different environmental variables (Jubair and Domaratzki 2023).

GWAS are an advanced technique for identifying specific gene loci linked to particular traits (Tibbs Cortes et al. 2021). AI can enhance GWAS by integrating high-dimensional genomic data with phenotypic observations and uncovering subtle associations that traditional breeding and statistical methods may fail to achieve (Tibbs Cortes et al. 2021). Various challenges are also associated with using genomic datasets that need to be addressed in future research on AI, including datasets being noisy, incomplete, or biased toward specific plant species or geographic regions (Jubair and Domaratzki 2023), which requires robust data preprocessing and imputation techniques (Xu et al. 2022). Translating massive genomic datasets using AI algorithms also consumes high performance and excessive computational power and resources that must be considered while using these methods (Farooq et al. 2024). The prospects in the genomic selection using AI are diverse.

They may include integrating multi-omics data, such as transcriptomics and proteomics (Rao et al. 2018), to provide a diverse and comprehensive understanding of plant trait development, which can substantially accelerate the development of high-yielding and climate-resilient crop varieties for the future.

3.11.4 Enviromics

Enviromics is a field of genetics that deals with analyzing and integrating plant genomic data with environmental data, such as temperature, precipitation, and soil conditions, to comprehend and predict the interactions of plants with the surrounding environment (Resende et al. 2024). Recent advancements in AI have played a pivotal role in enhancing enviromics studies by enabling the processing of large-scale environmental datasets and linking them to crop performance (Xu et al. 2022), which can truly enable researchers and plant breeders to develop climate-resilient crop varieties in the face of modern challenges of global climate change. The ML models are widely used to predict and simulate crop responses to various abiotic stressors, i.e., drought, heat, and salinity (Gill et al. 2022). Enviromics can also be combined with AI algorithms to facilitate spatial analysis to integrate environmental data with remote sensing imagery to map areas prone to various stresses (Resende et al. 2024), which enables researchers and plant breeders to prioritize regions for field trials and identify location-specific traits to enhance crop productivity.

An innovative application use case scenario is generative adversarial networks (GANs) (Saxena and Cao 2021), which can simulate future climate scenarios, generate synthetic environmental data (Jubair and Domaratzki 2023), and predict their impacts on crop performance, which enables crop breeders to simulate and explore future scenarios and develop crop varieties suitable for future (Xu et al. 2022). The challenges associated with economics include the heterogeneity and inconsistency of environmental datasets (Genaev et al. 2012; Jubair and Domaratzki 2023), and AI models and algorithms need to be improvised to account for these variations to ensure reliable predictions (Saxena and Cao 2021). Moreover, sophisticated data fusion techniques are required to properly integrate diverse enviromic, genomic, and phenomic datasets (Genaev et al. 2012; Jubair and Domaratzki 2023). AI-powered enviromics is expected to play a critical role in developing climate-resilient crop varieties and climate-smart agricultural practices.

3.11.5 Integration of Genomic, Phenomic, and Enviromic Data

It is the cornerstone of modern crop improvement programs. It is a process of combining genomic, phenomic, and enviromic data for a comprehensive and holistic understanding of the interactions of genes with the environment that influence crop performance, which can help create phenotypic outcomes (Xu et al. 2022; Resende et al. 2024). The input datasets for this purpose are complex, huge, heterogeneous, and multi-dimensional, which can be appropriately handled using AI models and algorithms (Jubair and Domaratzki 2023). The data integration for AI begins with the preprocessing of multi-omics datasets consisting of different formats, scales, and levels of granularity (Xu et al. 2022; Jubair and Domaratzki 2023). These datasets can be harmonized to extract meaningful relationships using algorithms, i.e., autoencoders (Gill et al. 2022) and multi-view learning models (Xu et al. 2024). The application use case scenario is data integration using predictive modeling, combining genomic markers, phenotypic traits, and environmental variables using AI models to predict crop performance (Sheikh et al. 2024). The commonly used DL architectures for this purpose include multilayer perceptrons (MLPs) (Xu et al. 2024) and RNNs (Khaki et al. 2020). AI models can be used to discover hidden traits by analyzing integrated datasets to identify previously unknown interactions between genes and environmental factors (Xu et al. 2022). These insights can pave the way for developing multi-trait breeding programs to simultaneously optimize crops for yield, stress tolerance, and nutrient contents.

Currently, several challenges and hurdles are associated with multi-omics integration due to the sheer volume and complexity of datasets that are difficult to handle and can consume considerable computational resources (Xu et al. 2022). Data quality and consistency remain challenging (Jubair and Domaratzki 2023), in addition to the interpretability of various AI models and algorithms (Feuerriegel et al. 2024) because breeders and researchers need to understand the biological basis to trust the predictions and subsequently apply them effectively.

3.12 Symbolic AI

3.12.1 Expert Systems

Expert systemsare logic-based reasoning and rule-driven algorithms that harness the power of AI to achieve decision-making capabilities similar to human experts to solve complex problems in many fields of life, i.e., medicine, agriculture, and manufacturing (Sharma 2021). These systems rely on a predefined set of rules and knowledge bases to mimic human expertise, offering precise recommendations for various agricultural systems, i.e., pest and disease management (Ibrahim et al. 2022). These systems typically consist of three or four essential and main components, i.e., knowledge base, inference engine, if-then rules, and user interface (Saleem et al. 2021). The knowledge base comprises domain-specific knowledge, i.e., pest lifecycles, crop characteristics, and environmental conditions (Ibrahim et al. 2022), where the inference engine applies logical reasoning and a set of if-then rules according to the knowledge base to drive conclusions or recommendations (Saleem et al. 2021), and user interface allows users to input data and receive actionable insights or information (Sharma 2021).

Expert systems can function properly in resource-constrained environments and do not require large datasets for training (Saleem et al. 2021), which enables smallholder farmers and breeders to benefit from these systems equally and improve efficiency (Farooq et al. 2024). Expert systems also have the ability to incorporate region-specific knowledge and tailor recommendations according to local agricultural and environmental conditions (Sharma 2021). However, the rigidity of rule-based systems hinders their ability to be implemented in novel situations because they rely heavily on the completeness and accuracy of the available knowledge base, which can sometimes become outdated with time, the introduction of new pests and diseases, or climate condition changes (Davis and King 1984; Sharma 2021). Hybrid systems combined with symbolic AI and ML are being developed to address these issues (Gill et al. 2022), and they can learn from the induction of new datasets while retaining the interpretability of rule-based approaches. There is further potential in using expert systems in combination with the data achieved from various drones, cameras, IoT devices, mobile applications, etc., to provide real-time monitoring and assessment capabilities, which can enable researchers and breeders to automate multiple tasks in agricultural experiments, thereby minimizing resource consumption and external support.

3.12.2 Inductive Logic Programming

Inductive Logic Programming (ILP) is an advanced subfield of ML that relies on first-order logic to represent data and hypotheses, learn rules and patterns from examples, and incorporate background knowledge within a structured data framework to create logical reasoning that can explain positive while rejecting negative instances (Cropper and Dumančić 2022). Traditional ML algorithms rely on and operate based on numerical data. At the same time, ILP works based on logical data (Zhang et al. 2023), making it particularly useful for agricultural and crop breeding applications involving qualitative and relational data (Farooq et al. 2024). ILP can be used to discover relationships between genetic traits and various environmental factors (Jubair and Domaratzki 2023) by analyzing datasets of various crop genotypes and phenotypes and subsequently identifying rules to explain specific gene–environment interactions (Xu et al. 2022) to develop climate-resilient crop varieties for the future. These techniques are particularly applicable to plant breeding programs that intend to target multiple traits simultaneously (Jubair and Domaratzki 2023).

Hypothesis generation is another perspective where ILP can propose new rules or relationships by analyzing experimental datasets that researchers and breeders may have overlooked (Cropper and Dumančić 2022). Interpretability of ILP is more straightforward as it consists of human-readable rules that can be directly understood and applied by researchers and crop breeders compared to other techniques (Zhang et al. 2023). Your strengths can sometimes become weaknesses, as ILP relies on logical rules and struggles with noisy or incomplete data, making it less flexible than statistical methods. Moreover, ILP algorithms are computationally intensive and require significant processing power in the case of large datasets (Cropper and Dumančić 2022). ILP can be used in crop improvement for multi-omics integration and precision agriculture (Zhang et al. 2023). It has the potential to be integrated with real-time data streams from IoT sensors and remote sensing platforms to develop dynamic and rule-based systems capable of adapting to changing environmental conditions.

3.13 Conclusions and Future Perspectives

There is substantial potential in using AI technologies for agriculture and crop improvement programs. It can be a game-changer due to its considerable ability to process and analyze large datasets, surpassing far beyond the traditional approaches. AI is already defining and transforming crop improvement and management methods, ranging from genomic selection, management strategies and practices, and yield predictions. AI tools and algorithms, i.e., ML, DL, CV, etc., have enabled high-throughput phenotyping, real-time decision-making, and designing climate-resilient crop varieties. Various AI techniques are being investigated for their potential to accelerate crop breeding cycles, optimize resource use, and increase agricultural productivity to meet emerging global challenges, i.e., climate change, population growth, and food insecurity. AI also has the potential to facilitate sustainable farming practices by optimizing resource allocation, reducing waste, and minimizing the environmental footprint in the agriculture sector. The current challenges faced by the use of AI in crop improvement in general and agriculture specifically include higher costs, resource allocation, infrastructure creation, data quality and accessibility, the digital divide, and ethical concerns related to labor displacement and transparency. Addressing these issues or creating a way forward for these challenges can benefit all stakeholders and contribute to a more sustainable and equitable agricultural system.

There is a need for synergy between modern AI and traditional crop improvement and agricultural practices because AI tools will continue to evolve and potentially become more integrated into diverse farming practices. For example, integrating live datasets from various cameras, drones, IoT devices, and edge computing can enable real-time monitoring and adaptive management of research trials and crops. In the meantime, various labor-intensive tasks like planting, irrigation, weeding, and harvesting can be automized seamlessly with robotics. These accomplishments are data- and budget-intensive, requiring collaborations across the board from multiple sectors, i.e., government institutions, academic institutions, private companies, crop breeders, and farmers, to make AI tools accessible, affordable, and tailored to the needs of all stakeholders in diverse agricultural systems. AI can empower crop breeders and farmers if incentives, i.e., open-source AI platforms and data accessibility, are provided to increase productivity, sustainability, and resilience in the agriculture sector.

References

Attri, I., Awasthi, L.K., Sharma, T.P. et al. (2023). A review of deep learning techniques used in agriculture. *Ecological Informatics* 102217.

Badapanda, K., Mishra, D.P., and Salkuti, S.R. (2022). Agriculture data visualization and analysis using data mining techniques: application of unsupervised machine learning. *TELKOMNIKA (Telecommunication Computing Electronics and Control)* 20: 98–108.

Bailey-Serres, J., Parker, J.E., Ainsworth, E.A. et al. (2019). Genetic strategies for improving crop yields. *Nature* 575: 109–118.

Balamurali, M. (2023). T-Distributed stochastic neighbor embedding. In: *Encyclopedia of mathematical geosciences*, 1527–1535. Springer.

Ben Ayed, R. and Hanana, M. (2021). Artificial intelligence to improve the food and agriculture sector. *Journal of Food Quality* 2021: 5584754.

Bentéjac, C., Csörgő, A., and Martínez-Muñoz, G. (2021). A comparative analysis of gradient boosting algorithms. *Artificial Intelligence Review* 54: 1937–1967.

Berahmand, K., Daneshfar, F., Salehi, E.S. et al. (2024). Autoencoders and their applications in machine learning: a survey. *Artificial Intelligence Review* 57: 28.

Biswas, R. and Goel, D.N. (2023). Intelligent chatbot assistant in agriculture domain. In: *Agriculture-Centric Computation: First International Conference, ICA 2023,* Chandigarh, India, May 11–13, 2023, Revised Selected Papers. Springer Nature p. 180.

Boukerche, A., Zheng, L., and Alfandi, O. (2020). Outlier detection: methods, models, and classification. *ACM Computing Surveys (CSUR)* 53: 1–37.

Bwambale, E., Abagale, F.K., and Anornu, G.K. (2023). Data-driven model predictive control for precision irrigation management. *Smart Agricultural Technology* 3: 100074.

Chapagain, R., Huth, N., Remenyi, T.A. et al. (2023). Assessing the effect of using different APSIM model configurations on model outputs. *Ecological Modelling* 483: 110451.

Cropper, A. and Dumančić, S. (2022). Inductive logic programming at 30: a new introduction. *Journal of Artificial Intelligence Research* 74: 765–850.

Das, S., Christopher, J., Apan, A. et al. (2021). UAV-Thermal imaging and agglomerative hierarchical clustering techniques to evaluate and rank physiological performance of wheat genotypes on sodic soil. *ISPRS Journal of Photogrammetry and Remote Sensing* 173: 221–237.

Davis, R. and King, J.J. (1984). *The Origin of Rule-based Systems in AI*. Rule-based expert systems: The MYCIN experiments of the Stanford Heuristic Programming Project.

De Ville, B. (2013). Decision trees. *Wiley Interdisciplinary Reviews: Computational Statistics* 5: 448–455.

Divyanth, L., Guru, D., Soni, P. et al. (2022). Image-to-image translation-based data augmentation for improving crop/weed classification models for precision agriculture applications. *Algorithms* 15: 401.

Domínguez, A.G., Roig-Tierno, N., Chaparro-Banegas, N. et al. (2024). Natural language processing of social network data for the evaluation of agricultural and rural policies. *Journal of Rural Studies* 109: 103341.

Dora, M., Kumar, A., Mangla, S.K. et al. (2022). Critical success factors influencing artificial intelligence adoption in food supply chains. *International Journal of Production Research* 60: 4621–4640.

D'Souza, J. (2024). Agriculture named entity recognition—towards FAIR, reusable scholarly contributions in agriculture. *Knowledge* 4: 1–26.

Dubois, D., Prade, H., and Ughetto, L. (1999). Fuzzy logic, control engineering and artificial intelligence. *Fuzzy Algorithms for control* 17–57.

Dwivedi, R., Dave, D., Naik, H. et al. (2023). Explainable AI (XAI): core ideas, techniques, and solutions. *ACM Computing Surveys* 55: 1–33.

Elango, E., Hanees, A., Shanmuganathan, B. et al. (2024). Precision agriculture: a novel approach on AI-driven farming. In: *Intelligent Robots and Drones for Precision Agriculture,* 119–137. Springer.

Ellis, E. and Paustian, K. (2024). Importance of on-farm research for validating process-based models of climate-smart agriculture. *Carbon Balance and Management* 19: 16.

Falcon, W.P., Naylor, R.L., and Shankar, N.D. (2022). Rethinking global food demand for 2050. *Population and Development Review* 48: 921–957.

Farooq, M.A., Gao, S., Hassan, M.A. et al. (2024). Artificial intelligence in plant breeding. *Trends in Genetics* .

Feuerriegel, S., Hartmann, J., Janiesch, C. et al. (2024). Generative ai. *Business & Information Systems Engineering* 66: 111–126.

Fuller, D.Q. and Stevens, C.J. (2019). Between domestication and civilization: the role of agriculture and arboriculture in the emergence of the first urban societies. *Vegetation History and Archaeobotany* 28: 263–282.

Gautron, R., Maillard, O.-A., Preux, P. et al. (2022). Reinforcement learning for crop management support: review, prospects and challenges. *Computers and Electronics in Agriculture* 200: 107182.

Genaev, M., Doroshkov, A., Morozova, E. et al. (2012). WheatPGE: a system for analysis of relationships among the phenotype, genotype, and environment in wheat. *Russian Journal of Genetics: Applied Research* 2: 262–269.

Ghazal, S., Munir, A., and Qureshi, W.S. (2024). Computer vision in smart agriculture and precision farming: techniques and applications. In: *Artificial Intelligence in Agriculture*.

Gill, T., Gill, S.K., Saini, D.K. et al. (2022). A comprehensive review of high throughput phenotyping and machine learning for plant stress phenotyping. *Phenomics* 2: 156–183.

GNAFC (2024). *Global Report on Food Crises (GRFC)*. Food Security Information Network (FSIN).

Gnanasaravanan, S., Tharani, B., and Sahu, M. (2021). Long short-term memory recurrent neural networks for plant disease identification. *Plant Pathology* 3.

González-Camacho, J.M., Ornella, L., Pérez-Rodríguez, P. et al. (2018). Applications of machine learning methods to genomic selection in breeding wheat for rust resistance. *The Plant Genome* 11: 170104.

Haider, H. and Lavanya, B. (2024). Revolutionizing agritech with deep learning-enhanced remote sensing for precision agricultural management. *PatternIQ Mining* 1: 63–76.

Hajjarpoor, A., Nelson, W.C., and Vadez, V. (2022). How process-based modeling can help plant breeding deal with G x E x M interactions. *Field Crops Research* 283: 108554.

Hamany Djande, C.Y., Piater, L.A., Steenkamp, P.A. et al. (2021). A metabolomics approach and chemometric tools for differentiation of barley cultivars and biomarker discovery. *Metabolites* 11: 578.

Hamed, M.A., El-Habib, M.F., Sababa, R.Z. et al. (2024). Artificial intelligence in agriculture: enhancing productivity and sustainability. *International Journal of Engineering and Information Systems* 8: 1–5.

Hamilton, J.D. (2020). *Time series analysis*. Princeton: Princeton university press.

Hijazi, O., Schoo, D., Schweiger, A. et al. (2023). Comparative life cycle assessment of FarmBot technology with conventional and organic agriculture. In: *2023 ASABE Annual International Meeting*, 1. American Society of Agricultural and Biological Engineers.

Hoogenboom, G., Porter, C., Shelia, V. et al. (2024). *Decision Support System for Agrotechnology Transfer (DSSAT) Version 4.8.5* (www.DSSAT.net). Gainesville, Florida, USA: DSSAT Foundation.

Hosna, A., Merry, E., Gyalmo, J. et al. (2022). Transfer learning: a friendly introduction. *Journal of Big Data* 9: 102.

Ibrahim, E.A., Salifu, D., Mwalili, S. et al. (2022). An expert system for insect pest population dynamics prediction. *Computers and Electronics in Agriculture* 198: 107124.

Javaid, M., Haleem, A., Khan, I.H. et al. (2023). Understanding the potential applications of artificial intelligence in agriculture sector. *Advanced Agrochem* 2: 15–30.

Jiang, J., Johansen, K., Tu, Y.H., and McCabe, M.F. (2022). Multi-sensor and multi-platform consistency and interoperability between UAV, Planet CubeSat, Sentinel-2, and Landsat reflectance data. *GIScience & Remote Sensing* 59 (1): 936–958.

Jiang, C., Yeddula, S.D., and Ku, W.-S. (2024). A convolutional neural network model for accurate short-term leaf area index prediction. *Modeling Earth Systems and Environment* 10: 6397–6412.

Jonquet, C., Toulet, A., Arnaud, E. et al. (2018). AgroPortal: a vocabulary and ontology repository for agronomy. *Computers and Electronics in Agriculture* 144: 126–143.

Jubair, S. and Domaratzki, M. (2019). Ensemble supervised learning for genomic selection. In: *2019 IEEE International Conference on Bioinformatics and Biomedicine (BIBM)*, 1993–2000. IEEE.

Jubair, S. and Domaratzki, M. (2023). Crop genomic selection with deep learning and environmental data: a survey. *Frontiers in Artificial Intelligence* 5: 1040295.

Kazerouni, A., Zhao, Q., Xie, J. et al. (2020). Active learning for skewed data sets. *arXiv preprint arXiv:2005.11442* .

Ketkar, N., Moolayil, J., Ketkar, N. et al. (2021). Convolutional neural networks. *Deep Learning with Python: Learn Best Practices of Deep Learning Models with PyTorch* 197–242.

Khaki, S., Wang, L., and Archontoulis, S.V. (2020). A CNN-RNN framework for crop yield prediction. *Frontiers in Plant Science* 10: 1750.

Kiliçarslan, S. and Celik, M. (2021). RSigELU: a non-linear activation function for deep neural networks. *Expert Systems with Applications* 174: 114805.

Kumar, A., Shumayla, and Upadhyay, S.K. (2024). CRISPR/Cas-Mediated Genome Editing in Plants: a Historical Perspective. *Applications of Genome Engineering in Plants* 1–19: https://doi.org/10.1002/9781394183913.ch1.

Lee, C.-S. and Baek, H.-J. (2023). A study on the need for AI literacy according to the development of artificial intelligence chatbot. *The Journal of the Korea Institute of Electronic Communication Sciences* 18: 421–426.

Li, Y., Zhan, X., Liu, S. et al. (2023). Self-supervised plant phenotyping by combining domain adaptation with 3D plant model simulations: application to wheat leaf counting at seedling stage. *Plant Phenomics* 5: 0041.

Liu, H., Chen, W., Li, Y. et al. (2022). CRISPR/Cas9 technology and its utility for crop improvement. *International Journal of Molecular Sciences* 23: 10442.

Liu, X., Wang, Z., Gmitter, F.G. Jr., et al. (2023). Effects of different rootstocks on the metabolites of Huanglongbing-affected sweet orange juices using a novel combined strategy of untargeted metabolomics and machine learning. *Journal of Agricultural and Food Chemistry* 71: 1246–1257.

Maillard, D. (2021). The obsolescence of man in the digital society. *International Journal for Applied Information Management* 1: 99–124.

Meivel, S. and Maheswari, S. (2021). Remote sensing analysis of agricultural drone. *Journal of the Indian Society of Remote Sensing* 49: 689–701.

Mienye, I.D., Swart, T.G., and Obaido, G. (2024). Recurrent neural networks: A comprehensive review of architectures, variants, and applications. *Information* 15: 517.

Mirhaji, H., Soleymani, M., Asakereh, A. et al. (2021). Fruit detection and load estimation of an orange orchard using the YOLO models through simple approaches in different imaging and illumination conditions. *Computers and Electronics in Agriculture* 191: 106533.

Montesinos López, O.A., Montesinos López, A., and Crossa, J. (2022). Random forest for genomic prediction. In: *Multivariate Statistical Machine Learning Methods for Genomic Prediction*, 633–681. Springer.

Naeem, S., Ali, A., Anam, S. et al. (2023). An unsupervised machine learning algorithms: comprehensive review. *International Journal of Computing and Digital Systems.*

Parida, P.K., Somasundaram, E., Krishnan, R. et al. (2024). Unmanned aerial vehicle-measured multispectral vegetation indices for predicting LAI, SPAD Chlorophyll, and yield of maize. *Agriculture* Basel 14.

Patel, V.K., Abhishek, K., and Selvarajan, S. (2024). Optimized recurrent neural network-based early diagnosis of crop pest and diseases in agriculture. *Discover Computing* 27: 43.

Peters, B., Blume-Werry, G., Gillert, A. et al. (2023). As good as human experts in detecting plant roots in minirhizotron images but efficient and reproducible: The convolutional neural network "RootDetector". *Scientific Reports* 13: 1399.

Pfeifer, B., Alachiotis, N., Pavlidis, P. et al. (2020). Genome scans for selection and introgression based on k-nearest neighbour techniques. *Molecular Ecology Resources* 20: 1597–1609.

Pisner, D.A. and Schnyer, D.M. (2020). Support vector machine. In: *Machine Learning*, 101–121. In: Elsevier.

Rao, M., Rao, V.R., and Angadi, S. (2018). Crop Improvement in the World-Past. In: *Present and Future*, 22. Asian Agri-History.

Rattan, P., Penrice, D.D., and Simonetto, D.A. (2022). Artificial intelligence and machine learning: what you always wanted to know but were afraid to ask. *Gastro Hep Advances* 1: 70–78.

Razavi, S. (2021). Deep learning, explained: fundamentals, explainability, and bridgeability to process-based modelling. *Environmental Modelling & Software* 144: 105159.

Resende, R.T., Hickey, L., Amaral, C.H. et al. (2024). Satellite-enabled enviromics to enhance crop improvement. *Molecular Plant* 17: 848–866.

Roy, S.K., Mondal, R., Paoletti, M.E. et al. (2021). Morphological convolutional neural networks for hyperspectral image classification. *IEEE Journal of Selected Topics in Applied Earth Observations and Remote Sensing* 14: 8689–8702.

Saadane, R., Chehri, A., and Jeon, S. (2022). AI-based modeling and data-driven evaluation for smart farming-oriented big data architecture using IoT with energy harvesting capabilities. *Sustainable Energy Technologies and Assessments* 52: 102093.

Saleem, H., Khan, A.R., Jilani, T.A. et al. (2021). A comprehensive review on the application of diagnostic expert systems in the field of agriculture. *International Journal on Emerging Technologies (Research Trend)* 12: 304–316.

Samal, I., Bhoi, T.K., Pradhan, A.K. et al. (2023). Plantix app: a success story of artificial intelligence in plant protection. *Van Sangyan* 24.

Sambo, F., Trifoglio, E., Di Camillo, B. et al. (2012). Bag of Naïve Bayes: biomarker selection and classification from genome-wide SNP data. *BMC Bioinformatics* 13: 1–10.

Saravanan, K.S. and Bhagavathiappan, V. (2024). AOQAS: ontology based question answering system for agricultural domain. *International Journal of Computer Information Systems and Industrial Management Applications* 16: 16–16.

Saxena, D. and Cao, J. (2021). Generative adversarial networks (GANs) challenges, solutions, and future directions. *ACM Computing Surveys (CSUR)* 54: 1–42.

Schirrmann, M., Landwehr, N., Giebel, A. et al. (2021). Early detection of stripe rust in winter wheat using deep residual neural networks. *Frontiers in Plant Science* 12: 469689.

Schoech, D., Jennings, H., Schkade, L.L. et al. (1985). Expert systems: artificial intelligence for professional decisions. *Computers in Human Services* 1: 81–115.

Shakya, A.K., Pillai, G., and Chakrabarty, S. (2023). Reinforcement learning algorithms: a brief survey. *Expert Systems with Applications* 231: 120495.

Sharma, R. (2021). Artificial intelligence in agriculture: a review. In: *2021 5th international conference on intelligent computing and control systems (ICICCS)*, 937–942. IEEE.

Sheikh, M., Iqra, F., Ambreen, H. et al. (2024). Integrating artificial intelligence and high-throughput phenotyping for crop improvement. *Journal of Integrative Agriculture* 23: 1787–1802.

Su, J., Zhu, X., Li, S. et al. (2023). AI meets UAVs: a survey on AI empowered UAV perception systems for precision agriculture. *Neurocomputing* 518: 242–270.

Tae, K.H. and Whang, S.E. (2021). Slice tuner: A selective data acquisition framework for accurate and fair machine learning models. In: *Proceedings of the 2021 International Conference on Management of Data*, 1771–1783.

Tibbs Cortes, L., Zhang, Z., and Yu, J. (2021). Status and prospects of genome-wide association studies in plants. *The Plant Genome* 14: e20077.

Tong, H. and Nikoloski, Z. (2021). Machine learning approaches for crop improvement: leveraging phenotypic and genotypic big data. *Journal of Plant Physiology* 257: 153354.

Tulbure, A.-A., Tulbure, A.-A., and Dulf, E.-H. (2022). A review on modern defect detection models using DCNNs–Deep convolutional neural networks. *Journal of Advanced Research* 35: 33–48.

Tyagi, K., Rane, C., Sriram, R. et al. (2022). Unsupervised learning. In: *Artificial Intelligence and Machine Learning for Edge Computing*, 33–52. Elsevier.

Upadhyay, S.K. (2021). *Genome Engineering for Crop Improvement*. Wiley. https://doi.org/10.1002/9781119672425.

Upadhyay, S.K. (2024). *Applications of Genome Engineering in Plants*. Wiley. https://doi.org/10.1002/9781394183913.

Wang, H., Zhang, L., and Zhao, J. (2023). Application of a fusion attention mechanism-based model combining bidirectional gated recurrent units and recurrent neural networks in soil nutrient content estimation. *Agronomy* 13: 2724.

Wenxia, B., Xuefeng, H., Gensheng, H. et al. (2021). Identification of maize leaf diseases using improved convolutional neural network. *Transactions of the Chinese Society of Agricultural Engineering* 37.

Wu, Z., Chen, Y., Zhao, B. et al. (2021). Review of weed detection methods based on computer vision. *Sensors* 21: 3647.

Xu, Y., Zhang, X., Li, H. et al. (2022). Smart breeding driven by big data, artificial intelligence, and integrated genomic-enviromic prediction. *Molecular Plant* 15: 1664–1695.

Xu, C., Si, J., Guan, Z. et al. (2024). Reliable conflictive multi-view learning. In: *Proceedings of the AAAI Conference on Artificial Intelligence*, 16129–16137.

Xuan, Y., Small, E., Sokol, K. et al. (2023). Can users correctly interpret machine learning explanations and simultaneously identify their limitations? *arXiv preprint arXiv:2309.08438*.

Yan, J. and Wang, X. (2022). Unsupervised and semi-supervised learning: the next frontier in machine learning for plant systems biology. *The Plant Journal* 111: 1527–1538.

Yu, D., Zha, Y., Sun, Z. et al. (2023). Deep convolutional neural networks for estimating maize above-ground biomass using multi-source UAV images: A comparison with traditional machine learning algorithms. *Precision Agriculture* 24: 92–113.

Zafar, A., Aamir, M., Mohd Nawi, N. et al. (2022). A comparison of pooling methods for convolutional neural networks. *Applied Sciences* 12: 8643.

Zhang, Z., Yilmaz, L., and Liu, B. (2023). A critical review of inductive logic programming techniques for explainable AI. *IEEE Transactions on Neural Networks and Learning Systems* 35 (8): 10220–10236.

Zou, Z., Chen, K., Shi, Z. et al. (2023). Object detection in 20 years: A survey. *Proceedings of the IEEE* 111: 257–276.

4

Advancements in Phenotyping and High-Throughput Imaging

Shrishti, Avani Vasudeva, and Papiya Mukherjee

Department of Botany, Panjab University, Sector 14, Chandigarh 160014, India

4.1 Introduction

Plant phenotyping, the thorough examination of observable plant traits determined by genetic and environmental factors, has become a cornerstone of modern agriculture and crop improvement research. It integrates advanced tools and methods for quantifying morphological, physiological, and pathological characteristics of plants, aiding in understanding their growth, development, and stress responses (Zhang and Zhang 2018). By combining high-throughput imaging systems with advanced data processing techniques, researchers can capture detailed, multi-modal data at unprecedented scales, offering new insights into plant behavior and performance under diverse conditions (Sheikh et al. 2024). The introduction of high-throughput phenotyping platforms, both ground-based and aerial, has revolutionized data collection. Ground-based systems, such as stationary platforms, vehicle-mounted units, and autonomous robots, provide high-resolution imaging for precise trait analysis. Meanwhile, aerial platforms, including unmanned aerial vehicles (UAVs), manned aircraft, and satellites, facilitate large-scale monitoring of crop fields, overcoming logistical challenges and providing rapid, non-invasive assessments (Zhang and Zhang 2018). These platforms employ cutting-edge imaging technologies, such as visible band imaging, thermal imaging, fluorescence imaging, spectral imaging, stereovision imaging, and tomographic imaging, to capture diverse aspects of plant morphology, physiology, and pathology. The combination of artificial intelligence (AI), machine learning (ML), and computer vision has further improved the efficacy and accuracy of phenotyping (Cembrowska-Lech et al. 2023). Advanced algorithms now enable the segmentation of plant components, extraction of complex traits, and the prediction of yield and stress responses (Singh et al. 2018). These technological advancements support the development of climate-resilient crops, improved resource use efficiency, and the identification of stress-tolerant genotypes, aligning with the global demand for sustainable agricultural practices. This article delves into the innovative methods, platforms, and imaging modalities driving the field of high-throughput phenotyping, underscoring their transformative impact on crop improvement and agricultural research.

4.2 Crop Phenotyping

Plant phenotyping is the qualitative and quantitative analysis of plant traits, such as morphology, physiology, and biochemical properties, and their interaction with genetic and environmental variables. It plays a vital role in crop improvement by allowing the identification of traits associated with yield, stress tolerance, disease resistance, and resource use efficiency (Das Choudhury et al. 2019). Modern phenotyping methods incorporate advanced technologies, including high-throughput imaging, sensor-based platforms, robotics, and AI, offering precise and efficient measurement of plant traits across spatial and temporal scales (Naqvi et al. 2024). These tools allow researchers to study plant responses under both controlled and field circumstances, bridging the genotype–phenotype gap (Park et al. 2024). Phenotyping across diverse genetic backgrounds and environments generates robust datasets essential for genomic selection, gene discovery, and precision breeding (Xu et al. 2022). Ultimately, this leads to the development of climate-resilient cultivars with enhanced agronomic performance, nutritional quality, and adaptability, contributing to global food security and sustainable agriculture.

4.3 Phenotyping Platforms

Phenotyping platforms are systems equipped with various sensors and technologies designed to capture and analyze physical and biochemical traits of a plant in a standardized manner for research and breeding applications. These platforms are categorized primarily into "Ground-based" and "Aerial" types, each tailored to specific observational needs (Zhang and Zhang 2018). Ground-based platforms capture detailed and high-resolution data. Aerial platforms, on the other hand, offering broader field coverage for large-scale crop assessments (Zhang and Zhang 2018; Abebe et al. 2023). Each platform type is optimized to monitor traits like growth, stress responses, and yield under varying conditions, providing comprehensive data to enhance research and breeding efforts (Table 4.1).

4.3.1 Ground-Based Platforms

Ground-based phenotyping platforms offer precise, high-resolution data collection at a close range and are instrumental for plant trait analysis directly in the field (Andrade-Sanchez et al. 2013). Operating at a close range, these platforms provide superior spatial resolution compared to aerial platforms (Davis et al. 2024). This category encompasses stationary systems, vehicle-mounted units, self-propelled robots, and portable devices, each facilitating non-invasive data acquisition through advanced sensors, imaging tools, and software to monitor plant growth, physiology, and health (Zhang and Zhang 2018). Overall, ground-based platforms provide adaptable, accurate phenotyping options; however, each type has trade-offs in terms of mobility, field coverage, cost, and slow speed of covering large areas (Abebe et al. 2023). Balancing these factors allows for tailored approaches to meet diverse crop assessment needs in agricultural research.

Table 4.1 Various plant crops were phenotyped for different traits using ground-based and aerial-based phenotyping platforms.

Platform	Designation	Sensors/imaging techniques	Plants	Traits measured	References
Ground-based platforms					
	Self-developed	Terrestrial laser scanner	Maize, soybean, wheat	Canopy height	Friedli et al. (2016)
	Self-developed	RGB, NIR	Rice	Shoot biomass panicle number, grain weight	Naito et al. (2017)
	CropQuant	RGB, NIR	General crop	Crop growth rate	Zhou et al. (2017)
Stationary	PhenoCam	RGB	Ecosystem	Canopy greenness	Richardson et al. (2018)
	Self-developed	Hemispherical video camera	Wheat, oat, barley	Crop lodging	Susko et al. (2018)
	BreedVision	Hyperspectral, 3D ToF	Triticales	Moisture content, growth stage, tiller density, and nitrogen content	Busemeyer et al. (2013)
	Self-developed	Ultrasonic, RGB, thermal IR, spectrometers	Soybean, Wheat	Canopy height, NDVI, canopy temperature	Bai et al. (2016)
Vehicle-based	GPhenoVision	RGB-D, thermal, hyperspectral	Cotton	Canopy height, canopy width, growth rate, projected leaf area, volume, yield	Jiang et al. (2018)
	Phenomobile lite	LiDAR, RGB, NDVI	Wheat	Plant height, biomass, ground cover	Jimenez-Berni et al. (2018)
	Self-developed	LiDAR	Maize	Plant height	Qiu et al. (2019)

(Continued)

Table 4.1 (Continued)

Platform	Designation	Sensors/imaging techniques	Plants	Traits measured	References
Self-propelled	Phenocart	Infrared, thermal imaging	Wheat	NDVI, canopy temperature	Crain et al. (2016)
	Field scanalyzer	RGB, FLUO, thermal IR, hyperspectral, 3D laser scanner	Wheat	Canopy height, spike number, canopy closure	Virlet et al. (2017)
	Self-developed	RGB	Indian bread wheat	Spike architecture based on rain fed conditions	Abdi et al. (2024)
	Pheno-Robot	LiDAR	General crops	Crop density and color	Pan et al. (2024)
	Self-developed	RGB	Tartary buckwheat	Seed traits	Park et al. (2024)
	Husky	3D-LiDAR, FARO, LiDAR	General crops	Crop density	Sanchez (2024)
Aerial-based platforms					
Unmanned aerial platforms (UAPs)	Anaconda	RGB, multispectral	Sorghum, maize	Plant height	Shi et al. (2016)
	Phantom DJI 3	RGB	Soybean	Weed detection	dos Santos Ferreira et al. (2017)
	UAV	Multispectral imaging	Citrus	Canopy area, NDVI, tree gaps	Ampatzidis and Partel (2019)
	Phantom 4	RGB	Induced lettuce lines	Carotenoid content	Maciel et al. (2019)
	3DR Solo quadcopter	Multispectral	Maize	NDVI, chlorophyll red-edge index, high conical reflectance factor	Fawcett et al. (2020)
	Matrice 600 Pro	RGB, multispectral camera, infrared thermal	Cotton	Yield	Feng et al. (2020)

	Self-developed octorotor	RGB, multispectral	Rice	Canopy height, canopy coverage	Wan et al. (2020)
	Matrice M300-RTK quadcopter	RGB, NIR, multispectral	Bread wheat	Spike and kernel count	Mohammed et al. (2024)
	DJI M300	RGB	Rice	Yield	Lu et al. (2024)
	Self-developed	Multispectral	Sugarbeet	Soil plant analysis development	Gao et al. (2024)
	Self-developed	LiDAR	Maize	Biomass	Li et al. (2015)
	AT-402B	RGB	General crop	Pest severity	Yang and Hoffmann (2015)
Manned aerial platforms (MAPs)					
Satellite	RapidEye	Multispectral, hyperspectral	Wheat	Nitrogen content	Basso et al. (2016)
	GeoEye-1	Multispectral	Turfgrasses	Nitrogen content	Caturegli et al. (2015)
	FLEX (Fluorescence explorer)	Fluorescence imaging spectrometer (FLORIS)	General crops	Photosynthesis	Drusch et al. (2016)
	Sentinel-1	Synthetic aperture radar	General crops	Enhancing agricultural management practices	Christiansen et al. (2019)

4.3.1.1 Stationary Platforms

Stationary platforms are pole- or tower-based systems in which sensors are mounted at the top of a fixed-height pole made of materials like aluminum, steel, or plastic fiber. For instance, Friedli et al. (2016) utilized a pole-based terrestrial laser scanning platform for monitoring the canopy height growth in plants of maize, soybean, and wheat. This setup involved a laser scanner mounted upside down on a 3.8 m high aluminum elevator tripod. Stationary systems are limited by their fixed positioning and reduced field coverage.

4.3.1.2 Vehicle-Based Platforms

Vehicle-based platforms comprise of high-efficiency sensors and cameras mounted on or behind tractor-like systems that provide mobility, cover large areas, and enable data collection across plots (Stentz et al. 2002). BreedVision platform is a tractor-mounted system to assess traits such as moisture content, plant height, and nitrogen content across 25 Triticale genotypes and proved highly efficient (Busemeyer et al. 2013). Despite offering broader range coverage, vehicle-based platforms can be strictly restricted by terrain and crop structure (Li et al. 2021).

4.3.1.3 Self-Propelled Platforms

Self-propelled platforms represent the recent advancements, integrating robotics for autonomous navigation and sensor-based imaging. For instance, PhenoCart is a system developed by Crain et al. (2016) that employed infrared and thermal sensors such as GreenSeeker for spectral reflectance to measure canopy temperature and normalized difference variation index (NDVI) in 10 different wheat breeding lines. These platforms offer high flexibility and precision but may involve high initial costs and complex management (Paul et al. 2022).

4.3.2 Aerial Platforms

Aerial platforms comprising UAVs and satellites have revolutionized high-throughput plant phenotyping by offering rapid, non-destructive monitoring over large areas that offer advantages over ground-based systems by covering extensive fields quickly, overcoming accessibility issues, and minimizing environmental interference (Zhang and Zhang 2018). Equipped with advanced sensors like RGB (red, green, and blue), hyperspectral, and thermal cameras, aerial systems can assess various plant traits, including stress responses and biomass (Abebe et al. 2023). However, limitations include lower spatial resolution and payload constraints, requiring careful calibration to achieve high precision.

4.3.2.1 Unmanned Aerial Platforms

Unmanned aerial platforms (UAPs) include UAVs, commonly known as drones, which operate at low altitudes to capture high-resolution imagery with advanced sensors like RGB, hyperspectral, multispectral, light detection and ranging (LiDAR), and thermal cameras (Zhang and Zhu 2023). They offer flexibility, cost-effectiveness, and capability for targeted phenotyping, making them ideal for evaluating crop health, biomass, and stress responses with exceptional spatial resolution (Fu et al. 2024). Guo et al. (2022) utilized UAVs mounted with LiDAR sensors to estimate canopy biomass and plant height in winter wheat (*Triticum* spp.), aiding in the assessment of nitrogen fertilizer effects.

4.3.2.2 Manned Aerial Platforms

Manned aerial platforms (MAPs) involve piloted aircraft equipped with sophisticated imaging systems for large-scale phenotyping. Although they can cover broader areas than UAVs, MAPs are more expensive and less efficient for frequent or localized data collection, making them more suitable for regional-scale agricultural studies (Li et al. 2021). Yang and Hoffmann (2015) used the AT-402B based on an RGB imaging system, to measure traits such as pest severity in crops.

4.3.2.3 Satellites

Satellites provide phenotyping data on a global scale through multispectral, hyperspectral, and radar sensors (Zhang et al. 2020). Although they facilitate monitoring of extensive fields, satellite imagery is constrained by lower spatial resolution, atmospheric interference, and limited revisitation frequency, making them advantageous for broad agricultural and environmental analyses (Li et al. 2021). Christiansen et al. (2019) operated Sentinel-1 by the European Space Agency, which provides synthetic aperture radar data valuable for monitoring broadacre fields and supports the observations of local conditions, enhancing agricultural management practices.

4.4 Plant Traits to Be Phenotyped

Plant phenotyping aims to analyze observable traits influenced by genetic and environmental factors. High-throughput imaging techniques revolutionize phenotyping by allowing non-invasive and rapid data collection (Sheikh et al. 2024). Traits commonly studied include morphological (e.g., plant height, biomass, leaf area), physiological (e.g., photosynthetic efficiency, water status, nutrient content), and pathological (e.g., disease symptoms, pest damage) (Zhang and Zhang 2018). Modern platforms integrate imaging modalities such as RGB cameras, thermal imaging, hyperspectral imaging (HSI), and LiDAR to provide comprehensive datasets that support breeding programs and environmental studies (Li et al. 2022).

4.4.1 Morphological Traits

Morphological traits refer to structural attributes of plants, including plant height, leaf area, root architecture, and canopy structure (Ghestem et al. 2014). These morphological traits are critical for understanding growth patterns, light interception, and overall biomass production. For example, under "plant canopy," canopy height and width are crucial for estimating plant growth and biomass. This is measured by LiDAR and stereo imaging systems enable precise 3D reconstructions of plant canopies, aiding in detailed morphological studies (Zhang and Zhang 2018). Singhvi et al. (2024) utilized computer vision and ML techniques to process imaging of *Populus trichocarpa,* extracting morphological traits and predicting treatment conditions (control or drought). The application of optical character recognition achieved 94.31% accuracy in reading labels, while classification models identified leaf characteristics and treatment conditions with average accuracies of 62.82% and 60.08% respectively (Singhvi et al. 2024).

4.4.2 Physiological Traits

Physiological traits provide insight into the metabolic and functional activities of plants, such as water use efficiency, photosynthesis, and nutrient assimilation (Singh et al. 2024). Pauli et al. (2016) utilized field-based high-throughput phenotyping to reveal temporal patterns of quantitative trait loci related to stress-responsive traits in cotton that provide insights into traits like stomatal conductance and transpiration rates under stress situations. Basener et al. 2024 developed an interpretable neural network trained on spectral data that learned spectral indicators related to chemical and physiological properties such as chlorophyll composition, aiding in species identification and health assessment.

4.4.3 Pathological Traits

Pathological traits involve identifying disease symptoms caused by biotic and abiotic stresses. Imaging techniques have significantly advanced disease detection through early and precise identification. Moghadam et al. (2017) proposed the use of HSI combined with ML algorithms to detect Tomato Spotted Wilt Virus in plants. This approach enables early and accurate identification of the virus, facilitating timely intervention and management. Natarajan et al. (2024) introduced a deep CNN model capable of quickly and precisely identifying various plant diseases from images. This model demonstrated high accuracy of 99.95% across multiple plant species and disease types, showcasing the potential of deep learning (DL) in plant pathology diagnostics.

4.5 High-Throughput Imaging Systems

High-throughput imaging systems are indispensable tools in modern plant phenotyping, enabling non-invasive, accurate, and comprehensive analysis of plant traits across several growth stages (Fahlgren et al. 2015). These systems utilize diverse imaging modalities, including RGB, hyperspectral, multispectral, thermal, and 3D imaging, to capture structural, physiological, and biochemical properties of plants. Each modality provides unique insights (Zhang and Zhang 2018). Advancements in imaging technologies have revolutionized plant phenotyping. Machine learning and computer vision techniques now enable automated analysis of complex datasets generated by high-throughput systems (Singhvi et al. 2024). These advancements include plant segmentation algorithms for identifying individual plants or organs within dense canopies (Kim et al. 2021) and DL models that predict traits like biomass, leaf area, and yield with high precision (Lube et al. 2022). Moreover, the integration of robotics with imaging systems, such as gantry-based platforms, has facilitated continuous, automated data collection in large-scale phenotyping platforms (Chaudhury et al. 2018). The usefulness of imaging systems lies in their capacity to address the growing demand for sustainable agriculture. These systems provide rapid, high-resolution data critical for evaluating genotype performance under diverse environmental conditions (Zhang and Zhang 2018). They are instrumental in studying plant responses to biotic and abiotic stresses, identifying stress-resilient genotypes, and optimizing resource use (Li et al. 2022). Additionally, imaging systems bridge the gap between lab-based studies and field applications, offering scalable solutions for breeding programs (Singhvi et al. 2024).

4.5.1 Visible Band Imaging

Visible band imaging is a non-destructive imaging in plant phenotyping, utilizing the visible spectrum light of approx. 400–700 nm to evaluate various plant traits such as chlorophyll content, leaf color and plant health. This technique typically involves the use of RGB sensors or cameras to capture light reflected off plants at specific wavelengths. The light reflected by plants is influenced by their biochemical and structural properties, which vary depending on plant health, growth, and stress levels (Li et al. 2014). By analyzing the reflected light in visible bands, visible band imaging enables the extraction of vital information about a plant's physiological status. For instance, healthy plants predominantly reflect green light due to the chlorophyll present in their leaves, while stressed or diseased plants may reflect different amounts of light in the red or blue regions of the spectrum, which can be indicative of specific physiological conditions (Banerjee et al. 2020). This imaging technique works by capturing RGB images of plants using specialized sensors or cameras and then analyzing the resulting data to derive various phenotypic results. For instance, visible band imaging has been employed in drought studies where visible spectrum data were used to monitor water stress in rice plants, offering real-time insights into their physiological responses to environmental conditions (Cho and Yang 2023). Visible band imaging has been proven to be advantageous due to its non-destructive nature, which enables continuous, repeated measurements over time without harming the plants, making it ideal for longitudinal studies and field applications (Gill et al. 2022). Furthermore, it is a cost-effective solution, as the equipment required is relatively affordable and accessible, making it a feasible choice for high-throughput phenotyping in both laboratory and field settings (Esser et al. 2023). However, visible band imaging does have some limitations. Its narrow spectral range restricts its ability to detect plant traits that are best observed in other parts of the spectrum, such as near-infrared or ultraviolet regions. For instance, visible band imaging can effectively assess aboveground traits, but it may fall short in analyzing belowground structures like root systems, which require more specialized techniques like X-ray tomography (Hou et al. 2022). Additionally, environmental factors such as lighting conditions, sensor noise, and background clutter can affect the accuracy and reliability of data collected. Variations in natural light during the day or shadows in the plant canopy can introduce noise that may confound results in field biology (Cho and Yang 2023).

The amalgamation of ML with visible band imaging has led to the development of automated phenotyping systems, which can analyze large datasets quickly and accurately, reducing the need for manual labor and minimizing human error (Singhvi et al. 2024). The continuous advancements in imaging systems, such as the development of mobile platforms equipped with visible band imaging sensors, have further broadened the application of this technique in field-based phenotyping. These systems enable collection of high-resolution, real-time data in outdoor environments, facilitating large-scale monitoring of crops across different conditions and stress scenarios (Lube et al. 2022).

As visible band imaging technologies improve, it is likely that their applications in crop monitoring and genetic studies and precision agriculture will expand, offering enhanced tools for plant researchers worldwide.

4.5.2 Thermal Imaging

Thermal imaging, an advanced technique in plant phenotyping, utilizes infrared radiation emitted from plants to assess their physiological conditions, including water stress, temperature variations, and overall health. Unlike visible band imaging, which captures light in the visible spectrum, thermal imaging detects infrared radiation (8–14 μm wavelength), which correlates with the plant's leaf surface temperature. This is particularly useful in high-throughput phenotyping because it allows researchers to monitor a wide array of plants in real-time without disrupting growth. The working principle of thermal imaging is based on the fact that plants emit infrared radiation that varies depending on their surface temperature, which, in turn, is influenced by environmental conditions, transpiration rates, and water availability. Thermal cameras detect this radiation and convert it into thermal images, where the temperature variations can indicate plant stress, such as water deficiency or disease. Recent studies have highlighted the use of UAVs equipped with thermal cameras for large-scale field phenotyping, providing high-resolution thermal images from aerial views (Ludovisi et al. 2017). Thermal imaging's main advantage lies in its ability to non-destructively monitor plant stress and health in real time, which makes it an ideal technique for large-scale phenotyping, especially under field conditions. It can be used to detect early signs of drought stress, disease, or other environmental stresses that may not yet be visible through other imaging techniques, such as visible light or HSI. Moreover, it offers the advantage of being rapid and capable of covering large areas, making it highly scalable for high-throughput phenotyping. For example, infrared thermography has been applied in field phenotyping of cotton, where it helped identify drought resistance traits (Pabuayon et al. 2019). However, there are also some limitations to thermal imaging. One of the primary disadvantages is that thermal signals can be influenced by external influences, such as ambient temperature, humidity, or soil type, which can affect the accuracy of the data. In addition, thermal imaging often requires careful calibration to account for environmental variables and ensure that the temperature variations detected are solely due to plant physiological processes, not external interference. For instance, in cases where the soil moisture content is high, it can influence the leaf temperature, creating false signals of water stress (Pineda et al. 2020). The principle behind thermal imaging is based on the emission of infrared radiation from plant surfaces. The amount of radiation emitted is directly proportional to the plant's surface temperature, which is in turn affected by factors like water availability, transpiration, and overall plant health. When a plant experiences water stress, for instance, its transpiration rate decreases, causing its surface temperature to rise. These changes in temperature can be captured and visualized through thermal cameras, enabling researchers to assess plant health and stress levels in real time. This technique has been successfully applied in high-throughput phenotyping systems for various crops, such as rice, maize, and cotton, enabling researchers to evaluate stress tolerance and growth efficiency across multiple plant species (Fan et al. 2022). Examples of recent applications of thermal imaging include its use in monitoring the drought responses of black poplar trees, where UAV-based thermal sensors were deployed to assess water stress in a large-scale field trial. This approach allowed researchers to monitor changes in plant canopy temperatures over large distances, offering a high-throughput solution for assessing plant stress responses to changing environmental situations (Ludovisi et al. 2017).

4.5.3 Fluorescence Imaging

Fluorescence imaging is a non-invasive technique used to monitor plant health, stress, and physiological processes by detecting emitted light from fluorescent molecules in response to excitation by a specific light wavelength. This technique primarily focuses on chlorophyll fluorescence, which provides insights into plant photosynthetic activity and overall health. Fluorescence imaging works by illuminating plant samples with a light of specific wavelengths, typically in the blue or ultraviolet spectrum (Abebe et al. 2023). When plants absorb this energy, they re-emit it at longer wavelengths, often in the red or far-red spectrum. This emitted light is captured by specialized imaging systems, which allow the quantification of fluorescence intensity, giving researchers a window into the plant's physiological condition, including stress responses. The working principle behind fluorescence imaging is based on the ability of chlorophyll molecules to absorb light and then re-emit it in the form of fluorescence. This fluorescence signal is a reflection of the plant's photosynthetic efficiency, and changes in fluorescence intensity can indicate stress factors such as water shortage, nutrient deficiencies, or disease (Li et al. 2014). The main benefit of fluorescence imaging lies in its capability to provide real-time, non-destructive measurements of plant health and physiological status. It is particularly useful in high-throughput phenotyping, as it enables large-scale screening of plants for various traits, such as photosynthetic efficiency, drought tolerance, and disease resistance. Fluorescence imaging can be applied to a range of plants under controlled conditions or field environments, offering a scalable solution for crop improvement research (McAusland et al. 2019). Additionally, it can be combined with other imaging techniques, such as hyperspectral and multispectral imaging (MSI), to improve the accuracy and depth of phenotypic assessments. However, there are some limitations to fluorescence imaging. The fluorescence signal can be affected by environmental factors such as light intensity and temperature, which may introduce variability in measurements. Additionally, while fluorescence imaging is effective for studying photosynthesis and plant stress, it may not provide comprehensive insights into other aspects of plant health, such as nutrient status or root system architecture. Calibration and careful control of environmental conditions are essential to minimize such interference, especially in large-scale applications (Abebe et al. 2023). Recent studies demonstrate the application of fluorescence imaging in advancing plant research. For example, chlorophyll fluorescence has been used to identify drought-tolerant wheat varieties, highlighting the potential for fluorescence-based screening in crop breeding programs (McAusland et al. 2019).

4.5.4 Spectral Imaging

Spectral imaging is an advanced technique that captures images across multiple wavelengths of light, enabling the identification of various plant traits that are often imperceptible to the human eye. It works by utilizing the reflectance or emission spectra of plant surfaces, providing a detailed snapshot of plant health, stress responses, and biochemical compositions. Spectral imaging systems, such as HSI or MSI, typically operate by recording light in narrow bands across the visible and infrared spectra, allowing the differentiation of plant tissues and the detection of subtle physiological changes. Hyperspectral imaging, for example, can capture hundreds of spectral bands, providing a high-resolution spectral signature of plants that can be analyzed to discern various stress factors like drought or disease (Banerjee et al. 2020). The working principle of spectral imaging relies on capturing the spectral reflectance or emission of plant surfaces and processing it through advanced algorithms to extract meaningful plant phenotypic data (Das Choudhury et al. 2019). Spectral imaging offers several advantages, including its ability to analyze plants without physical contact, making it a non-destructive and high-throughput tool. It can assess a variety of plant characteristics simultaneously and is applicable under field conditions, providing detailed understandings into the overall health and biochemical composition of plants. Models like the "Field Robot" and "Prospector" systems utilize spectral imaging to generate high-resolution 3D phenotypic models, which can be utilized to assess plant growth and yield potential (Rife et al. 2021). However, spectral imaging does have its drawbacks, including its reliance on specific light conditions and calibration to minimize environmental interference. Variability in environmental situations, such as lighting or temperature, can affect the accuracy of spectral data, making it essential to control these factors during imaging. Additionally, while spectral imaging provides valuable information on plant health, it may not capture deeper physiological parameters like soil moisture or root dynamics, which require complementary techniques (Tayade et al. 2022). Recent developments in spectral imaging for high-throughput phenotyping include combining spectral data with ML algorithms to improve the accuracy and prediction of plant performance, as seen in the use of aerial hyperspectral imagery for yield phenotyping in wheat (Moghimi et al. 2020).

4.5.5 Stereovision Imaging

Stereovision imaging is an advanced 3D imaging technique that captures depth information of plant structures by analyzing two or more images taken from slightly different angles, mimicking human binocular vision (Yin et al. 2024). This technique works by triangulating the disparity between the corresponding points in stereo pairs to reconstruct three-dimensional models of plant morphology. Stereovision allows for accurate measurements of plant traits such as leaf area, height, and canopy structure, which are essential for high-throughput phenotyping. Unlike 2D imaging methods, stereovision can offer depth perception, making it particularly valuable for assessing the spatial arrangement of plant organs and their growth dynamics (Xiang et al. 2021). The working principle of stereovision involves capturing two or more images of a plant from different viewpoints and comparing the pixel differences between the images. This disparity is then used to calculate depth information, generating a 3D point cloud or surface model. The resulting data can be processed to evaluate plant architecture and quantify characteristics such as biomass, stem growth, and root systems (Li et al. 2014). It is also highly accurate for measuring plant volumes and growth, which is crucial for precision agriculture.

The technique can be integrated with other imaging modalities, like thermal or HSI, to provide complementary data for a more detailed analysis of plant health (Cho and Yang 2023). However, there are some disadvantages, such as sensitivity to environmental factors like lighting conditions and plant movement, which may introduce inaccuracies in depth estimation. Additionally, stereovision systems require precise calibration and high computational power to process large datasets effectively, which can limit their efficiency in field applications (Guo et al. 2023). Recent work in stereovision imaging includes the development of systems like the "PhenoStereo" platform, which utilizes stereo vision for high-throughput phenotyping of field-grown plants, and the usage of low-cost stereo cameras integrated with AI for autonomous plant monitoring in various agricultural settings (Xu and Li 2022). These models enable efficient, scalable phenotyping of large crop populations under real-world field conditions.

4.5.6 Tomographic Imaging

Tomographic imaging is a non-invasive, high-resolution technique used in plant phenotyping to reconstruct three-dimensional images of internal plant structures by analyzing multiple two-dimensional cross-sectional images captured from various angles (Li et al. 2014). This technique operates on the principle of capturing projection data and reconstructing it into 3D models using algorithms like filtered back projection or iterative reconstruction methods. Common modalities in tomographic imaging comprise X-ray computed tomography (CT), magnetic resonance imaging, and positron emission tomography (PET). Each modality offers distinct advantages for visualizing and quantifying internal plant traits, such as root architecture, vascular systems, and biomass distribution (Zhang et al. 2024). The working principle of tomographic imaging involves passing energy waves, such as X-rays, magnetic fields, or radioactive tracers, through plant tissues. Detectors capture the resulting signals, which vary based on tissue density or chemical composition. These signals are processed computationally to generate detailed 3D reconstructions. For example, X-ray CT relies on the attenuation of X-rays by plant tissues to visualize structural traits, while PET tracks radiotracers to study dynamic processes like carbon allocation and water uptake (Wang et al. 2015). Tomographic imaging offers unparalleled advantages, such as the ability to study internal plant structures and physiological processes non-destructively. It allows for dynamic monitoring of plant responses to environmental conditions and provides high spatial resolution for detailed trait analysis. Additionally, the integration of tomographic imaging with optical projection tomography further enhances the resolution of smaller structures like fine roots (Lee et al. 2017). However, the technique has limitations, including high costs, the necessity for specialized equipment, and extensive computational requirements. Environmental factors like movement and sample preparation challenges also affect imaging accuracy, particularly for field-grown plants. Moreover, the ionizing radiation used in X-ray CT and PET can limit repeated imaging of the same specimen. Recent advancements include the use of models like the Siemens Biograph Vision PET/CT scanner and Zeiss Xradia 620 Versa micro-CT system, which offer improved resolution and faster imaging for high-throughput phenotyping (Karahara et al. 2023). Applications include analyzing drought tolerance traits and carbon allocation patterns in root systems under stress conditions. These innovations continue to push the boundaries of plant phenotyping, enabling more precise and dynamic studies of plant biology (Duncan et al. 2022).

4.6 High-Throughput Image Data Processing

High-throughput image data processing (HTIDP) in plant phenotyping has emerged as a critical tool for accelerating agricultural research and improving crop productivity. This process involves the usage of advanced computational techniques, such as AI, ML, and DL, to efficiently analyze vast datasets obtained from imaging platforms. HTIDP transforms raw phenotypic data into meaningful insights that aid in plant breeding, stress analysis, and crop improvement (Das Choudhury et al. 2019; Upadhyay 2026). The initial phase of HTIDP includes preprocessing raw images obtained from high-throughput phenotyping platforms. This step encompasses image normalization, noise reduction, and feature enhancement to prepare the data for subsequent analysis. Advanced convolutional neural networks (CNNs) have been employed for precise segmentation of plant components, such as leaves and stems, facilitating accurate trait extraction (Mostafa et al. 2023). Feature extraction is a critical stage where quantitative traits, including plant height, leaf area, and biomass, are measured. Machine learning models, including support vector machines and decision trees, have been utilized for this purpose (Ubbens and Stavness 2017). Notably, DL frameworks such as VGG-16 (Mostafa et al. 2023) and YOLOv5 (Cardellicchio et al. 2023) have demonstrated superior performance in tasks like stand count estimation and bloom detection, owing to their ability to capture intricate patterns in image data. The scalability of HTIDP allows for the processing of millions of images in a relatively short timeframe, a feat unattainable through manual analysis. Cloud-based platforms, such as Microsoft Azure (Debauche et al. 2022), AutoML (Koh et al. 2021), provide the necessary computational resources to handle large-scale phenotypic datasets, enabling real-time processing and analysis. Despite these developments, challenges persist in standardizing data formats and ensuring interoperability among diverse imaging systems. The development of open-source software and publicly available datasets is crucial in addressing these issues, promoting collaboration and reproducibility in the research community. Recent studies have focused on integrating multi-modal data, compounding information from various imaging modalities to provide a comprehensive phenotypic profile (Stumpe et al. 2024). For example, the fusion of thermal, hyperspectral, and fluorescence imaging has been employed to assess plant health and stress responses, offering a holistic view of plant physiology. HTIDP in plant phenotyping encompasses several critical steps to efficiently analyze large volumes of phenotypic data. A comprehensive framework for HTIDP includes firstly data acquisition, in which utilizing multimodal, multi-view, and time-lapse imaging systems is done to capture diverse aspects of plant morphology and function (Das Choudhury et al. 2019). After which, image pre-processing is done, implementing techniques such as noise reduction, normalization, and enhancement to improve image quality and prepare data for analysis (Hartmann et al. 2011). After that, segmentation is done employing ML-based methods to accurately delineate plant structures from the background, facilitating precise trait extraction (Lee et al. 2018). Then feature extraction is done, quantifying phenotypic traits like plant height, leaf area, and biomass using advanced image analysis algorithms (Wang et al. 2020). Then, data integration and analysis are done for combining extracted features with environmental and genotypic data to perform comprehensive statistical analyses, enabling the identification of trait associations and patterns (Danilevicz et al. 2021). And at last, data management and storage are done for implementing robust information systems to handle large datasets, ensuring data integrity, accessibility, and facilitating subsequent analyses (Klukas et al. 2014). By systematically executing these steps, HTIDP enables the efficient and accurate assessment of plant phenotypes, thereby advancing research in plant science and agriculture.

4.7 Future Prospectives

Future trends point towards the continued integration of AI and advanced data analytics to enhance the predictive power of phenotyping systems. The development of multi-modal imaging platforms, capable of combining data from different imaging modalities (e.g., hyperspectral, thermal, and fluorescence), will enable researchers to generate holistic phenotypic profiles. Additionally, the use of autonomous robotics and drones equipped with advanced imaging systems will further optimize field phenotyping, making it more efficient and accessible for large-scale applications. Another key trend lies in the evolution of real-time data processing pipelines supported by cloud computing and edge AI, ensuring rapid and actionable insights for crop management and breeding programs. The creation of standardized, open-access databases and software frameworks will promote global collaboration and reproducibility in phenotyping research. Moreover, innovations in quantum computing and neural network architectures hold the potential to process and analyze phenotypic data at unprecedented speeds, addressing challenges posed by the ever-increasing scale of plant phenotyping studies. The integration of environmental monitoring systems with phenotyping platforms is another future direction, enabling dynamic analysis of plant-environment interactions under diverse climatic scenarios. This will support the breeding of crops with enhanced resilience to climate change. Lastly, personalized phenotyping platforms tailored for specific crops and regions will emerge, allowing targeted solutions to local agricultural challenges and further advancing the goal of sustainable agriculture.

4.8 Conclusion

The advancements in phenotyping and high-throughput imaging have revolutionized plant science by allowing accurate, non-destructive, and large-scale assessment of plant traits. The integration of cutting-edge imaging technologies, such as thermal imaging, spectral imaging, stereovision, and tomographic imaging, has provided detailed insights into plant morphology, physiology, and responses to environmental stress. These advancements have bridged the gap between phenotype and genotype, offering a clearer understanding of plant behavior and aiding in the development of climate-resilient and resource-efficient crops. Furthermore, the application of AI, ML, and computer vision has streamlined data processing, improving accuracy and scalability in extracting complex phenotypic traits from vast datasets.

References

Abdi, H., Alipour, H., Bernousi, I. et al. (2024). Morpho-colorimetric diversity and genome-wide association study of wheat spike architecture based on image processing under well-watered and rain-fed conditions. *Journal of Plant Growth Regulation* 44 (2): 850–867.

Abebe, A.M., Kim, Y., Kim, J. et al. (2023). Image-based high-throughput phenotyping in horticultural crops. *Plants* 12 (10): 2061.

Ampatzidis, Y. and Partel, V. (2019). UAV-based high throughput phenotyping in citrus utilizing multispectral imaging and artificial intelligence. *Remote Sensing* 11 (4): 410.

Andrade-Sanchez, P., Gore, M.A., Heun, J.T. et al. (2013). Development and evaluation of a field-based high-throughput phenotyping platform. *Functional Plant Biology* 41 (1): 68–79.

Bai, G., Ge, Y., Hussain, W. et al. (2016). A multi-sensor system for high throughput field phenotyping in soybean and wheat breeding. *Computers and Electronics in Agriculture* 128: 181–192.

Banerjee, B.P., Joshi, S., Thoday-Kennedy, E. et al. (2020). High-throughput phenotyping using digital and hyperspectral imaging-derived biomarkers for genotypic nitrogen response. *Journal of Experimental Botany* 71 (15): 4604–4615.

Basener, W., Basener, A., and Luegering, M. (2024). An interpretable neural network for vegetation phenotyping with visualization of trait-based spectral features. *arXiv preprint* arXiv 2407:10333. https://doi.org/10.48550/arXiv.2407.10333.

Basso, B., Fiorentino, C., Cammarano, D., and Schulthess, U. (2016). Variable rate nitrogen fertilizer response in wheat using remote sensing. *Precision Agriculture* 17: 168–182.

Busemeyer, L., Mentrup, D., Möller, K. et al. (2013). BreedVision—a multi-sensor platform for non-destructive field-based phenotyping in plant breeding. *Sensors* 13 (3): 2830–2847.

Cardellicchio, A., Solimani, F., Dimauro, G. et al. (2023). Detection of tomato plant phenotyping traits using YOLOv5-based single stage detectors. *Computers and Electronics in Agriculture* 207: 107757.

Caturegli, L., Casucci, M., Lulli, F. et al. (2015). GeoEye-1 satellite versus ground-based multispectral data for estimating nitrogen status of turfgrasses. *International Journal of Remote Sensing* 36 (8): 2238–2251.

Cembrowska-Lech, D., Krzemińska, A., Miller, T. et al. (2023). An integrated multi-omics and artificial intelligence framework for advance plant phenotyping in horticulture. *Biology* 12 (10): 1298.

Chaudhury, A., Ward, C., Talasaz, A. et al. (2018). Machine vision system for 3D plant phenotyping. *IEEE/ACM Transactions on Computational Biology and Bioinformatics* 16 (6): 2009–2022.

Cho, W.J. and Yang, M. (2023). High-throughput plant phenotyping system using a low-cost camera network for plant factory. *Agriculture* 13 (10): 1874.

Christiansen, M.P., Teimouri, N., Laursen, M.S. et al. (2019). Preprocessed sentinel-1 data via a web service focused on agricultural field monitoring. *IEEE Access* 7: 65139–65149.

Crain, J.L., Wei, Y., Barker, J. III, et al. (2016). Development and deployment of a portable field phenotyping platform. *Crop Science* 56 (3): 965–975.

Danilevicz, M.F., Bayer, P.E., Nestor, B.J. et al. (2021). Resources for image-based high-throughput phenotyping in crops and data sharing challenges. *Plant Physiology* 187 (2): 699–715.

Das Choudhury, S., Samal, A., and Awada, T. (2019). Leveraging image analysis for high-throughput plant phenotyping. *Frontiers in Plant Science* 10: 508.

Davis, J.M., Gaillard, M., Tross, M.C. et al. (2024). 3D reconstruction enables high-throughput phenotyping and quantitative genetic analysis of phyllotaxy. *bioRxiv* 2024-10. https://doi.org/10.1101/2024.10.03.616344.

Debauche, O., Mahmoudi, S., Manneback, P., and Lebeau, F. (2022). Cloud and distributed architectures for data management in agriculture 4.0: review and future trends. *Journal of King Saud University, Computer and Information Sciences* 34 (9): 7494–7514.

Drusch, M., Moreno, J., Del Bello, U. et al. (2016). The fluorescence explorer mission concept—ESA's earth explorer 8. *IEEE Transactions on Geoscience and Remote Sensing* 55 (3): 1273–1284.

Duncan, K.E., Czymmek, K.J., Jiang, N. et al. (2022). X-ray microscopy enables multiscale high-resolution 3D imaging of plant cells, tissues, and organs. *Plant Physiology* 188 (2): 831–845.

Esser, F., Rosu, R.A., Corneließen, A. et al. (2023). Field robot for high-throughput and high-resolution 3D plant phenotyping. *arXiv preprint* arXiv 2310.11516. https://doi.org/10.48550/arXiv.2310.11516.

Fahlgren, N., Gehan, M.A., and Baxter, I. (2015). Lights, camera, action: high-throughput plant phenotyping is ready for a close-up. *Current Opinion in Plant Biology* 24: 93–99.

Fan, Z., Sun, N., Qiu, Q. et al. (2022). In situ measuring stem diameters of maize crops with a high-throughput phenotyping robot. *Remote Sensing* 14 (4): 1030.

Fawcett, D., Panigada, C., Tagliabue, G. et al. (2020). Multi-scale evaluation of drone-based multispectral surface reflectance and vegetation indices in operational conditions. *Remote Sensing* 12: 514.

Feng, A., Zhou, J., Vories, E.D. et al. (2020). Yield estimation in cotton using UAV-based multi-sensor imagery. *Biosystems Engineering* 193: 101–114.

Friedli, M., Kirchgessner, N., Grieder, C. et al. (2016). Terrestrial 3D laser scanning to track the increase in canopy height of both monocot and dicot crop species under field conditions. *Plant Methods* 12: 1–15.

Fu, H., Lu, J., Cui, G. et al. (2024). Advanced plant phenotyping: Unmanned aerial vehicle remote sensing and CimageA software technology for precision crop growth monitoring. *Agronomy* 14 (11): 2534.

Gao, W., Zeng, W., Li, S. et al. (2024). Remote sensing estimation of sugar beet SPAD based on un-manned aerial vehicle multispectral imagery. *PLoS One* 19 (6): e0300056.

Ghestem, M., Veylon, G., Bernard, A. et al. (2014). Influence of plant root system morphology and architectural traits on soil shear resistance. *Plant and Soil* 377: 43–61.

Gill, T., Gill, S.K., Saini, D.K. et al. (2022). A comprehensive review of high throughput phenotyping and machine learning for plant stress phenotyping. *Phenomics* 2 (3): 156–183.

Guo, Y., He, J., Huang, J. et al. (2022). Effects of the spatial resolution of UAV images on the prediction and transferability of nitrogen content model for winter wheat. *Drones* 6 (10): 299.

Guo, X., Qiu, Y., Nettleton, D., and Schnable, P.S. (2023). High-throughput field plant phenotyping: a self-supervised sequential CNN method to segment overlapping plants. *Plant Phenomics* 5: 0052.

Hartmann, A., Czauderna, T., Hoffmann, R. et al. (2011). HTPheno: an image analysis pipeline for high-throughput plant phenotyping. *BMC Bioinformatics* 12: 1–9.

Hou, L.H., Gao, W., Weng, Z.H. et al. (2022). Use of X-ray tomography for examining root architecture in soils. *Geoderma* 405: 115405.

Jiang, Y., Li, C., Robertson, J.S. et al. (2018). GPhenovision: a ground mobile system with multi-modal imaging for field-based high throughput phenotyping of cotton. *Scientific Reports* 8 (1): 1213.

Jimenez-Berni, J.A., Deery, D.M., Rozas-Larraondo, P. et al. (2018). High throughput determination of plant height, ground cover, and above-ground biomass in wheat with LiDAR. *Frontiers in Plant Science* 9: 237.

Karahara, I., Yamauchi, D., Uesugi, K., and Mineyuki, Y. (2023). Three-dimensional visualization of plant tissues and organs by X-ray micro–computed tomography. *Microscopy* 72 (4): 310–325.

Kim, M., Lee, C., Hong, S. et al. (2021). High-throughput phenotyping methods for breeding drought-tolerant crops. *International Journal of Molecular Sciences* 22 (15): 8266.

Klukas, C., Chen, D., and Pape, J.M. (2014). Integrated analysis platform: an open-source information system for high-throughput plant phenotyping. *Plant Physiology* 165 (2): 506–518.

Koh, J.C., Spangenberg, G., and Kant, S. (2021). Automated machine learning for high-throughput image-based plant phenotyping. *Remote Sensing* 13 (5): 858.

Lee, K.J., Calder, G.M., Hindle, C.R. et al. (2017). Macro optical projection tomography for large-scale 3D imaging of plant structures and gene activity. *Journal of Experimental Botany* 68 (3): 527–538.

Lee, U., Chang, S., Putra, G.A. et al. (2018). An automated, high-throughput plant phenotyping system using machine learning-based plant segmentation and image analysis. *PLoS One* 13 (4): e0196615.

Li, L., Zhang, Q., and Huang, D. (2014). A review of imaging techniques for plant phenotyping. *Sensors* 14 (11): 20078–20111.

Li, W., Niu, Z., Huang, N. et al. (2015). Airborne LiDAR technique for estimating biomass components of maize: a case study in Zhangye City, Northwest China. *Ecological Indicators* 57: 486–496.

Li, D., Quan, C., Song, Z. et al. (2021). High-throughput plant phenotyping platform (HT3P) as a novel tool for estimating agronomic traits from the lab to the field. *Frontiers in Bioengineering and Biotechnology* 8: 623705.

Li, M., Frank, M.H., and Migicovsky, Z. (2022). ColourQuant: a high-throughput technique to extract and quantify color phenotypes from plant images. In: *High-Throughput Plant Phenotyping: Methods and Protocols* (ed. A. Lorence and K. Medina Jimenez), 77–85. New York, NY: Springer US.

Lu, X., Shen, Y., Xie, J. et al. (2024). Phenotyping of panicle number and shape in rice breeding materials based on unmanned aerial vehicle imagery. *Plant Phenomics* 6: 0265.

Lube, V., Noyan, M.A., Przybysz, A. et al. (2022). MultipleXLab: a high-throughput portable live-imaging root phenotyping platform using deep learning and computer vision. *Plant Methods* 18 (1): 38.

Ludovisi, R., Tauro, F., Salvati, R. et al. (2017). UAV-based thermal imaging for high-throughput field phenotyping of black poplar response to drought. *Frontiers in Plant Science* 8: 1681.

Maciel, G.M., de Araújo Gallis, R.B., Barbosa, R.L. et al. (2019). Image phenotyping of inbred red lettuce lines with genetic diversity regarding carotenoid levels. *International Journal of Applied Earth Observation and Geoinformation* 81: 154–160.

McAusland, L., Atkinson, J.A., Lawson, T., and Murchie, E.H. (2019). High throughput procedure utilising chlorophyll fluorescence imaging to phenotype dynamic photosynthesis and photoprotection in leaves under controlled gaseous conditions. *Plant Methods* 15: 1–15.

Moghadam, P., Ward, D., Goan, E. et al. (29 November–1 December 2017). Plant disease detection using hyperspectral imaging. In 2017 International Conference on Digital Image Computing: Techniques and Applications (DICTA), Sydney, NSW, Australia, 1–8. IEEE. https://doi.org/10.1109/DICTA.2017.8227476.

Moghimi, A., Yang, C., and Anderson, J.A. (2020). Aerial hyperspectral imagery and deep neural networks for high-throughput yield phenotyping in wheat. *Computers and Electronics in Agriculture* 172: 105299.

Mohammed, A., Ali, N., Bais, A. et al. (2024). From fields to pixels: UAV multispectral and field-captured RGB imaging for high-throughput wheat spike and kernel counting. *IEEE Journal of Selected Topics in Applied Earth Observations and Remote Sensing* 17: 17806–17819.

Mostafa, S., Mondal, D., Panjvani, K. et al. (2023). Explainable deep learning in plant phenotyping. *Frontiers in Artificial Intelligence* 6: 1203546.

Naito, H., Ogawa, S., Valencia, M.O. et al. (2017). Estimating rice yield related traits and quantitative trait loci analysis under different nitrogen treatments using a simple tower-based field phenotyping system with modified single-lens reflex cameras. *ISPRS Journal of Photogrammetry and Remote Sensing* 125: 50–62.

Naqvi, R.Z., Farooq, M., Zahid, R. et al. (2024). Recent advances in plant phenomics and speed breeding for climate-smart agriculture. In: *Plant Speed Breeding and High-throughput Technologies* (ed. J.-t. Chen), 82–103. CRC Press.

Natarajan, S., Chakrabarti, P., and Margala, M. (2024). Robust diagnosis and meta visualizations of plant diseases through deep neural architecture with explainable AI. *Scientific Reports* 14 (1): 13695.

Pabuayon, I.L.B., Sun, Y., Guo, W., and Ritchie, G.L. (2019). High-throughput phenotyping in cotton: a review. *Journal of Cotton Research* 2: 1–9.

Pan, Y., Hu, K., Liu, T. et al. (2024). Pheno-robot: an auto-digital modelling system for in-situ phenotyping in the field. *arXiv preprint* arXiv 2402.09685. https://doi.org/10.48550/arXiv.2402.09685.

Park, J.E., Kim, B.H., Kim, W. et al. (2024). Correction: High-throughput image data analysis shows the genetic diversity in Tartary buckwheat (*Fagopyrum tataricum* Gaertn). *Plant Biotechnology Reports* 18 (6): 811.

Paul, K., Chatterjee, S.S., Pai, P. et al. (2022). Viable smart sensors and their application in data-driven agriculture. *Computers and Electronics in Agriculture* 198: 107096.

Pauli, D., Andrade-Sanchez, P., Carmo-Silva, A.E. et al. (2016). Field-based high-throughput plant phenotyping reveals the temporal patterns of quantitative trait loci associated with stress-responsive traits in cotton. *G3: Genes, Genomes, Genetics* 6 (4): 865–879.

Pineda, M., Barón, M., and Pérez-Bueno, M.L. (2020). Thermal imaging for plant stress detection and phenotyping. *Remote Sensing* 13 (1): 68.

Qiu, Q., Sun, N., Bai, H. et al. (2019). Field-based high-throughput phenotyping for maize plant using 3D LiDAR point cloud generated with a "Phenomobile". *Frontiers in Plant Science* 10: 554.

Richardson, A.D., Hufkens, K., Milliman, T. et al. (2018). Tracking vegetation phenology across diverse North American biomes using PhenoCam imagery. *Scientific Data* 5 (1): 1–24.

Rife, T.W., Courtney, C., Hershberger, J. et al. (2021). Prospector: a mobile app for high-throughput NIRS phenotyping. *arXiv preprint* arXiv 2104.07071. https://doi.org/10.48550/arXiv.2104.07071.

Sanchez, J.R. (2024). Automated unmanned systems and data analytics for in-field digital crop phenotyping Doctoral dissertation, University of Georgia.

dos Santos Ferreira, A., Freitas, D.M., da Silva, G.G. et al. (2017). Weed detection in soybean crops using ConvNets. *Computers and Electronics in Agriculture* 143: 314–324.

Sheikh, M., Iqra, F., Ambreen, H. et al. (2024). Integrating artificial intelligence and high-throughput phenotyping for crop improvement. *Journal of Integrative Agriculture* 23 (6): 1787–1802.

Shi, Y., Murray, S.C., Rooney, W.L. et al. (2016). Corn and sorghum phenotyping using a fixed-wing UAV-based remote sensing system. In: *Autonomous Air and Ground Sensing Systems for Agricultural Optimization and Phenotyping*, vol. 9866, 46–53. SPIE.

Singh, A.K., Ganapathysubramanian, B., Sarkar, S., and Singh, A. (2018). Deep learning for plant stress phenotyping: trends and future perspectives. *Trends in Plant Science* 23 (10): 883–898.

Singh, H., Kumar, N., and Kumar, A. (2024). Enhancing resource use efficiency in crops through plant functional traits. In: *Plant Functional Traits for Improving Productivity* (ed. N. Kumar and H. Singh), 97–117. Singapore: Springer Nature Singapore.

Singhvi, V., Lunga, L., Nidhi, P. et al. (2024). High-throughput phenotyping using computer vision and machine learning. *arXiv preprint* arXiv:2407.06354. https://doi.org/10.48550/arXiv.2407.06354.

Stentz, A., Dima, C., Wellington, C. et al. (2002). A system for semi-autonomous tractor operations. *Autonomous Robots* 13: 87–104.

Stumpe, E., Bodner, G., Flagiello, F., and Zeppelzauer, M. (2024). 3D multimodal image registration for plant phenotyping. *arXiv preprint* arXiv:2407.02946. https://doi.org/10.48550/arXiv.2407.02946.

Susko, A.Q., Gilbertson, F., Heuschele, D.J. et al. (2018). An automatable, field camera track system for phenotyping crop lodging and crop movement. *HardwareX* 4: e00029.

Tayade, R., Yoon, J., Lay, L. et al. (2022). Utilization of spectral indices for high-throughput phenotyping. *Plants* 11 (13): 1712.

Ubbens, J.R. and Stavness, I. (2017). Deep plant phenomics: a deep learning platform for complex plant phenotyping tasks. *Frontiers in Plant Science* 8: 1190.

Upadhyay, S.K. (2026). In: *Designer Crops: Implications of Artificial Intelligence and Modern Biotechnological Methods*. In: Elsevier Science.

Virlet, N., Sabermanesh, K., Sadeghi-Tehran, P., and Hawkesford, M.J. (2017). Field scanalyzer: an automated robotic field phenotyping platform for detailed crop monitoring. *Functional Plant Biology* 44: 143–153.

Wan, L., Cen, H., Zhu, J. et al. (2020). Grain yield prediction of rice using multi-temporal UAV-based RGB and multispectral images and model transfer–a case study of small farmlands in the South of China. *Agricultural and Forest Meteorology* 291: 108096.

Wang, Q., Komarov, S., Mathews, A.J. et al. (2015). Combined 3D PET and optical projection tomography techniques for plant root phenotyping. *arXiv preprint* arXiv:1501.00242. https://doi.org/10.48550/arXiv.1501.00242.

Wang, R., Qiu, Y., Zhou, Y. et al. (2020). A high-throughput phenotyping pipeline for image processing and functional growth curve analysis. *Plant Phenomics* 2020: 7481687.

Xiang, L., Tang, L., Gai, J., and Wang, L. (2021). Measuring stem diameter of sorghum plants in the field using a high-throughput stereo vision system. *Transactions of the ASABE* 64 (6): 1999–2010.

Xu, R. and Li, C. (2022). A review of high-throughput field phenotyping systems: focusing on ground robots. *Plant Phenomics* 2022: 9760269.

Xu, Y., Zhang, X., Li, H. et al. (2022). Smart breeding driven by big data, artificial intelligence, and integrated genomic-enviromic prediction. *Molecular Plant* 15 (11): 1664–1695.

Yang, C. and Hoffmann, W.C. (2015). Low-cost single-camera imaging system for aerial applicators. *Journal of Applied Remote Sensing* 9 (1): 096064.

Yin, H., Luo, S., Tang, J., and Hua, J. (2024). RGB camera-based monocular stereo vision applied in plant phenotype: a survey. *Computers and Electronics in Agriculture* 227: 109523.

Zhang, Y. and Zhang, N. (2018). Imaging technologies for plant high-throughput phenotyping: a review. *Frontiers of Agricultural Science and Engineering* 5 (4): 406–419.

Zhang, Z. and Zhu, L. (2023). A review on unmanned aerial vehicle remote sensing: platforms, sensors, data processing methods, and applications. *Drones* 7 (6): 398.

Zhang, C., Marzougui, A., and Sankaran, S. (2020). High-resolution satellite imagery applications in crop phenotyping: an overview. *Computers and Electronics in Agriculture* 175: 105584.

Zhang, Q., Luan, R., Wang, M. et al. (2024). Research progress of spectral imaging techniques in plant phenotype studies. *Plants* 13 (21): 3088.

Zhou, J., Reynolds, D., Websdale, D. et al. (2017). CropQuant: an automated and scalable field phenotyping platform for crop monitoring and trait measurements to facilitate breeding and digital agriculture. *BioRxiv [Preprints]* 161547: https://doi.org/10.1101/161547.

5

Applications of AI in the Genomic Analysis of Crop Plants for Improved Agricultural Outcomes

Satwi Shah[1], *Nancy Vora*[1], *and Manan Shah*[2]

[1] *Nirma University, Department of Computer Science and Engineering, Sarkhej Gandhinagar Highway, Ahmedabad 380055, Gujarat, India*
[2] *Pandit Deendayal Energy University, Department of Chemical Engineering, School of Energy Technology, Gandhinagar, Knowledge Corridor, Raisan Village, 382426, Gujarat, India*

5.1 Introduction

Over the centuries, transcendental transformation has been observed in the field of agriculture, not only pertained by population growth factor but has also derived its emergence from contribution from adversaries like climate change and ever-increasing food security demand. Traditional cultivation methods fall short to commensurate with increasing complexities of global challenge, including reduction in arable land, capricious weather patterns, and evolving pests and diseases, hence as whole blighting the crop production. These challenges necessitate innovative approaches which could be benevolent for enhancing crop resilience and yield, while maintaining environmental sustainability (Crossa et al., 2017). The integration of modern scientific advancements with traditional practices has become indispensable in addressing these demands (Ma et al., 2025).

The field of plant breeding has witnessed revolutionization with the introduction of genomics which allows precise genetic manipulation accounting for improved crop traits. With the use of sophisticated molecular tools that includes GWAS and QTL mapping, in the field of GS, it helps to identify and introduce productive traits into crop populations (Kumar et al., 2024). The technologies used in this process allow the selection of such traits that would be drought resistant, higher resistance to pest infection, higher nutritional content, and therefore making the whole breeding cycle shorter. The latest advancements in gene-editing technologies, particularly CRISPR-Cas9, have streamlined the process to an extent that has facilitated the alteration of targets particularly at the molecular level and is addressing agricultural challenges (Ma et al., 2025).

The integration of genomics with recent advancements in technologies like AI, ML, and big data analytics begets the novel era in agriculture with precision. AI-driven systems have made room for the possibility of optimizing data collection not only from high-throughput phenotyping but also included predictive modeling which has largely changed the breeding program for better. Unmanned aerial vehicles (UAVs) are being used to help track the growth of crops and keep check of their health by equipping UAVs with multi-spectral imaging sensors which has greatly helped by providing real-time monitoring. The analysis drawn from the vast datasets which be accounted as a result from big data integration has enabled genomics, phenomics, and fostering predictive breeding strategies that are responsible for genotype-environment interactions (Crossa et al., 2017). These tools have contributed in streamlining the efficiency of trait selection, and in addition to it, they have also played an important role by contributing in the development of climate-resilient crop varieties (Ma et al., 2025).

One of the recently fast paced developing discipline, rooted in computational algorithm and data-driven problem-solving, is now being widely recognized and accepted in majority of areas as a solution and a driving force for innovation, has played a paramount role in revolutionizing the domains including but not restricted to the genomics and agriculture. With the help of techniques such as ML and deep learning, it leverages them by analyzing complex datasets which can later be useful to discover patterns and accordingly predict its outcomes, drawing resemblance for decision-making capabilities from humans. The computational power and data acquisition field have been subjected to advancements, enabling AI to process huge datasets of genomic and phenotypic data and thereby creating opportunities for breakthroughs in research and innovation (Yue et al., 2023).

Artificial intelligence successfully plays the role of catalyst in the domain comprising genomic research by putting forward the data analysis process at a speedy pace and also aims at improving precision. Genomics dataset being characterized by its complexities and size calls for sophisticated tools to help us gain meaningful insights. Neural network and predictive modeling framework have effectively increased the efficacy in gene identification, functional annotation, and sequence analysis. These methodologies have been proved to be instrumental in handling challenges in areas like genome assembly and regulatory sequence prediction. AI helps in linking genetic markers to phenotypic traits, thus expediting GS and improving crop varieties (Varshney et al., 2021; van Dijk et al., 2021).

Artificial intelligence leverages genomic data in optimizing crop yield, resilience, and sustainability by its convergence in precision agriculture. From the derivation of correlation of genomic and environmental data, AI has helped effectively in enabling real-time decision-making for allocating resources, management of pests, and adaptation to climate change. For instance, AI-driven identification has helped us in recognizing drought-resistant genes in maize and salt-tolerance genes in rice which has ultimately contributed to the development of crop varieties which constitute resilient nature. Advancements have been considered utilitarian for addressing global food security amidst environmental and population pressures (Wójcik-Gront et al. 2024).

Limitation of AI couldn't be merely contained by singling out its only role, the very usefulness in GS by undertaking strategic development of high-yield and climate-resilient crops but has scope to expand further. Many algorithms of ML like Random Forest (RF) and Convolutional Neural Networks (CNNs) can efficiently make prediction of breeding values by the analysis of genetic markers. AI can target interventions for improvement in crops by refining phenotypic trait identification in genomic studies (Wójcik-Gront et al. 2024). Reduction in breeding cycles numbers and consumption of resources have been observed by adaption of such tools, thereby facilitating hasty acclimatization to the environmental challenges.

The main objective of this chapter will be to zero in on the changes in plant breeding, influenced by the changing face of AI and ML because it can transcend other challenges such as non-linear genetic effects, overfitting, and computation inefficiency. The models covered by this review are Reproducing Kernel Hilbert Space (RKHS), Bayesian lasso (BL), RF, and Artificial Neural Networks (ANNs) among others and their applicability to handle large genomic datasets for improved prediction. This chapter explores the possibility of AI and ML in the selection and genomic process as this has several more advantages compared to the evolved selection methods to develop solutions towards the sustainability of agricultural practices for most of the global challenges.

5.2 AI Methods

5.2.1 AI-Powered Multi-Omics Integration in Crop Science

Integration of multi-omics data which includes genomics, transcriptomics, proteomics, metabolomics, and epigenomics has drastically brought advancement in crop science with the aid of comprehensive knowledge of biological systems. AI emerges as a pivotal tool to integrate these multiple datasets by inferring meaningful patterns and untangling complex interrelationships among molecular interactions. Multi-omics integration enabled by AI results in the prediction of precise traits, understanding of genotype-environment-phenotype interactions, and building of resilient and productive crops (Figure 5.1).

Figure 5.1 Various AI techniques.

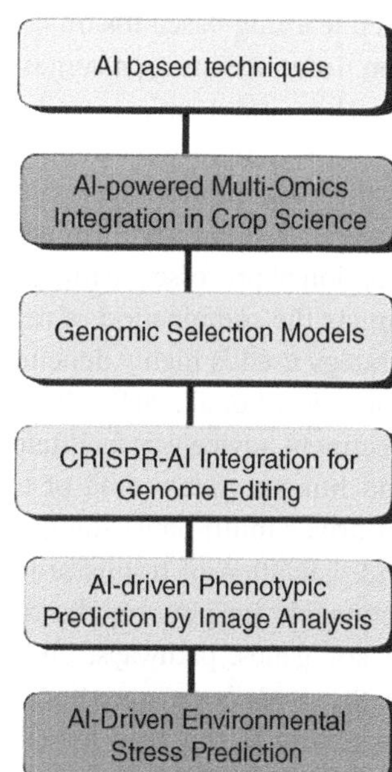

The first step in implementing AI-powered multi-omics integration is formed by data acquisition and preprocessing. High-throughput technologies like next-generation sequencing (NGS), mass spectrometry, and microarrays are used to obtain data. The noisy and heterogeneous nature of these datasets necessitates cleaning and normalization to account for batch effects, missing values, and platform-specific variability. High-dimensional data is commonly compressed into interpretable formats using dimensionality reduction methods, such as Principal Component Analysis (PCA) and t-distributed Stochastic Neighbor Embedding (t-SNE), without losing essential information. For instance, a framework for integrating multi-omics data to model genotype-environment-phenotype relationships was designed by (Subramanian et al., 2020). Better capabilities in predictive modeling in plant systems were shown by applying AI-driven preprocessing methods. Feature extraction and engineering are involved in the preparation of the data for AI models. Thousands of variables that are complex and may or may not be relevant to the biological phenomenon being studied are contained in omics data. The most informative features are isolated by feature selection methods while hierarchical representations of the data are learned by models through embedding layers for neural networks. An example is the model Cross-omics Linked unified embedding with Contrastive Learning and Self Attention (CLCLSA), which was proposed by (Murmu et al., 2024), with contrastive learning and self-attention, to be effectively learned from the incomplete multi-omics data. This allows for the participation of missing values in a dataset in making accurate predictions. Once the data is preprocessed and features are extracted, model development and learning are followed. The AI algorithms used in normal practice are Support Vector Machines (SVMs), RFs, CNNs, and GNNs. The strength of these models in capturing non-linear relations and high-order interactions amongst omics layers is exploited. For instance, the multi-omics fusion strategy is typically considered state-of-the-art as reported in CustOmics by (Benkirane et al. 2023), enabling the use of deep learning-based fusion between multiple omics data for phenotypic outcome prediction. Intermediate representations of different types of data were developed by this model; therefore, cross-omics correlations were able to be identified, and thus, phenotypic outcome predictions were improved. The most challenging part of multi-omics is constituted by the integration strategies. In the early integration approach, all the omics layers data are integrated in the raw form prior to analysis. Intermediate integration, involving the combination of processed features of each layer, is recognized as another field. Late integration targets the combination of results from each omics type in a single analysis. The kind of strategy used is highly dependent on the background information and the research question (Picard et al., 2021). It was observed in their work on precision oncology that better treatment advice was facilitated by the late integration of the individual analyses in plant and human system data of transcriptomics and metabolomics. The commencement of AI-driven multi-omics integration occurs with model development, but the interpretation and validation of results are considered equally important. In biological interpretation, actionable insights can be derived from the model outputs through the identification of key genes, pathways, and metabolites associated with specific traits. The reliability of the model's computational predictions can be bolstered by experimental validation. For example, the need for laboratory work to validate genotype–phenotype interactions discovered via AI models was underlined by Ritchie et al. (2015). Such validation ensures that statistical robustness can be historically confirmed with biological interpretation. Despite its promise, challenges are faced in multi-omics integration, including missing data, scalability, and computational complexity. Some of these issues have been addressed by recent advances in AI. ML techniques for handling incomplete datasets are provided by (Flores et al. 2023) to ensure that the absence of certain omics layers will not compromise overall analyses. A comparison of bioinformatics and ML approaches for the processing of large-scale omics data (Wang et al., 2014; Farooq et al., 2024), allowing these methods to be scaled for real-world applications. The nature of crop science is set to be altered by the synergy between AI and multi-omics integration. Different datasets can be integrated, providing an integrated portrait of plant biology to garner insights on stress resilience, nutrient efficiencies, and disease resistance. Examples of the use of AI-assisted systems in aiding the understanding of plant defense mechanisms are provided by (Murmu et al. 2024), showing how attention is directed toward key regulatory networks by the integrated analysis of transcriptomics and proteomics data. Varieties with the capacity to withstand the impact of climate change and other environmental challenges are developed through these approaches, thus assuring food and environmental safety and sustainability.

Transcriptomics and proteomics developments explained the crops' drought tolerance using AI inputs and concluded that important genes and proteins in crop plants had been identified regarding their potential functionality regarding providing osmotic regulation, reactive oxygen species (ROS) scavenging pathways, and signaling through) abscisic acid (ABA) in drought (Murmu et al. 2024), thereby aiding in upholding sustainable agriculture.

Using a framework called deep learning that assimilates the genomic, transcriptomic, and metabolomic data known as CustOmics advanced the idea for enhancing the nitrogen use efficiency (NUE) in maize (Benkirane et al. 2023). Through intermediate integration approaches, the analysis was able to capture cross-omics correlations aiding in the discovery of the pertinent genes and metabolites that intervene in nitrogen uptake, assimilation, and recycling. Missing values are addressed using imputation algorithms which help in elucidating metabolic pathways that affect the NUE. This would provide the breeding programs with good targets to achieve high NUE genotypes. This research therefore contributes to the reduction of nitrogen fertilizer dependence, enhancing crop yield, and promotion of sustainable agriculture for environmental and economic sustainability concerns.

The CLCLSA model of predicting plant resistance to fungal pathogens, integrating genomics and epigenomics data through contrastive learning and self-attention mechanisms, was proposed by (Murmu et al., 2024). The challenge of missing omics layers was alleviated in the AI-based model by employing appropriate imputation strategies, and some of the significant epigenetic markers associated with resistance to fungi included deoxyribonucleic acid (DNA) methylation sites. This model demonstrated a much higher accuracy of predicting resistance phenotypes when traditional methods were compared. Molecular markers that were integrated into breeding programs provided an avenue for basis in breeding crops that are more resistant to fungi than the levels of chemical fungicides in use at the moment and promote sustainable agricultural practices of crop protection.

In summary, profound changes in modern crop science are being brought about through AI-tailored multi-omics integration. From data preprocessing to model development, each workflow step is contributing towards the revelation of plant systems' complexity.

Challenges such as missing data and computational complexity are being addressed in order for full utilization of multi-omics integration to be enabled. With evolving technologies, it is expected that the understanding of plant biology will continue to be informed through this approach in the endeavor towards innovative solutions for global agricultural problems.

Artificial intelligence-powered integrative multi-omics is reshaping agricultural research to capture the whole equation of plant biology. The selection of proper AI model indicates advanced data acquisition, preprocessing, and feature engineering to enable the AI model to capture these complex molecular interactions and predict traits very accurately. Genomics, transcriptomics, proteomics, metabolomics, and epigenomics provide a broader scope for understanding the relationship among genotype, environment, and phenotype to promote crop resilience and yield. The AI services hold promise, as there are only the disadvantages of missing data and computational complexity to consider, which the persistent and timely advances in AI tackle head-on toward sustainable, efficient agriculture. This approach will likely change crop improvement and global food security completely.

5.3 Genomic Selection Models: Implementation in Plant Breeding

GS stands for a revolutionary breeding approach in which the genome-wide markers are utilized to predict genetic potential. Some key steps have to be considered in implementing GS. First of all, creating a diverse and representative training population forms the backbone of GS. Relevant traits, including yield, drought tolerance, and disease resistance, are assessed by genotyping and phenotyping. The size and genetic diversity of the training population greatly influence the accuracy of GS models, as a wide genetic variation is necessary for developing strong models (Crossa et al. 2017). Subsequently, genotyping is performed with high-throughput technologies to acquire genome-wide marker data. SNP markers are most frequently used because of their abundance and ease of detection. With NGS being developed further, genotyping is becoming more economical, hence facilitating genome-wide studies (Heffner et al., 2010). High-quality data of key traits across a wide range of environmental conditions is collected under phenotyping for key traits. Correctly captured phenotypic data is quite crucial, because small errors might hinder the GS models from producing predictions effectively (Jannink et al. 2010). Subsequently, the developed models are trained with estimates of association between genome-wide markers and the appropriate phenotypic traits. Among them are best linear unbiased prediction, ridge regression best linear unbiased prediction, and Bayes-based methodologies (Meuwissen et al. 2001), that effectively capture both additive and non-additive genetic effects for optimal predictive precision (Su et al. 2012). Validation of the developed predictive models involves ensuring their reliability and generalizability. Usually, cross-validation methods are employed wherein the training dataset is divided into subsets for the purpose of training and testing the model. Thus, overfitting is avoided as well as achieving good performance with unseen datasets (Heslot et al. 2012). Thereafter, for individuals lacking information on phenotypic data, their GEBVs are calculated. GEBVs depend purely on genotypic data.

That means there is no need for extensive phenotyping, and it allows breeders to efficiently find candidates with the best genetic potential (Bernardo 2014).

International Maize and Wheat Improvement Center (CIMMYT) used GS models to improve the yield of wheat under diverse environments. Using the genomic and phenotypic information of a wide training population, predictive models were developed by using Ridge Regression Best Linear Unbiased Prediction (RR-BLUP). Genomic-wide markers related to stability in yield and tolerance to abiotic stresses were identified. The yield-related traits were determined quickly across different environments through the combining of information with high-throughput phenotyping technologies. This ensured robust predictions from the models. The models were evidenced to be good predictors of the yields of untested genotypes by cross-validation. This would ensure faster selection of high-yielding wheat lines adapted to drought and heat stress. CIMMYT demonstrated that it could half the breeding cycle with no loss in genetic diversity, and the results are much better efficiency in breeding and food security of stress-prone areas.

Genomic selection is a modeling technique with good potential for improving the resistance of maize to Northern Corn Leaf Blight (NCLB). Genotyping 2 million SNP markers in these training populations, researchers assess this wealth of data based on various disease severities from elsewhere. Genomic breeding values for resistance traits are estimated using different prediction models, including one of the Bayesian methods (BayesB). By tracking how the model responded to interactions with the environment, one could understand how and under what conditions the resistance genes performed best. Prediction accuracy was good even with a small training population set and was supported through cross-validation. This entailed a reliance on fewer chemicals. Conclusively, the application of GS in maize breeding programs has led to the generation of improved genotypes resistant to NCLB that can reduce yield losses and, thus, lower production costs.

Genomic selection in soybean breeding provides support for enhancing oil content in rice and other grains (Bernardo 2014). Detailed modeling of oil-concentration changes combined with determination of genetic markers was employed upon the population suitable for analysis. These models estimate not only the contribution by genes but also the direct contribution of the genes involved in oil production. Such models give more accurate predictions. In contrast to the classical MAS, GS has been shown to outperform other contexts in which gene interactions are relevant. What is even more exciting is that the study revealed the accumulation of important alleles for oil content in many of the tested lines, although they were not directly involved in the study-for they are high-very-high-potential candidates for future breeding. (Bernardo 2014) showed how GS could help breeders get faster advances toward high oil production and steer on reducing the time taken to obtain superior-soybean varieties.

Finally, high GEBVs individuals are selected as parents for the next breeding cycle. The genetic gain can accelerate with favorable alleles combined in subsequent generations. Adding GS to breeding pipelines was demonstrated to dramatically shorten breeding cycles, especially to that needed for long-generation crops such as wheat and maize (Voss-Fels et al. 2019). These steps outline how GS can be efficient and highly beneficial in revolutionizing plant breeding.

At its core, GS has transformed plant breeding by genome-wide markers predicting genetic potential. The text speaks directly about the many important steps included in

GS: diverse training populations, high-throughput genotyping technologies, and strong predictive models. All of these approaches have been useful in highlighting the right traits for yield, drought tolerance, and disease resistance and civilization in the number of breeding cycles through the reduced consumption of resources. In that reproductive methodologies of GS have transformed coefficients, CIMMYT has been recognized as a successful institution applying them in wheat, maize, and soybean breeding. For example, GS delivered drought-tolerant wheat and heat-resistant maize varieties and high-oil-content soybeans. Such instances indicate the potential of this application to address the global challenges of climate change, food security, and sustainable agriculture. In my opinion, GS is an association of advanced genomic tools and statistical modeling that helps expedite crop genetic improvement. One is sure that GS does not merely accelerate the breeding cycle period but also assures genetic diversity. It is a powerful tool and part of the bright future of agriculture. With very good effects on addressing challenges to the environment and asserting that crops are resilient while optimizing the investment of resources, this will occupy an overall focused position in trying to enhance global food security through the exploration of sustainable farming solutions.

5.3.1 CRISPR-AI Integration for Genome Editing in Plant Breeding

Scientists have first identified those traits they aim to improve the plants for-having higher yield or better tolerance against stress, then searched for which genes control the expression of the said traits. Traits are therefore selected based on the objectives that are set upon the breeding programs. AI algorithms analyze large chunks of genomic data to find those genes related to the traits they target. ML models can predict how these genes function and are regulated. Thus, the researchers will find the best genes to edit based on such models. For instance, the complex genetic data studied with deep learning techniques, including CNNs, assist in identifying patterns within the genes related to the desired traits (van Dijk et al., 2021). This process ensures that the editing efforts are targeted to the genes with the highest impact.

First, they begin with identification of specific traits they are willing to enhance. In the case of a breeding program, researchers want either high yield or high tolerance to stresses; these researchers have identified genes in charge of the same traits. Then, the algorithm determines what characteristics will be more valuable in that specific breeding. From these sets, large genomic data will be run through an algorithm, identifying specific genes responsible for such targeted traits. Researchers predict how such genes work and how they react to one another and to external control. As a result, they are enabled to find which genes to select for editing in the first place. For example, deep techniques of learning with convolutional networks can be deployed to scan big genetic information datasets for patterns across genes associated with the traits which are desired (van Dijk et al., 2021). This step is important because it ensures that genome editing focuses on the genes that will make the biggest difference. On a basic level, the targeting of genes follows after this: guide RNAs (gRNA) target the Cas9 nuclease to specific sequences to edit DNA; the efficiency of gRNAs for specificity determines whether the CRISPR system was used successfully (Kumar et al., 2024). Tools for predicting gRNA activity include, among others, DeepCRISPR and CRISPRscan; both predict gRNA activity based on DNA accessibility sequence context and potential off-target sites. These tools use large datasets from previous experiments to predict gRNAs with high target efficiency and minimal risk of off-target edits Chuai, G., et al. (2018). The optimization through AI in this way ensures that the designed gRNAs are cutting at correct genomic locations very accurately, making the editing process more reliable (Chuai et al. 2018). Plant species with complex genomes or recalcitrant types of tissue constitute significant challenges when it comes to the delivery of CRISPR components into plant cells (Mao et al., 2019). AI comes in handy at optimizing delivery techniques specific to given plant species; it can base a decision on experimental data (Hesami and Jones 2020) and recommend suitable delivery techniques in the form of Agrobacterium-mediated transformation, electroporation, or particle bombardment, among others (Kocsisova et al., 2023). An example is the utilization of AI algorithms to predict transformation protocol success rates in monocots and dicot species (Niedbała et al., 2021). In addition, the use of AI-guided techniques helps optimize temperature conditions, tissue types, and the reagents in transformation to boost the efficiency of delivery by CRISPR, consequently enhancing the expression and uptake of the editing machinery. Plant cells edit their genomes at target loci with desired genetic changes upon the introduction of CRISPR components, which AI tools make it possible to detect and validate the edits. Through the use of AI algorithms on high-throughput sequencing data, precise edits can be confirmed, as well as possible off-target effects. CRISPResso2 and GUIDE-seq tools process sequencing reads to map edits with single-nucleotide resolution, thereby providing insights into the success and accuracy of the genome editing process (Narra et al., 2025). AI processing of large-scale sequencing data makes it possible to identify edited lines very quickly, accelerating the validation phase considerably. Beyond editing, AI is engaged in phenotypic evaluation of genetic modification. The incorporation of omics data, including transcriptomics, proteomics, and metabolomics, into AI models allows for prediction of the eventual physiological effect of gene editing on specific traits. This step is important in determining whether the edits meet their intended goal of agronomic improvement. For instance, AI frameworks have been applied to predict the effects of CRISPR-induced mutations on drought tolerance through modeling gene expression changes under stress conditions (Farooq et al., 2024). The predictions guarantee that the edited plants exhibit desirable traits in real-world agricultural environments.

Using the CRISPR-AI tool, the researchers defined zinc uptake transporters and other transporters for the movement of zinc and successfully raised its level in wheat. Molecular modeling-based prediction of zinc bioavailability targeted a transporter in the plant ZIP family, TaZIP, and highly specific gRNA design implemented CRISPR edits. High-throughput sequencing was then performed using AI tools to check for off-target effects. The prediction of zinc bioaccumulation in the edited wheat varieties considered the results of PCR tests and AI-augmented multi-omics analysis. The edited wheat lines contained 30% more zinc and did not have reduced yield. The ability of the CRISPR–AI integration to remediate micronutrient deficiencies in staple crops offers a scalable solution for ameliorating malnutrition.

Recently, scientists applied CRISPR-AI to provide improvement of drought tolerance in rice using key genes involved in ABA signaling. AI combed through large genomic data to identify key genes, which included OsPYL1, a major regulator of the plant's drought response machinery. They introduced specific mutations into OsPYL1 using

CRISPR-Cas9, which improved the plant's ability to utilize water. They then used tools such as DeepCRISPR to design highly specific gRNAs that reduce off-target effects to ensure that the gene editing was correct. The crew went to verify the edits via CRISPResso2 to verify that the changes were on level after the edits were made. This is the place ML, which includes gene expression and metabolite levels in response to drought conditions, came into play, to predict how the plants might react. When tested in a controlled environment, the altered rice plant life had mind-blowing effects, growing a quarter extra underneath drought situations as compared to ordinary crops. This study shows how the union of CRISPR plus AI can create a revolution in agriculture that will enable plants to better cope with pressure from their environment.

Finally, CRISPR-AI is scaled by propagating a breeding program through the high-level application of validated edits onto lines that bear good traits or even across some selected promising candidates, where extensive evaluation in heterogeneous environments tests such traits for efficiency and stability (Saini, 2025). This integration of CRISPR and AI accelerates the development of improved crop varieties while also ensuring their adaptability to changing climatic and environmental conditions.

The combined effects of CRISPR and AI are changing modern agriculture and genome editing, allowing particular crop traits to be improved quickly and decisively. From the identification of desirable traits, such as high yield, stress tolerance, or nutritional benefits, AI algorithms allow for maximum juxtaposition of extensive genomic data, pinning down the primacy of the responsible genes. Using CNNs, ML models can predict gene function and interactions that will allow researchers to target high-impact genes for editing. Tools such as DeepCRISPR and CRISPRscan endorse the proper design of gRNAs, minimizing off-target effects and providing for a more efficient function of the CRISPR-Cas9 system. AI technology fosters delivery of CRISPR components into challenging plant species and validates edits by conducting high-throughput sequencing analysis. The integration is no longer limited to perform the former gene-editing operations. AI models now could numerically predict the physiological effects of gene edits using multi-omics data to ensure that the modified plants would perform satisfactorily under real-world conditions. Case studies involving augmenting zinc levels in wheat or promoting drought tolerance in rice show that CRISPR-AI may respond successfully to malnutrition and climate resilience challenges. In my opinion, the collaboration between AI and CRISPR is an unprecedented breakthrough in the enduring sustainability of agriculture. They define a breeding program profile, allow for ever more precise breeding, and offer unrivalled potential for breaching food security matters, thus being a potent tool for a world rapidly changing.

5.3.2 AI-Driven Phenotypic Prediction Through Image Analysis

Advancements in AI has changed the approach to deriving phenotypic prediction by enabling high-throughput image processing technologies. Phenotyping is the evaluation of quantitative plant traits identifiable by function; possible traits include plant height, leaf area, and responses associated with stresses, such as wilting. In general, the operation of measuring phenotypic traits is laborious, with human subjectivity often limiting the scale of evaluation. AI has automated this process to some extent, particularly with ML and deep learning algorithms, so that the speed and accuracy increases are multiples of the original time (Fahlgren et al. 2015).

Different forms of imaging, such as red, green, and blue (RGB), hyperspectral, and thermal imaging can also provide accurate phenotypic information on plants. For example, a form of hyperspectral imaging can be used to analyze reflected wavelengths not in the visible spectrum; the information can be utilized to make assessments of plant stress and health. 3D imagers can create a 3D printed representation of the architecture of plants, which allows for evaluation of height growth and canopy structure. When employed in conjunction with AI-based support structures, imaging can enhance the accuracy and consistency of phenotypic prediction and allow the quantitative examination of individual phenotypic traits (Yang et al. 2020). Some of the more successful modeled approaches with image-based phenotyping have been based on ML, and, in particular, the CNN approach, as CNNs can learn a tremendous amount of image-based features, such as patterns in the leaf venation and the symptoms associated with diseases. This enables prediction accuracy that has not been fully explored until recently, based on actual data values for traits such as biomass, yield, and resistance to diseases. Additionally, methods of transfer learning have allowed for the transfer of pretrained models onto new datasets and thus reduced burden of bias to limited training data (Singh et al. 2016). However, data sharing and standardization issues continue to be a significant challenge within AI-based phenotyping. Comprehensive, unique datasets are required to effectively train and validate models in this area. Fragmented datasets and inconsistent formats continue to hinder this from happening. For example, European Plant Phenotyping Network (EPPN) and International Plant Phenotyping Network (IPPN) in Europe developed standardized protocols, that stimulated institutions to share their data and collaborate on joint-use platforms (Tardieu et al. 2017). Agriculture and image analysis and AI present applications that are absolutely game-changing for agriculture. The use of AI speeds breeding programs to find superior genotypes more quickly and cost effectively. In the context of precision agriculture, AI-based tools enable real-time monitoring of crops along with optimizing resources such as water and fertilizer. This capability allows AI-based phenotyping to assist in crop improvements for climate resilient traits by assessing linkage to drought and/or heat tolerance for agricultural sustainability (Yang et al. 2020).

In a comprehensive study to develop an AI model to evaluate the phenotypic characteristics of lettuce at or during its growth stage, reviewed by Maraveas (2024) described RGB images of the lettuce plants at various growth stages and trained deep learning model(s) to predict both geometric indices such as leaf area, and texture and color traits. In an extensive analysis, the errors were found to be <0.55% for geometric and < 1.7% for texture and color indices. This robust system detected differences in phenotypes, with, allow for the favorable and healthy production of lettuce. Overall, results shown that AI can accurately predict crop growth traits and enable optimization of the management of crops, when undertaking critical decision-making periods (Maraveas 2024). AI approaches were developed to assess the laborious and time-consuming task of counting maize plants on large sites (Maraveas 2024).

High-resolution cameras fitted on drones took multi-spectral images of maize fields. These multi-spectral images were then processed using digital image processing and features extracted using this processing. An SVM model was developed to classify the spectral data obtained from the multi-spectral images, with a classification rate of approximately 80–88%. This algorithm that leverages AI and is scalable makes it practical for the application of large-scale agriculture, the automation process of counting really mitigated manual labor and at the same time enable real-time information to come in this way to inform better field management, and thereby yield, (Maraveas 2024).

Maraveas recently reviewed the latest development of the root phenotyping via Anti-Gravity Stem-Seeking, or AGSS, Algorithm (2024). It was able to correct distortion in images of maize root and determine important traits like root, length, and diameter; however, its detection of number of lateral roots was less sensitive with the reliability being greater than 90%. It has speed and power that makes it suitable for use in embedded systems. It proved to be an impressive demonstration of how AI can solve the manual root measurement problems; so therefore, this is a faster and more accurate answer to allow breeders to select for stronger root traits (Maraveas 2024).

Artificial intelligence-enabled image analysis is going to revolutionize phenotypic prediction, thus enhancing the field of agricultural research and crop breeding. With AI application, specifically through the implementation of deep learning methods (i.e., CNN), traditional plant phenotyping processes have evolved from very simple observations to completely automated and quantitative analyses. The characterization of plant traits, such as yield, biomass, and resistance to disease, has increased in speed and accuracy to an unprecedented level. In combination with AI, other high-throughput imaging techniques have also become available, such as hyperspectral and 3D imaging, that allow for highly precise data collection related to evaluating plant growth, health, and canopy structure. The capacity for transfer learning allows for the differentiation of smaller and lower level data through the inclusion of pretrained models into new datasets. International programs like EPPN and IPPN facilitate collaboration and enact agreements for data sharing under current issues regarding the sharing and standardizing of data. AI-enabled image analysis is definitely going to create a new paradigm in the world of phenotyping prediction, which will further expand research in agricultural and crop breeding research. This evolution is happening by taking plant phenotyping processes from vague observations to fully automated quantitative analysis, thanks to AI, specifically deep learning methodologies (i.e., CNNs). Plant trait characterization speed and accuracy (yield, biomass, and disease resistance) has accelerated at tremendous rates. With the help of AI, high-throughput imaging methods, hyperspectral and 3D imaging, allow for highly sophisticated data collection to measure plant growth, health, and canopy structure. Transfer learning separates small- and low-level data, with pretrained models in new dataset. To date, they are presenting possibilities that satisfy imminent food security challenges through smarter, more productive, sustainable, viable, and climate change-resilient agricultural solutions.

5.3.3 AI for Predictive Modeling of Environmental Stress Response

Drought, salinity, and extreme temperatures are among the biggest stresses of crop productivity and food security. AI predictive modeling is working on changing the way in which researchers think to study and mitigate such environmental stresses on crops. With data integration from sensors, remote sensing, and genomics, the AI models would be capable of predicting the exact plant responses to environmental stress, providing sufficient time for important intervention before agricultural production has suffered any overt implications on a larger scale.

Despite the complexities of the models, they can effectively analyze complex datasets ranging from weather cues, soil characteristics, and plant phenotypes with ML, thus unveiling patterns that shall be helpful in predicting stress responses. Usually, deep learning models like CNNs and Recurrent Neural Networks (RNNs) would find enhanced robustness while performing non-linear correlations in such datasets. For ruptured drought stress, the CNN tracks spatial data using hyperspectral images, while the RNN temporalizes the usage of plant response over time (Fahlgren et al. 2015). Integration of multi-source data is very fundamental within predictive modeling. Stress detection via hyperspectral imaging is accomplished by detecting spectral reflectance through wavelengths out of the vision range, thereby being able to visualize small physiological displacement in a plant. Integration of data sources such as canopy temperature from thermal imaging and genomic data enables AI models to predict stress-tolerant phenotypes with improved predictive performance (Singh et al. 2016). Strength of predictive modeling thrives on quality data preprocess and good feature extraction. Image segmentation and normalization techniques thus improve data quality for AI algorithms, leading to good model performance. AI may use remote sensing technologies to bring fast monitoring of wide-scale impacts of environmental stress on agricultural crops in extensive regions (Yang et al. 2020). Predictive modeling is not without its problems. These include standardization of the data, heavy computational requirements, and interpretability. Issues are being addressed in international collaborative efforts and global phenotyping networks aiming towards data sharing and standardization protocols (Tardieu et al. 2017). The application of predictive modeling in precision agriculture is also becoming more popular in optimizing irrigation schedules, early stress detection, and selecting tolerant crop varieties with AI-based tools by farmers. These technologies contribute to climate-smart agriculture able to provide productive solutions under changing climatic conditions.

The cyclic stress ratio (CSR) of soil was predicted using ML models like Decision Tree, RF, KNN (K-Nearest Neighbor), and Extra Tree. The dataset was obtained from LEAP-2020, which contained the outcomes of cyclic triaxial and direct shear experiments. The major input parameters were cycle count, axial strain, axial stress, load in KPa, and void ratio. These models were trained to predict CSR using these parameters following standard ASTM D 3999–3991 and ASTM D 6528–6507. The outcomes revealed that classifiers such as RF and Extra Tree show great cross-validation accuracy of 0.998 and 0.987 respectively by predicting the robustness of the CSR (Pawar et al. 2023). The end conclusion reached was that AI models could sufficiently replace traditional, very labor-intensive soil testing methods by allowing proactive measures against liquefaction-related soil threats.

A deep learning-based ANN model has been developed that predicts the potential of soil liquefaction by computing CSR values using cyclic triaxial and cyclic direct simple shear tests. The ANN model is composed of two hidden layers with 32 neurons in each and was trained using experimental data from LEAP-2020. The model classified the soil samples into different CSR classes. It reached 98.12% accuracy in cyclic direct shear tests and 96.13% in cyclic triaxial tests (Pawar et al. 2023). It successfully recognized soil behavior under dynamic stress conditions. Therefore, there was an enhancement in more precise predictions of liquefaction potential. The authors argued that deep learning techniques can enhance geotechnical assessments drastically by saving much time and costs involved with laboratory-based traditional testing.

The soil behavior under dynamic environmental conditions has been analyzed with an emphasis on cyclic stress, focusing on its influence on the stability of the soil. The dataset contained the results of cyclic triaxial and cyclic direct shear tests with the critical factors including axial strain, major and minor principal stress, deviatoric stress, and void ratio. The study concluded that the increase of confining pressure and relative density reduced excess pore water pressure, thereby reducing liquefaction risk (Pawar et al. 2023). The AI model predicted CSR values well, which enabled the assessment of soil stability over earthquake-prone territories. The final conclusion was made that AI-assisted predictive models could be involved in civil engineering projects for greater risk assessment with mitigation strategies.

The advent of AI-driven predictive modeling of environmental stress responses brings to the world the revolutionary change through which one tackles modern issues involving agriculture. Diverse datasets coupled with the implementation of sophisticated algorithms take the ache away since such a method merely allows the management and monitoring of crop stress by paving way to resilient, productive, and sustainable farming systems.

In summary, predictive modeling powered by AI is redefining the management of environmental stress in agriculture, permitting far more proactive and efficient responses to the more significant challenges posed by drought, salinity, and extreme temperatures. Drawing upon a diverse range of datasets, such as weather, soil parameters, plant phenotypes, and genomic information, models based on AI predict plant responses to stress and provide insights before irreversible damage occurs. Deep learning models gain robustness from CNNs and RNNs because they are designed to detect both spatial and temporal stress-responses. The combination of these technologies allows predictions of stress-tolerant phenotypes with high confidence by integrating data and information from hyperspectral imaging and thermal sensing. International collaboration is prospering in unison to overcome copytrend difficulties in standardization, computational burdens, and model interpretability. Currently, AI-based tools are helping with the irrigation schedule, early stress detection, and resistant crop selection, all under climate-smart agriculture. Under the umbrella of soil behavior prediction, there are parallels drawn between the application of AI for soil behavior prediction and similar works in geotechnical inference, adding another dimension of adaptability. In summary, AI-based modeling has raised productivity and sustainability in agriculture and has consequently created space for building climate-resilient farming systems responsible for facing challenges in the future.

5.4 Challenges in AI Techniques for Genomic Selection in Plants

Due to use of AI techniques in GS in plants, the future scope for further development in this field has increased enormously but we also face different types of challenges for it. These challenges include different types of difficulties ranging from accessibility of data to quality of it, we have to also consider ethical reasons. By solving these challenges only, we can explore the maximum potential of AI techniques for GS in plants.

One major obstacle is the availability of high-quality phenotypic and genotypic datasets. Genome-wide marker information and associated phenotypes are required for GS models, which need to be extensive. Yet, acquiring such datasets is a time-consuming, labor-intensive, and costly process, particularly for complex traits such as drought tolerance or disease resistance. This especially applies to phenotypic data, which are sensitive to more noise caused by environmental fluctuations and measurement errors that also influence the accuracy of the models and prediction results (Jannink et al. 2010). These datasets include transcriptomics, proteomics, and metabolomics data, and their integration along with genomic data is a non-trivial problem. Heterogeneity plays an important role in these datasets, making the preprocessing and analytical step very challenging due to their high-dimensional nature. Given the complexity of the data, AI models need considerable computational resources, and improper integration of AI models can yield poor prediction performance or wrong biological interpretation of the valuable insights it generates (Farooq et al., 2024). Deep learning and ML algorithms used in GS often suffer in terms of overfitting, especially on small datasets. Overfitting is when a model does well on training data but not on unseen data. This can result in misleading predictions, particularly when models are used on different populations or in different environmental conditions. Meanwhile, the lack of interpretability, an issue known as the "black-box" problem, is another major limitation especially for complex AI models (van Dijk et al., 2021). Running deep learning models or doing simulation-based GWAS needs a high-performance computing infrastructure that is not available in resource-constrained settings. That limits the broad application of AI methods in GS, especially in low- and middle-income countries (Burrow et al., 2021). Another challenge is the validation of AI models in the context of plant breeding programs. Cross-validation techniques provide some assurance of performance, but it's the application in practice that often reveals the differences in predicted versus actual results. Training AI models on just one crop species, genetic population, or geographic area is not enough to generalize them to new datasets, and those with poor phenotype-to-genotype compatibility will not be useful due to their suboptimal prediction performance (Crossa et al. 2017). The use of AI in genome editing opens up ethical and regulatory challenges, particularly when applied in conjunction with technologies such as CRISPR. If there are efforts to use AI to guide the editing of genes, there could be unforeseen socio-economic fractures such as big wigs controlling crop innovations. Moreover, the regulations governing genome-edited crops differ by region, leading to ambiguity in the commercial viability of AI-powered GS methods (Bernardo 2014). Another barrier is a lack of expertise in both AI and plant genomics among plant breeders. Implementing AI tools in conventional breeding programs demands extensive training and capacity development. Plant breeders have little chances to interpret the AI model outputs or fit them into breeding pipelines without sufficient knowledge and resources (Voss-Fels et al. 2019). There has been much progress in genomic selection in major crops like wheat, rice, and maize but little with minor and underutilized crops. The limited genomic resources for these crops hinder the advancement and utilization of AI-based methods, which may lead to ignoring the key food and economic crops.

5.4.1 Potential of AI Techniques in Future Genomic Selection for Plants

The use of AI in GS for plants has the potential to reshape plant breeding by focusing on the development of crop varieties that are rapid, precise, and efficient. In the future, we see growth potentials in the field that will be reliant on solving current challenges and expanding trajectories of AI opportunities in plant genomics and breeding.

We space advances in high-throughput phenotyping and genotyping technologies generating larger, more complete datasets for AI models to leverage. Combining multimodal-omics data will allow for high-dimensional information pertaining to the relationship between the genotype and phenotype with the help of sophisticated AI approaches, particularly deep learning, with the help of data from multiple temporal scales. The use of "integrative predictive tools" that combines heterogeneous data types can enhance the accuracy of predictions, that may help breeders effectively target traits of more complexity (Farooq et al., 2024). Explainable AI model development is key, in advancing breeders' trust, and ensuring benefits in practice.

There are many explanations for the AI model decisions using Explainable AI, like SHAP (Shapley Additive Explanations) or LIME (Local Interpretable Model-agnostic Explanations), which provide informative explanations about AI model decision making (van Dijk et al., 2021; Azodi et al., 2020) that enhances model transparency and helps confirm model consistency.

Artificial intelligence will increasingly be used in conjunction with genome editing tools (e.g., CRISPR). gRNAs that are designed with the help of AI models will be more efficient and accurate, and they can develop better off-target predictions and better-designed delivery systems, which will improve genome editing payload. Breeders will therefore be able to more accurately define disease, stress tolerance, and nutritional composition, et cetera (Iksat et al. 2025).

Artificial intelligence will help automate menial tasks (phenotyping, genotyping, and data analysis) which will allow breeding programs to be more efficient and focus on high-value tasks. Robotic systems that are integrated with AI will facilitate high-throughput field phenotyping, which will decrease labor costs and improve efficiency. Automated systems for decision-making will also be a benefit for breeders selecting appropriate breeding candidates for breeding cycles (Farooq, M. A., et al. 2024).

Climate change poses increasing challenges to agriculture, and genomic selection combined with advanced data-driven models is expected to play an important role in developing climate-resilient crop varieties. Data from the AI models will thus be useful in predicting how different genetic modifications affect plant traits across diverse environmental conditions, informing breeders in their quest of developing crops that can withstand extreme climatic events, pests, and diseases (Voss-Fels et al. 2019). The increasing availability of user-friendly, open-source AI tools could democratize GS, empowering breeders across the globe to gainfully utilize recent advances in genomic technologies. New initiatives such as Galaxy and CyVerse have emerged and been established to give researchers easy access to computing resources and analytical tools. Such platforms will enable greater contributions in GS applications that will benefit breeders operating under a resource-constrained environment (Crossa et al. 2017).

Minor crops and orphan traits will increasingly be pursued with AI methods as a way to address threats to food security and nutrition in diverse agricultural systems. Genomic resources and AI tools can be used to pursue the potential of under-utilized crops in sustainable agriculture. AI-enabled GS will require coordination among a global research network and commitment by public and private sectors. Outlining common resources, datasets, and protocols with partners will increase efficiency and fairness (Bernardo 2014).

Artificial intelligence will work in combination with synthetic biology approaches where AI could be useful to predict how synthetic gene constructs will influence plant phenotypes and ultimately lead to the establishment of new genetic pathways for traits like enhanced photosynthesis, nitrogen fixation, or pathogen resistance, all of which could accelerate the velocity of crop improvements (Saini, 2025). It will be important to create ethical prescriptions and reconcile whichever governance structure for AI-enabled GS technologies will be accepted. Such a governance structure will have to address data privacy, equity of access to technology, and socio-economic considerations.

5.5 Conclusions

This chapter highlighted the integration of traditional knowledge with advances in sciences, genomics, and AI for adaptation in agriculture. These technologies have indeed revolutionized the nature of plant breeding, by allowing drought tolerance and pest resistance to be added along with quality of nutrients while also being able to achieve precision agriculture. The choices of traits which are slowly induced by selection could gain renewed acceleration through tools such as CRISPR-Cas9 and UAVs with multi-spectral imaging to help develop climate-resilient crops against this threat to food security among the world.

It then further explored about the progressing applications of AI in multi-omics integration, GS, and phenotypic prediction. AI-enabled multi-omics integration to enhance crop science: $G \times E \times P$ interactions. Models such as CLCLSA and CustOmics are using AI to intelligently combine several datasets for the purpose of optimizing crop resilience and yield. Recently, GS, which uses AI, works to shorten breeding cycles while emphasizing certain traits in wheat, maize, and soybean. We have achieved new best-in-class levels of precision by integrating AI with CRISPR to automate genome editing, and we can design traits in which new engineered genomes can be scaled up for sustainable agricultural production, such as CRISPR-enhanced drought tolerance and zinc biofortification. AI also ensures climate-smart agriculture through predictive modeling of environmental stress responses and phenotypic prediction by advanced imaging.

Finally, the challenges in the implementation of AI for GS, which include data quality, computing requirements, and ethical problems, were talked about in this chapter. Challenges such as data heterogeneity, generalization, and validation in different environments are limiting wider adoption. Progress in these areas is hampered by the ethical questions to be addressed as well as inequalities of resources. Addressing these challenges will further improve the use of AI techniques in GS for plants.

Concluding the chapter, it seems that both these areas of application will bring about the transformation of agriculture, resilience, and sustainability. Combating existing issues and widening AI applications to underused crops and orphan characteristics will enable fairer, scalable, and climate-resilient farming systems able to satisfy future worldwide food security necessities.

References

Azodi, C.B., Tang, J., and Shiu, S.-H. (2020). Opening the black box: interpretable machine learning for geneticists. *Trends in Genetics* 36 (6): 442–455. https://doi.org/10.1016/j.tig.2020.03.005.

Benkirane, H., Pradat, Y., Michiels, S., and Cournède, P.-H. (2023). CustOmics: a versatile deep-learning based strategy for multi-omics integration. *PLoS Computational Biology* 19 (3): e1010921. https://doi.org/10.1371/journal.pcbi.1010921.

Bernardo, R. (2014). Genomewide selection when major genes are known. *Crop Science* 54: 68–75. https://doi.org/10.2135/cropsci2013.05.0315.

Burrow, H.M., Mrode, R., Mwai, A.O. et al. (2021). Challenges and opportunities in applying genomic selection to ruminants owned by smallholder farmers. *Agriculture* 11 (11): 1172. https://doi.org/10.3390/agriculture11111172.

Chuai, G., Ma, H., Yan, J. et al. (2018). DeepCRISPR: optimized CRISPR guide RNA design by deep learning. *Genome Biology* 19: 80. https://doi.org/10.1186/s13059-018-1459-4.

Crossa, J., Pérez-Rodríguez, P., Cuevas, J. et al. (2017). Genomic selection in plant breeding: methods, models, and perspectives. *Trends in Plant Science* 22 (11): 961–975. https://doi.org/10.1016/j.tplants.2017.08.011.

van Dijk, A.D.J., Kootstra, G., Kruijer, W., and de Ridder, D. (2021). Machine learning in plant science and plant breeding. *iScience* 24 (1): 101890. https://doi.org/10.1016/j.isci.2020.101890.

Fahlgren, N., Gehan, M.A., and Baxter, I. (2015). Lights, camera, action: high-throughput plant phenotyping is ready for a close-up. *Current Opinion in Plant Biology* 24: 93–99. https://doi.org/10.1016/j.pbi.2015.02.006.

Farooq, M.A., Gao, S., Hassan, M.A. et al. (2024). Artificial intelligence in plant breeding. *Trends in Genetics* 40 (10): 891–905. https://doi.org/10.1016/j.tig.2024.07.001.

Flores, J.E., Claborne, D.M., Weller, Z.D. et al. (2023). Missing data in multi-omics integration: recent advances through artificial intelligence. *Frontiers in Artificial Intelligence* 9 (6): 1098308. https://doi.org/10.3389/frai.2023.1098308.

Heffner, E., Lorenz, A., Jannink, J.-L., and Sorrells, M. (2010). Plant breeding with genomic selection: gain per unit time and cost. *Crop Science* 50: https://doi.org/10.2135/cropsci2009.11.0662.

Hesami, M. and Jones, A.M.P. (2020). Application of artificial intelligence models and optimization algorithms in plant cell and tissue culture. *Applied Microbiology and Biotechnology* 104 (22): 9449–9485. https://doi.org/10.1007/s00253-020-10888-2.

Heslot, N., Yang, H.-P., Sorrells, M.E., and Jannink, J.-L. (2012). Genomic selection in plant breeding: a comparison of models. *Crop Science* 52: 146–160. https://doi.org/10.2135/cropsci2011.06.0297.

Iksat, N., Madirov, A., Zhanassova, K., and Masalimov, Z. (2025). Artificial intelligence-assisted CRISPR/Cas systems for targeting plant viruses. *Genes (Basel)* 16 (11): 1258. https://doi.org/10.3390/genes16111258.

Jannink, J.-L., Lorenz, A.J., and Iwata, H. (2010). Genomic selection in plant breeding: from theory to practice. *Briefings in Functional Genomics* 9 (2): 166–177. https://doi.org/10.1093/bfgp/elq001.

Kocsisova, Z. and Coneva, V. (2023). Strategies for delivery of CRISPR/Cas-mediated genome editing to obtain edited plants directly without transgene integration. *Frontiers in Genome Editing* 5: 1209586. https://doi.org/10.3389/fgeed.2023.1209586.

Kumar, R., Das, S.P., Choudhury, B.U. et al. (2024). Advances in genomic tools for plant breeding: harnessing DNA molecular markers, genomic selection, and genome editing. *Biological Research* 57 (1): 80. https://doi.org/10.1186/s40659-024-00562-6.

Ma, Z., Ren, J., Liu, Q. et al. (2025). Integrating traditional breeding and modern biotechnology for advanced forest tree improvement. *International Journal of Molecular Sciences* 26 (17): 8591. https://doi.org/10.3390/ijms26178591.

Mao, Y., Botella, J.R., Liu, Y., and Zhu, J.-K. (2019). Gene editing in plants: progress and challenges. *National Science Review* 6 (3): 421–437. https://doi.org/10.1093/nsr/nwz005.

Maraveas, C. (2024). Image analysis artificial intelligence technologies for plant phenotyping: current state of the art. *AgriEngineering.* 6 (3): 3375–3407. https://doi.org/10.3390/agriengineering6030193.

Meuwissen, T.H., Hayes, B.J., and Goddard, M.E. (2001). Prediction of total genetic value using genome-wide dense marker maps. *Genetics* 157 (4): 1819–1829. https://doi.org/10.1093/genetics/157.4.1819.

Montesinos-López, O.A., Montesinos-López, A., Pérez-Rodríguez, P. et al. (2021). A review of deep learning applications for genomic selection. *BMC Genomics* 22: 19. https://doi.org/10.1186/s12864-020-07319-x.

Murmu, S., Sinha, D., Chaurasia, H. et al. (2024). A review of artificial intelligence-assisted omics techniques in plant defense: current trends and future directions. *Frontiers in Plant Science* 15: 1292054. https://doi.org/10.3389/fpls.2024.1292054.

Narra, M., Ray, A., Polley, B. et al. (2025). AI-driven advances in plant biotechnology: sharpening the edge of plant tissue culture and genome editing. *Frontiers in Plant Science* 16: 1718810. https://doi.org/10.3389/fpls.2025.1718810.

Niedbała, G., Niazian, M., and Sabbatini, P. (2021). Modeling *Agrobacterium*-mediated gene transformation of tobacco (*Nicotiana tabacum*)—A model plant for gene transformation studies. *Frontiers in Plant Science* 12: 695110. https://doi.org/10.3389/fpls.2021.695110.

Pawar, A., Jolly, A., Pandey, V. et al. (2023). Artificial Intelligence algorithms for prediction of cyclic stress ratio of soil for environment conservation. *Environmental Challenges* 12: 100730. https://doi.org/10.1016/j.envc.2023.100730.

Picard, M., Scott-Boyer, M.-P., Bodein, A. et al. (2021). Integration strategies of multi-omics data for machine learning analysis. *Computational and Structural Biotechnology Journal* 19: 3735–3747. https://doi.org/10.1016/j.csbj.2021.06.030.

Ritchie, M.D., Holzinger, E.R., Li, R. et al. (2015). Methods of integrating data to uncover genotype–phenotype interactions. *Nature Reviews Genetics* 16 (2): 85–97. https://doi.org/10.1038/nrg3868.

Saini, H. (2025). Integration of artificial intelligence with genome editing: a new era for molecular plant breeding. *Progress in Molecular Plant Physiology and Pathology* 102952. https://doi.org/10.1016/j.pmpp.2025.102952.

Singh, A., Ganapathysubramanian, B., Singh, A.K., and Sarkar, S. (2016). Machine learning for high-throughput stress phenotyping in plants. *Trends in Plant Science* 21 (2): 110–124.

Su, G., Christensen, O.F., Ostersen, T. et al. (2012). Estimating additive and non-additive genetic variances and predicting genetic merits using genome-wide dense single nucleotide polymorphism markers. *PLoS One* 7 (9): e45293. https://doi.org/10.1371/journal.pone.0045293.

Subramanian, I., Verma, S., Kumar, S. et al. (2020). Multi-omics data integration, interpretation, and its application. *Bioinformatics and Biology Insights* 14: 1–24. https://doi.org/10.1177/1177932219899051.

Tardieu, F., Cabrera-Bosquet, L., Pridmore, T., and Bennett, M. (2017). Plant phenomics: from sensors to knowledge. *Current Biology* 27 (15): R770–R783. https://doi.org/10.1016/j.cub.2017.05.055.

Varshney, R.K., Roorkiwal, M., and Sorrells, M.E. (2021). Genomic selection for crop improvement: advances, challenges, and opportunities. *The Plant Genome* 14 (2): e20149. https://doi.org/10.1002/tpg2.20149.

Voss-Fels, K.P., Cooper, M., and Hayes, B.J. (2019). Accelerating crop genetic gains with genomic selection. *Theoretical and Applied Genetics* 132 (3): 669–686. https://doi.org/10.1007/s00122-018-3270-8.

Wang, L., Wang, G., and Alexander, C.A. (2014). Big data in bioinformatics: a survey of distributed computing frameworks. *Bioinformatics* 31 (17): 2792–2798.

Wójcik-Gront, E., Zieniuk, B., and Pawełkowicz, M. (2024). Harnessing AI-powered genomic research for sustainable crop improvement. *Agriculture* 14 (12): 2299. https://doi.org/10.3390/agriculture14122299.

Yang, W., Feng, H., Zhang, X. et al. (2020). Crop phenomics and high-throughput phenotyping: past decades, current challenges, and future perspectives. *Molecular Plant* 13 (2): 187–214. https://doi.org/10.1016/j.molp.2020.01.008.

Yue, T., Wang, Y., Zhang, L. et al. (2023). Deep learning for genomics: from early neural nets to modern large language models. *International Journal of Molecular Sciences* 24 (21): 15858. https://doi.org/10.3390/ijms242115858.

6

Current Applications of Artificial Intelligence and Machine Learning in Plant Functional Genomics

Siddharth Singh[1], Sanjana Mishra[1], Amaan Arif[1], Prekshi Garg[2], and Prachi Srivastava[1]

[1] *Amity Institute of Biotechnology, Amity University Uttar Pradesh, Lucknow, 226028, India*
[2] *BioinfoCore Solutions (OPC) Pvt. Ltd., A-15 Mall Avenue Awas-Vikas Colony, Lucknow, 226001, India*

6.1 Introduction

The emergence of genomic technology has brought plant biology research into the post-genomics age. A new "green revolution" in agriculture is being propelled by the outstanding advancements in genetics and agricultural applications (Wang et al. 2023). Many significant crops have superior genomes, which have allowed for the identification of many genes governing agronomically significant features and the disclosure of the underlying processes. The knowledge of plant evolution and adaptability at the microscopic level has greatly expanded as a result of the study of plant genomics (Song et al. 2023; Wang et al. 2024). By gaining molecular-level insights into plants' ecological adaptation, stress tolerance, and pest and disease resistance, researchers have been able to make significant progress in plant breeding owing to the plenty of plant genomic data (Abbas et al. 2024). Plants face a range of abiotic stresses, such as drought, salinity, and temperature fluctuations, as well as biotic stresses from pathogens and pests (Singh et al. 2021a,b). By integrating omics data, researchers can identify molecular markers and pathways associated with stress resilience. This information aids in the development of stress-tolerant crop varieties that can thrive in changing environmental conditions, ensuring food security and sustainable agriculture (Cembrowska-Lech et al. 2023). Furthermore, the rapid advancements (shown in figure 1) in high-throughput sequencing technologies have led to an explosion of omics data, necessitating the development of sophisticated analytical tools for effective data integration.

Advanced bioinformatics platforms and ML algorithms are increasingly being utilized to analyze large-scale multi-omics datasets (shown in figure 2), enhancing the accuracy and efficiency of genomic research (Jamil et al. 2020). By synthesizing data from these diverse sources, researchers can gain deeper insights into the intricate regulatory networks that

Table 6.1 An overview of popular machine learning techniques and their main applications.

Methods	Main applications	References
Linear regression (LR)	• Classification. Functional causal modelling. • Metabolomics. • Genotype-phenotype associations.	Xi et al. (2014); Ogutu et al. (2012); Xu et al. (2008)
Support vector machines (SVM)	• Cancer genomics classification. • Outliers detection. • Discovery of new biomarkers and new drug targets.	(Kircher (2014); Zhang et al. (2013); Wu et al. (2011)
Random decision forest (RDF)	• Genome-wide association (GWA). • Epistasis detection. • Pathway analysis. • Visualization of decision processes.	Ernst and Kellis (2015); Chen and Ishwaran (2012)
Naive Bayes	• Short-sequences classification. • Multi-class prediction. • DNA barcoding. Biomarker selection.	Nielsen and Matz (2006); Silla and Freitas (2009)
K nearest neighbors (KNN)	• Cancer genomics classification. • Gene expression analysis.	Garrow et al. (2005); Yao and Ruzzo (2006); Parry (2010)
Principal component analysis	• Dimensionality reduction. • Cancer classification. SNPs tagging. • Visualization of genetic distances. • Proteomic analysis.	Maisuradze et al. (2009); Taguchi and Okamoto (2011); Worley and Powers (2013); Hsu et al. (2014)
Dynamic Bayesian networks (DBNs)	• Gene regulation analysis. • Epigenetic data integration. • Protein sequencing.	Gusmao et al. (2014); Zhang et al. (2016)
Linear discriminant analysis (LDA)	• Data pre-processing. • Motifs identification. • Cancer genomics classification.	Yuan et al. (2008); Huang et al. (2009)
k-means clustering	• Genome clustering. • Gene expression pattern recognition. • Image segmentation.	Handhayani and Hiryanto (2015); Oyelade (2016)

Table 6.2 An overview of the most pertinent deep learning architectures and their uses.

Methods	Main applications	References
Deep neural networks (DNNs)	• Cancer genomics. • Protein sequence classification. • Phenotype from genotype prediction.	Angermueller et al. (2017); Sundaram (2018); Date and Kikuchi (2018); Zhang et al. (2019)
Multilayer perceptron (MLP)	• Protein structure prediction. • Molecular classification. • Cancer genomics.	Mojarad et al. (2010); Oh (2013)
Convolutional neural networks (CNNs)	• Modelling regulatory elements. • Detection of DNA accessibility. • Finding binding sites sequences.	Min et al. (2016); Kelley et al. (2018); Zhou (2018); Quang and Xie (2019)
Recurrent neural networks (RNNs)	• Transcription factor binding sites prediction. • Mutation and variants identification. • Protein homology detection.	Quang and Xie (2019); Liu (2019); Shen et al. (2018)
Long short term memory (LSTM)	• Splicing prediction. • Gene expression regulation. • Detection of genomic long-term correlations.	Tavakoli (2019); Lee et al. (2020)
Deep Boltzman machines (DBMs)	• Protein function prediction. • SNPs pattern recognition. • Cancer genomics.	Hess et al. (2017); Nivaashini and Soundariya (2018)
Deep belief networks (DBNs)	• Enhancers prediction. • Gene expression pattern recognition. • Cancer classification. • Drug discovery.	Bu et al. (2017); Karabulut and Ibrikci (2017)
Autoencoders (AEs)	• Dimensionality Reduction. • Mutation and variants identification. • Methylation analysis. • Drug discovery.	Wang and Gu (2018); Heje Grønbech et al. (2019); Levy (2020)
Generative adversarial networks (GANs)	• Data denoising. • Data augmentation. • Missing data imputation. • Genome editing.	Ghahramani et al. (2018); Gupta and Zou (2019); Marouf (2020)

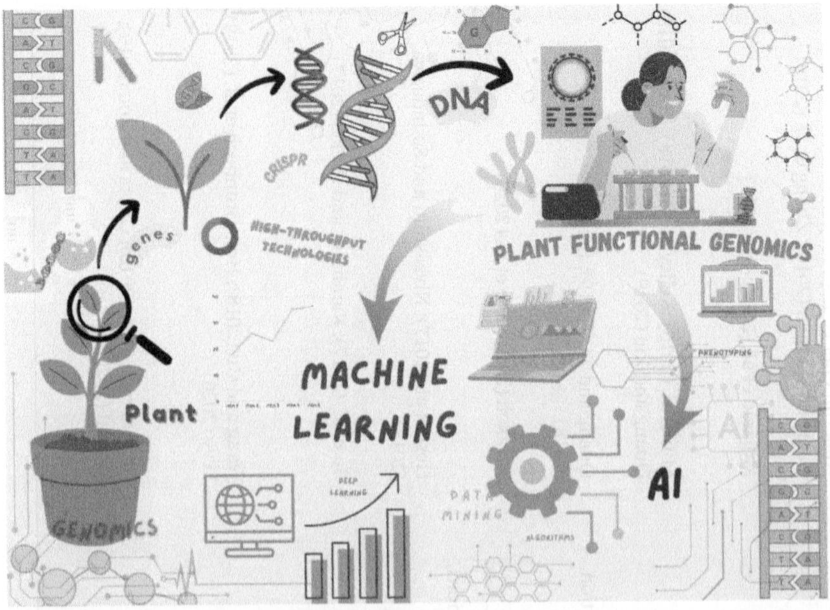

Figure 6.1 An overview of plant functional genomics, integrating high-throughput technologies, and machine learning (ML) with artificial intelligence (AI). These approaches enhance data mining, DL, and phenotyping, enabling precise gene function analysis and trait prediction. This interdisciplinary framework accelerates crop improvement and advances sustainable agriculture.

govern plant development, growth, and responses to environmental stimuli. This integrative approach bridges the gap between genetic information and phenotypic traits, enabling scientists to decipher complex interactions at multiple biological levels. Understanding these interactions is essential for advancing plant genomics research and improving agricultural productivity (Huang et al. 2017; Subramanian et al. 2020).

Utilizing transcriptomics, epigenomics, proteomics, interactomics, metabolomics, and phenomics (Nabwire et al. 2021), functional genomics research is essential to comprehending plant signal transduction and metabolic processes (Lu and Zou 2024). Furthermore, using gene deletion, overexpression, and transgenic technologies, scientists may examine the roles of certain genes in plant growth, development, and environmental adaptability (Roy et al. 2022). The complexity and size of omics data might quickly become too massive for statistical correlations or visual analysis. As a result, the use of AI or machine intelligence (ML) has increased (McCarthy et al. 2006). AI is capable of managing data sets that are too large for human brains to process, extracting information that goes beyond what we currently know about the system being studied, and most importantly improving itself automatically via dealing with training data.

The rapid advancement of high-throughput technology in the biological sciences over the past several decades has resulted in the generation of tremendous amounts of data. In order to comprehend how biological molecules contribute to the structure, function, and dynamics of a living system, such as a cell, tissue, or organism, these data are intended to quantify and characterize specific ensembles of biological molecules, including proteins, metabolites, DNA, and RNA (McKusick and Ruddle 1987).

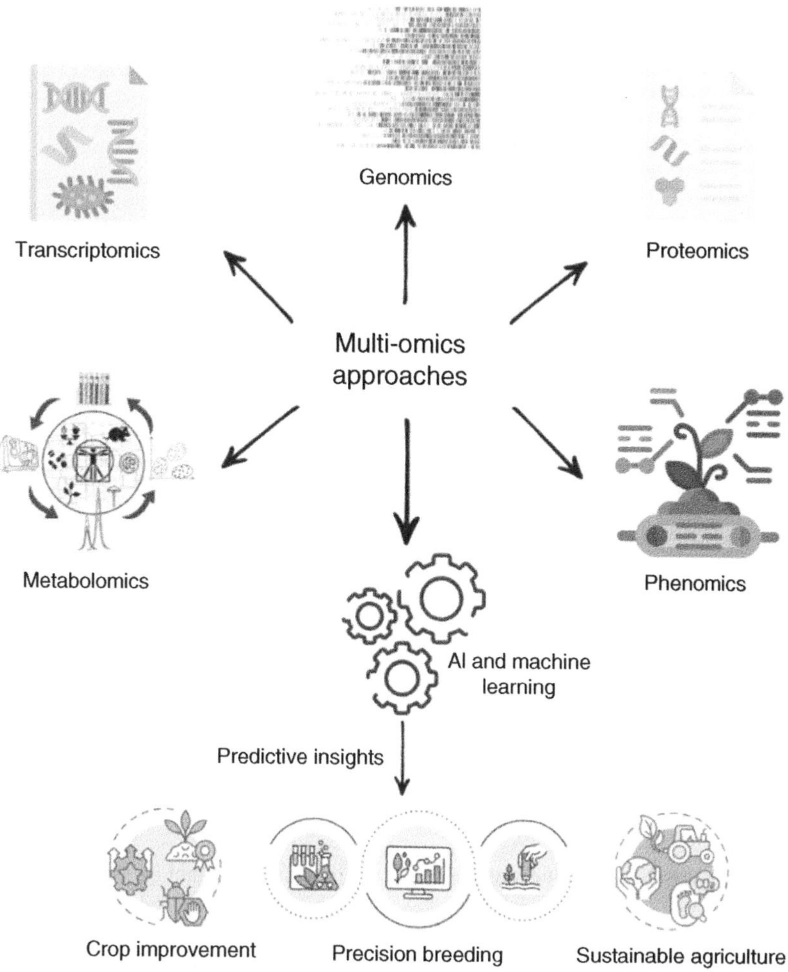

Figure 6.2 Multi-omics data integration framework.

The various fields of study that contribute to this enormous amount of biological data are named after the primary focus of their research, whether it is the DNA information content of an organism or system (genomics), the reversible modifications that DNA can undergo (epigenomics), the RNA transcripts produced by a genome (transcriptomics), the complete set and dynamics of RNA modifications (epitranscriptomics), the translational outcomes of protein-coding transcripts (proteomics), or the metabolites (metabolomics) that can exist in a particular organism or system at a specific time and condition in both physiological and pathological conditions (Caudai et al. 2021). Although each of these disciplines is a separate field of study, the information and data they generate come together to form the ambitious objective of functional genomics, a field of study that aims to describe the behavior and interactions of all key players that connect a set of evident features of a cell or oneself to the functional relationship within the environmental factors and the fundamental genetic makeup.

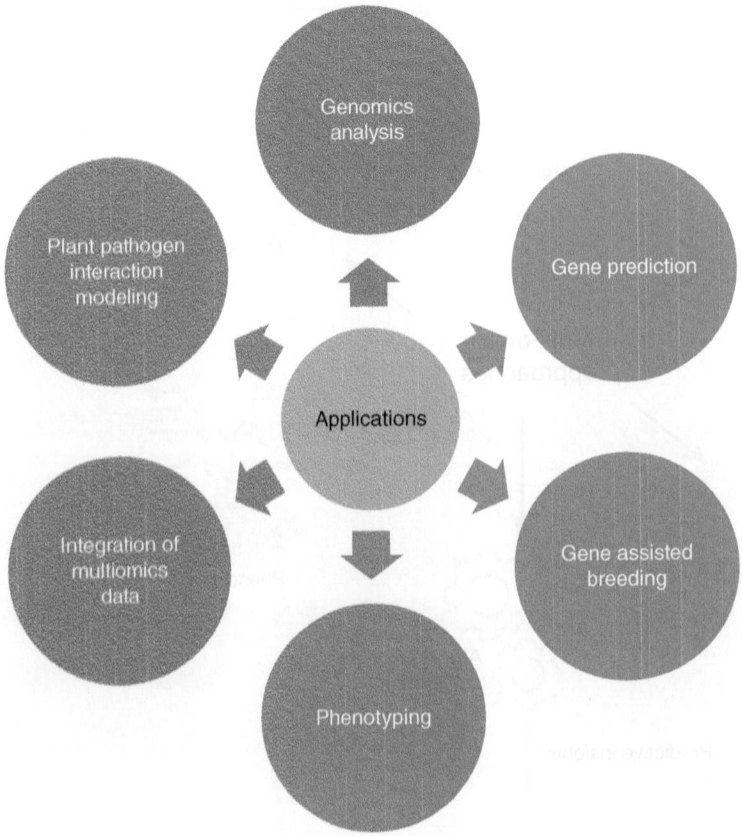

Figure 6.3 Illustrating the various applications of AI and ML in enhancing plant genomics.

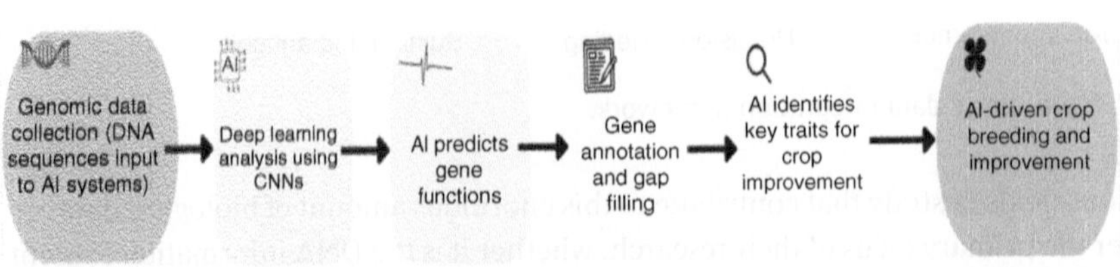

Figure 6.4 Genomic data analysis and crop improvement.

6.2 Artificial Intelligence and Machine Learning in Plant Genomics

AI is revolutionizing the way machines perform tasks traditionally associated with human intelligence, such as interpreting natural language, recognizing patterns, and making informed decisions. At the core of AI lies ML, a subset of AI that empowers systems to learn from vast amounts of data, improving autonomously over time without being explicitly programmed. By leveraging algorithms, ML can detect patterns within datasets and

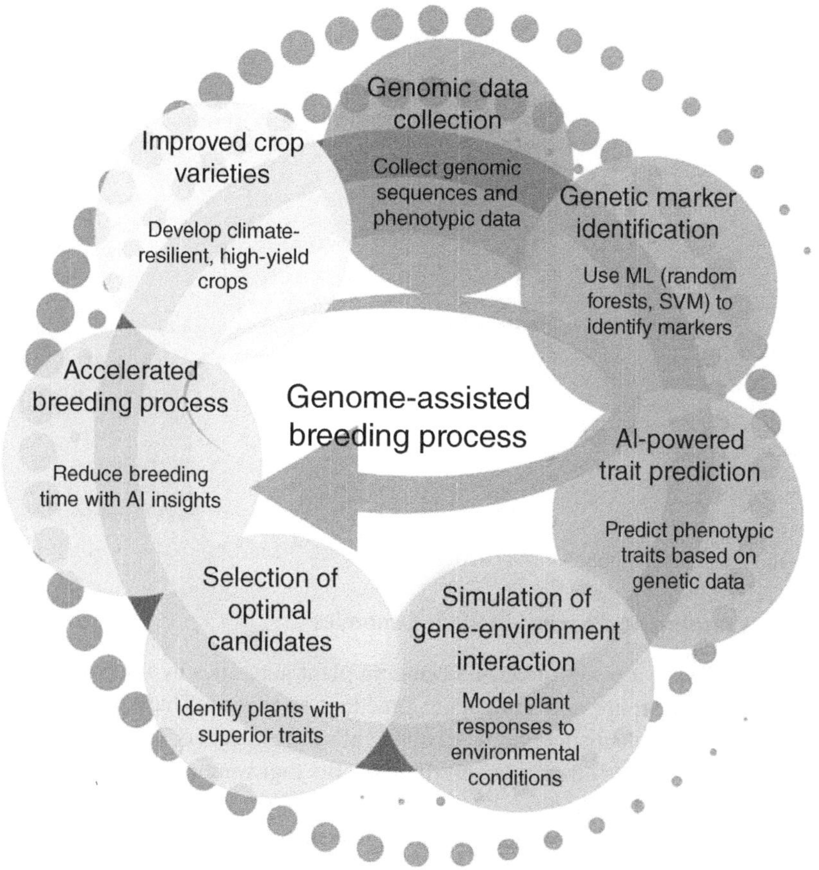

Figure 6.5 Genome-assisted breeding integrates genomics, AI, and machine learning to enhance crop traits and accelerate breeding.

predict outcomes, adapting to new data inputs as they arise. The power of ML is derived from its various subfields, such as supervised learning, where algorithms are trained using labeled data to predict outcomes, and unsupervised learning, which identifies hidden structures within unlabeled data. Reinforcement learning, another facet of ML, enables systems to learn through feedback—receiving rewards or penalties based on actions taken within an environment. A more advanced dimension of ML, DL, utilizes neural networks with multiple layers to process and interpret complex data such as images and text, driving innovations in areas like image recognition and natural language processing (Schneider 2020; Ferrario and Loi 2021).

AI has become a powerful tool for deciphering large omics datasets and comprehending the complex systems that underlie plant stress responses. This paper explores the use of AI techniques, including ML and DL, as viable methods for interpreting complicated omics data in plant functional genomics. Also explores the challenges posed by AI in plant genomes as well as potential avenues for plant functional genomics research.

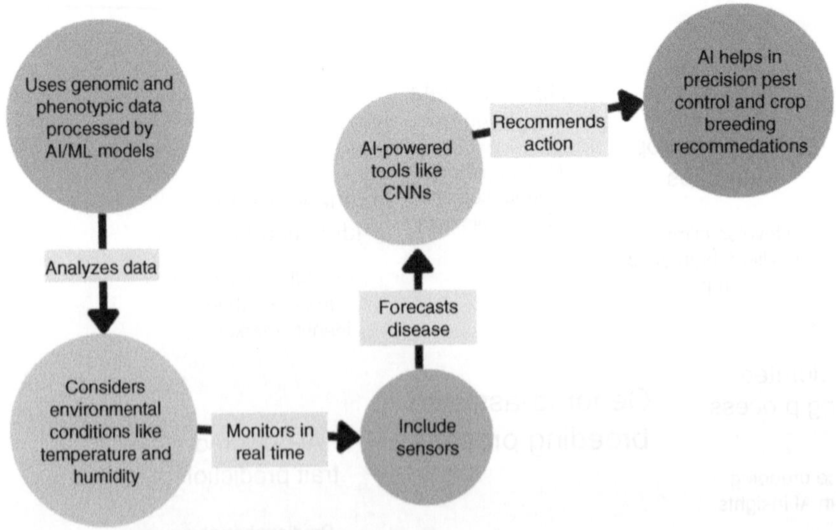

Figure 6.6 AI and ML plant–pathogen interaction modelling.

6.2.1 Machine Learning Approaches in Plant Genomics

ML algorithms are playing a pivotal role in advancing plant genomics by improving our understanding of genetic traits and enhancing breeding strategies. These techniques, divided into supervised and unsupervised learning, allow researchers to analyze vast amounts of genomic data more efficiently, offering insights that were previously difficult to obtain through traditional methods.

In supervised learning, models are trained on labeled datasets, where input data is paired with the correct output. Among the key supervised learning algorithms, Random Forest (RF) stands out as an ensemble learning method that builds multiple decision trees during training and outputs the most frequent prediction. This approach is particularly effective in capturing the complex, non-linear relationships between genotype and phenotype, making it a powerful tool for predicting complex traits. Similarly, the support vector machine (SVM) is widely used for classification tasks, where it identifies the optimal hyperplane that separates different classes. Its ability to operate efficiently in high-dimensional spaces makes SVM an ideal candidate for genomic data analysis, which is often highly multidimensional (Lourenço et al. 2022).

The distinctive characteristic of ML algorithms, such as linear regression, clustering, and Bayesian networks, within AI is their independence from the requirement to be manually programmed to complete a specific job (refer Table 1). In the early 1980s, ML techniques were initially used in the field of biology (Stormo et al. 1982). All fields of study associated with functional genomics, including genomics (de Ridder et al. 2013; Angermueller et al. 2016; Camacho et al. 2018), transcriptomics (Zhang 2019), proteomics (Park and Kellis 2015; Ragoza et al. 2017), and metabolomics (Grapov et al. 2018; Zampieri et al. 2019), have used ML systems in recent times.

ML algorithms are also playing a pivotal role in analyzing complex genomic data. Algorithms like RF and XGBoost are being used to predict crop yields and identify desirable

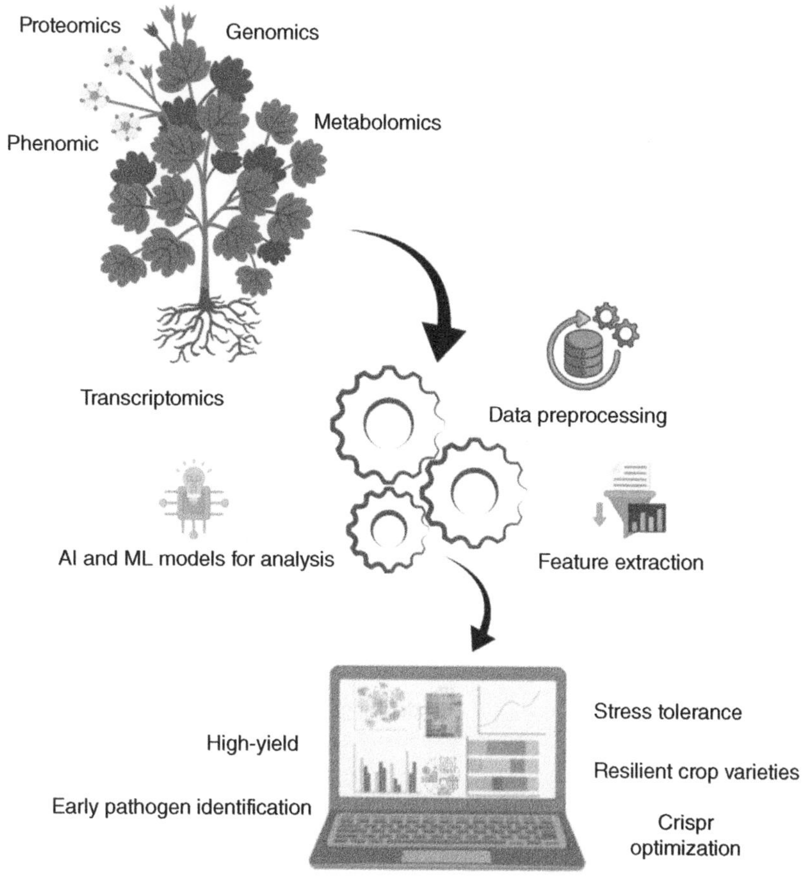

Figure 6.7 Showing use of AI and ML models for modern plant breeding.

genotypes by analyzing high-throughput phenotyping data. These algorithms excel in processing large datasets, making them ideal for handling the immense volume of genetic and phenotypic data generated in plant genomics. Additionally, support vector regression (SVR) has been employed in studies that relate hyperspectral measurement data to plant genotypes. SVR automatically selects relevant features from the data, which improves the understanding of the relationships between genotype and phenotype, ultimately contributing to more accurate predictions of plant traits (Gabur et al. 2022).

6.2.2 Deep Learning Approaches

DL techniques are the most promising in addressing the complexity of omics and data. These techniques use mathematical operations organized in several layers that are related to one another for processing information. Despite the fact that the first neural network models were put into practice over 60 years ago, their initial computational and financial expenses made them an intriguing but unsustainable resource. Frank Rosenblatt originally presented the Perceptron, the first neural network design, to the scientific world

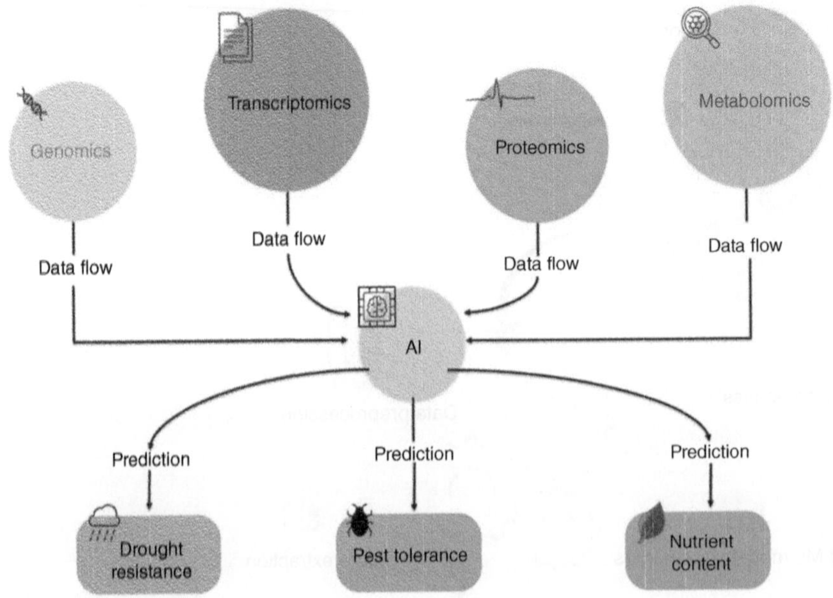

Figure 6.8 Omics data integration with AI.

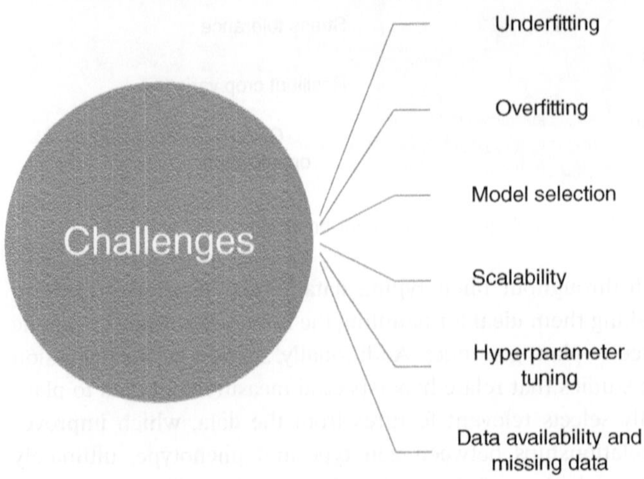

Figure 6.9 Challenges and considerations in AI and ML applications for plant genomics.

in 1958 (Rosenblatt 1958), but it had a limited capacity for learning. Furthermore, the computer that was executing the Perceptron algorithm had relatively little computing capacity, even though it was the size of a vast room. Effective use of DL techniques (refer Table 2) has generally only been made feasible in the past ten years due to a sharp rise in processor performance that fulfilled their high computational demands, particularly when gaming-focused Graphical Processing Units (GPUs) were repurposed (Dettmers 2015). During those years, a massive drop in sequencing costs made an abundance of

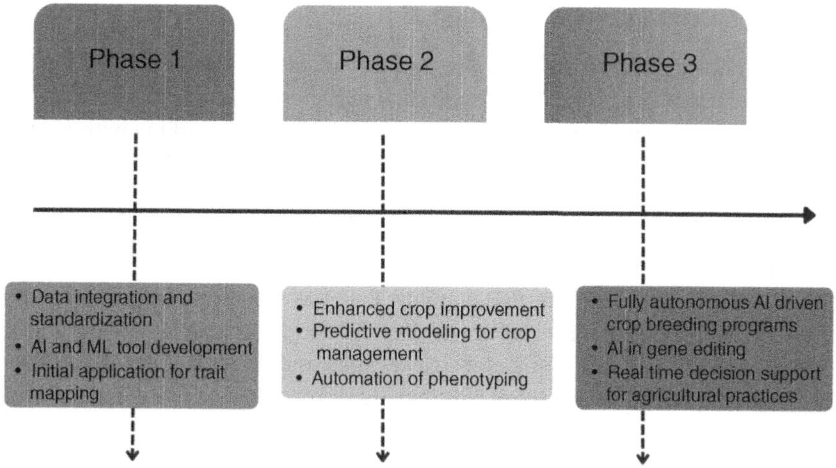

Figure 6.10 Roadmap diagram for AI and machine learning in plant genomics.

big genome-scale datasets available, which in turn made functional genomics a rich environment for DL applications (Zou 2018; Eraslan et al. 2019; Esteva 2019).

DL techniques, particularly CNNs, have also revolutionized plant breeding, especially in image-based analysis. CNNs are capable of automating the analysis of plant images, such as identifying and counting morphological features in embryos with remarkable precision (Kelley et al. 2018). This ability significantly enhances crop propagation practices by providing more accurate assessments of plant characteristics, facilitating better decision-making in breeding programs.

6.2.3 Benefit of Deep Learning Over Machine Learning

The primary benefit of DL over ML techniques is end-to-end learning, which allows for the immediate extraction of categorization or prediction outcomes from the unprocessed data. End-to-end learning gains from eliminating the potential bias imposed by manual involvement in the different data processing steps, even though the approach is not immune to potential sources of bias (such as input data selection for the network training phase). Additionally, DL techniques make it easier to integrate many input data formats, including text, numbers, photos, and audio files. Lastly, when it comes to abstraction, DL architectures are far more capable than conventional ML methods.

6.3 Applications of Artificial Intelligence and Machine Learning in Plant Genomics

AI and ML have brought transformative changes to plant genomics (figure 3), driving significant advancements in genome sequencing, gene expression prediction, trait identification, and crop improvement. These technologies provide powerful tools for analysing vast amounts of biological data, offering insights that were previously beyond the reach of traditional methods.

6.3.1 Genomic Analysis and Gene Prediction

AI and ML have become crucial tools in the analysis of plant genomes, significantly improving the accuracy of gene functions. Traditional methods of gene identification and annotation can be time-consuming and incomplete, especially when working with poorly understood or complex genomes. AI, particularly DL techniques like CNNs, has proven to be extremely effective in processing large-scale genomic data. By recognizing patterns in DNA sequences, these algorithms can predict gene functions and provide new insights into previously unexplored areas of the genome.

The ability of AI to analyze and interpret genomic data has significant implications for crop improvement as depicted in figure 4. For example, AI-driven systems can help identify genes that confer resistance to environmental stresses such as drought or pests. This capability is especially valuable for developing crops that are better suited to changing climates and global food security challenges. Moreover, AI can enhance the process of genomic sequencing by automating data analysis, reducing the time required for gene identification and annotation. This accelerates the development of new crop varieties with desirable traits, such as higher yield, disease resistance, or improved nutritional content. By filling gaps in poorly annotated genomes, AI contributes to creating a more complete and accurate understanding of plant genetics.

Several studies illustrate the importance of AI and ML in genomic research. For instance, research by Gao et al. explored the incorporation of gene interaction data into genomic prediction, offering more precise tools for plant breeding (Gao et al. 2017). Similarly, the work by Pook et al. demonstrated the potential of local CNNs for improving genomic prediction, enabling more accurate assessments of genetic traits across diverse plant species. The integration of DL techniques for plant genomics is not just about improving accuracy but also about enhancing the speed and efficiency of genetic research (Pook et al. 2020). As AI continues to evolve, its role in plant genomics will only expand, offering exciting possibilities for the future of crop breeding and agricultural sustainability. In this regard, studies like those by Wang et al. emphasize the significant potential of AI in driving crop improvement, especially in the context of global challenges like climate change and food security (Wang et al. 2020).

6.3.2 Genomic-Assisted Breeding

Building upon the advancements in genomic analysis and gene prediction, AI and ML are further transforming plant genomics through genomic-assisted breeding as represented in figure 5. This approach leverages advanced algorithms like marker-assisted selection (MAS) and the simulation of gene–environment interactions to accelerate crop improvement and help breeders develop plants with superior traits, such as higher yields, pest resistance, and enhanced climate resilience (Mir et al. 2019).

ML techniques, particularly RFs and SVMs, are now essential tools for linking genetic variations to specific phenotypic traits. These models analyze complex genomic datasets, enabling breeders to identify genetic markers associated with desirable traits early in the breeding process (Montesinos López et al. 2022). For instance, AI can pinpoint genetic variations that contribute to drought tolerance or pest resistance, facilitating the selection of plants with optimal genetic profiles before they are grown in the field. By using AI and ML to correlate genetic data with phenotypic outcomes, breeders can significantly speed up the process of developing improved crop varieties. Traditional breeding methods often take years, as breeders must cross-pollinate plants and wait for multiple generations to assess phenotypic traits. However, AI-driven genomic-assisted breeding reduces this time significantly, allowing for faster identification of the best candidates and improving breeding efficiency.

Furthermore, AI's ability to simulate gene–environment interactions plays a critical role in developing climate-resilient crops. Through ML models, breeders can predict how plants will respond to different environmental conditions, such as changes in temperature, soil moisture, and other stress factors. This allows breeders to identify plants that are more likely to thrive under unpredictable and extreme weather conditions, ultimately helping to secure food production in the face of climate change. Research by Montesinos López et al. demonstrated the power of statistical ML methods, using tools like the SKM library to improve the accuracy of genomic prediction models (Montesinos López et al. 2023). Ogutu et al. also compared various ML techniques, including RFs, boosting, and SVMs, finding that these models significantly outperformed traditional methods in genomic selection tasks (Ogutu et al. 2011). By incorporating AI and ML into genomic-assisted breeding, plant breeders can now more accurately and efficiently select plants with the desired traits, reducing time and resource expenditures. This powerful combination of technology is not only enhancing agricultural productivity but also facilitating the development of crops that can better adapt to the challenges posed by climate change, ultimately leading to more sustainable and resilient agriculture.

6.3.3 Plant–Pathogen Interaction Modelling

The integration of AI and ML into plant–pathogen interaction research has revolutionized the field, enhancing our understanding of the complex relationships between plants and pathogens (shown in figure 6). By leveraging these advanced computational techniques, researchers can predict pathogen effectors, identify plant resistance genes, and model how plants respond to pathogen attacks across diverse environmental conditions. These insights are crucial for developing strategies that enhance crop resilience and reduce the impact of plant diseases on food security.

AI and ML models can analyze vast amounts of genomic and phenotypic data, enabling them to make predictions about plant responses to pathogens. For instance, through the identification of specific genetic markers, AI models can predict which plants are more likely to resist certain diseases. These predictions are particularly important when considering the wide range of pathogens that affect crops and the unpredictable nature of plant diseases, influenced by factors such as climate change and agricultural practices. By understanding these interactions, researchers can improve breeding programs to create more disease-resistant crops. Furthermore, AI models are capable of simulating plant–pathogen interactions under varying environmental conditions. This capability allows for a deeper understanding of how environmental factors like temperature, humidity, and soil conditions impact disease progression and plant responses. As a result, farmers can make better decisions about crop management practices, optimizing interventions based on real-time weather patterns and other environmental variables.

The application of AI and ML in plant–pathogen interaction modeling also plays a crucial role in early disease detection and monitoring. Real-time data from sensors, drones, and satellites can be processed using AI algorithms to track the spread of diseases across agricultural regions. This proactive monitoring allows for timely interventions, preventing outbreaks from becoming widespread and helping mitigate losses before they occur. For example, AI-powered diagnostic tools, such as CNNs, can analyze images of plant leaves to detect early symptoms of disease. This technology is especially useful in regions where resources are limited, as it can work with minimal data and still provide accurate predictions of plant health (Afifi et al. 2021).

Moreover, AI's role extends beyond disease detection. It is also instrumental in understanding the molecular mechanisms underlying plant immunity and pathogen virulence. For example, studies have utilized AI to predict how specific pathogens interact with plant immune receptors and how these interactions can lead to disease or resistance. By analyzing these molecular interactions at a high level, AI models can identify potential targets for genetic modification or breeding to enhance disease resistance in crops.

In addition to its predictive capabilities, AI can also optimize disease management strategies by recommending personalized treatments based on data from various sources. These strategies can include tailored use of fungicides or the selection of resistant plant varieties. AI also helps to manage and reduce pesticide use, offering more sustainable solutions to plant disease control. By analyzing large datasets from agricultural fields, AI can suggest precise interventions that minimize environmental impact while maximizing crop yield and health.

6.3.4 High-Throughput Phenotyping and Trait Prediction

Following the advancements in genomic-assisted breeding, AI-powered systems are now enhancing high-throughput phenotyping and trait prediction, providing breeders with tools to more accurately predict and evaluate complex traits in plants. Traits such as yield, drought resistance, and pathogen resilience are often influenced by multiple genes and environmental factors, making them difficult to measure through traditional methods. However, with the help of ML models, particularly deep CNNs, AI can now predict these traits with greater precision (figure 7).

High-throughput phenotyping is a critical component of modern plant breeding. It involves the use of advanced imaging technologies and ML algorithms to collect and analyze large volumes of phenotypic data. For example, drones, sensors, and cameras are used to capture images of plants, which are then processed using AI to extract detailed information about plant growth, stress responses, and other phenotypic characteristics (Barbedo 2019; Gill et al. 2022). This ability to quickly and accurately measure complex traits enables breeders to assess plant performance across large fields, streamlining the process of identifying plants with desirable traits. AI-driven phenotyping systems can rapidly assess how plants respond to different environmental conditions, such as drought or pathogen attacks, and predict how these responses correlate with genetic data. This is particularly valuable when working with traits that are difficult to measure manually or that take time to manifest, like drought tolerance or pathogen resistance. By combining genomic data with phenotypic data, AI can generate highly accurate predictions of how a plant will perform under various environmental stressors, enabling breeders to make informed decisions early in the breeding process.

The integration of AI with phenotyping has a direct impact on the development of robust crop varieties that are tailored to specific agricultural challenges. This technology accelerates the breeding of crops that are not only higher yielding but also more resilient to the challenges posed by climate change, such as water scarcity or extreme weather conditions. Research in this field highlights the power of AI and ML in improving phenotyping accuracy and efficiency. Yue et al. demonstrated the use of AI in predicting complex plant traits based on genomic and phenotypic data, highlighting the role of DL models in improving trait prediction accuracy (Yue et al. 2023). Similarly, Mir et al. explored how high-throughput phenotyping can be combined with AI to assess plant health, growth patterns, and stress responses, providing a comprehensive understanding of how plants adapt to environmental challenges (Mir et al. 2019). Further, studies by some scientists underscored the transformative impact of AI in plant breeding. Like Farooq et al. reviewed the various ways AI is reshaping plant breeding, from trait prediction to improving breeding strategies (Farooq et al. 2024), while Crossa et al. discussed how ML algorithms are being used to analyze large genomic and phenotypic datasets, offering increased predictive breeding accuracy (Crossa et al. 2024). These advancements allow breeders to predict plant performance more accurately, ensuring that new crop varieties meet the needs of farmers and are better suited to both current and future agricultural challenges.

The continued integration of AI-powered phenotyping systems with genomic data is paving the way for more precise and efficient plant breeding programs (Silva et al. 2019; Zingaretti et al. 2020). With the ability to rapidly assess plant traits and predict how plants will respond to various stressors, AI is not only enhancing our understanding of plant biology but also accelerating the development of crops that are more resilient, productive, and adaptable to changing environmental conditions.

6.3.5 Integration of Multi-Omics Data

The integration of multi-omics data has become a cornerstone of modern plant research, and AI is playing a crucial role in unlocking its full potential (figure 8). The multi-omics approach encompasses genomics, transcriptomics, proteomics, and metabolomics, each offering valuable insights into the complex biological processes that govern plant traits. AI-driven models have the ability to analyze and synthesize these diverse data sources, facilitating a deeper understanding of how genes interact with environmental factors to influence traits such as growth, yield, and stress resilience.

In particular, the combination of these omics layers enables the identification of key molecular pathways and biomarkers associated with important traits, such as drought tolerance, pest resistance, and adaptation to extreme temperatures (Yang et al. 2021; Vasupalli et al. 2024). AI's ability to process and analyze large-scale, high-dimensional datasets allows for the prediction of complex traits that are often influenced by a combination of genetic, environmental, and phenotypic factors (Xu et al. 2022; Murmu et al. 2024). These predictions are invaluable for crop breeding, where the goal is to develop plant varieties that can thrive under adverse conditions and ensure global food security. For example, the integration of genomics with transcriptomics data helps to uncover gene expression patterns that correspond to specific traits, while proteomics and metabolomics provide insights into the functional outputs of these genes. This multi-layered approach enables the identification of genes and proteins that may play pivotal roles in plant stress responses. AI techniques, particularly ML, help distill this information, uncovering hidden patterns and complex relationships that would be impossible to identify manually. This process improves the accuracy of trait prediction models, making it possible to develop plants with enhanced resilience to environmental challenges (Mansoor et al. 2024).

ML algorithms, such as DL and ensemble learning methods, are particularly adept at handling the complexity of multi-omics data. By leveraging vast datasets, these models can predict how plants will respond to different environmental conditions and stresses. Van Dijk et al. highlighted the importance of these AI-driven approaches in plant science and breeding. Through the use of ML, scientists are able to integrate data from multiple omics platforms, providing a more comprehensive picture of plant biology (Van Dijk et al. 2021). This integration also facilitates the discovery of novel genetic variants and pathways that contribute to desirable traits, accelerating the development of crops with improved stress resistance. Moreover, the application of AI to multi-omics integration is transforming crop breeding by enabling more precise and efficient selection of traits. Traditional breeding methods often involve a trial-and-error approach, which can be time-consuming and resource-intensive (Aziz and Masmoudi 2025). However, with AI, plant breeders can simulate various breeding scenarios, predicting which genetic combinations are most likely to result in the desired traits. This precision reduces the time required to develop new varieties and ensures that crops are tailored to meet the challenges of a changing climate.

6.3.5.1 Case Study Showcasing Successful Integration of Multi-Omics in Plant Research

The integration of multi-omics data with ML has led to remarkable advancements in plant research. A notable case study involves *Arabidopsis thaliana*, where researchers combined genomic, transcriptomic, and metabolomic data to identify regulatory networks influencing flowering time (Danilevicz et al. 2022). By utilizing a multi-omics approach, they were able to pinpoint specific genes that play crucial roles in environmental responses, improving the understanding of plant adaptation mechanisms. In maize breeding, ML models have been successfully applied to integrate genomic and phenotypic data, significantly improving the accuracy of yield-related trait predictions. This approach has allowed breeders to make more informed selections, accelerating the development of high-yield maize varieties (Schrag et al. 2018; Hu et al. 2022). Similarly, in wheat research, the integration of multi-omics data has been instrumental in identifying genetic markers associated with disease resistance. By leveraging ML techniques, researchers have facilitated the development of wheat varieties with enhanced resistance to pathogens, ensuring greater crop stability and food security (Hernández et al. 2012; Wu et al. 2022).

6.4 Challenges in AI and ML Applications in Plant Genomics

6.4.1 Complexity of Plant Genomes and Data Management

The complexity of plant genomes presents significant challenges in applying AI and ML technologies to plant genomics. Plants exhibit intricate genetic architectures, including polyploidy (multiple sets of chromosomes) and abundant repetitive DNA sequences, which make genome sequencing and analysis particularly difficult (Hudson 2021). These complexities complicate the task of identifying genetic variations that influence important traits, such as disease resistance or drought tolerance. Moreover, the vast, multidimensional data generated through plant genomics, including genomic sequences, transcriptomic data, and phenotypic observations, require robust data management systems to handle the scale and heterogeneity of the information (Heckenberger et al. 2008; Kuriakose et al. 2020).

Efficient data storage, sharing, and integration are paramount for AI applications to be successful in this field. Standardized data formats and comprehensive platforms for data integration are essential to facilitate collaboration and ensure the usability of data across research groups. This challenge is particularly relevant in the context of agricultural research, where interdisciplinary efforts are needed to bridge gaps between genomics, phenomics, and environmental factors (Zhao et al. 2019; Kuriakose et al. 2020). Data management systems that enable seamless integration and analysis of complex datasets can help unlock the potential of AI in understanding plant genetic diversity and its relationship to phenotypic traits.

6.4.2 Scarcity of Annotated Genomic Data

A significant challenge in applying AI to plant genomics is the scarcity of well-annotated genomic data, especially for non-model plants that are important for food security. While model organisms like Arabidopsis and rice have been extensively studied and their genomes annotated, many crops, particularly those in developing regions or less-studied species, lack comprehensive genomic annotations. This limits the ability to train AI models on sufficient datasets, which is essential for creating accurate and reliable predictions about gene function and trait associations (Tauro et al. 2022).

The lack of annotated data impedes progress in identifying genetic variations that are linked to valuable agronomic traits such as yield, pest resistance, or drought tolerance. However, emerging AI techniques such as transfer learning offer a promising solution to this problem. Transfer learning allows models trained on well-characterized species to be adapted for use with less-studied crops, overcoming the data scarcity issue by leveraging the knowledge gained from model organisms (Zhao et al. 2024). This approach can significantly reduce the amount of annotated genomic data required to make meaningful predictions in crop improvement, thus expanding the applicability of AI across a wider range of plant species.

6.4.3 Interpretability and Validation of AI Models

One of the most critical challenges in using AI models for plant genomics is the interpretability of the results (refer figure 9). Many AI models, particularly DL algorithms, operate as "black boxes," meaning that their decision-making processes are not easily understood by humans. In the context of plant genomics, this lack of transparency makes it difficult for researchers to validate the biological relevance of AI-driven insights and apply them with confidence in practical breeding programs (Cobianchi et al. 2022; Collins et al. 2022; Walker et al. 2022). Interpretability is especially important in genomics, where the objective is often to link genetic variants with phenotypic traits, such as plant growth or disease resistance. If an AI model provides a useful prediction but the underlying biological mechanism is not understood, it becomes difficult to trust or validate the model's predictions. To address this, researchers are working on methods to improve the transparency of AI models through techniques like explainable AI (XAI), which aims to provide insights into how and why certain predictions are made (Mostafa et al. 2023; Tjoa and Guan 2021).

Moreover, rigorous validation processes are needed to assess whether AI models can generalize across different datasets and experimental conditions. This is critical for ensuring that the models remain reliable when applied to diverse plant species and breeding environments. In order to ensure the validity of AI models, validation strategies such as cross-validation, independent testing sets, and field trials must be incorporated into the development pipeline. By enhancing both interpretability and validation, researchers can increase the trust and reliability of AI models, making them more widely applicable to plant genomics research.

6.5 Future Prospects

The future of AI and ML in plant genomics holds immense potential, with these technologies poised to revolutionize the field through advanced, innovative approaches as shown in figure 10. These technologies are increasingly being employed to process and analyze vast and complex genomic data, enabling researchers to gain deeper insights into plant biology and genetics. By improving the accuracy, efficiency, and scalability of plant research, AI and ML are set to drive significant breakthroughs in crop improvement, disease resistance, climate resilience, and overall agricultural productivity. As these methods integrate with emerging tools and methodologies, the impact on plant science will be transformative, paving the way for smarter, more sustainable agricultural practices in the years ahead.

6.5.1 AI-Driven Precision Breeding

Precision breeding is one of the most exciting future approaches in plant genomics, powered by AI. Advanced ML algorithms will increasingly be able to analyze genetic data, predict desirable traits, and design breeding strategies that are more efficient and targeted. AI systems can process vast amounts of data from genome-wide association studies (GWAS), transcriptomic analyses, and phenotypic data to identify the best breeding candidates. This technology will facilitate the rapid development of crops that are better suited to changing environmental conditions and evolving agricultural needs (Yano et al. 2016; Sun et al. 2024).

6.5.2 AI for Synthetic Biology in Plants

Synthetic biology, which combines biological and engineering principles to design new biological parts, systems, and organisms, is another emerging area where AI and ML can play a pivotal role. In plant genomics, AI can help design novel genetic circuits, optimize metabolic pathways, and predict the effects of gene editing (Jagadeesh Chandra Bose et al. 2024).

This will open new avenues for creating plants with custom traits, such as enhanced resistance to climate extremes or increased productivity. By utilizing AI to model and simulate the effects of genetic modifications at the molecular level, scientists can accelerate the development of genetically engineered plants that address global challenges such as food security, biofuel production, and sustainable agriculture. AI's ability to model complex biological networks will be key to unlocking the full potential of synthetic biology in plants (Zhang et al. 2024).

6.5.3 Automated Discovery of Functional Genomic Elements

One of the most challenging aspects of plant genomics is understanding the function of each gene and its contribution to complex traits. AI technologies, particularly DL algorithms, hold great promise in automatically identifying functional elements within plant genomes (Chen et al. 2024). These elements include promoters, enhancers, and other regulatory sequences that play critical roles in gene expression. By automating the process of functional annotation of plant genomes, AI will enable a deeper understanding of how genes interact within networks to influence phenotypic traits. This will facilitate the discovery of new genetic pathways and improve our ability to manipulate plant genomes for specific purposes, such as increasing disease resistance or improving nutritional value (Zhao et al. 2023).

6.5.4 AI-Enhanced Crop Phenotyping in Complex Environments

AI also plays a key role in enhancing crop phenotyping, especially in complex environments where traditional phenotyping methods fall short. In such settings, AI can be used to analyze data collected from various platforms, such as high-throughput field phenotyping systems, mobile apps, and remote sensing technologies. By combining machine vision, AI, and advanced data analytics, researchers will be able to analyze subtle phenotypic variations that are often overlooked in traditional breeding programs (Yan et al. 2017). These phenotypic traits are often critical to crop performance, such as resistance to abiotic stressors like drought, salinity, and heat. AI's ability to automate and scale phenotyping will make it possible to analyze vast populations of plants under real-world conditions, accelerating the pace of crop improvement (Yang et al. 2020; Su et al. 2023).

6.6 Conclusion

The integration of AI and ML in plant functional genomics is revolutionizing agricultural research and breeding practices. By leveraging advanced computational tools, researchers can analyze vast genomic datasets, predict gene functions, and enhance trait selection with unprecedented accuracy. AI-driven approaches such as DL, CNNs, and ensemble learning models have significantly improved GWAS and MAS, facilitating the identification of key genetic markers.

Furthermore, the incorporation of multi-omics data—spanning genomics, transcriptomics, proteomics, and metabolomics—has provided a comprehensive framework for linking genetic variations with phenotypic traits. AI-powered high-throughput phenotyping has further streamlined plant breeding by enabling rapid assessment of traits such as yield, stress tolerance, and disease resistance. This has accelerated the development of resilient crops capable of withstanding environmental challenges, thus contributing to food security and sustainable agriculture. Despite these advancements, challenges remain in AI-driven plant genomics, including the complexity of plant genomes, data integration issues, and model interpretability. Addressing these limitations requires interdisciplinary collaborations, improved algorithm transparency, and enhanced validation methodologies. Overall, the continued evolution of AI and ML will play a transformative role in plant genomics, driving agricultural innovation, improving breeding efficiency, and ensuring crop sustainability in the face of global challenges.

References

Abbas, Q., Wilhelm, M., Kuster, B. et al. (2024). Exploring crop genomes: assembly features, gene prediction accuracy, and implications for proteomics studies. *BMC Genomics* 25: 619.

Afifi, A., Alhumam, A., and Abdelwahab, A. (2021). Convolutional neural network for automatic identification of plant diseases with limited data. *Plants* 10 (1): 28. https://doi.org/10.3390/plants10010028.

Angermueller, C., Lee, H.J., Reik, W., and Stegle, O. (2017). DeepCpG: accurate prediction of single-cell dna methylation states using deep learning. *Genome Biology* 18: https://doi.org/10.1186/s13059-017-1189-z.

Angermueller, C., Pärnamaa, T., Parts, L., and Stegle, O. (2016). Deep learning for computational biology. *Molecular Systems Biology* 12: https://doi.org/10.15252/msb.20156651.

Aziz, M.A. and Masmoudi, K. (2025). Molecular breakthroughs in modern plant breeding techniques. *Horticultural Plant Journal* 11 (1): 15–41. https://doi.org/10.1016/j.hpj.2024.01.004.

Barbedo, J.G.A. (2019). A review on the use of unmanned aerial vehicles and imaging sensors for monitoring and assessing plant stresses. *Drones* 3 (2): 40. https://doi.org/10.3390/drones3020040.

Bu, H., Gan, Y., Wang, Y. et al. (2017). A new method for enhancer prediction based on deep belief network. *BMC Bioinformatics* 18: https://doi.org/10.1186/s12859-017-1828-0.

Camacho, D.M., Collins, K.M., Powers, R.K. et al. (2018). Next-generation machine learning for biological networks. *Cell* 173: 1581–1592. https://doi.org/10.1016/j.cell.2018.05.015.

Caudai, C., Galizia, A., Geraci, F. et al. (2021). AI applications in functional genomics. *Computational and Structural Biotechnology Journal* 11 (19): 5762–5790. https://doi.org/10.1016/j.csbj.2021.

Cembrowska-Lech, D., Krzemińska, A., Miller, T. et al. (2023). An integrated multi-omics and artificial intelligence framework for advance plant phenotyping in horticulture. *Biology* 12: https://doi.org/10.3390/biology12101298.

Chen, X.D. and Ishwaran, H. (2012). Random forests for genomic data analysis. *Genomics* 99 (6): 323–329. https://doi.org/10.1016/j.ygeno.2012.04.003.

Chen, Z., Ain, N.U., Zhao, Q., and Zhang, X. (2024). From tradition to innovation: conventional and deep learning frameworks in genome annotation. *Briefings in Bioinformatics* 25 (3): bbae138. https://doi.org/10.1093/bib/bbae138.

Cobianchi, L., Verde, J.M., Loftus, T.J. et al. (2022). Artificial intelligence and surgery: ethical dilemmas and open issues. *Journal of the American College of Surgeons* 235 (2): 268–275. https://doi.org/10.1097/XCS.0000000000000242.

Collins, J.W., Marcus, H.J., Ghazi, A. et al. (2022). Ethical implications of AI in robotic surgical training: a Delphi consensus statement. *European Urology Focus* 8 (2): 613–622. https://doi.org/10.1016/j.euf.2021.04.006.

Crossa, J., Montesinos-Lopez, O.A., Costa-Neto, G. et al. (2024). Machine learning algorithms translate big data into predictive breeding accuracy. *Trends in Plant Science* 30: https://doi.org/10.1016/j.tplants.2024.09.011.

Danilevicz, M.F., Gill, M., Anderson, R. et al. (2022). Plant genotype to phenotype prediction using machine learning. *Frontiers in Genetics* 13: 822173. https://doi.org/10.3389/fgene.2022.822173.

Date, Y. and Kikuchi, J. (2018). Application of a deep neural network to metabolomics studies and its performance in determining important variables. *Analytical Chemistry* 90: 1805–1810. https://doi.org/10.1021/acs.analchem.7b03795.

de Ridder, D., de Ridder, J., and Reinders, M.J.T. (2013). Pattern recognition in bioinformatics. *Briefings in Bioinformatics* 14: 633–647. https://doi.org/10.1093/bib/bbt020.

Dettmers, T. (2015). Deep learning in a nutshell: history and training. https://devblogs.nvidia.com/parallelforall/deeplearning-nutshell-history-training/

Eraslan, G., Avsec, Ž., Gagneur, J., and Theis, F.J. (2019). Deep learning: new computational modelling techniques for genomics. *Nature Reviews Genetics* 20: 389–403. https://doi.org/10.1038/s41576-019-0122-6.

Ernst, J. and Kellis, M. (2015). Large-scale imputation of epigenomic datasets for systematic annotation of diverse human tissues. *Nature Biotechnology* 33: 364–376. https://doi.org/10.1038/nbt.3157.

Esteva, A. (2019). A guide to deep learning in healthcare. *Nature Medicine* 25: 24–29. https://doi.org/10.1038/s41591-018-0316-z.

Farooq, M.A., Gao, S., Hassan, M.A. et al. (2024). Artificial intelligence in plant breeding. *Trends in Genetics : TIG* 40 (10): 891–908. https://doi.org/10.1016/j.tig.2024.07.001.

Ferrario, A. and Loi, M. (2021). *Algorithm, Machine Learning and Artificial Intelligence*. Social Science Research Network. https://doi.org/10.2139/SSRN.3817377.

Gabur, I., Simioniuc, D., Snowdon, R., and Cristea, D. (2022). Machine learning applied to the search for nonlinear features in breeding populations. *Frontiers in Artificial Intelligence* 5: https://doi.org/10.3389/frai.2022.876578.

Gao, N., Martini, J.W.R., Zhang, Z. et al. (2017). Incorporating gene annotation into genomic prediction of complex phenotypes. *Genetics* 207 (2): 489–501. https://doi.org/10.1534/genetics.117.300198.

Garrow, A.G., Agnew, A., and Westhead, D.R. (2005). Tmb-hunt: a web server to screen sequence sets for transmembrane beta-barrel proteins. *Nucleic Acids Research* 33: W188–W192. https://doi.org/10.1093/nar/gki384.

Ghahramani, A., Watt, F.M., and Luscombe, N.M. (2018). Generative adversarial networks simulate gene expression and predict perturbations in single cells. *BioRxiv* 262501. https://doi.org/10.1101/262501.

Gill, T., Gill, S.K., Saini, D.K. et al. (2022). A comprehensive review of high throughput phenotyping and machine learning for plant stress phenotyping. *Phenomics (Cham, Switzerland)* 2 (3): 156–183. https://doi.org/10.1007/s43657-022-00048-z.

Grapov, D., Fahrmann, J., Wanichthanarak, K., and Khoomrung, S. (2018). Rise of deep learning for genomic, proteomic, and metabolomic data integration in precision medicine. *OMICS* 22: 630–636. https://doi.org/10.1089/omi.2018.0097.

Gupta, A. and Zou, J. (2019). Feedback GAN for DNA optimizes protein functions. *Nature Machine Intelligence* 1: 105–111.

Gusmao, E.G., Dieterich, C., Zenke, M., and Costa, I.G. (2014). Detection of active transcription factor binding sites with the combination of dnase hypersensitivity and histone modifications. *Bioinformatics* 30: 3143–3151. https://doi.org/10.1093/bioinformatics/btu519.

Taguchi, Y. and Okamoto, A. (2011). Principal component analysis for bacterial proteomic analysis. *2011 IEEE International Conference on Bioinformatics and Biomedicine Workshops (BIBMW)*, pp. 961–963.

Handhayani, T. and Hiryanto, L. (2015). Intelligent kernel k-means for clustering gene expression. *Procedia Computer Science* 59: 171–177.

Heckenberger, M., Maurer, H.P., Melchinger, A.E., and Frisch, M. (2008). The Plabsoft database: a comprehensive database management system for integrating phenotypic and genomic data in academic and commercial plant breeding programs. *Euphytica* 161 (2): 173–179. https://doi.org/10.1007/s10681-007-9478-3.

Heje Grønbech, C., Maximillian Fornitz Vording, Pascal N Timshel et al. (2019). scVAE: variational auto-encoders for single-cell gene expression data. https://www.biorxiv.org/content/10.1101/318295v3.

Hernández, M., Hernández, M., Crossa, J. et al. (2012). Multi-trait and multi-environment QTL analyses for resistance to wheat diseases. *PLoS One* 7: https://doi.org/10.1371/journal.pone.0038008.

Hess, M., Lenz, S., Blätte, T.J. et al. (2017). Partitioned learning of deep Boltzmann machines for SNP data. *Bioinformatics* 33: 3173–3180. https://doi.org/10.1093/bioinformatics/btx408.

Hsu, Y.-L., Huang, P.-Y., and Chen, D.-T. (2014). Sparse principal component analysis in cancer research. *Translational Cancer Research* 3 (3): 182–190. https://doi.org/10.3978/j.issn.2218-676X.2014.05.06.

Hu, H., Meng, Y., Liu, W. et al. (2022). Multi-trait genomic prediction improves accuracy of selection among doubled haploid lines in maize. *International Journal of Molecular Sciences* 23: https://doi.org/10.3390/ijms232314558.

Huang, D., Quan, Y., He, M., and sen Zhou B. (2009). Comparison of linear discriminant analysis methods for the classification of cancer based on gene expression data. *Journal of Experimental & Clinical Cancer Research* 28: 149. https://doi.org/10.1186/1756-9966-28-149.

Huang, S., Chaudhary, K., and Garmire, L. (2017). More is better: recent progress in multi-omics data integration methods. *Frontiers in Genetics* 8: https://doi.org/10.3389/fgene.2017.00084.

Hudson, I.L. (2021). Data integration using advances in machine learning in drug discovery and molecular biology. In: *Artificial Neural Networks*, vol. 2190 (ed. H. Cartwright), 137–157). Methods in Molecular Biology. Humana. https://doi.org/10.1007/978-1-0716-0826-5_7.

Jagadeesh Chandra Bose, K., Kaur, S., Sharma, S. et al. (2024). Methods from the field of synthetic biology that aim to improve plant growth and resistance to stress through the use of genetic engineering. In: *Metabolomics, Proteomics and Gene Editing Approaches in Biofertilizer Industry* (ed. S. Kaur, V. Dwibedi, and P.K. Sahu), 127–145. Springer. https://doi.org/10.1007/978-981-97-2910-4_6.

Jamil, I., Remali, J., Azizan, K. et al. (2020). Systematic multi-omics integration (MOI) approach in plant systems biology. *Frontiers in Plant Science* 11: https://doi.org/10.3389/fpls.2020.00944.

Karabulut, E.M. and Ibrikci, T. (2017). Discriminative deep belief networks for microarray based cancer classification. *Biomedical Research-Tokyo.* 28: 1016–1024.

Kelley, D.R., Reshef, YA., Bileschi, M. et al. (2018). Sequential regulatory activity prediction across chromosomes with convolutional neural networks. https://genome.cshlp.org/content/early/2018/03/27/gr.227819.117.

Kircher, M. (2014). A general framework for estimating the relative pathogenicity of human genetic variants. *Nature Genetics* 46: 310–315. https://doi.org/10.1038/ng.2892.

Kuriakose, S.V., Pushker, R., and Hyde, E.M. (2020). Data-driven decisions for accelerated plant breeding. In: *Accelerated Plant Breeding*, vol. 1 (ed. S. Gosal and S. Wani), 63–78. Springer. https://doi.org/10.1007/978-3-030-41866-3_4.

Lee, D., Zhang, J., Liu, J., and Gerstein, M.B. (2020). Epigenome-based splicing prediction using a recurrent neural network. *PLoS Comput Biol. 2020 Jun 25;16(6):e1008006*. https://doi.org/10.1371/journal.pcbi.1008006. PMID: 32584815; PMCID: PMC7343189.

Levy, J.J. (2020). Methylnet: an automated and modular deep learning approach for dna methylation analysis. *BMC Bioinformatics* 21: https://doi.org/10.1186/s12859-020-3443-8.

Liu, Q. (2019). Detection of dna base modifications by deep recurrent neural network on oxford nanopore sequencing data. *Nature Communications* 10: https://doi.org/10.1038/s41467-019-10168-2.

Lourenço, V., Ogutu, J., Rodrigues, R., and Piepho, H. (2022). Genomic prediction using machine learning: a comparison of the performance of regularized regression, ensemble, instance-based and deep learning methods on synthetic and empirical data. *BMC Genomics* 25: https://doi.org/10.1101/2022.06.09.495423.

Lu, Y. and Zou, Q. (2024). Functional genomics and comparative genomics analysis in plants. *Current Issues in Molecular Biology* 46 (12): 13780–13782. https://doi.org/10.3390/cimb46120823.

Maisuradze, G.G., Liwo, A., and Scheraga, H.A. (2009). Principal component analysis for protein folding dynamics. *Journal of Molecular Biology* 385 (1): 312–329. https://doi.org/10.1016/j.jmb.2008.10.018.

Mansoor, S., Karunathilake, E.M.B.M., Tuan, T.T., and Chung, Y.S. (2024). Genomics, phenomics, and machine learning in transforming plant research: advancements and challenges. *Horticultural Plant Journal*. https://doi.org/10.1016/j.hpj.2023.09.005.

Marouf, M.M. (2020). Realistic in silico generation and augmentation of single-cell rna-seq data using generative adversarial networks. *Nature Communications* 11: https://doi.org/10.1038/s41467-019-14018-z.

McCarthy, J., Minsky, M., Rochester, N., and Shannon, C.E. (2006). A proposal for the Dartmouth summer research project on artificial intelligence, August 31, 1955. *AI Magazine* 27: 12–14.

McKusick, V.A. and Ruddle, F.H. (1987). Editorial: a new discipline, a new name, a new journal. *Genomics* 1: 1–2.

Min, X., Zeng, W., Chen, S. et al. (2016). Predicting enhancers with deep convolutional neural networks. *BMC Bioinformatics* 18: https://doi.org/10.1186/s12859-017-1878-3.

Mir, R.R., Reynolds, M., Pinto, F. et al. (2019). High-throughput phenotyping for crop improvement in the genomics era. *Plant Science* 282: 60–72. https://doi.org/10.1016/j.plantsci.2019.01.007.

Mojarad, S.A., Dlay, S.S., W.L., Woo and Sherbet, G.V. (2010). Breast cancer prediction and cross validation using multilayer perceptron neural networks. *2010 7th International Symposium on Communication Systems, Networks & Digital Signal Processing (CSNDSP 2010)*, pp. 760–764.

Mojarad, S.A., Dlay, S.S., Woo, W.L., and Sherbet, G.V. (2011). Cross Validation Evaluation for Breast Cancer. Prediction Using Multilayer Perceptron Neural Networks. *American J. of Engineering and Applied Sciences* 4 (4): 576–585, 1941–7020.

Montesinos López, O.A., Montesinos López, A., and Crossa, J. (2022). Random forest for genomic prediction. In: *Multivariate Statistical Machine Learning Methods for Genomic Prediction*, 257–278. Springer. https://doi.org/10.1007/978-3-030-89010-0_15.

Montesinos López, O.A., Mosqueda González, B.A., Montesinos López, A., and Crossa, J. (2023). Statistical machine-learning methods for genomic prediction using the SKM library. *Genes* 14 (5): 1003. https://doi.org/10.3390/genes14051003.

Mostafa, S., Mondal, D., Panjvani, K. et al. (2023). Explainable deep learning in plant phenotyping. *Frontiers in Artificial Intelligence* 6: 1203546. https://doi.org/10.3389/frai.2023.1203546.

Murmu, S., Sinha, D., Chaurasia, H. et al. (2024). A review of artificial intelligence-assisted omics techniques in plant defense: current trends and future directions. *Frontiers in Plant Science* 15: 1292054. https://doi.org/10.3389/fpls.2024.1292054.

Nabwire, S., Suh, H., Kim, M. et al. (2021). Review: application of artificial intelligence in phenomics. *Sensors (Basel, Switzerland)* 21: https://doi.org/10.3390/s21134363.

Nielsen, R. and Matz, M.V. (2006). Statistical approaches for DNA barcoding. *Systematic Biology* 55 (1): 162–169. https://doi.org/10.1080/10635150500431239.

Nivaashini, M. and Soundariya, R. (2018). Deep Boltzmann machine based breast cancer risk detection for healthcare systems. *International Journal of Pure and Applied Mathematics* 119: 581–590.

Ogutu, J.O., Schulz-Streeck, T., and Piepho, H.-P. (2012). Genomic selection using regularized linear regression models: ridge regression, lasso, elastic net and their extensions. *BMC Proceedings* 6: S10. https://doi.org/10.1186/1753-6561-6-S2-S10.

Ogutu, J.O., Piepho, H.P., and Schulz-Streeck, T. (2011). A comparison of random forests, boosting and support vector machines for genomic selection. *BMC Proceedings* 5 (Suppl 3): S11. https://doi.org/10.1186/1753-6561-5-S3-S11.

Oh, S.-H. (2013). Protein disorder prediction using multilayer perceptrons. *International Journal of Contents.* 9: 11–15.

Oyelade, J. (2016). Clustering algorithms: their application to gene expression data. *Bioinformatics and Biology Insights* 10: 237–253. https://doi.org/10.4137/BBI.S38316.

Park, Y. and Kellis, M. (2015). Deep learning for regulatory genomics. *Nature Biotechnology* 33: 825–826. https://doi.org/10.1038/nbt.3313.

Parry, R.M. (2010). k-nearest neighbor models for microarray gene expression analysis and clinical outcome prediction. *The Pharmacogenomics Journal* 10: 292–309. https://doi.org/10.1038/tpj.2010.56.

Pook, T., Freudenthal, J., Korte, A., and Simianer, H. (2020). Using local convolutional neural networks for genomic prediction. *Frontiers in Genetics* 11: 561497. https://doi.org/10.3389/fgene.2020.561497.

Quang, D. and Xie, X. (2019). Factornet: a deep learning framework for predicting cell type specific transcription factor binding from nucleotide-resolution sequential data. *Methods* 166: 40–47. https://doi.org/10.1016/j.ymeth.2019.03.020.

Ragoza, M., Hochuli, J., Idrobo, E. et al. (2017). Protein-ligand scoring with convolutional neural networks. *Journal of Chemical Information and Modeling* 57: 942–957. https://doi.org/10.1021/acs.jcim.6b00740.

Rosenblatt, F. (1958). The perceptron: a probabilistic model for information storage and organization in the brain. *Psychological Review* 65: 386–408. https://doi.org/10.1037/h0042519.

Roy, R., Debashree, I., Srivastava, S. et al. (2022). Clustered Regularly Interspaced Short Palindromic Repeats (CRISPR)/Cas9 off-targets: computational analysis of causes, prediction, detection and overcoming strategies. *Current Bioinformatics* 17: 119–132.

Schneider, D. (2020). Machine learning and artificial intelligence. In: *Health Services Research* (ed. V. Jain and J.M. Chatterjee). Springer. https://doi.org/10.1007/978-3-030-28357-5_14.

Schrag, T., Westhues, M., Schipprack, W. et al. (2018). Beyond genomic prediction: combining different types of omics data can improve prediction of hybrid performance in maize. *Genetics* 208: 1373–1385. https://doi.org/10.1534/genetics.117.300374.

Shen, Z., Bao, W., and Huang, D.-S. (2018). Recurrent neural network for predicting transcription factor binding sites. *Scientific Reports* 8: https://doi.org/10.1038/s41598-018-33321-1.

Silla C.N., Freitas A.A. (2009). A global-model naive Bayes approach to the hierarchical prediction of protein functions, *2009 9th IEEE International Conference on Data Mining*, pp. 992–997.

Silla Jr., Carlos, N., and Freitas, Alex A. (2009) A global-model naive Bayes approach to the hierarchical prediction of protein functions. In: Wang, Wei and Kargupta, Hillol and Ranka, Sanjay and Yu, Philip S. and Wu, Xindong, eds. *2009 Ninth IEEE International Conference on Data Mining. IEEE*, pp. 182–196. ISBN 978-1-4244-5242-2. (doi:10.1109/ICDM.2009.85)

Silva, J., Teixeira, R., Silva, F. et al. (2019). Machine learning approaches and their current application in plant molecular biology: a systematic review. *Plant Science : An International Journal of Experimental Plant Biology* 284: 37–47. https://doi.org/10.1016/J.PLANTSCI.2019.03.020.

Singh, S., Koyama, H., Bhati, K.K., and Alok, A. (2021a). Biotechnological importance of NAC family genes in improvement of food crop. *Journal of Plant Research* 134 (3): 475–495.

Singh, S., Kudapa, H., Garg, V., and Varshney, R.K. (2021b). Comprehensive analysis and identification of drought-responsive candidate NAC genes in three semi-arid tropics (SAT) legume crops. *BMC Genomics* 22: 289.

Song, B., Ning, W., Wei, D. et al. (2023). Plant genome resequencing and population genomics: current status and future prospects. *Molecular Plant* 16: 1252–1268.

Stormo, G.D., Schneider, T.D., Gold, L., and Ehrenfeucht, A. (1982). Use of the 'perceptron' algorithm to distinguish translational initiation sites in *E. coli*. *Nucleic Acids Research* 10: 2997–3011. https://doi.org/10.1093/nar/10.9.2997.

Su, D., Qiao, Y., Jiang, Y. et al. (2023). Editorial: AI, sensors and robotics in plant phenotyping and precision agriculture, volume II. *Frontiers in Plant Science* 14: 1215899. https://doi.org/10.3389/fpls.2023.1215899.

Subramanian, I., Verma, S., Kumar, S. et al. (2020). Multi-omics data integration, interpretation, and its application. *Bioinformatics and Biology Insights* 14: https://doi.org/10.1177/1177932219899051.

Sun, L., Lai, M., Ghouri, F. et al. (2024). Modern plant breeding techniques in crop improvement and genetic diversity: from molecular markers and gene editing to artificial intelligence—a critical review. *Plants* 13 (19): 2676. https://doi.org/10.3390/plants13192676.

Sundaram, L. (2018). Predicting the clinical impact of human mutation with deep neural networks. *Nature Genetics* 50: 1161–1170. https://doi.org/10.1038/s41588-018-0167-z.

Taguchi, Y.h. and Okamoto, A. (2012). Principal Component Analysis for Bacterial Proteomic Analysis. In: Shibuya, T., Kashima, H., Sese, J., Ahmad, S. (eds) Pattern Recognition in Bioinformatics. PRIB 2012. *Lecture Notes in Computer Science*, vol 7632. Springer, Berlin, Heidelberg. https://doi.org/10.1007/978-3-642-34123-6_13.

Tauro, F., Maltese, A., Giannini, R., and Harfouche, A. (2022). Latent heat flux variability and response to drought stress of black poplar: a multi-platform multi-sensor remote and proximal sensing approach to relieve the data scarcity bottleneck. *Remote Sensing of Environment* 268: 112771. https://doi.org/10.1016/j.rse.2021.112771.

Tavakoli, N. (2019, July). Modeling genome data using bidirectional LSTM. In *2019 IEEE 43rd annual computer software and applications conference* (*COMPSAC*) (Vol. 2, pp. 183–188). IEEE.

Tjoa, E. and Guan, C. (2021). A survey on explainable artificial intelligence (XAI): toward medical XAI. *IEEE Transactions on Neural Networks and Learning Systems* 32 (11): 4793–4813. https://doi.org/10.1109/TNNLS.2020.3027314.

Van Dijk, A.D.J., Kootstra, G., Kruijer, W., and de Ridder, D. (2021). Machine learning in plant science and plant breeding. *iScience* 24 (1): 101890. https://doi.org/10.1016/j.isci.2020.101890.

Vasupalli, N., Bhat, J.A., Jain, P. et al. (2024). Omics big data for crop improvement: opportunities and challenges. *The Crop Journal* 12 (6): 1517–1532. https://doi.org/10.1016/j.cj.2024.10.007.

Walker, K.A., Basisty, N., Wilson, D.M. 3rd, and Ferrucci, L. (2022). Connecting aging biology and inflammation in the omics era. *The Journal of Clinical Investigation* 132 (14): e158448. https://doi.org/10.1172/JCI158448.

Wang, D. and Gu, J. (2018). Vasc: dimension reduction and visualization of single-cell RNA-seq data by deep variational autoencoder. *Genomics, Proteomics & Bioinformatics* 16: 320–331. https://doi.org/10.1016/j.gpb.2018.08.003.

Wang, Y., Zeng, J., Yang, G. et al. (2023). Harnessing knowledge from plant functional genomics and multi-omics for genetic improvement. *International Journal of Molecular Sciences* 24 (12): 10347. https://doi.org/10.3390/ijms241210347. PMID: 37373493; PMCID: PMC10299164.

Wang, H., Cimen, E., Singh, N., and Buckler, E. (2020). Deep learning for plant genomics and crop improvement. *Current Opinion in Plant Biology* 54: 34–41. https://doi.org/10.1016/j.pbi.2019.12.010.

Wang, J., Kan, S., Liao, X. et al. (2024). Plant organellar genomes: much done, much more to do. *Trends in Plant Science* 29: 754–769.

Worley, B. and Powers, R. (2013). Multivariate analysis in metabolomics. *Current Metabolomics* 1: 92–107. https://doi.org/10.2174/2213235X11301010092.

Wu, S., Szilagyi, A., and Zhang, Y. (2011). Improving protein structure prediction using multiple sequence-based contact predictions. *Structure* 19: 1182–1191. https://doi.org/10.1016/j.str.2011.05.004.

Wu, F., Zhou, Y., Shen, Y. et al. (2022). Linking multi-omics to wheat resistance types to *Fusarium* head blight to reveal the underlying mechanisms. *International Journal of Molecular Sciences* 23: https://doi.org/10.3390/ijms23042280.

Xi, B., Gu, H., Baniasadi, H., and Raftery, D. (2014). Statistical analysis and modeling of mass spectrometry-based metabolomics data. *Methods in Molecular Biology* 2014: 333–353. https://doi.org/10.1007/978-1-4939-1258-2_22.

Xu, H., Yang, L., and Freitas, M.A. (2008). A robust linear regression based algorithm for automated evaluation of peptide identifications from shotgun proteomics by use of reversed-phase liquid chromatography retention time. *BMC Bioinformatics* 9: 347. https://doi.org/10.1186/1471-2105-9-347.

Xu, Y., Zhang, X., Li, H. et al. (2022). Smart breeding driven by big data, artificial intelligence, and integrated genomic-enviromic prediction. *Molecular Plant* 15 (11): 1664–1695. https://doi.org/10.1016/j.molp.2022.09.001.

Yan, J., Risacher, S., Shen, L., and Saykin, A. (2017). Network approaches to systems biology analysis of complex disease: integrative methods for multi-omics data. *Briefings in Bioinformatics* 19: 1370–1381. https://doi.org/10.1093/bib/bbx066.

Yang, W., Feng, H., Zhang, X. et al. (2020). Crop phenomics and high-throughput phenotyping: past decades, current challenges, and future perspectives. *Molecular Plant* 13 (2): 187–214. https://doi.org/10.1016/j.molp.2020.01.008.

Yang, Y., Saand, M.A., Huang, L. et al. (2021). Applications of multi-omics technologies for crop improvement. *Frontiers in Plant Science* 12: 563953. https://doi.org/10.3389/fpls.2021.563953.

Yano, K., Yamamoto, E., Aya, K. et al. (2016). Genome-wide association study using whole-genome sequencing rapidly identifies new genes influencing agronomic traits in rice. *Nature Genetics* 48 (8): 927–934. https://doi.org/10.1038/ng.3596.

Yao, Z. and Ruzzo, W.L. (2006). A regression-based k nearest neighbor algorithm for gene function prediction from heterogeneous data. *BMC Bioinformatics* 7: S11. https://doi.org/10.1186/1471-2105-7-S1-S11.

Yuan, Y., Liang, Y., Yi, L. et al. (2008). Uncorrelated linear discriminant analysis (ULDA): a powerful tool for exploration of metabolomics data. *Chemometrics and Intelligent Laboratory Systems* 93: 70–79.

Yue, T., Wang, Y., Zhang, L. et al. (2023). Deep learning for genomics: from early neural nets to modern large language models. *International Journal of Molecular Sciences* 24 (21): 15858. https://doi.org/10.3390/ijms242115858.

Zampieri, G., Vijayakumar, S., Yaneske, E., and Angione, C. (2019). Machine and deep learning meet genome-scale metabolic modeling. *PLoS Computational Biology* 15. https://doi.org/10.1371/journal.pcbi.1007084.

Zhang, F., Kaufman, H.L., Deng, Y., and Drabier, R. (2013). Recursive SVM biomarker selection for early detection of breast cancer in peripheral blood. *BMC Medical Genomics* 6: S4. https://doi.org/10.1186/1755-8794-6-S1-S4.

Zhang, Y., An, L., Yue, F., and Hardison, R.C. (2016). Jointly characterizing epigenetic dynamics across multiple human cell types. *Nucleic Acids Research* 44: 6721–6731. https://doi.org/10.1093/nar/gkw278.

Zhang, Z. (2019). Deep learning in omics: a survey and guideline. *Briefings in Functional Genomics* 18: 41–57. https://doi.org/10.1093/bfgp/ely030.

Zhang, D., Xu, F., Wang, F. et al. (2024). Synthetic biology and artificial intelligence in crop improvement. *Plant Communications* 5 (1): 101220. https://doi.org/10.1016/j.xplc.2024.101220.

Zhang, F., Song, H., Zeng, M. et al. (2019). Deepfunc: a deep learning framework for accurate prediction of protein functions from protein sequences and interactions. *Proteomics* 19: https://doi.org/10.1002/pmic.201900019.

Zhao, C., Zhang, Y., Du, J. et al. (2019). Crop phenomics: current status and perspectives. *Frontiers in Plant Science* 10: 714. https://doi.org/10.3389/fpls.2019.00714.

Zhao, L., Walkowiak, S., and Fernando, W.G.D. (2023). Artificial intelligence: a promising tool in exploring the phytomicrobiome in managing disease and promoting plant health. *Plants* 12 (9): 1852. https://doi.org/10.3390/plants12091852.

Zhao, Z., Alzubaidi, L., Zhang, J. et al. (2024). A comparison review of transfer learning and self-supervised learning: definitions, applications, advantages, and limitations. *Expert Systems with Applications* 242: 122807. https://doi.org/10.1016/j.eswa.2023.122807.

Zhou, J. (2018). Deep learning sequence-based ab initio prediction of variant effects on expression and disease risk. *Nature Genetics* 50: 1171–1179. https://doi.org/10.1038/s41588-018-0160-6.

Zingaretti, L., Gezan, S., Ferrão, L. et al. (2020). Exploring deep learning for complex trait genomic prediction in polyploid outcrossing species. *Frontiers in Plant Science* 11: https://doi.org/10.3389/fpls.2020.00025.

Zou, J. (2018). A primer on deep learning in genomics. *Nature Genetics* 51: 12–18. https://doi.org/10.1038/s41588-018-0295-5.

7

AI Models for Studying and Integrating Plant Multiple Omics

Maliheh Eftekhari and Mohammad Reza Naghavi

University of Tehran, Division of Biotechnology, Department of Agronomy and Plant Breeding, College of Agricultural and Natural Resources, Daneshkadeh Crossroad, Karaj, 31587-77871, Iran

7.1 Introduction

Rapid advancements in artificial intelligence (AI) and multi-omics technologies are changing crop improvement strategies. Traditional breeding methods have always used a combination of selecting traits based on phenotype and various genetic markers. However, these methods are now being increasingly supported by AI-based techniques that can handle large amounts of data from different sources such as genomics, transcriptomics, proteomics, metabolomics, and phenomics. The combination of AI and multi-omics has the potential to give us a better understanding of how plants work as a whole. This knowledge can then be used to create models that predict complex traits and speed up breeding programs (Zhang et al. 2022).

AI techniques like machine learning (ML), deep learning (DL), and network-based models have already shown success in analyzing and integrating multi-omics data. These techniques have provided valuable insights into gene regulation, responses to stress, and predicting crop yields. For example, ML algorithms such as random forests (RFs) and support vector machines (SVMs) have been applied to classify gene expression patterns and predict phenotypic traits based on input from multi-omics data (Yang et al. 2021). DL models, especially convolutional neural networks (CNNs) and recurrent neural networks (RNNs), are very efficient in extracting complex patterns from multi-omics datasets, thus improving trait prediction accuracy for crops. For example, DL models employing both RNNs and CNNs have turned out to be successful in analyzing environmental and genomic data with better predictive accuracy in yields by bridging spatial and temporal dependencies (Khaki et al. 2020). These models are better than classical statistical approaches because of the use of ML frameworks to merge high-dimensional omics data, thus demonstrating their potential for precision breeding (Vidyarthi et al. 2024).

The integration of multi-omics data allows for potent insights into plant responses to environmental conditions, genetic variability, and metabolic pathways. AI-driven models provide information about the complex interaction between genotype, environment, and phenotype, enabling scientists to uncover causal links that were previously difficult to

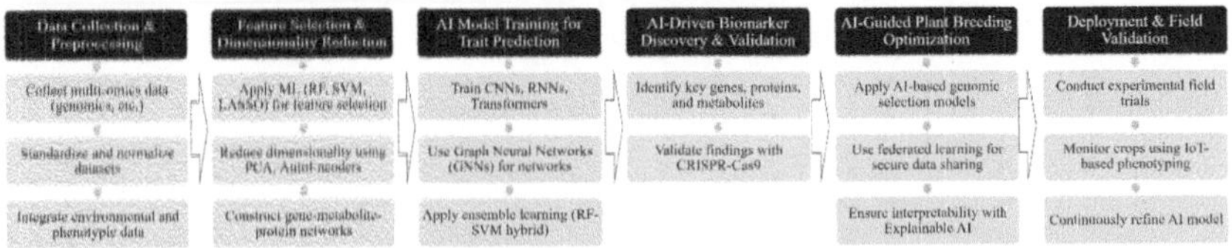

Figure 7.1 AI-driven multi-omics integration for optimizing plant breeding strategy.

establish with classical methods (Wu and Xie 2025). Such technological advances have produced more appreciable breeding strategies and developed climate-resilient varieties of crops, all of which are going to push global food security (Yang et al. 2021; Upadhyay 2026). The high-level algorithmic framework presented in Figure 7.1 shows AI-driven plant improvement using multi-omics integration. The algorithm follows a stepwise approach, integrating genomics, transcriptomics, proteomics, metabolomics, and phenomics with ML and DL to enhance trait prediction, stress tolerance, and crop improvement.

7.2 Plant Multi-Omics: Data Types and Challenges

7.2.1 Genomics: DNA Sequencing, SNP Analysis, and Structural Variations

Whole-genome sequencing provides complete representations of organisms' genetic blueprints, thereby facilitating the identification of genes associated with agronomic characteristics such as drought tolerance and yield optimization. Reduced-representation sequencing is the genotyping-by-sequencing and exome sequencing, which gives a cost-effective approach for analyzing targeted genomic regions (Gunasekaran et al. 2024). Long-read sequencing (e.g., PacBio and Oxford Nanopore) has significantly enhanced genome assembly, particularly when the target plants have large, highly repetitive, and polyploid genomes (Pokrovac and Pezer 2022).

Single-nucleotide polymorphisms (SNPs) have been described as frequently showing genetic variation and provide important markers in Genome-Wide Association Study (GWAS), quantitative trait locus mapping and genomic selection. Advances in the high-density SNP and next-generation sequencing have made it possible to perform genome-wide SNP discovery to better show allele variations associated with desirable traits (Mandaliya et al. 2010).

These include large-scale genetic changes such as insertions, deletions (indels), duplications, inversions, and translocations that affect gene expression and phenotypic characteristics in plants to a great extent. Technologies such as optical mapping, Hi-C sequencing, and third-generation sequencing have improved the identification and characterization of these structural variants (SVs), which would otherwise be hidden in traditional SNP-based studies (Feuk et al. 2006).

Many crops, including wheat and sugarcane, have complex genomes that contain polyploidy and repetitive elements, adding great complexities to genome assembly and annotation (Pokrovac and Pezer 2022; Taneja and Upadhyay 2025). High-throughput sequencing will generate terabytes of genomic data, necessitating a lot of storage space, computational efforts, and an in-depth understanding of bioinformatics.

Considerations toward using cloud computing and AI bioinformatics pipelines for improving scalability and efficiency are sought (Gunasekaran et al. 2024). Many plant genomes are still poorly annotated with functional gaps in the interpretation of genomes. Pan-genome sequencing, functional genomics initiatives, and other research proposals aim to fill such gaps (Mandaliya et al. 2010). Genomic data alone cannot give a complete picture of complex plant traits. An interfacing of genomics with transcriptomics, metabolomics, and phenomics continues to be hampered by incompatibilities between the data structures, scales, and analytical approaches employed in transcriptomics, metabolomics, and phenomics (Feuk et al. 2006).

7.2.2 Transcriptomics: RNA Sequencing, Gene Expression Analysis, and Regulatory RNA Analysis

RNA-seq is the premier approach to transcriptomic analyses and displays great advantages over microarrays by being able to detect both known and novel transcripts without any prior knowledge of the gene sequence. It is thus very useful in obtaining a snapshot of gene expression across various tissues, developmental stages, and environmental conditions (Upadhyay et al. 2015; Sharma et al. 2025a,b; Deepika et al. 2024, 2025a,b; Upton et al. 2023). Single-cell RNA sequence is becoming a promising avenue to investigate gene expression at the cellular level by recognizing rare cell types and dynamic regulatory mechanisms (Grones et al. 2024).

The transcriptome-based approaches are for determining differential gene expression that enables one to identify which genes are either up- or down-regulated in response to stress, developmental cues, or genetic modifications. Use of advanced methodologies such as 3′ RNA-seq and Simplified Poly(A) Anchored Sequencing fashioned modern techniques that significantly improved the efficiency, sensitivity, and cost-effectiveness of gene expression quantification (Wang et al. 2021).

Small RNAs (miRNAs, siRNAs) essentially take part in post-transcriptional gene regulation by silencing target genes. The growing advances in transcriptomics allow increasingly high-resolution profiling of regulatory RNAs (Sun et al. 2014). Long non-coding RNAs have been established as major regulators in plant stress responses and development (Bhatia et al. 2019, 2021; Chen et al. 2020; Jha et al. 2020; Kaur and Upadhyay 2020; Taneja and Upadhyay 2020; Upadhyay 2020).

RNA-seq generates large datasets, which need extensive computational time and power to process, align, and analyze. The handling of large transcriptomic datasets remains a bottleneck for many institutions (Han et al. 2015). Unlike genomic sequences, transcriptomes are much more dynamic and vary with tissues and conditions. The de novo transcript assembly is difficult, especially in polyploid plant species with a complex genome (Pucker and Schilbert 2019). Single-cell RNA-seq has better resolution; however, its implementation in plant models is novel and still requires optimization in sample preparation, sequencing depth, and computational tools (Islam et al. 2024).

To determine functional qualities on those genes that were differently expressed by the gene expression studies, strict validation has to be applied. Functional analysis generally involves gene knockout, overexpression, and biochemical assays to confirm gene function and its impact on plant phenotypes (Madhu et al. 2025; Ishu et al. 2025; Deepika et al.

2025c; Chauhan et al. 2024). One prevalent approach is virus-induced gene silencing, which provides transient and rapid gene silencing for functional analysis without using stable transgenic lines. In barley, studies on this method have proved very successful in the identification of disease resistance genes with a fast and convenient validation approach (Gunupuru et al. 2018).

Besides, CRISPR/Cas9 has long emerged as the most powerful tool to create targeted gene knockouts for functional validation in an ever-expanding variety of plant species (Upadhyay 2021, 2024). Extending this very capability, CRISPR activation (CRISPRa) allows one to target specific genes for upregulation, serving as an alternative approach to overexpression studies. Unlike the conventional transgenic approach, which is limited by vector capacity and transformation efficiency, CRISPRa allows the simultaneous activation of multiple genes through the design of guide RNAs placed in a single construct (Ding et al. 2022; Kumar and Shumayla 2024). Overexpression studies using CRISPRa were applied to study stress-responsive genes, providing new insights into their functions and regulatory networks. By linking these validation methods with transcriptomic data, researchers may be able to bridge the gap between gene expression profiles and observable plant traits, and ensure that transcriptomic findings provide meaningful biological insights.

7.2.3 Proteomics: Protein Profiling, Interactions, and Functions

Shotgun proteomics and two-dimensional electrophoresis allow large-scale protein profiling to study differentially expressed proteins in plant tissues under various conditions. Label-free quantification (LFQ) and stable isotope labeling techniques enable accurate protein abundance measurements, providing insights into metabolic regulation (Chen and Harmon 2006).

Shotgun proteomics and 2D-electrophoresis are two standard methods for large-scale protein profiling to identify differentially expressed proteins in plant tissues under many conditions. LFQ and stable isotope labeling enable accurate protein quantifications to investigate metabolic regulation (Elmore et al. 2023). Understanding protein–protein interactions (PPIs) is important for understanding signaling pathways and regulatory networks. The common methodologies used for PPI studies in plants include Y2H, AP-MS, and proximity labeling (Uhrig and Moorhead 2013).

The plant proteome is very complex, presenting a wide range of expression levels, modifications, and interactions between proteins. Low-abundance regulatory proteins and membrane proteins are still difficult to detect because of technical limitations (Baginsky 2009). In contrast with nucleic acids, proteins cannot be amplified, making it challenging to detect low-abundance proteins. Improvement in fractionation techniques and enrichment methods is required to extend proteome coverage (Dwevedi and Kayastha 2013). The wealth of proteomic data requires sophisticated bioinformatics tools and databases for annotation and pathway analysis. Functional validation of identified proteins is a bottleneck that necessitates integrating experiments based on either gene knockout or overexpression (Bertone and Snyder 2005). Though proteomics presently provides functional insight, its integration with genomics, transcriptomics, and metabolomics is challenging because of notable differences in data formats, scales, and analytical methodologies (Elmore et al. 2023).

7.2.4 Metabolomics: Small Molecules and Metabolic Pathways

These metabolites are directly involved in growth and development; examples include carbohydrates such as glucose and sucrose, amino acids, lipids, and organic acids. They provide the essential building blocks for biosynthetic pathways and energy metabolism (Sukumaran et al. 2020). Plants produce vast numbers of secondary metabolites that help in defense against pathogens, herbivores, and abiotic stress. Examples include flavonoids, alkaloids, phenolics, and terpenoids. Recent advances in metabolic engineering have enabled the targeted enhancement of beneficial secondary metabolites for crop improvement (Zhou et al. 2022a). The processes that map metabolic pathways provide insight into those biochemical networks which govern plant growth, stress adaptation, and nutrient utilization. High-throughput metabolomics techniques, including mass spectrometry and nuclear magnetic resonance (NMR) spectroscopy, allow the identification of the metabolic rearrangements occurring in response to environmental signals (Wu et al. 2023).

Plants produce a range of metabolites, many of which are species-specific and vary based on tissue, development phase, and environmental conditions. Identifying and characterizing novel metabolites remains a major challenge due to their structural diversity and low abundance. (Razzaq et al. 2022). Unlike DNA and proteins, metabolites do not have a universal amplification method, which means that their detection relies heavily on the sensitivity of the instruments used. To accurately measure metabolites in various biological processes, advanced techniques such as liquid chromatography-mass spectrometry and gas chromatography-mass spectrometry are necessary (Collins et al. 2021). Variability in metabolomics workflows, such as sample preparation, extraction methods, and analytical platforms, impacts data reproducibility. Combining metabolomics data with other omics layers is still computationally difficult because of differences in data structure and biological interpretation (Manickam et al. 2023). Metabolomic studies often find important changes in metabolite profiles, but connecting these changes to specific genes or enzymes needs further confirmation. The use of CRISPR-based metabolic engineering has emerged as a promising tool for confirming metabolic pathways and improving plant biochemical characteristics (Dixit et al. 2021; Zhou et al. 2022a).

7.2.5 Phenomics: High-Throughput Phenotyping and Its Role in AI-Driven Crop Improvement

RGB (Red, Green, and Blue), hyperspectral, and thermal imaging enable non-destructive monitoring of plant shape, growth patterns, and reactions to stress. ML algorithms examine imaging data to identify drought stress, disease signs, and nutrient deficiencies (Murphy et al. 2024). Lidar scanning and stereo imaging create 3D models of plant structure, allowing for accurate analysis of canopy shape and biomass estimation. These techniques are commonly used to evaluate drought tolerance and plant health in outdoor settings (Rahaman et al. 2015). Sensors that measure chlorophyll fluorescence, stomatal conductance, and transpiration rates provide valuable information about how well plants are photosynthesizing and using water. In precision agriculture, spectral indices like Normalized

Difference Vegetation Index are commonly used to monitor the health of plants (Tayade et al. 2022). Rhizotrons, X-ray CT scanning, and shovelomics are used to analyze root structure and how efficiently plants take up water and nutrients. Understanding root characteristics is crucial for breeding crops that can withstand climate challenges (White et al. 2012).

While collecting genomic data has become much easier, gathering phenotypic data is still a big challenge because it is complicated and can be affected by the environment. High-throughput phenotyping seeks to solve this problem by making it possible to collect data on a large scale and automatically (Mir et al. 2019). The large amount of phenotypic data generated needs strong systems for storing, processing, and integrating data in order to gain valuable insights. To make data analysis easier, AI-powered phenomic databases and cloud computing platforms are being created (Nabwire et al. 2021). Phenotypic traits are greatly affected by environmental factors such as temperature, humidity, and soil type. This makes it difficult to have consistent measurements in different places. However, methods like multi-environment trials and field-based robotic phenotyping are helping to improve the analysis of trait stability (Ma et al. 2022). The high cost of phenotyping platforms, imaging systems, and AI infrastructure restricts accessibility, mostly in developing countries. Affordable, portable phenotyping tools such as the Phenocart are being developed to expand accessibility (Crain 2016; Crain et al. 2016).

7.2.6 Challenges in Data Integration: Heterogeneity, Dimensionality, and Missing Data

Integrating multi-omics data is still a major challenge because of the differences, large number of variables, and missing information in various datasets. These issues make it difficult to effectively combine genomic, transcriptomic, proteomic, metabolomic, and phenotypic data, which in turn affects the precision and trustworthiness of predictive models (Flores et al. 2023). Different omics layers use different measurement technologies, which leads to incompatible data formats, scales, and distributions. Differences in experimental protocols and platforms create systematic biases that make it difficult to integrate data from different omics layers. To address this challenge, researchers are developing AI-driven models like multi-level feature contrast clustering and self-attention networks. These models aim to extract informative features from diverse datasets and enhance the performance of cross-omics integration (Zhang et al. 2024).

Omics datasets often have many variables (thousands to millions) but only a few samples. This can lead to problems like overfitting and data sparsity. To tackle these issues, we can use feature selection and dimensionality reduction techniques such as autoencoders (AEs) and matrix factorization. These methods help us find important patterns in complex data with many dimensions (Lee et al. 2020). Missing data in omics studies occur due to factors such as cost constraints, instrument sensitivity, or data loss during processing. AI-based imputation techniques, including cross-omics AEs and contrastive learning, have enhanced data recovery and model performance. The bwm-R package has been created to handle block-wise missing data, showing improved accuracy in binary classification and regression tasks (Baena-Miret et al. 2024).

7.3 AI Models for Analyzing Individual Omics

7.3.1 Genomics: Deep Learning for Variant Prediction and CNNs for Sequence Analysis

DL has furthered advances in predicting genomic variants and has thus aided in identifying important genetic elements, disease-causing mutations, and regulatory variants. Classical methods like GWAS and statistical models frequently struggle to cope with complex and unpredictable genetic interactions. However, DL models such as CNNs and graph neural networks (GNNs) have emerged to classify variants, model regulatory sequences, and interpret non-coding variants with considerable success (Voytetskiy et al. 2022).

CNNs are excellent tools for predicting the function of genetic variants. They work by analyzing DNA sequences and recognizing any patterns in the data. CNNs are best suited to finding specific kinds of mutations, namely, SNPs, indels, and SVs, which are further useful for understanding regulatory and coding variants. One of the most prominent frameworks for CNNs is DeepSEA, which has been widely utilized to predict the impact of non-coding variants on gene regulation. DeepSEA learns by training on extensive datasets containing information on chromatin structure, the material that constitutes chromosomes. It enables the accurate prediction of a region of DNA's accessibility, modifications present on histones (proteins aiding the packaging of DNA), and binding sites for transcription factors. This technique has been reported to outperform older statistical methods in pressing functional variant prioritization and uncovering SNPs related to diseases (Zhou and Troyanskaya 2015).

Recent advancements in GNNs have further improved genomic predictions by integrating sequence data with other biological data types (omics layers). GNN-based methods have shown better performance compared to CNNs and RNNs in modeling long-range genomic dependencies and complex effects of genetic variants (Voytetskiy et al. 2022).

CNNs are extensively used for detecting functional regulatory elements within non-coding DNA sequences, improving the prediction of enhancers, promoters, and transcription factor binding sites. ExplaiNN is a CNN-based model that is interpretable and has been created to accurately predict transcription factor binding and chromatin accessibility. Unlike conventional DL models, ExplaiNN offers biologically understandable insights into analyzing genomic sequences, making it easier to discover motifs and study gene regulation (Novakovsky et al. 2023).

DL technology has transformed the work of analyzing genomic sequences. However, the tedious process of manual tuning of CNNs would remain, the requirement of which will take a lot of time and expertise. To remedy this situation, Automated Modelling for Biological Evidence-based Research (AMBER) already created an automated framework for neural architectures specifically to optimize CNN architectures toward genomic analysis. AMBER employs a technology called Neural Architecture Search, introducing what it considers to be the best model for a specific biological question that is user-defined. The above work was a huge step toward achieving higher accuracy in allele-specific binding predictions, yielding better classifications of variants, and model regulatory sequences better. The main advantages of AMBER over the traditional manually designed CNN models are that it exhibits higher prediction accuracy and improves computational efficiencies. All these advantages together allow more DL techniques that enable genomic research on a different level and offer possibilities for deeper study of biological data (Zhang et al. 2021).

7.3.2 Transcriptomics: RNNs/LSTMs for Time-Series Gene Expression and Attention Models

Transcriptomics provides insight into how genes respond developmentally, often as a function of environmental changes, or due to modified genetic background. Unlike genome sequences that remain static, transcriptomic data are regularly changing and time-dependent, calling for advanced AI strategies to discover patterns and make predictions. DL models like RNNs, long short-term memory (LSTM) networks, and attention-based architectures have been successful in capturing the complex time-dependent relationships in gene expression data (Zhou et al. 2022b).

Conventional ML models find it hard to handle time-related dependencies of gene expression profiles because of their complex patterns over time. RNN, composed mainly of the LSTM architecture, can capture long-term dependencies in the sequential data and solve the vanishing gradient problem therefore supplementing gene expression analysis over time (Monti et al. 2022). Multi-LSTM expands using empirical mode decomposition, which stabilizes and makes the data more model-friendly. The inherent strength of the method makes the LSTMs more powerful in predicting gene expression patterns at time points that might not be known (Zhou et al. 2022b). Common LSTM models are time-step agnostic, which can downplay critical regulatory events. Attention mechanisms provide innate time points for the model to concentrate on during training; thus enhancing the understanding and precision of the model prediction (Monti et al. 2022). Transcriptional activity's row-randomization RNNs rebuild the gene regulatory networks (GRNs) based on knowledge of how genes connect through time. This provides enhanced GRN study possibilities that enable the direct identification of regulatory structures in plant transcriptomics (Monti et al. 2022).

7.3.3 Proteomics: ML-Based Protein Function Prediction, Deep Learning for Structure Modeling

Thus, protein function prediction is significant in plant proteomics to figure out enzyme activity, signaling cascades, and stress response machinery. However, experimental real cognition for protein functions is resource-intensive, making ML-based computational approaches highly essential. A host of ML approaches, such as SVMs, RFs, and deep neural networks (DNNs), have been successfully employed to predict proteins' functions based on sequence features, structural properties, and interaction networks (Bonetta and Valentino 2020).

Early ML methods used physical and chemical properties, amino acid composition, and evolutionary characteristics to categorize proteins into functional groups. Recently, new techniques like AEs and deep representation learning have been developed to automatically identify complex features in protein sequences (Le 2022). The prPred-DRLF web server uses DL and natural language processing techniques to classify plant resistance proteins, which are important for defending against pathogens. The model performed better than existing methods in cross-validation tests, showing that DL is effective for predicting protein functions (Le 2022).

To predict a protein's function, interaction with other proteins, and regulation of metabolism, it is very important to know about the protein structure. However, conventional ways of determining protein structures, like X-ray crystallography and NMR spectroscopy, are bound to be slow and very expensive, which necessitates the development of AI-based computational approaches. The newest developments in DL, especially with AlphaFold2 and other AI-based protein structure prediction programs, have greatly increased the accuracy and speed of tertiary protein structure predictions (Pakhrin et al. 2021). DL models like AlphaFold2 have changed the game in protein structure prediction by using DNNs and evolutionary sequence alignments. These models can accurately predict how different parts of a protein interact, how it folds, and its secondary structure, even for proteins that do not have any known relatives. AlphaFold2 has had a huge impact on structural biology, offering detailed protein models that help with understanding functions and designing new proteins (Pakhrin et al. 2021). Recent methods combine CNNs and DL architectures to refine distance maps and improve 3D structural modeling. CNN-based frameworks significantly enhance protein structure similarity detection and improve fold recognition in newly sequenced proteins (Torrisi et al. 2020).

PPIs are essential for various biological processes, including metabolism, gene expression, and plant responses to stress. With the help of AI, researchers can now predict PPIs more accurately by using models that analyze the specific sequences and structures of proteins involved in these interactions. DL techniques, such as graph-based learning and attention mechanisms, have proven to be particularly effective in analyzing PPI networks. These advanced methods allow for a more detailed understanding of how different proteins communicate with each other, which in turn leads to better predictions of plant signaling pathways and their ability to adapt to stressful conditions (Torrisi et al. 2020).

7.3.4 Metabolomics: Random Forests and SVMs for Metabolite Classification

Metabolomics produces high-dimensional biochemical datasets that capture metabolic profiles, stress responses, and phenotype-associated alterations in plants. ML models, specially RF and SVMs, have been widely used for metabolite classification, biomarker discovery, and predictive modeling (Ghosh et al. 2020). RF is an extensively used ensemble learning method that makes multiple decision trees and aggregates predictions, making it stout against overfitting and well-suited for intricate metabolomics datasets. RF has been used to classify metabolomics profiles with high accuracy, particularly in distinguishing between different plant genotypes and stress responses (Mendez et al. 2019). In MS-based metabolomics, RF models identify important metabolites associated with specific plant traits, improving our understanding of how plants metabolize and adapt (Enot et al. 2006). SVMs are robust classification models that separate metabolite classes using hyperplanes in high-dimensional space. Kernel-based SVMs outperform traditional statistical models in metabolic biomarker discovery and classification of stress-responsive metabolites (Huang et al. 2015).

7.3.5 Phenomics: Image-Based Deep Learning Models for Trait Prediction

Phenomics is important for connecting genetic variations to visible plant characteristics. Image-based DL models make it possible to analyze plant traits quickly and without causing any harm. Unlike traditional methods of studying plants, which are slow, require a lot of manual work, and can be affected by human mistakes, this new approach uses AI to analyze images (Murphy et al. 2024).

CNNs extract structural and spectral features from plant images, enabling accurate measurement of leaf area, height, biomass, and chlorophyll content. Transfer learning has improved CNN performance in predicting traits from hyperspectral and RGB images, minimizing the requirement for extensive labeled datasets (Adams et al. 2021).

Multitask deep learning allows for the simultaneous extraction of multiple plant traits, making the process more efficient. A modified ResNet50 model was able to accurately predict leaf count, projected leaf area, and genotype classification in *Arabidopsis thaliana*, reducing the need for manual annotation by 75% (Dobrescu et al. 2020).

Transformer architectures, like Vision Transformers (ViTs), are better at extracting features for tasks such as disease detection, growth monitoring, and stress response assessment. Methods using ViTs for phenotyping are more effective than CNNs in recognizing small plant characteristics in various environmental settings (Weng et al. 2019).

7.4 AI Approaches for Multi-Omics Integration

7.4.1 Deep Learning Approaches

Integrating multi-omics data is complex, diverse, and may have missing information, so an arduous challenge. Nevertheless, it is reported that some DL methodologies like AEs and multi-modal neural networks have achieved rigorous extraction of valuable biological insights on those data types. These methods contribute to a supervised learning process of improving prediction accuracy and understanding biological processes through feature compression via interactions amongst different omics layers (Benkirane et al. 2023).

AEs are DL models that learn to represent complex, high-dimensional data in a more compact, lower-dimensional form. Stacked AEs (SAEs) are a type of AE that consists of multiple layers of encoders and decoders stacked on top of each other. SAEs have been widely used in multi-omics studies, which involve analyzing different types of biological data such as genomics, transcriptomics, proteomics, and metabolomics. These studies often face the challenge of dealing with a large number of features (variables) compared to the number of samples, known as the curse of dimensionality. SAEs help overcome this challenge by learning hierarchical representations of the data and capturing important underlying patterns or features. One specific application of SAEs in cancer research is the SAEsurv-net model, which combines multi-omics data for predicting cancer survival outcomes. This model has shown better performance compared to traditional ML models when it comes to predicting patient survival based on various omics measurements (Wu and Fang 2022). Variational AEs have also been explored for unsupervised clustering and feature extraction, enabling improved classification of multi-omics profiles (Zhang et al. 2019).

Multi-modal neural networks combine various omics layers such as genomics, transcriptomics, and proteomics to discover connections between different omics fields. CustOmics, a DL-based multi-omics integration method, uses a two-phase strategy where each omics type is trained separately before learning interactions between them. This approach leads to improved feature extraction (Benkirane et al. 2023). Multi-view Factorization AEs combine biological interaction networks with omics data, leading to better predictions of disease progression and phenotype associations (Ma and Zhang 2019). Hierarchical AE-based integration has been shown to perform better than traditional models in multi-omics survival prediction by focusing on the most informative omics layers (Wissel et al. 2021).

7.4.2 Machine Learning Models for Feature Selection Across Omics and Improving Predictive Accuracy

Multi-omics data integration needs strong feature selection methods to find useful biomarkers and remove unnecessary variables. RFs and SVMs are commonly used for feature selection in multi-omics because they can work with high-dimensional datasets and non-linear relationships (Wang 2023). RF uses ensemble decision trees to rank features according to their importance, making it efficient for multi-omics classification tasks. The Dynamic Correlated Regularized Random Forest model combines RF feature selection with data reduction techniques, boosting computational efficiency (Wang 2023). SVM-based Recursive Feature Elimination is a technique that repeatedly removes the least useful features from a dataset, leading to better model performance. Ensemble SVM-RFE takes this a step further by improving the accuracy of gene selection in complex omics datasets with many variables, using a method called bootstrap sampling (Anaissi et al. 2016).

Ensemble learning combines multiple ML models to enhance predictive accuracy, especially in multi-omics integration. RF-SVM hybrid models integrate RF feature selection with SVM classification, enhancing robustness and precision (Zheng and Jin 2019). Bagging-based SVM ensembles enhance classification stability and address class imbalances in omics datasets (Anaissi et al. 2016). Random Generalized Linear Models integrate RF with feature selection techniques to improve interpretability while keeping predictive accuracy (Song et al. 2013).

7.4.3 Network-Based Models for Gene-Protein-Metabolite Interactions

Graph-based AI methods are now crucial for modeling intricate biological networks that involve gene, metabolite, and protein interactions. These techniques employ GNNs and knowledge graphs (KGs) to reveal concealed connections in multi-omics datasets, making it easier to identify important molecular interactions (Tran and Gao 2023). GNNs effectively capture the relationships between genes, metabolites, and proteins, leading to better functional annotations and disease predictions. LATTE2GO, a new GNN-based framework for integrating multiple omics data types, improves protein function prediction by leveraging diverse interaction networks (Tran and Gao 2023). DeepMOCCA, a graph-based AI model, combines PPI networks with multi-omics data to enhance survival predictions in cancer research (Althubaiti et al. 2021).

Graph-based embeddings have been successfully used to create multi-omics disease networks, which represent the relationships between biomolecules and phenotypic traits. A new model for learning graph representations has enhanced the classification of multi-omics patients, especially in cases where biological datasets are sparse (Hussein et al. 2022).

KGs combine various biological datasets by representing connections between genes, proteins, and metabolites in organized networks. These models improve the study of biological pathways, drug development, and functional genomics. OncoNet Ontology (ONO) uses BioBERT-based text mining and extensive KGs to identify relationships among genes, cancer types, and biomarkers, leading to better predictions of phenotypes (Karim et al. 2023). MPK-GNN, a GNN that uses multiple sources of prior knowledge, has performed better than other AI-based methods in classifying cancer subtypes (Xiao et al. 2023).

7.5 Applications of AI-Driven Multi-Omics in Crop Improvement

7.5.1 Genetic Trait Prediction: Using Integrated Omics for Breeding Strategies

AI-driven multi-omics methods greatly improve trait prediction accuracy by combining genomics, transcriptomics, and metabolomics. ML models, like DL-based genomic selection, enhance the selection of high-yielding and stress-resilient crop varieties (Zhang et al. 2022). Multi-omics-assisted genomic prediction models are better than traditional genomic best linear unbiased prediction at predicting agronomic traits (Ali et al. 2024).

7.5.2 Disease Resistance and Stress Tolerance: AI Models for Stress-Responsive Gene Discovery

Multi-omics AI frameworks combine various biological data such as gene regulatory networks, epigenetic modifications, and metabolomic markers to find traits in crops that help them adapt to stress. Using AI to model gene expression has uncovered important transcription factors that play a role in tolerance to drought, heat, and salinity (Roychowdhury et al. 2023). Using CRISPR-Cas9 to validate genes identified by AI improves precision breeding for climate-resilient crops (Razzaq et al. 2022).

7.5.3 Yield Prediction and Quality Improvement: Multi-Omics AI Frameworks for Phenotype Prediction

AI-driven phenomics combines image-based DL models with genomic prediction to improve crop yield estimation. Multi-omics approaches enhance seed nutritional quality by identifying important metabolic pathways that control protein and oil content (Hu et al. 2021). Multi-trait genomic selection using AI has resulted in significant yield improvements in breeding programs for rice, wheat, and maize (Yang et al. 2021).

7.5.4 Precision Agriculture: AI-Guided Multi-Omics Insights for Sustainable Farming

AI models combine remote sensing data with omics-driven insights to optimize fertilizer application, pest control, and irrigation management (Thriveni et al. 2024). Predictive AI models using soil microbiomics and metabolomics enhance soil health monitoring and promote sustainable farming practices (Zenda et al. 2021).

7.6 Challenges and Future Directions in AI-Driven Multi-Omics for Crop Science

7.6.1 Data Quality, Standardization, and Interoperability

Discrepancies in omics data formats, missing values, and batch effects obstruct AI-driven multi-omics analysis. Standardization efforts, for instance FAIR (Findable, Accessible, Interoperable, and Reusable) data principles, enhance interoperability across multi-omics datasets (Barash 2016). Multi-omics data harmonization methods allow seamless integration of genomic, transcriptomic, proteomic, and phenotypic data (Cembrowska-Lech et al. 2023).

7.6.2 Model Interpretability and Explainability in Plant Science

AI models need to be understandable for regulatory approval and scientific validation in plant research. Explainable AI (XAI) methods like SHAP (Shapley Additive Explanations) and LIME (Local Interpretable Model-agnostic Explanations) enhance clarity in multi-omics predictions (Hussein et al. 2024). Graph-based learning models provide a systematic way to visualize gene-metabolite-protein interactions, which can be useful for functional annotation and generating hypotheses (Zampieri 2018).

7.6.3 Computational Efficiency and Scalability of AI Models

Multi-omics datasets are large and diverse, needing high-performance computing (HPC) and distributed learning frameworks to scale AI models. RandomSCM, a sparse classifier ensemble model, balances predictive precision and computational efficiency, making it appropriate for large-scale omics applications (Godon et al. 2022). Federated learning and cloud-based AI platforms enable collaborations between multiple institutions while keeping data private (Wu and Xie 2025).

7.6.4 Ethical and Regulatory Aspects of AI in Crop Genomics

Ethical concerns about AI-driven plant breeding include data ownership, algorithmic bias, and equitable access to genomic resources. Regulatory frameworks must evolve to address AI-driven genomic selection and CRISPR-Cas9 applications in agriculture (Reznikov and Turlakova 2024). Clear AI governance and responsible AI use policies will be essential for ensuring the fair and sustainable use of AI in crop improvement (Wu and Xie 2025).

7.6.5 Future Directions in Multi-Omics Approaches for Plant Science

- Self-supervised learning for automated multi-omics feature extraction to enhance AI precision in genomic selection.
- Hybrid AI models combining DL, graph-based learning, and probabilistic AI for better biological network inference.
- Ethical AI frameworks for plant biotechnology, ensuring fair AI applications in genomics and precision agriculture.

AI-powered multi-omics research has great potential for sustainable agriculture, precision breeding, and climate-resilient crop development, as long as data quality, model interpretability, computational scalability, and ethical considerations are properly addressed (Cembrowska-Lech et al. 2023).

7.7 Conclusion

The employability of AI in multi-omics has revolutionized crop improvement. It is used for precise prediction of genetic traits, stress tolerance analysis, and yield optimization. Some prominent AI techniques employed in the field of crop improvement include:

- DL methods (CNNs, RNNs, and transformers) performing sequence-based analysis and omics feature extraction.
- ML models (RFs, SVMs, and ensemble learning) for multi-omics classification and biomarker discovery.
- Graph-based AI methods for modeling gene-protein-metabolite interactions in complex biological networks.
- Knowledge graphs and XAI for improving model interpretability in plant breeding.
- Federated learning and HPC to scale AI-driven crop genomics while ensuring data privacy (Cembrowska-Lech et al. 2023).

In the future, AI-driven multi-omics research will be crucial for sustainable agriculture, precision breeding, and climate resilience. Some of the new exploring areas include:

- Self-supervised learning and generative AI models for automated feature extraction in genomic selection (Yang et al. 2021).
- Integration of AI with IoT-driven phenotyping for real-time crop monitoring and smart farming (Konar 2024).
- Ethical AI frameworks for responsible genomic selection and regulatory compliance in plant biotechnology (Viduedo et al. 2024).
- AI systems that combine genetic, phenotypic, and environmental information to enhance predictions for how crops will adapt (Murmu et al. 2024).
- With ongoing advancements in AI, high-throughput sequencing, and phenotyping technologies, AI-driven multi-omics will transform crop improvement, leading to sustainable food production and increased agricultural resilience (Zhang et al. 2022).

References

Adams, J., Qiu, Y., Posadas, L. et al. (2021). Phenotypic trait extraction of soybean plants using deep convolutional neural networks with transfer learning. *Big Data and Information Analytics* 6 (1): 26–40.

Ali, B., Huguenin-Bizot, B., Laurent, M. et al. (2024). High-dimensional multi-omics measured in controlled conditions are useful for maize platform and field trait predictions. *Theoretical and Applied Genetics* 137 (7): 175.

Althubaiti, S., Kulmanov, M., Liu, Y. et al. (2021). DeepMOCCA: a pan-cancer prognostic model identifies personalized prognostic markers through graph attention and multi-omics data integration. *BioRxiv* 2003–2021.

Anaissi, A., Goyal, M., Catchpoole, D.R. et al. (2016). Ensemble feature learning of genomic data using support vector machine. *PLoS One* 11 (6): e0157330.

Baena-Miret, S., Reverter, F., and Vegas, E. (2024). A framework for block-wise missing data in multi-omics. *PLoS One* 19 (7): e0307482.

Baginsky, S. (2009). Plant proteomics: concepts, applications, and novel strategies for data interpretation. *Mass Spectrometry Reviews* 28 (1): 93–120.

Barash, C.I. (2016). Omics challenges and unmet translational needs. *Applied & Translational Genomics* 10: 1.

Benkirane, H., Pradat, Y., Michiels, S. et al. (2023). CustOmics: a versatile deep-learning based strategy for multi-omics integration. *PLoS Computational Biology* 19 (3): e1010921.

Bertone, P. and Snyder, M. (2005). Prospects and challenges in proteomics. *Plant Physiology* 138 (2): 560–562.

Bhatia, G., Sharma, S., Upadhyay, S.K., and Singh, K. (2019). Long non-coding RNAs coordinate developmental transitions and other key biological processes in grapevine. *Scientific Reports* 9 (1): https://doi.org/10.1038/s41598-019-38989-7.

Bhatia, G., Upadhyay, S.K., Upadhyay, A., and Singh, K. (2021). Investigation of long non-coding RNAs as regulatory players of grapevine response to powdery and downy mildew infection. *BMC Plant Biology* 21 (1): https://doi.org/10.1186/s12870-021-03059-6.

Bonetta, R. and Valentino, G. (2020). Machine learning techniques for protein function prediction. *Proteins: Structure, Function, and Bioinformatics* 88 (3): 397–413.

Cembrowska-Lech, D., Krzemińska, A., Miller, T. et al. (2023). An integrated multi-omics and artificial intelligence framework for advance plant phenotyping in horticulture. *Biology* 12 (10): 1298.

Chauhan, H., Alok, A., Aiana et al. (2024). CRISPR/Cas9 edited StbHLH47 lines exhibit altered expression profiling of iron regulating genes and increased iron content in Solanum tuberosum. *Current Plant Biology* 38: 100354. https://doi.org/10.1016/j.cpb.2024.100354.

Chen, S. and Harmon, A.C. (2006). Advances in plant proteomics. *Proteomics* 6 (20): 5504–5516.

Chen, L., Zhu, Q.-H., and Kaufmann, K. (2020). Long non-coding RNAs in plants: emerging modulators of gene activity in development and stress responses. *Planta* 252 (5): 92.

Collins, S.L., Koo, I., Peters, J.M. et al. (2021). Current challenges and recent developments in mass spectrometry–based metabolomics. *Annual Review of Analytical Chemistry* 14 (1): 467–487.

Crain, J.L. (2016). *Leveraging the Genomics Revolution with High-Throughput Phenotyping for Crop Improvement of Abiotic Stresses*. Kansas State University.

Crain, J.L., Wei, Y., Barker, J. III, et al. (2016). Development and deployment of a portable field phenotyping platform. *Crop Science* 56 (3): 965–975.

Deepika, Madhu, and Upadhyay, S.K. (2024). Deciphering the features and functions of serine/arginine protein kinases in bread wheat. *Plant Gene* 38: 100451. https://doi.org/10.1016/j.plgene.2024.100451.

Deepika, Shekhawat, J., Madhu et al. (2025a). Genome-wide analysis unveiled the characteristic features and functions of AFC genes of bread wheat (*Triticum aestivum* L.). *Journal of Plant Growth Regulation* 44 (5): 2466–2484. https://doi.org/10.1007/s00344-024-11559-6.

Deepika, Madhu, Shekhawat, J. et al. (2025b). Pre-mRNA processing factor 4 kinases (PRP4Ks): exploration of molecular features, interaction network and expression profiling in bread wheat. *Journal of Plant Growth Regulation* 44 (2): 806–820. https://doi.org/10.1007/s00344-024-11489-3.

Deepika, Madhu, Ishu, and Upadhyay, S.K. (2025c). A LysM domain-containing protein TaLysM1-A confers salinity and drought tolerance in transgenic Arabidopsis. *Plant Physiology and Biochemistry* 229: 110629. https://doi.org/10.1016/j.plaphy.2025.110629.

Ding, X., Yu, L., Chen, L. et al. (2022). Recent progress and future prospect of CRISPR/Cas-derived transcription activation (CRISPRa) system in plants. *Cells* 11 (19): 3045.

Dixit, S., Shukla, A., Singh, V., and Upadhyay, S.K. (2021). *Engineering of Plant Metabolic Pathway for Nutritional Improvement*. Wiley. https://doi.org/10.1002/9781119672425.ch20.

Dobrescu, A., Giuffrida, M.V., and Tsaftaris, S.A. (2020). Doing more with less: a multitask deep learning approach in plant phenotyping. *Frontiers in Plant Science* 11: 141.

Dwevedi, A. and Kayastha, A.M. (2013). Soybean proteomics: applications and challenges. In: *A Comprehensive Survey of International Soybean Research–Genetics, Physiology* (ed. J.E. Board), 395–412. Rijeka, Croatia: Agronomy and Nitrogen Relationships. InTech.

Elmore, J.M., Velásquez-Zapata, V., and Wise, R.P. (2023). Next-generation yeast two-hybrid screening to discover protein–protein interactions. In: *Protein-Protein Interactions: Methods and Protocols* (ed. S. Mukhtar), 205–222. Springer.

Enot, D.P., Beckmann, M., Overy, D. et al. (2006). Predicting interpretability of metabolome models based on behavior, putative identity, and biological relevance of explanatory signals. *Proceedings of the National Academy of Sciences* 103 (40): 14865–14870.

Feuk, L., Carson, A.R., and Scherer, S.W. (2006). Structural variation in the human genome. *Nature Reviews Genetics* 7 (2): 85–97.

Flores, J.E., Claborne, D.M., Weller, Z.D. et al. (2023). Missing data in multi-omics integration: recent advances through artificial intelligence. *Frontiers in Artificial Intelligence* 6: 1098308.

Ghosh, T., Zhang, W., Ghosh, D. et al. (2020). Predictive modeling for metabolomics data. *Computational Methods and Data Analysis for Metabolomics* 313–336.

Godon, T., Plante, P. L., Bauvin, B. et al. (2022). RandomSCM: interpretable ensembles of sparse classifiers tailored for omics data, *arXiv preprint* arXiv:2208.06436 [Preprint].

Grones, C., Eekhout, T., Shi, D. et al. (2024). Best practices for the execution, analysis, and data storage of plant single-cell/nucleus transcriptomics. *The Plant Cell* 36 (4): 812–828.

Gunasekaran, D., Ardell, D.H., and Nobile, C.J. (2024). SNP-SVant: a computational workflow to predict and annotate genomic variants in organisms lacking benchmarked variants. *Current Protocols* 4 (5): e1046.

Gunupuru, L.R., Perochon, A., Ali, S.S. et al. (2018). Virus-induced gene silencing (VIGS) for functional characterization of disease resistance genes in barley seedlings. In: *Barley: Methods and Protocols* (ed. W.A. Harwood), 95–114. Springer.

Han, Y., Gao, S., Muegge, K. et al. (2015). Advanced applications of RNA sequencing and challenges. *Bioinformatics and Biology Insights* 9: BBI-S28991.

Hu, H., Campbell, M.T., Yeats, T.H. et al. (2021). Multi-omics prediction of oat agronomic and seed nutritional traits across environments and in distantly related populations. *Theoretical and Applied Genetics* 134: 4043–4054.

Huang, X., Xu, Q.S., Yun, Y.H. et al. (2015). Weighted variable kernel support vector machine classifier for metabolomics data analysis. *Chemometrics and Intelligent Laboratory Systems* 146: 365–370.

Hussein, S., Vu, T., Langel, L. et al. (2022). Effective subject representation based on multi-omics disease networks using graph embedding, In *2022 IEEE International Conference on Bioinformatics and Biomedicine (BIBM)*, (6–8 December 2022), Las Vegas, USA & Changsha, China. IEEE, pp. 1911–1918.

Hussein, A., Prasad, M. and Braytee, A. (2024). Explainable AI methods for multi-omics analysis: a survey, *arXiv preprint* arXiv:2410.11910 [Preprint].

Ishu, Shumayla, Madhu, and Upadhyay, S.K. (2025). Complementation with TaNCL2-A reinstates growth and abiotic stress response in *atncl* mutant of Arabidopsis. *Plant Science* 353: 112411. https://doi.org/10.1016/j.plantsci.2025.112411.

Islam, M.T., Liu, Y., Hassan, M.M. et al. (2024). Advances in the application of single-cell transcriptomics in plant systems and synthetic biology. *Biodesign Research* 6: 29.

Jha, U.C., Nayyar, H., Jha, R. et al. (2020). Long non-coding RNAs: emerging players regulating plant abiotic stress response and adaptation. *BMC Plant Biology* 20: 1–20.

Karim, M.R., Comet, LM., Shajalal, M. et al. (2023). From large language models to knowledge graphs for biomarker discovery in cancer, *arXiv preprint* arXiv:2310.08365 [Preprint].

Kaur, A. and Upadhyay, S.K. (2020). Application of lncRNA in plants improvement: a future perspective. In: *Long Noncoding RNAs in Plants* (ed. S.K. Upadhyay). Elsevier. https://doi.org/10.1016/B978-0-12-821452-7.00021-0.

Khaki, S., Wang, L., and Archontoulis, S.V. (2020). A CNN-RNN framework for crop yield prediction. *Frontiers in Plant Science* 10: 1750.

Konar, M. (2024). Integrating IoT and AI for automated crop with smart Urban farming, In *2024 International Conference on Emerging Innovations and Advanced Computing (INNOCOMP)*, (25–26 May 2024), Sonipat, India. IEEE, pp. 454–462.

Kumar, A., Shumayla, Upadhyay, S.K. (2024). CRISPR/Cas-mediated genome editing in plants: a historical perspective. In: *Applications of Genome Engineering in Plants* (ed. S.K. Upadhyay), 1–19. Wiley. https://doi.org/10.1002/9781394183913.ch1.

Le, N.Q.K. (2022). Potential of deep representative learning features to interpret the sequence information in proteomics. *Proteomics* 22 (1–2): 2100232.

Lee, B., Zhang, S., Poleksic, A. et al. (2020). Heterogeneous multi-layered network model for omics data integration and analysis. *Frontiers in Genetics* 10: 1381.

Ma, T. and Zhang, A. (2019). Integrate multi-omics data with biological interaction networks using Multi-view Factorization Auto Encoder (MAE). *BMC Genomics* 20 (Suppl 11): 944.

Ma, Z., Rayhana, R., Feng, K. et al. (2022). A review on sensing technologies for high-throughput plant phenotyping. *IEEE Open Journal of Instrumentation and Measurement* 1: 1–21.

Madhu, Deepika, Sharma, A. et al. (2025). Expression of *Glutathione Reductase* 2-D (TaGR2-D) From *Triticum aestivum* facilitates arsenic tolerance by modulating GSH pool and arsenic transporters in Transgenic Arabidopsis. *Physiologia Plantarum* 177 (4): e70439. https://doi.org/10.1111/ppl.70439.

Mandaliya, V.B., Pandya, R.V., and Thaker, V.S. (2010). Single nucleotide polymorphism (SNP): a trend in genetics and genome analyses of plants. *General and Applied Plant Physiology* 36 (3–4): 159–166.

Manickam, S., Rajagopalan, V.R., Kambale, R. et al. (2023). Plant metabolomics: current initiatives and future prospects. *Current Issues in Molecular Biology* 45 (11): 8894–8906.

Mendez, K.M., Reinke, S.N., and Broadhurst, D.I. (2019). A comparative evaluation of the generalised predictive ability of eight machine learning algorithms across ten clinical metabolomics data sets for binary classification. *Metabolomics* 15: 1–15.

Mir, R.R., Reynolds, M., Pinto, F. et al. (2019). High-throughput phenotyping for crop improvement in the genomics era. *Plant Science* 282: 60–72.

Monti, M., Fiorentino, J., Milanetti, E. et al. (2022). Prediction of time series gene expression and structural analysis of gene regulatory networks using recurrent neural networks. *Entropy* 24 (2): 141.

Murmu, S., Sinha, D., Chaurasia, H. et al. (2024). A review of artificial intelligence-assisted omics techniques in plant defense: current trends and future directions. *Frontiers in Plant Science* 15: 1292054.

Murphy, K.M., Ludwig, E., Gutierrez, J. et al. (2024). Deep learning in image-based plant phenotyping. *Annual Review of Plant Biology* 75: 771–795.

Nabwire, S., Suh, H.K., Kim, M.S. et al. (2021). Application of artificial intelligence in phenomics. *Sensors* 21 (13): 4363.

Novakovsky, G., Fornes, O., Saraswat, M. et al. (2023). ExplaiNN: interpretable and transparent neural networks for genomics. *Genome Biology* 24 (1): 154.

Pakhrin, S.C., Shrestha, B., Adhikari, B. et al. (2021). Deep learning-based advances in protein structure prediction. *International Journal of Molecular Sciences* 22 (11): 5553.

Pokrovac, I. and Pezer, Ž. (2022). Recent advances and current challenges in population genomics of structural variation in animals and plants. *Frontiers in Genetics* 13: 1060898.

Pucker, B. and Schilbert, H.M. (2019). Genomics and transcriptomics advance in plant sciences. In: *Molecular Approaches in Plant Biology and Environmental Challenges* (ed. S.P. Singh, S.K. Upadhyay, A. Pandey, and S. Kumar), 419–448. Springer.

Rahaman, M.M., Chen, D., Gillani, Z. et al. (2015). Advanced phenotyping and phenotype data analysis for the study of plant growth and development. *Frontiers in Plant Science* 6: 619.

Razzaq, A., Wishart, D.S., Wani, S.H. et al. (2022). Advances in metabolomics-driven diagnostic breeding and crop improvement. *Metabolites* 12 (6): 511.

Reznikov, R.B. and Turlakova, S.S. (2024). Data science methods and models in modern economy. *Economic Space* 191: 104–113.

Roychowdhury, R., Das, S.P., Gupta, A. et al. (2023). Multi-omics pipeline and omics-integration approach to decipher plant's abiotic stress tolerance responses. *Genes* 14 (6): 1281.

Sharma, Y., Deepika, Kaur, A., and Upadhyay, S.K. (2025a). Exploration of mitochondrial calcium uniporters suggested their role in abiotic stress response in bread wheat. *Plant Physiology and Biochemistry* 227: 110180. https://doi.org/10.1016/j.plaphy.2025.110180.

Sharma, Y., Dixit, S., Singh, K., and Upadhyay, S.K. (2025b). Vascular plant one-zinc finger transcription factors: exploration of characteristic features, expression, coexpression and interaction suggested their diverse role in bread wheat (*Triticum aestivum* L.). *Plant Growth Regulation* 105 (1): 71–87. https://doi.org/10.1007/s10725-024-01264-5.

Song, L., Langfelder, P., and Horvath, S. (2013). Random generalized linear model: a highly accurate and interpretable ensemble predictor. *BMC Bioinformatics* 14: 1–22.

Sukumaran, S.T., Sugathan, S., and Abdulhameed, S. (2020). *Plant Metabolites: Methods, Applications and Prospects*. Springer.

Sun, X., Zhang, Y., Zhu, X. et al. (2014). Advances in identification and validation of plant microRNAs and their target genes. *Physiologia Plantarum* 152 (2): 203–218.

Taneja, M. and Upadhyay, S.K. (2025). Heavy Metal-Associated (HMA) domain-containing proteins: insight into their features and roles in bread wheat (*Triticum aestivum* L.). *Biology* 14 (7): 818. https://doi.org/10.3390/biology14070818.

Taneja, M. and Upadhyay, S.K. (2020). An overview of long non-coding RNA in plants. In: *Long Non-coding RNAs in Plants* (ed. S.K. Upadhyay). Elsevier. https://doi.org/10.1016/B978-0-12-821452-7.00001-5.

Tayade, R., Yoon, J., Lay, L. et al. (2022). Utilization of spectral indices for high-throughput phenotyping. *Plants* 11 (13): 1712.

Thriveni, V., Teotia, J., Hazra, S. et al. (2024). A review on integrating bioinformatics tools in modern plant breeding. *Archives of Current Research International* 24 (9): 293–308.

Torrisi, M., Pollastri, G., and Le, Q. (2020). Deep learning methods in protein structure prediction. *Computational and Structural Biotechnology Journal* 18: 1301–1310.

Tran, N.C. and Gao, J.X. (2023). Integrating heterogeneous biological networks and ontologies for improved protein function prediction with graph neural networks, In *2023 IEEE International Conference on Bioinformatics and Biomedicine (BIBM)*, (5–8 December 2023), Istanbul, Turkey. IEEE, pp. 43–48.

Uhrig, R.G. and Moorhead, G.B. (2013). Plant proteomics: current status and future prospects. *Journal of Proteomics* 88: 34–36.

Upadhyay, S.K. (2020). *Long Non-Coding RNAs in Plants: Roles in Development and Stress*. Elsevier. https://doi.org/10.1016/C2019-0-03200-1.

Upadhyay, S.K. (2021). *Genome Engineering for Crop Improvement*. Wiley 10.1002/9781119672425.

Upadhyay, S.K. (2024). *Applications of Genome Engineering in Plants*. Wiley. https://doi.org/10.1002/9781394183913.

Upadhyay, S.K. (2026). *Designer Crops: Implications of Artificial Intelligence and Modern Biotechnological Methods*. In: Elsevier Science.

Upadhyay, S.K., Sharma, S., Singh, H. et al. (2015). Whitefly genome expression reveals host-symbiont interaction in amino acid biosynthesis. *PLoS One* 10 (5): e0126751.

Upton, R.N., Correr, F.H., Lile, J. et al. (2023). Design, execution, and interpretation of plant RNA-seq analyses. *Frontiers in Plant Science* 14: 1135455.

Viduedo, P.H.Z., Candeia, V.C., Martins, V.B. et al. (2024). Harnessing the power of ai and machine learning for next-generation sequencing data analysis: a comprehensive review of applications, challenges, and future directions in precision oncology. *Revista Ibero-Americana de Humanidades, Ciências e Educação* 10 (8): 2898–2904.

Vidyarthi, V.M., Bhandari NS, Goyal S et al. (2024). Design and implementation of deep learning CN-RNN algorithm for crop analysis, In *2024 5th International Conference on Image Processing and Capsule Networks (ICIPCN)*, 3–4 July 2024, Dhulikhel, Nepal, IEEE, pp. 614–619.

Voytetskiy, A., Herbert, A. and Poptsova, M. (2022). Graph neural networks for Z-DNA prediction in genomes, In *2022 IEEE International Conference on Bioinformatics and Biomedicine (BIBM)*, (6–8 December 2022), Las Vegas, USA & Changsha, China. IEEE, pp. 3173–3178.

Wang, Z. (2023). Research on feature selection methods based on random forest. *Tehnički vjesnik* 30 (2): 623–633.

Wang, X., Tu, M., Wang, Y. et al. (2021). Whole-genome sequencing reveals rare off-target mutations in CRISPR/Cas9-edited grapevine. *Horticulture Research* 8: 114.

Weng, Y., Zeng, R., Wu, C. et al. (2019). A survey on deep learning based plant phenotype research in agriculture. *Scientia Sinica (Vitae)* 49 (6): 698–716.

White, J.W., Andrade-Sanchez, P., Gore, M.A. et al. (2012). Field-based phenomics for plant genetics research. *Field Crops Research* 133: 101–112.

Wissel, D., Rowson, D., and Boeva, V. (2021). Hierarchical autoencoder-based integration improves performance in multi-omics cancer survival models through soft modality selection. *BioRxiv* 2009–2021.

Wu, X. and Fang, Q. (2022). Stacked autoencoder based multi-omics data integration for cancer survival prediction, *arXiv preprint* arXiv:2207.04878 [Preprint].

Wu, Y. and Xie, L. (2025). AI-driven multi-omics integration for multi-scale predictive modeling of genotype-environment-phenotype relationships. *Computational and Structural Biotechnology Journal* 27: 265–277.

Wu, M., Northen, T.R., and Ding, Y. (2023). Stressing the importance of plant specialized metabolites: omics-based approaches for discovering specialized metabolism in plant stress responses. *Frontiers in Plant Science* 14: 1272363.

Xiao, S., Lin, H., Wang, C. et al. (2023). Graph neural networks with multiple prior knowledge for multi-omics data analysis. *IEEE Journal of Biomedical and Health Informatics* 27 (9): 4591–4600.

Yang, Y., Saand, M.A., Huang, L. et al. (2021). Applications of multi-omics technologies for crop improvement. *Frontiers in Plant Science* 12: 563953.

Zampieri, G. (2018). Prioritisation of candidate disease genes via multi-omics data integration.

Zenda, T., Liu, S., Dong, A. et al. (2021). Omics-facilitated crop improvement for climate resilience and superior nutritive value. *Frontiers in Plant Science* 12: 774994.

Zhang, X., Zhang J, Sun K et al. (2019). Integrated multi-omics analysis using variational autoencoders: application to pan-cancer classification, In *2019 IEEE International Conference on Bioinformatics and Biomedicine (BIBM)*, (18–21 November 2019), San Diego, CA, USA. IEEE, pp. 765–769.

Zhang, Z., Park, C.Y., Theesfeld, C.L., and Troyanskaya, O.G. (2021). An automated framework for efficiently designing deep convolutional neural networks in genomics. *Nature Machine Intelligence* 3 (5): 392–400.

Zhang, R., Zhang, C., Yu, C. et al. (2022). Integration of multi-omics technologies for crop improvement: status and prospects. *Frontiers in Bioinformatics* 2: 1027457.

Zhang, J., Ren, H., Jiang, Z. et al. (2024). Strategic multi-omics data integration via multi-level feature contrasting and matching. *IEEE Transactions on NanoBioscience* 23: 579–590.

Zheng, W. and Jin, M. (2019). FTA: a novel feature training approach for classification, In *Proceedings of the 33rd Pacific Asia Conference on Language, Information and Computation*, (13–15 September 2019), Tokyo, Japan. Waseda University, pp. 316–323.

Zhou, J. and Troyanskaya, O.G. (2015). Predicting effects of noncoding variants with deep learning–based sequence model. *Nature Methods* 12 (10): 931–934.

Zhou, S., Ma, Y., Shang, Y. et al. (2022a). Functional diversity and metabolic engineering of plant-specialized metabolites. *Life Metabolism* 1 (2): 109–121.

Zhou, Y., Jia, E., Shi, H. et al. (2022b). Prediction of time-series transcriptomic gene expression based on long short-term memory with empirical mode decomposition. *International Journal of Molecular Sciences* 23 (14): 7532.

8

Artificial Intelligence and Synthetic Biology for Crop Breeding

Arukshita Chandra[1], Sakshi Singh[1], Divya Mohanty[1], Charu Sharma[2], Shrishti[2], Papiya Mukherjee[2], and Nupur Mondal[3]

[1] *University of Delhi, Department of Botany, Hindu College, Sudhir Bose Marg, Delhi, 110007, India*
[2] *Panjab University, Department of Botany, Sector 14, Chandigarh, 160014, India*
[3] *University of Delhi, Department of Botany, Atma Ram Sanatan Dharma College, Dhaula Kuan, Delhi, 110021, India*

8.1 Introduction

The artificial intelligence (AI) revolution is reforming every aspect of our lives, from work to living. It is changing industries, boosting innovation, and opening up new avenues. In sectors like agriculture, healthcare, and education, AI is increasing efficiency and is entirely redefining how things get done. This is not just a technology shift; it is a paradigm shift in which creativity, intelligence, and technology combine to tackle both emerging and conventional challenges. In agriculture, AI is increasing productivity, sustainability, and precision. Using machine learning (ML), data analytics, and robotics, AI is helping solve some of the greatest agricultural challenges. It unlocks the potential for the future, reforming everything with each new algorithm and ML model.

AI, especially tools like ML and deep learning (DL), is playing a bigger and more exciting role in areas like genome research and agriculture. These technologies are helping farmers grow more food, use resources more wisely, and make agriculture sustainable (Wójcik-Gront et al. 2024). When AI is combined with synthetic biology (SynBio), we see major advances in plant genetics. This is extremely important for creating stronger, disease-resistant crops. One standout technique is SynBio, which has the ability to precisely edit genes. CRISPR/Cas9, a powerful gene editing tool that can add, remove, or interchange specific genes (Chen et al. 2019). When AI works together with these tools, it enhances biofortification, where crops are improved to contain more essential nutrients that people need for good health (Zhu et al. 2020). This not only helps fight malnutrition but also makes our food safer and more nutritious (Sticklen 2015).

As modern breeding programs collect more and more data, AI and ML have become valuable tools for plant breeders. These technologies help sort through huge amounts of genetic, physical, and environmental information, making the work faster and decisions more accurate (Cembrowska-Lech et al. 2023). AI helps in phenomic selection by analyzing genetic data to identify markers linked to valuable traits in crops. This process, known as marker-assisted breeding, helps breeders pick the best plants with desirable genetic traits. AI can handle complex data and uncover patterns that traditional methods might miss (Sood et al. 2021; Kumar 2023). When combined with genomic selection tools, AI also improves our ability to predict how plants will perform in different environments, a major advantage in the face of climate change and shifting growing conditions (Jeon et al. 2023).

AI is also making a big transformation in agriculture in the form of high-throughput phenotyping, which is a quick and accurate measurement of different traits in plants. With the arrival of AI-powered tools like imaging drones and satellite remote sensors, farmers and researchers can now easily track parameters like plant height, leaf size, and biomass (Santangeli et al. 2020). This technological advancement is helping to make farming more precise, efficient, and environmentally friendly.

In this chapter, we'll explore how AI is changing the face of agriculture, from boosting crop yields to making farming more sustainable and efficient. We'll look at how tools like ML, SynBio, and advanced genetics are helping farmers grow smarter and more resilient crops. By exploring these innovations, we'll see how AI is assisting in tackling some of the major challenges in farming today and shaping the future of food production.

8.2 History

Global climate change, combined with the COVID-19 pandemic, has hit the agricultural sector badly, causing a noticeable rise in hunger levels by 2020. Hence, it has become more important than ever to explore new plant breeding technologies that can boost genetic diversity and help find new genes to make crops more resistant (Rai 2022). The agriculture, fishery, and forestry industries have greatly benefited from developments in genetic research. In fact, nearly half of the 1–3% yearly growth in crop and livestock productivity is credited to advances in genetic engineering (Goold et al. 2018).

SynBio is a relatively new and fast-growing field (Figure 8.1) that really started gaining momentum in the early 2000s (Dhar and Srinivas 2021). Although the term was first used by Dr. Barbara Hobom, it became widely recognized after the first official SynBio Meeting held at MIT in 2004 (Das and Dhar 2021). SynBio is sometimes also called biological engineering, since both use engineering approaches to redesign genes and organisms (Das and Dhar 2021). Over the years, this field has had a huge impact on agriculture around the world. For instance, advances in genetic engineering have led to a significant increase in global production of maize and soybean between the years 1996 and 2013, with over 195 million tons of maize and 110 million tons of soybean being produced (Goold et al. 2018).

AI is playing an important role in improving agriculture. By using ML and data analysis, AI is helping farmers and scientists process large amounts of information quickly and efficiently (Malik and Pradhan 2024). These technologies are speeding up crop breeding programs, making it easier to develop high-yield, climate-resilient plants, detect diseases early, and selective breeding (Pandey et al. 2024). A technique called speed breeding, driven by next-generation breeding technologies using AI, is now being used to fasten crop improvement by shortening the time it takes to develop better plant varieties (Rai 2022). In addition, next-generation sequencing (NGS) is helping improve AI tools for agricultural applications (Shendure et al. 2017). The combination of AI and SynBio for crop breeding

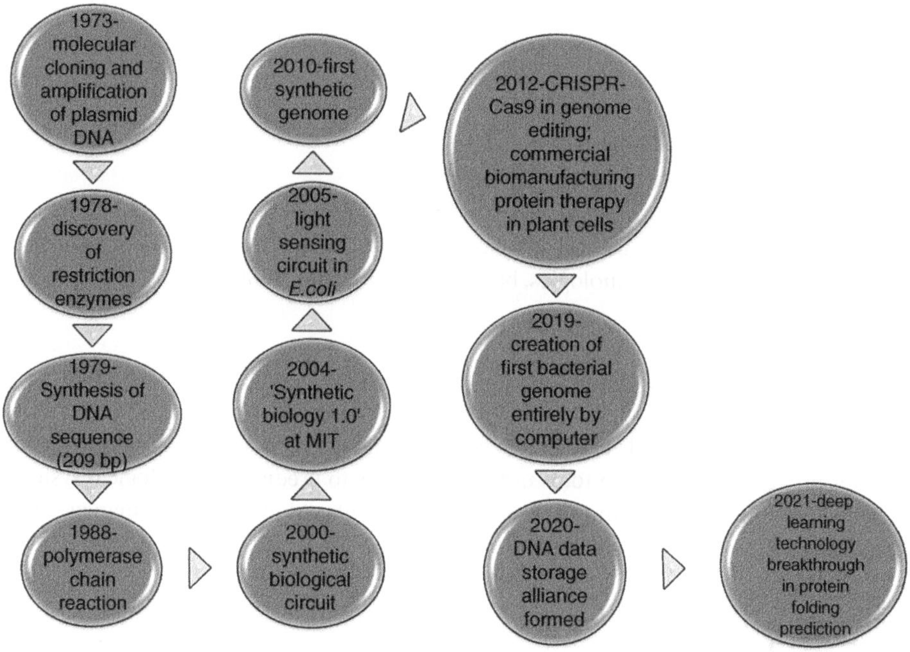

Figure 8.1 A brief timeline of synthetic biology.

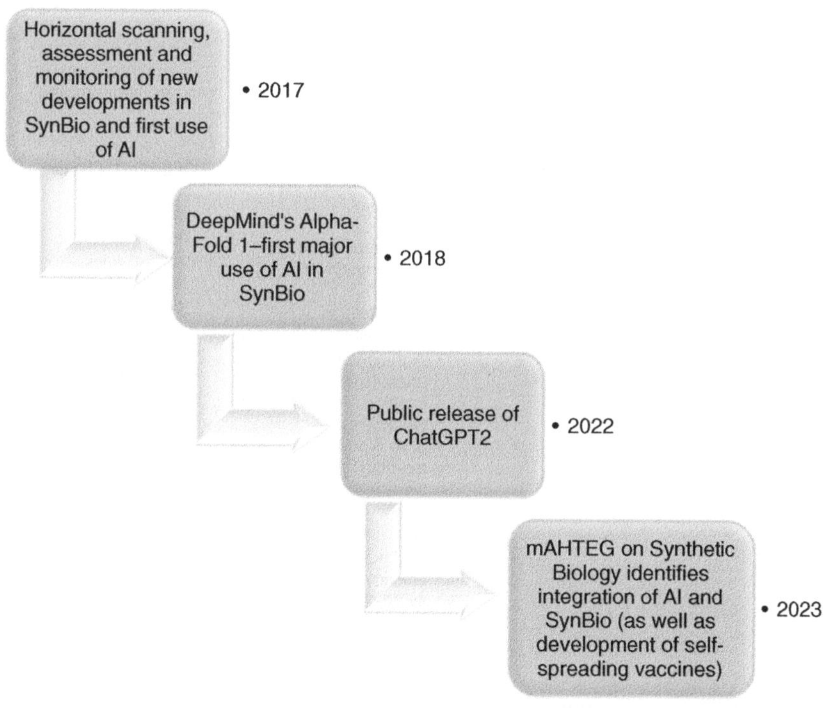

Figure 8.2 An integrated timeline of synthetic biology and AI after the emergence of AI.

(Figure 8.2) has led to exciting new innovations in agricultural biotechnology (Malik and Pradhan 2024). These fields together have significantly boosted the agricultural economy, benefiting markets both nationally and globally.

8.3 Methods, Models, and Algorithms

Major tools and technologies used in SynBio can be divided into four categories, Figure 8.3 (Meige et al. 2024): design technologies, building technologies, analysis tools, and enabling technologies.

8.3.1 Design Technologies

Gene and protein design tools help create genetic sequences and protein structures using computer programs. Simulation tools use math models to predict how biological systems will behave, saving time compared to wet-lab experiments. AI and ML help manage large data sets and uncover hidden patterns (Meige et al. 2024).

8.3.2 Build Technologies

This involves creating custom DNA sequences, gene delivery systems, genome editing tools, and fermentation technologies. New DNA can be sequenced using NGS to confirm

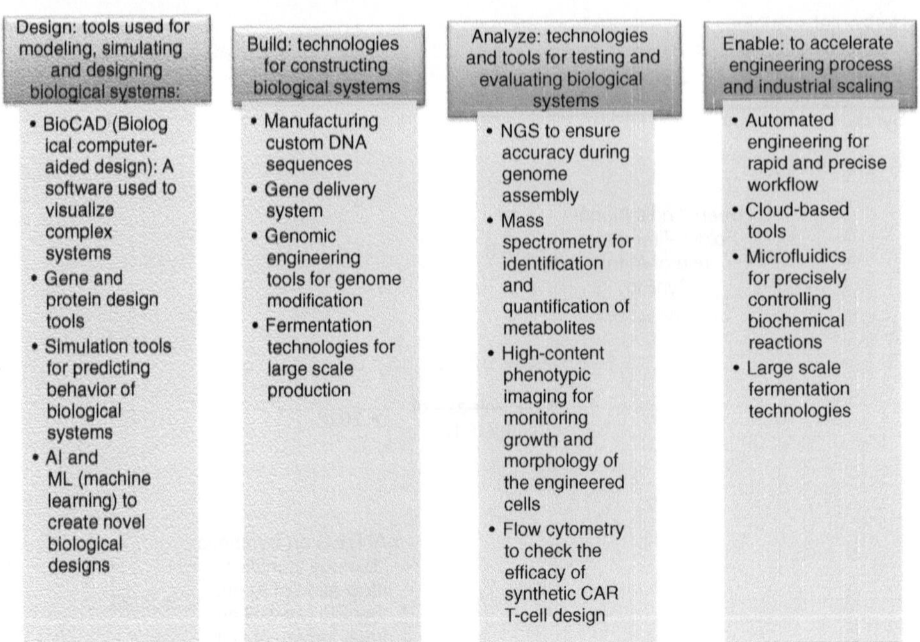

Figure 8.3 Workflow methods of synthetic biology.

its accuracy. The DNA can then be used to make proteins that perform specific functions. Efficient DNA delivery is crucial for SynBio, and methods like viral vectors (viruses used to carry DNA), liposomes, and physical techniques can help. Genome editing tools, like CRISPR-Cas, can change the genetic code, helping with gene regulation. Fermentation, an old technique, is now used to produce mRNA or cells on a large scale (Meige et al. 2024).

8.3.3 Analysis Tools

NGS is a fast technology used to analyze genetic material. PCR is a quick, precise method to measure target genes. Mass spectrometry helps identify and measure proteins and other substances. Advanced imaging techniques are used to monitor engineered cells (Meige et al. 2024).

8.3.4 Enable Technologies

This includes automation, cloud labs, microfluidics, and large-scale fermentation. Automation and robotics allow SynBio processes to be done faster and more consistently with less human involvement. Cloud labs are centralized, automated spaces for conducting experiments remotely. Microfluidics gives precise control over chemical reactions, overcoming the limitations of traditional methods. Large-scale fermentation uses bioreactors to produce bioengineered products cost-effectively (Meige et al. 2024).

8.4 Synthetic Genomics

Synthetic genomics involves designing and creating entire genomes. Genome minimization helps generate simpler organisms by identifying and removing unnecessary genetic elements, making them easier to study. This process can reduce the metabolic load and direct resources to more beneficial pathways, such as in plants (Goold et al. 2018). Genetic recording involves removing specific codons from a genome, which can then be used in other organisms to produce novel amino acids. For instance, replacing 1557 leucine codons in 176 genes of *Salmonella typhimurium* had no effect on cell growth (Goold et al. 2018). Genome editing, using technologies like CRISPR/Cas9, ZFN, and TALEN, allows precise genetic modifications, overcoming the limitations of traditional breeding methods. The CRISPR/Cas9 system uses sequence-specific nucleases and guide RNA, and tools like WheatCRISPR help design specific gRNA for wheat (Naik et al. 2024). CRISPR can also be used for treating various diseases (Bera 2019; Fernandez 2024).

8.5 Artificial Intelligence (AI)

AI is a key part of SynBio because it helps handle large amounts of data, making SynBio even more powerful when combined with engineering and biology (Fernandez 2024). AI uses technologies like ML and DL to analyze data, predict outcomes, and process information (Thomas 2024). For example, ChatGPT is a popular AI tool that generates text, images, and more, based on a large amount of internet data (Thomas 2024). AI has three main skills: learning, reasoning, and self-correction. It helps with analyzing plant traits, identifying genes that control plant growth, studying climate change effects on crops, and finding the best gene combinations to increase crop yields. ML also helps identify genomic regions by studying features like DNA methylation. One interesting area in plant science is digital plant phenotyping, which uses imaging systems to study plants (Naik et al. 2024). DL, a part of ML, uses neural networks, and models like XGBoost are already used in agriculture, like predicting maize yield (Nyeki et al. 2019; Naik et al. 2024). ML methods can be divided into supervised and unsupervised models, linear and non-linear algorithms, and shallow and DL models (Niazian et al. 2020).

8.5.1 AI and SynBio Working Together

AI is transforming SynBio by improving the precision of gene editing tools like CRISPR/Cas9. It helps analyze large data sets to predict which genes should be modified and suggests the best experiments. AI also helps predict protein structures and functions, which can be used in treatments and enzyme production. It can identify efficient metabolic pathways and improve their outputs, strengthening SynBio's role in solving challenges in fields like medicine and agriculture (Arif et al. 2024).

8.5.2 AI in Genome Data Analysis

AI, especially ML and DL, is becoming essential for genome data analysis and in agriculture to boost crop production, sustainability, and resource management (Wójcik-Gront et al. 2024). By combining AI with SynBio, researchers have made big strides in understanding plant genetics, which helps create resilient crop varieties and better disease management. Techniques like CRISPR/Cas9 allow for precise genetic modifications, and synthetic promoters help control gene expression, improving desired traits (Chen et al. 2019; Yasmeen et al. 2023). AI and SynBio also help in biofortification, where crops are designed to have more nutrients, addressing global nutrition challenges and improving food security (Zhu et al. 2020). AI and computational tools are also crucial for designing genetic systems and modeling biological processes, as follows (Sticklen 2015).

8.5.2.1 Predictive Analytics and Decision Support

AI, particularly ML algorithms, is being used to analyze large datasets from breeding programs, helping make better decisions about plant breeding. By using historical data, predictive analytics can identify the best genotypes, improving selection efficiency (Bhagat et al. 2022; Cembrowska-Lech et al. 2023; Rasool and Ammar 2023).

8.5.2.2 Genomic Selection and Marker-Assisted Breeding

AI is helping with genomic selection by analyzing genomic data to find genetic markers linked to important traits. This allows breeders to select plants with the best genetic profiles more efficiently. AI can identify patterns in genomic data that traditional methods might miss (Sood et al. 2021; Kumar 2023). Combining AI with genomic tools has also improved the accuracy of predicting plant performance under different conditions (Jeon et al. 2023).

Table 8.1 Summary of widely used databases in plant genetics and breeding research.

Database	Key features
AtMAD	High-quality multi-omics data of *Arabidopsis thaliana*
GoMapMan	Gene functional annotations in plant sciences
HapRice	SNP-haplotype database for rice
NPACT	Natural compounds with anti-cancerous activity
PGDD	Gene and genome duplication in plants
Phytozome	Plant comparative genomics
PlantGDB	Plant genome browsers
PMRD	Plant microRNA database
PLUTO	Information on plant varieties

The Genomics-FM model helps researchers analyze functional genomics more effectively, even with limited data (Kiranmai and Gurbani 2023).

8.5.2.3 Phenotyping and Trait Analysis

AI is playing a big role in high-throughput phenotyping, which involves measuring plant traits quickly. AI-powered imaging, such as drones and remote sensing, collects detailed data on plant features like height and leaf area (Santangeli et al. 2020). Platforms like GB3.0 link DNA elements to biological data, aiding in the design of new crop traits (Vázquez-Vilar et al. 2017).

8.5.2.4 Data Integration and Management

As genomic data grows, it's becoming more complex to manage. AI helps with organizing and analyzing this data. For example, the Global Pharmacopoeia Genome Database (GPGD) helps store and manage medicinal plant genomes, addressing data storage and integration challenges (Liao et al. 2021; Meng et al. 2022). Table 8.1 enlists several more databases that are widely used for plant genetics and breeding research.

8.6 Applications

SynBio, CRISPR, and AI are revolutionizing agriculture by accelerating breeding cycles and enhancing crop traits. For instance, genetically engineered bacteria by companies like Pivot Bio fix nitrogen in corn, reducing the need for chemical fertilizers, while SynBio techniques, as mentioned in Table 8.2, come with various benefits that enable the creation of disease-resistant, nutritionally rich crops, and AI supports these advancements by analyzing data for pest detection, irrigation timing, yield prediction, and climate adaptation (Figure 8.4). Together, these technologies have led to innovations such as robot-ready crops that signal AI-powered machinery for targeted interventions and speed breeding techniques that rapidly generate improved varieties in crops like wheat, barley, maize, and

Table 8.2 Synthetic biology applications in crops.

Synthetic biology technique	Crops applied	Outcomes
CRISPR-Cas9	Wheat, rice, tomato	Enhanced tolerance to abiotic stress, improved yield, and resistance to diseases.
RNA Interference (RNAi)	Cotton, maize, wheat	Resistance to pests and diseases, e.g., cotton bollworm and aphids.
TILLING (targeted induced local lesions in genomes)	Wheat, rice, tomato	Identification of novel allelic variations for nutritional and stress-tolerant traits.
Genomic selection	Multiple crop species	Accelerated identification of superior genotypes with desired traits.

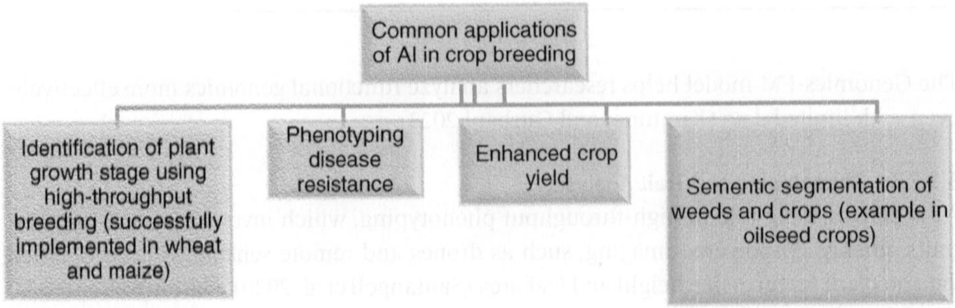

Figure 8.4 Applications of AI in crop breeding.

horticultural species (Table 8.3). AI algorithms have been successfully used for the analysis of gene expression data and forecasting gene alterations, thereby helping in the creation of safe and efficient genetically modified crops. In the field of crop management, predictive analytics, and imaging technologies help in the evaluation of environmental and crop data. It can forecast the best irrigation time and make weather predictions and water requirements of the crop by using data to analyze soil moisture, etc. AI improves the effectiveness of CRISPR edits and enhances the accuracy of gene alterations (Malik and Pradhan 2024). Robot-ready cropsare genetically engineered plants that work seamlessly with AI-powered farm machinery. These crops are modified to provide real-time signals or feedback that machines can detect and respond to, and this allows easy management of watering or pest control.

Many successful applications of AI and SynBio in crop breeding have been documented. For example, the use of speed breeding techniques combined with AI has significantly reduced the time required to develop new crop varieties. For instance, the use of genomic prediction models has significantly accelerated the breeding process, allowing for the selection of elite genotypes with superior traits (Zhang et al. 2022). This approach has been instrumental in enhancing maize productivity, particularly in regions facing climate variability (Bhagat et al. 2022). The development of drought-tolerant varieties of crops such as winter wheat has been facilitated by AI-driven phenotyping methods that accurately

Table 8.3 AI applications in cropbreeding.

Crop Name	AI/Machine Learning Technique Used	Traits Studied	Prediction Carried Out
Glycine max (Soybean)	Neural networks, random forest	Evaluation of yield traits	Identification of high-yielding soybean varieties.
Zea mays (Maize)	Neural networks, CNNs	Stress tolerance and productivity traits	Selection of optimal parental lines for breeding.
Oryza sativa (Rice)	Deep CNNs	Resistance to pests and diseases	Assessment of biotic stress tolerance levels.
Brassica rapa	Artificial neural networks	Seed production traits	Prediction of seed development efficiency.
Solanum tuberosum	Artificial neural networks	Key agronomic characteristics	Determination of superior-performing genotypes.
Triticum aestivum (Wheat)	ANN, Support Vector Machines (SVMs)	Salinity and drought stress tolerance	Identification of stress-tolerant wheat cultivars.
Daucus carota (Carrot)	Random forest regression	Yield traits using spectral and satellite data	Prediction of carrot yield with high accuracy.
Camellia sinensis (Tea Plant)	Least Squares-SVM (LS-SVM)	Nitrogen nutrition status	Classification of plants based on nitrogen levels.
Trachyspermum ammi	ANN, Multiple Linear Regression (MLR)	Seed yield and essential oil content	Prediction of optimal conditions for higher yield.
Arabidopsis thaliana	Decision Tree, SVM, Naïve Bayes	miRNA expression under abiotic stress	Prediction of plant responses to drought, salinity, and heat.

assess root system architecture and other critical traits (Chang-Brahim et al. 2024). The success stories extend to horticultural crops, where genetic engineering techniques, including CRISPR-Cas9, have been employed to enhance traits such as disease resistance and abiotic stress tolerance (Bhavanee et al. 2024).

8.7 Global and National Progress

Globally, the integration of AI in crop breeding is gaining momentum, with various countries investing in research and development to harness these technologies. The Food and Agriculture Organization (FAO) emphasizes the importance of innovative breeding strategies to address food security challenges (Ntsomboh-Ntsefong et al. 2023). The United States, for instance, has seen significant advancements in regulatory frameworks that support the use of genome editing in crop improvement (Pixley et al. 2019). In contrast, other regions may face hurdles such as the need for substantial computational resources, the complexity of data interpretation, and the integration of AI tools into existing breeding programs that slow down the adoption of these innovative techniques (Nales 2023).

Nationally, initiatives aimed at promoting sustainable agriculture through SynBio and AI are being implemented. For example, research institutions and universities are collaborating with agricultural industries to develop new crop varieties that are resilient to climate change and pests (Sargent et al. 2022). At the national level, regulatory frameworks surrounding the use of AI and genome editing in agriculture vary widely. The need for biosafety regulations is paramount to ensure that genetically modified organisms (GMOs) and genome-edited crops are safe for human consumption and the environment. However, the adoption of AI in agriculture is not without challenges.

8.8 Challenges

The use of AI and SynBio in agriculture faces several challenges that must be addressed for successful commercial deployment (Figure 8.5). One of the main hurdles is data complexity and quality, as AI-driven crop improvement relies heavily on accurate and comprehensive datasets (Kumar 2023). Agricultural data often suffers from missing information and biases, affecting the accuracy of AI predictions. Additionally, the cost of developing and commercializing AI and SynBio technologies is high, with significant investments required for resources, infrastructure, and skilled researchers (Ibrahim 2023). Ethical concerns also arise, with critics arguing that genetic modifications could disrupt the natural order and lead to unforeseen consequences (Roberts and Smith 2022). Despite the precision of AI and SynBio, biological and ecological complexities still pose risks, as

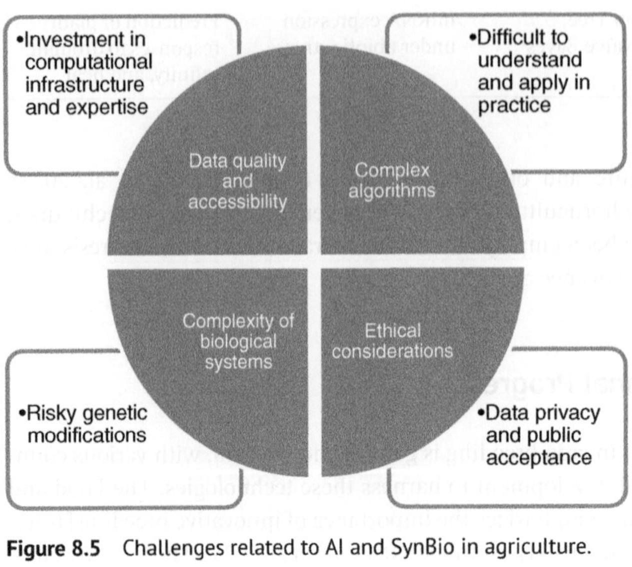

Figure 8.5 Challenges related to AI and SynBio in agriculture.

unintended effects on crop growth and environmental interactions may occur (Pixley et al. 2019;Sargent et al.2022). Public perception and acceptance of these technologies, shaped by the history of GMOs, may also hinder their adoption unless trust-building efforts are made through effective communication (Mohr and Kühl 2021). Moreover, biosafety and biosecurity concerns are central, as the deployment of genetically engineered organisms could lead to environmental or health risks, prompting the need for strong regulatory frameworks (Gómez-Tatay and Hernández-Andreu 2019; Zeng et al. 2022). Biosafety regulations like the Cartagena Protocol and national frameworks, such as those by the EPA, USDA, and FDA in the U.S., are essential to ensuring safe usage, alongside the more stringent European Union regulations (Sasani 2024).

8.9 Future Prospects

Leveraging advanced technologies for enhancing crop yield and productivity is one of the major goals of contemporary plant breeding (Farooq et al. 2024). AI is widely used in precision breeding methods. AI-driven genomics and phenomics tools are playing a significant role in selecting and breeding plants with desired traits more efficiently in comparison to traditional breeding methods. This can be helpful in the creation of high-yielding varieties resistant to climate change and diseases (Malik and Pradhan 2024;Toledo et al. 2024). Breeders are contemplating Next-Gen AI for comprehensive analysis of complex traits under variable environmental conditions (Rai 2022). AI can be used to trace artificial routes to generate biofuels, which in turn could reduce the use of fossil fuels. AI-driven drones are also being used for monitoring crops. This convergence could be extremely helpful for farmers, as it could accurately convey necessary information about the health of their crops, pests, soil health, etc. (Malik and Pradhan 2024). Advancements in AI-driven CRISPR technology are paving the way for crops that are resistant, productive, and rich in nutrients. These innovations are promising for improving global food security by introducing new and better crop varieties (Jennifer 2024). AI with SynBio could give a boost to the agricultural industry and can significantly aid sustainable farming by optimizing resource utilization, increasing productivity, and creating climatic change-resistant crops (Malik and Pradhan 2024). AI models and algorithms have the potential of comparative genomics and tracing the pathway of gene transfer from a model organism to a crop (Khan et al. 2022). It possesses the capability of screening genomes using several models and algorithms, analyzing the data, and using it for genetic improvements (Khan et al. 2022). DL models, imaging techniques algorithms, etc. play a significant role in automating the phenotypic analysis, thus improving the efficiency of crop breeding (Mushtaq et al. 2024). AI is an emerging field that harbors great potential in the agricultural sector. Its use has just begun and will continue as it matures due to the addition of new algorithms (Naik et al. 2024). Developing AI machines that incorporate moral and ethical considerations in breeding programs is necessary to ensure sustainable and socially responsible agricultural practices (Khan et al. 2022). This requires collaborative efforts of plant scientists, agronomists, regulatory bodies, AI, and SynBio researchers. Furthermore, hybrid intelligence, which is the combination of AI algorithms and human intelligence, can support such advancements in the field of agriculture (Mushtaq et al. 2024).

8.10 Conclusion

The realm of crop breeding has been revolutionized by the convergence of two groundbreaking technologies: AI and SynBio. As the global population continues to surge, reaching an estimated 9.7 billion by 2050, the need for innovative solutions to enhance agricultural productivity and sustainability has never been more pressing (Naresh et al. 2020). The integration of AI and SynBio holds immense potential in addressing the multifaceted challenges facing the agricultural sector (Arif et al. 2024). The methodologies and algorithms employed in this field are diverse, ranging from DL for predictive crop modelling to CRISPR-based gene editing for targeted trait enhancement (Malik and Pradhan 2024). The application of these technologies has resulted in the successful development of various crops with improved characteristics, such as disease-resistant maize, drought-tolerant wheat, and nutrient-fortified rice (Ghosh et al. 2018; Bhagat et al. 2022; Bhavanee et al. 2024; Chang-Brahim et al. 2024). As research continues to evolve, the future of AI and SynBio in crop improvement looks promising, with potential breakthroughs in climate-resilient and sustainable agricultural practices.

References

Arif, A., Garg, P., and Srivastava, P. (2024). Unlocking the transformative power of synthetic biology. *Archives of Biotechnology and Biomedicine* 8: 9–16.

Bera, R.K. (2019). *Synthetic Biology, Artificial Intelligence, and Quantum Computing (Synthetic Biology-New Interdisciplinary Science)*. IntechOpen.

Bhagat, P., Naz, F., and Magda, R. (2022). Artificial intelligence solutions enabling sustainable agriculture: a bibliometric analysis. *PLoS One* 17 (6): e0268989. https://doi.org/10.1371/journal.pone.0268989.

Bhavanee, K.D., Krishnamoorthi, A., Rathva, H.M. et al. (2024). Advancements in genetic engineering for enhanced traits in horticulture crops: a comprehensive review. *Journal of Advances in Biology & Biotechnology* 27: https://doi.org/10.9734/jabb/2024/v27i2702.

Cembrowska-Lech, D., Krzemińska, A., Miller, T. et al. (2023). An integrated multi-omics and artificial intelligence framework for advance plant phenotyping in horticulture. *Biology* 12: https://doi.org/10.3390/biology12101298.

Chang-Brahim, I., Koppensteiner, L.J., Beltrame, L. et al. (2024). Reviewing the essential roles of remote phenotyping, GWAS and explainable AI in practical marker-assisted selection for drought-tolerant winter wheat breeding. *Frontiers in Plant Science* 15: https://doi.org/10.3389/fpls.2024.1319938.

Chen, K., Wang, Y., Zhang, R. et al. (2019). Crispr/cas genome editing and precision plant breeding in agriculture. *Annual Review of Plant Biology* 70 (1): 667–697. https://doi.org/10.1146/annurev-arplant-050718-100049.

Das, M. and Dhar, P.K. (2021). What is synthetic biology? *Asian Biotechnology & Development Review* 23 (3): 1–12. https://openurl.ebsco.com/contentitem/gcd%3A156045932.

Dhar, P.K. and Srinivas, K.R. (2021). What is synthetic biology? *Asian Biotechnology and Development Review* 23 (3): 1–14. https://ris.org.in/sites/default/files/2022-03/ABDR%20Novermber%202021.pdf.

Farooq, M.A., Gao, S., Hassan, M.A. et al. (2024). Artificial intelligence in plant breeding. *Trends in Genetics* 40 (10): 891–908.

Fernandez, M. (2024). *Artificial Intelligence Powering Synthetic Biology: The Fundamentals*. S&P Global.

Ghosh, S., Watson, A., Gonzalez-Navarro, O.E. et al. (2018). Speed breeding in growth chambers and glasshouses for crop breeding and model plant research. *Nature Protocols* 13: https://doi.org/10.1038/s41596-018-0072-z.

Gómez-Tatay, L. and Hernández-Andreu, J.M. (2019). Biosafety and biosecurity in synthetic biology: a review. *Critical Reviews in Environmental Science and Technology* 49 (17): 1587–1621. https://doi.org/10.1080/10643389.2019.1579628.

Goold, H.D., Wright, P., Hailstones, D. et al. (2018). Emerging opportunities for synthetic biology in agriculture. *Genes (Basel)* 9 (7): 341.

Ibrahim, U. (2023). Artificial intelligence in agricultural extension for sustainable development. *IJASR* 1 (3): 259–268. https://doi.org/10.59890/ijasr.v1i3.740.

Jennifer,D. (2024). Combining AI and crispr will be transformational. WIRED.

Jeon, D., Kang, Y., Lee, S. et al. (2023). Digitalizing breeding in plants: a new trend of next-generation breeding based on genomic prediction. *Frontiers in Plant Science* 14: https://doi.org/10.3389/fpls.2023.1092584.

Khan, M.H., Wang, S., Wang, J. et al. (2022). Applications of artificial intelligence in climate-resilient smart crop breeding. *International Journal of Molecular Sciences* 23 (19): 11156.

Kiranmai, B. and Gurbani, R. (2023). Integrating phylogenetic analysis and machine learning for enhanced phenotype prediction in Oryza sativa. *International Journal of Agriculture and Biosciences* 12 (4): 277–283.

Kumar, J. (2023). Optimizing irrigation and nutrient management in agriculture through artificial intelligence implementation. *International Journal of Environment and Climate Change* 13 (10): 4016–4022. https://doi.org/10.9734/ijecc/2023/v13i103077.

Liao, B., Hu, H., Xiao, S. et al. (2021). Global pharmacopoeia genome database is an integrated and mineable genomic database for traditional medicines derived from eight international pharmacopoeias. *Science China Life Sciences* 65 (4): 809–817. https://doi.org/10.1007/s11427-021-1968-7.

Malik, P. and Pradhan, B.R. (2024). Agricultural biotechnology and AI: pioneering sustainable farming for the future. *ReseaproJournals* 1 (2): 19–22.

Meige,A., Eagar, R., Arthur D. Little et al. (2024). The brave new world of synthetic biology. *Blue Shift Report005: 12-23, 25-36, 46-64.*

Meng, F., Tang, Q., Chu, T. et al. (2022). Tcmpg: an integrative database for traditional Chinese medicine plant genomes. *Horticulture Research* 9: https://doi.org/10.1093/hr/uhac060.

Mohr, S. and Kühl, R. (2021). Acceptance of artificial intelligence in german agriculture: an application of the technology acceptance model and the theory of planned behavior. *Precision Agriculture* 22 (6): 1816–1844. https://doi.org/10.1007/s11119-021-09814-x.

Mushtaq, M.A., Ahmed, H.G., Zeng, Y. et al. (2024). Applications of artificial intelligence in wheat breeding for sustainable food security. *Sustainability* 16: 1–24.

Naik, Y.D., Zhao, C., Channale, S. et al. (2024). *Frontier Technologies for Crop Improvement*, vol. 37–54, 55–57. Springer.

Nales, P. (2023). Breeding by intervening: exploring the role of associations and deliberation in consumer acceptance of different breeding techniques. *Public Understanding of Science* 32 (7): 889–906. https://doi.org/10.1177/09636625231168087.

Naresh, R.K., Chandra, M.S., Charankumar, G.R. et al. (2020). The prospect of artificial intelligence (AI) in precision agriculture for farming systems productivity in sub-tropical India: a review. *Current Journal of Applied Science and Technology* 39 (48): 96–110.

Niazian, M., Kiani, S., and Shafiee-Jood, M. (2020). Machine learning for plant breeding and biotechnology. *Agronomy* 10 (10): 436. https://doi.org/10.3390/agronomy10100436.

Ntsomboh-Ntsefong, G., Toukam, G.M., and Likeng-Li-Ngue, B.C. (2023). A review of population genetics and the prospects for crop improvement. *Academia Biology*. https://doi.org/10.20935/acadbiol6117.

Nyeki, A., Kerepesi, C., Daróczy, B. et al. (2019). Maize yield prediction based on artificial intelligence using spatio-temporal data. In: *Precision Agriculture'19* (ed. J.V. Stafford), 1011–1017. Wageningen Academic.

Pandey, M.K., Bentley, A., Desmae, H. et al. (2024). *Frontier Technologies for Crop Improvement*, 50–52. Springer.

Pixley, K., Falck-Zepeda, J., Giller, K. et al. (2019). Genome editing, gene drives, and synthetic biology: will they contribute to disease-resistant crops, and who will benefit? *Annual Review of Phytopathology* 57 (1): 165–188. https://doi.org/10.1146/annurev-phyto-080417-045954.

Rai, K.K. (2022). Integrating speed breeding with artificial intelligence for developing climate-smart crops. *Springer Nature* 49 (12): 1–63.

Rasool, A. and Ammar, A. (2023). The dawn of AI in agriculture: From predictive analysis to autonomous farming. *Trends in Agricultural Practices and Sustainability* 23 (1): 73–83. https://www.trendsaps.com/articles/1-1-73-83-TAPS-23-021.pdf.

Roberts, L. and Smith, E. (2022). Ethical considerations in AI use for agricultural extension: a review of key issues. *Journal of Agricultural Ethics* 34 (4): 409–426.

Santangeli, A., Chen, Y., Kluen, E. et al. (2020). Integrating drone-borne thermal imaging with artificial intelligence to locate bird nests on agricultural land. *Scientific Reports* 10 (1): https://doi.org/10.1038/s41598-020-67898-3.

Sargent, D., Conaty, W., Tissue, D., and Sharwood, R. (2022). Synthetic biology and opportunities within agricultural crops. *Journal of Sustainable Agriculture and Environment* 1 (2): 89–107. https://doi.org/10.1002/sae2.12014.

Sasani, M.S. (2024). The importance of biosecurity in emerging biotechnologies and synthetic biology. *Avicenna Journal of Medical Biotechnology* 16 (4): 223.

Shendure, J., Balasubramanian, S., Church, G.M. et al. (2017). DNA sequencing at 40: past, present and future. *Nature* 550: 345–353.

Sood, A., Sharma, R., and Bhardwaj, A. (2021). Artificial intelligence research in agriculture: a review. *Online Information Review* 46 (6): 1054–1075. https://doi.org/10.1108/oir-10-2020-0448.

Sticklen, M. (2015). Advances in crop synthetic biology: a platform that bridges the expertise of crop molecular biologists, engineers, software developers and mathematical modelers. *Advances in Crop Science and Technology* 03 (3): https://doi.org/10.4172/2329-8863.1000171.

Thomas, J. (2024). Black box biotech: integration of artificial intelligence with synthetic biology: addressing the risks, hype and inequities underpinning generative biology. *African Centre for Biodiversity (ACB)* 7-13: 14–22.

Toledo, C.A., Crawford, M.M., and Tuinstra, M.R. (2024). Integrating multi-modal remote sensing, deep learning, and attention mechanisms for yield prediction in plant breeding experiments. *Frontiers in Plant Science* 15: 1408047.

Vázquez-Vilar, M., Quijano-Rubio, A., Fernández-del-Carmen, A. et al. (2017). GB3.0: a platform for plant bio-design that connects functional DNA elements with associated biological data. *Nucleic Acids Research* 45 (4): 2196–2209. https://doi.org/10.1093/nar/gkw1326.

Wójcik-Gront, E., Bartłomiej, Z., and Pawełkowicz, M. (2024). Harnessing AI-powered genomic research for sustainable crop improvement. *Agriculture* 14 (12): 2299. *ProQuest*. https://www.proquest.com/scholarly-journals/harnessing-ai-powered-genomic-research/docview/3149495959/se-2, https://doi.org/10.3390/agriculture14122299.

Yasmeen, E., Wang, J., Riaz, M. et al. (2023). Designing artificial synthetic promoters for accurate, smart, and versatile gene expression in plants. *Plant Communications* 4 (4): 100558. https://doi.org/10.1016/j.xplc.2023.100558.

Zeng, X., Jiang, H., Yang, G. et al. (2022). Regulation and management of the biosecurity for synthetic biology. *Synthetic and Systems Biotechnology* 7 (2): 784–790.

Zhang, Q., Zhang, Q., and Jensen, J. (2022). Association studies and genomic prediction for genetic improvements in agriculture. *Frontiers in Plant Science* 13: https://doi.org/10.3389/fpls.2022.904230.

Zhu, Q., Wang, B., Tan, J. et al. (2020). Plant synthetic metabolic engineering for enhancing crop nutritional quality. *Plant Communications* 1 (1): 100017. https://doi.org/10.1016/j.xplc.2019.100017.

9

Hub Gene Prediction by Machine Learning for Regulating Plant Stress Responses

Perumalla Srikanth[1], Ann Maxton[2], and Sam A. Masih[3]

[1]*Mohan Babu University, Agricultural Microbiology, School of Agriculture, Sree Sainath Nagar, Tirupati, Andhra Pradesh, 517102, India*
[2]*Sam Higginbottom University of Agriculture, Technology and Sciences, Genetics and Plant Breeding, Naini, Prayagraj, 211007, India*
[3]*Department of Molecular and Cellular Engineering, Sam Higginbottom University of Agriculture, Technology and Sciences, Prayagraj, 211007, India*

9.1 Introduction

In natural settings, a variety of biotic and abiotic stressors and their many combinations result in a unique stress response. This implies that our knowledge of plant stress response would be greatly expanded by the identification and characterization of important genes and their co-expression partners, which exhibit an expression profile that distinguishes between biotic and abiotic stress responses. Additionally, these genes would serve as targets for genetic modification to enhance plant stress tolerance.

Finding, contrasting, and comparing the stress-specific gene expression profile of one stress condition with other stresses is made possible by the availability of numerous genome-wide transcriptome data sets for the same stress condition (Singh et al. 2021a, b). A strong statistical foundation for reevaluating initial findings, increasing sensitivity with larger sample sizes, and testing novel hypotheses is provided by meta-analyses that aggregate related studies. To increase statistical robustness and identify weak signals, meta-analysis of microarray studies is frequently employed, particularly in clinical research (Liu et al., 2013).

Determining gene function becomes essential when reviewing this subject. According to Gerstein et al. (2007), a "gene" was formerly thought to be a site where sequence changes resulted in the inactivation of a desired feature, linking function to an assayable phenotype. On the other hand, a genomic region was considered to have a function in the more current ENCODE project if it "participated in at least one biochemical RNA and/or chromatin-associated event in at least one cell type." The authors' declaration that 80.4% of the human genome is functional was the result of their loose definition of function, which drew harsh criticism. However, thanks to programs like ENCODE and ever-growing genomics data, we now have a better understanding of the signatures linked to known functional sequences, which machine learning (ML) techniques used to identify genes and define their roles (Graur et al. 2013).

ML has been effectively applied in numerous biological domains in recent years. In addition to making few assumptions on the underlying probability distributions and generation techniques of the data, ML-based algorithms are effective enough to handle large data sets with significant levels of noise, dimensionality, and/or incompleteness. Although the line between ML and statistics is rather hazy in practice, the main goal of ML techniques is prediction, as opposed to the inferential goal of conventional statistical approaches. Generally speaking, there are two categories of ML algorithms: supervised and unsupervised. Often used for binary/multi-class classification of test instances or for numerical prediction of the trait values (regression), supervised algorithms like random forests (RFs), support vector machines (SVMs), and k-nearest neighbors (kNNs) require explicit label definitions, whereas unsupervised techniques like principal components analysis, k-means clustering, and self-organizing maps are label-free and mainly used for clustering and feature extraction. ML algorithms can be further divided into two categories based on the algorithm development process: feature-based and artificial neural network (ANN)-based. ANN techniques like convolutional neural networks (CNNs) and recurrent neural networks (RNNs) may extract features from the training data on their own, while feature-based techniques like RF and SVM require the explicit specification of distinct features. Moreover, ANN algorithms usually need hundreds to millions of input data points for accurate model construction, whereas feature-based approaches may only need a few dozen to a few hundred input training data points. ANN techniques like CNNs and RNNs may extract features from the training data on their own, while feature-based techniques like RF and SVM require the explicit specification of distinct features. Moreover, ANN algorithms usually need hundreds to millions of input data points for accurate model construction, whereas feature-based approaches may only need a few dozen to a few hundred input training data points.

The application of ANNs with several layers of neurons is known as deep learning. We mainly concentrate on well-known supervised techniques for gene function prediction. The majority of the solutions that have been presented employ a certain strategy for developing algorithms, which involves creating a model using training data and then cross-validating it. Furthermore, several of these methods additionally use entirely different "test" or "challenge" data to test their models, which the algorithm has not come across throughout the training or validation process (Bzdok et al. 2018).

9.2 Methods of Machine Learning for Gene Regulatory Network Prediction

9.2.1 Statistical Method

Statistics-based methods have been frequently used to predict gene regulatory networks (GRNs) from time-series transcriptome datasets because these datasets provide dynamic information that helps us understand the temporal dynamics of different biological processes. When combined with sparse estimate algorithms, the Autoregressive Exogenous Variables (ARX) model, a type of state-space model, allows us to reconstruct a GRN by

Table 9.1 Different gene prediction algorithms and their functions.

S. No.	Name of the algorithms	Function	References
1	CNN	Identifying cis-regulatory elements	Alipanahi et al. 2015
2	SVM, CNN	Expression of genes	Wang et al. 2018
3	kNN	Localization within cells	Washburn et al. 2019
4	RF	Characteristics	Lloyd et al. 2015
5	RF, logistic regression, decision trees, and naïve Bayes	Pathways of metabolism	Dale et al. 2010

describing time-varying processes seen in time-series records. GRNs were reconstructed using time-series expression datasets from *Escherichia coli, Mycobacterium TB*, and *Mus musculus* using the sparse estimation algorithm fused Lasso (Omranian et al. 2016).

Using a statistical sparse estimation method Group SCAD (Smoothly Clipped Absolute Deviation), a type of L_1 regularization technique, Koda et al. (2017) developed gene–gene temporal interactions for 3621 periodically expressed genes observed in a time-series RNA-Seq dataset based on ARX models. They also predicted GRNs with 2187 genes and 3107 directed edges. In order to understand the physiological response of five tropical Asian rice cultivars to high temperatures and water shortages in agricultural field conditions, an environmental gene regulatory influence network was also inferred from time-series transcriptome (RNA-Seq) and chromatin accessibility (ATAC-seq) datasets using the Inferelator algorithm 2, a type of sparse regression approach (Greenfield et al. 2013; Wilkins et al. 2016). Based on a dynamic Bayesian network technique, de Luis Balaguer et al. (2017) created GENIST3 and used it to predict GRN from time-series transcriptome data of Arabidopsis root stem cells as well as cell-type-specific data. Examples of statistics-based methods for determining GRNs from non-time-series transcriptome datasets are also available (Table 9.1). Using Context Likelihood of Relatedness (CLR4) (Faith et al. 2007), a mutual information (MI)-based GRN inference approach, Xiong et al. (2017) inferred GRNs to identify key genes in the maize seed development process using RNA-Seq data from various tissues: embryos, endosperms, whole seeds, and other tissues (Xiong et al. 2017). They created a GRN with 10,932 nodes and 48,740 edges and deduced the gene regulatory link between a tissue factor (TF) gene and either a non-TF or another TF gene based on the z-score of MI.

9.2.2 Machine Learning Method

In addition to showcasing its prowess in GRN inference (Table 9.1), ML, a branch of computer science that provides data-driven prediction, has garnered significant interest for its numerous applications in contemporary biology (Camacho et al. 2018; Webb 2018). Guo et al. (2016) used 166 transcriptome datasets obtained from the PLEXDB and 27 (9 trials with three biological replicates) to estimate the global GRN in *Fusarium graminearum* using the MinReg algorithm5, which is based on a greedy algorithm and a derivative of Bayesian networks (Guo et al. 2016).

By overlaying data from its protein–protein interaction network and the differently expressed genes of its mutant in *F. graminearum*, they were able to identify 968 potential regulators and depict a subnetwork for the regulatory gene FAC1. Recent GRN inference research using both static and dynamic transcriptome data from different species has made extensive use of GENIE36, a tree-based ML method (Huynh-Thu et al. 2010) (Redekar et al. 2017). Huang et al. (2018) developed four tissue-specific GRNs by using GENIE3 to infer GRNs from more than 1000 publicly accessible RNA-Seq data from different tissues, including leaf, root, shoot, apical meristem, and seed. They used publicly accessible ChIP-seq datasets to validate the anticipated regulatory networks for transcription factors KN1 (KNOTTED1), FEA4 (fasciated ear4), and O2 (Opaque2). In order to depict a GRN made up of nitrogen-responsive TF and non-TF genes, Varala et al. (2018) utilized dynamic factor graph models (Mirowski and LeCun 2009) to a fine-scale time-series transcriptome in response to nitrogen supply in *Arabidopsis* shoots and roots.

9.2.3 Combined Method

For GRN inference, combinatorial application of several algorithms may be a viable approach (Marbach et al. 2012). Using microarray-based time-series transcriptome data, Foo et al. (2018) used three different algorithms Inferelator, TIGRESS (Trustful Inference of Gene Regulation with Stability Selection7) (Haury et al. 2012), and GENIE3 to infer GRNs involved in *Arabidopsis* defense response. They also used the Y1H assay, data from an *Arabidopsis cistrome* map, and gene expression profiles from overexpressors of a related gene to confirm a specific subnetwork. Redekar et al. (2017) used 60 RNA-Seq datasets (three biological replicates, five stages of developing seeds, and four experimental lines) to infer the GRNs between TFs and co-expressed modules for seed development in soybean10 using five different algorithms: ARACNE8 (Margolin et al. 2006), GENIE3 (Huynh-Thu et al. 2010), TIGRESS (Haury et al. 2012), partial correlation (GeneTS9) (Schäfer and Strimmer 2005), and CLR (Faith et al. 2007). The resulting GRNs were then compared with published GRNs of *Arabidopsis*.10. In order to integrate different knowledge related to the regulation of gene expression, Banf and Rhee developed a novel GRN inference strategy called GRACE (Gene Regulatory network inference Accuracy Enhancement). This strategy generates GRNs through a number of steps, including initial network prediction from gene expression data using a RF regression model and integrating information related to gene regulation, meta-network construction based on information of functionally related genes for network module extraction, and ensembles of Markov Random Fields for further selection of regulatory links (Banf and Rhee 2017). The authors used experimentally identified cis-motifs for TFs and conserved sequence information in their promoter regions, along with gene expression data from 83 tissues and stages, to infer the developmental GRN in *Arabidopsis*. This resulted in an initial GRN with 10,098 links, 4305 targets, and 325 regulators (Figure 9.1).

9.3 Prediction of Different Biological Systems

Large-scale phenotypic data is making it possible for ML techniques to predict phenotypes based on genotype. Precision medicine, which aims to enhance clinical decision-making

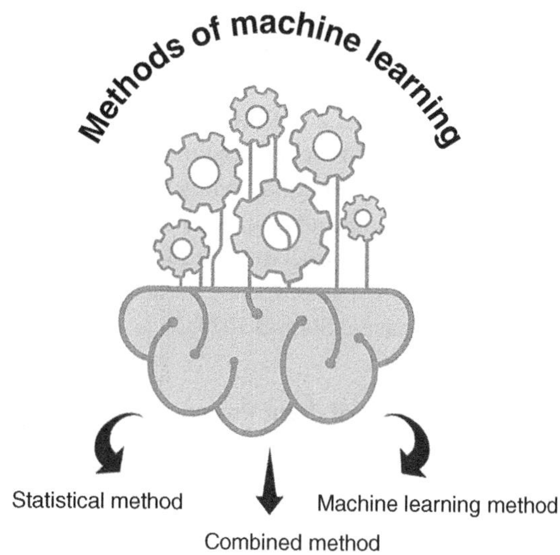

Figure 9.1 Methods for machine learning for gene regulatory network prediction.

based on patient genotypes, is the best illustration of this. In the plant sciences, there are surprisingly few phenotypic databases with well-documented metadata, which hinders dispersed innovation in extensive ML-based data analysis. However, ML techniques have been used to create predictive models and gain new biological insights. Here, we outline four uses of ML-based biological function prediction in plants: genomic prediction, phenotypic prediction from genotype, gene ontology (GO) category prediction, and gene association with metabolic pathways (Cruz and Wishart 2007; Kourou et al. 2015).

9.3.1 Prediction of Plant Genomics

Complex interactions between genetic, transport, and metabolic processes are all part of plants' molecular response to stress (Sahoo et al. 2020). Plant genomics was transformed by high-throughput sequencing technology, which made it possible to quickly and affordably generate large amounts of genetic data (Unamba et al. 2015). Genomic methods have been used to manage abiotic stress by identifying stress-responsive genes (SRGs), clarifying regulatory networks, and creating genetic markers for breeding initiatives (Kamali and Singh 2023). Plant stress is mostly controlled at the genomic level by cis-regulatory DNA sequences, transcription factors, and chromatin structure. Numerous facets of sequence analysis, including phylogenetic analysis, mapping chromosomal location, identifying orthologous genes, figuring out gene functions, studying introns, identifying cis-regulatory elements, and examining 5′ and 3′ untranslated regions, depend heavily on databases and computational tools (Joshi et al. 2021).

Numerous facets of sequence analysis, including phylogenetic analysis, mapping chromosomal location, identifying orthologous genes, figuring out gene functions, studying introns, identifying cis-regulatory elements, and examining 5′ and 3′ untranslated regions, depend heavily on databases and computational tools. While databases like riceDB, SoyBase, TAIR, and MaizeGDB offer useful resources that are specialized to a given plant species, databases like Gramene, Phytozome, Ensembl Plants, PlantGDB, and VISTA offer user-friendly interfaces that help retrieve plant genes and orthologous sequences (Joshi et al. 2021). The following tools help with gene mapping, structural analysis, gene–gene interactions, and cis-regulatory element identification: MapGene2Chromosome v2, MapChart, GSDS 2.0, PIECE, IUPred, ProtParam, Tbtools, and PlantCARE. Furthermore, a comprehensive understanding of genetic information and its applications is facilitated by QTL genomics, pangenomics, structural genomics, comparative genomics, epigenomics, genome editing, synthetic genomics, phylogenomics, functional annotation, and genome-wide association studies (GWAS). Finding genetic variations linked to crops' ability to withstand abiotic stress has been made possible in large part by GWAS (Nyirahabimana and Solmaz 2024).

9.3.2 Prediction of Plant Proteomics

The scientific community has already developed specialized bioinformatics resources that compile data from several "-omics" investigations in order to decipher the intricate proteome features of plant stress responses. This enables effective data mining and cooperative research (Ambrosino et al. 2020). Plant proteomics research is supported by a number of useful tools and databases (including Araport, Plant Proteome Database, UniProtKB/Swiss-Prot, etc.) (Bateman et al. 2023). Additionally, mass spectrometry-oriented proteomics data, including investigations conducted on plants, are stored in PRIDE Database and Massive (Deutsch et al. 2020). Distributing proteomics data across several repositories is made possible by ProteomeXchange. While SUBA4 provides information on the subcellular location of *Arabidopsis* proteins, supporting spatial proteomics studies, Plant PTM Viewer facilitates the investigation and display of post-translational protein modifications. Although there are many databases and bioinformatics tools accessible for plant study, their efficient use frequently necessitates specific expertise. Although proteomics databases like the Plant Proteome Database and UniProtKB/Swiss-Prot provide extensive resources for protein data, their utilization necessitates the ability to navigate intricate datasets and comprehend sophisticated proteomic concepts (Willems et al. 2022).

Simple tools like Plant PTM Viewer can assist novices in visualizing post-translational changes without requiring a great deal of prior knowledge, although they do not have the sophisticated data manipulation features of more specialized applications. Consequently, researchers, students, and professionals can find it easier to access and comprehend these tools and databases through user-friendly interfaces, online training materials, and educational platforms. Adopting and effectively utilizing bioinformatics resources can be facilitated by interactive tutorials, video demonstrations, and step-by-step instructions, which allow for smooth data analysis and interpretation (Lucaciu et al. 2019).

9.3.3 Prediction of Plant Metabolic Pathways

The two types of plant metabolic pathways are primary and specialized, with the genes involved in the latter frequently experiencing functional divergence and lineage-specific duplications. As a result, predicting the function of these genes based on sequence similarity as is usually done for gene function annotation following genome sequencing often results in inaccurate, insufficient, or no gene annotation. A metabolic gene can be more accurately linked to its pathway and even to particular enzyme activity by combining inputs from gene and protein structural studies, expression patterns, metabolic profiles, and ML homology research by Dale et al. (2010) found to predict metabolic pathways in an organism from its genome-wide gene complement, the authors used 123 features, such as reaction evidence, pathway holes (i.e., steps with unknown enzymes), pathway connectivity, gene chromosomal location and neighbors, pathway variants (i.e., alternative routes to an end product), and taxonomic range. These data were subjected to four widely used ML techniques: ensemble methods, logistic regression, decision trees (DTs), and naïve Bayes (NB). The performance of their prior non-ML approach, PathoLogic, was improved, according to the authors (Karp et al. 2010).

They point out how crucial it is for ML to be data-driven, scalable, and adjustable as opposed to the hard-coded rules that are a part of the fundamental PathoLogic algorithm and do not scale as the amount of training data increases. The availability of data sets that match enzymes to the reaction they catalyze remains a significant barrier to the effectiveness of ML-based pathway prediction techniques; the PlantCyc database helps to partially solve this issue (Schläpfer et al. 2017). A recent study (Toubiana et al. 2019) used a combination of transcriptomics and untargeted metabolomics data from tomatoes in an integrated RF framework to find and validate novel metabolic enzymes, whereas the PathoLogic program uses publicly available genetic data to approach pathway prediction. They specifically hypothesized and experimentally confirmed the existence of a new mechanism for the degradation of melibiose in plants. This use of a prediction method based on ML shows how ML may be applied to obtain new biological insights.

9.3.4 Prediction of Plant Transcriptomic Studies

Researchers can better understand the underlying molecular mechanisms by studying the entire set of RNA transcripts in particular cells or tissues under various conditions. This is known as transcriptomics, and it frequently reveals how plants regulate gene expression in response to oxidative stress, including genes involved in reactive oxygen species (ROS) production and scavenging. To discover important genes and regulatory processes involved in stress responses, for example, Adil et al. (2022) and Sehar et al. (2023) carried out a transcriptomics study to examine gene expression patterns in rice plants exposed to heavy metals (such as cadmium and arsenic). Cufflinks, which is used for transcript assembly, quantification, and differential expression analysis from RNA-Seq data and is extensively utilized in crops like rice and *Arabidopsis*, is one of the frequently used tools and databases for plants. TopHat maps RNA-Seq reads to reference genomes and identifies splice junctions in soybean and barley (Ghosh and Chan 2016). StringTie makes transcript assembly and quantification in tomatoes easier; DESeq2, which is frequently used in maize and wheat, helps with differential gene expression analysis (both DESeq2 and Cufflinks can help identify differentially expressed genes related to oxidative stress responses, giving insights into the plant's ability to manage ROS accumulation under various abiotic stresses) (Niedziela et al. 2022). For transcriptomics data and analysis across a variety of plant species, including *Arabidopsis*, rice, maize, soybean, barley, tomato, and poplar, the Plant Expression Database, PlantGenIE, *Arabidopsis* Information Resource, RiceXPro,

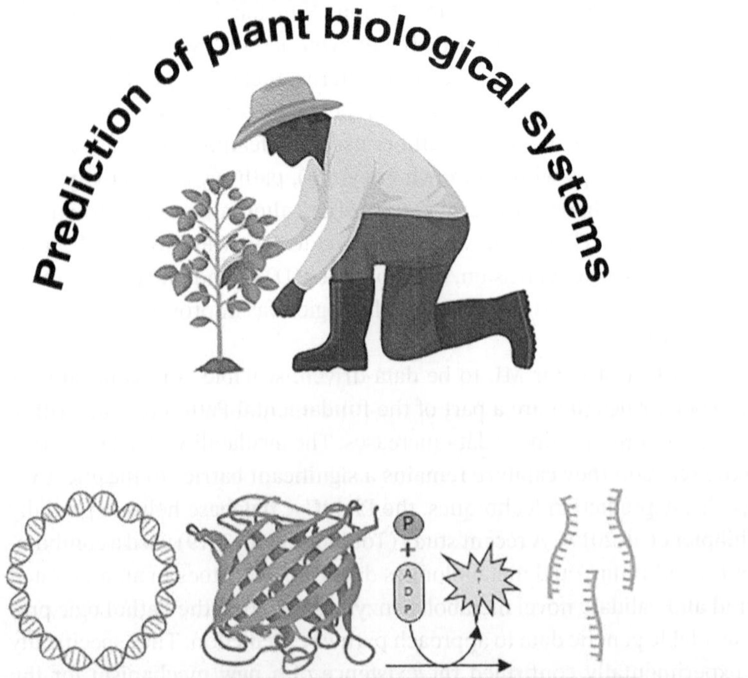

Figure 9.2 Prediction of different biological systems.

MaizeGDB, and Genevestigator are useful tools. Though they have steep learning curves and significant computational needs, transcriptomics tools like DESeq2 and StringTie are effective for transcript assembly and differential expression analysis. For novice researchers without much bioinformatics training, these difficulties may seem overwhelming. On the other hand, platforms like PlantGenIE and the *Arabidopsis* Information Resource (TAIR) offer more approachable interfaces that let novices examine transcriptomics data with less technical complexity, while they might not give as in-depth analysis as command-line tools (Hruz et al. 2008) (Figure 9.2).

9.4 Applications of Machine Learning for Regulating Plant Stress Responses

ML is tackling issues related to agriculture and food security and transforming plant research and crop resilience (Araus et al. 2022). Without labeled data, self-supervised deep learning models like ESM-2, DNABERT-2, and MSA Transformer can glean new information about the structure, function, and evolution of proteins and DNA (Datta et al. 2024). NetGO3 employs ESM for cutting-edge gene ontology prediction from sequences (Islam et al. 2024), and ESM-2 self-learns protein sequence features by masking and predicting masked sections.

Data gathering, such as RNA sequencing, is part of the first ML phase, and de-noising improves expression recovery. For training data representation, supervised ML employs a variety of features, including physicochemical characteristics and amino acid sequences (Sperschneider et al. 2018). Three methods filter, wrapper, and embedding are essential for feature selection (Khalid et al. 2014). A key component of ML is algorithm selection, which can be divided into three categories: semi-supervised (managing labeled and unlabeled data), unsupervised (identifying data patterns without known outcomes, such as clustering and dimension reduction), and supervised (establishing input–output relationships from training data) (Lac et al. 2024). SVM, DT, RF, ANN, and NB are examples of common supervised algorithms, whereas hierarchical clustering, k-means, and independent component analysis (ICA) are examples of unsupervised algorithms (Neeraj and Maurya 2020). High-dimensional data is reduced to fewer uncorrelated principal components (PCs) via principal component analysis (PCA) (Patel et al. 2024). Plant stress protein interactions can be better understood structurally with the help of AlphaFold2 and conformational subsampling. Ab initio 3D ligand structure prediction in proteins for de novo molecule design is made possible by the most recent work (Krishna et al. 2024).

9.5 Future Research and Challenges

Integration of artificial intelligence (AI) with Omics Data in Early Disease identification: To facilitate the early identification of plant illnesses, a promising future path is the smooth integration of AI with diverse omics data, including transcriptomics, proteomics, metabolomics, and genomics. Even before obvious symptoms appear, sophisticated AI algorithms may examine multi-dimensional omics datasets to find minute alterations in plant molecular profiles linked to the start of disease. High-Throughput Phenotyping with AI-Omics Fusion: This approach presents an intriguing way to combine high-throughput phenotyping approaches with AI-assisted omics methodologies. By connecting chemical reactions to detectable phenotypic characteristics, this fusion can allow for real-time plant health monitoring. Our knowledge of plant defense systems is improved by the incorporation of phenomics data into AI-driven analytical pipelines.

Predictive Modeling of Disease Dynamics: To anticipate disease dynamics in plant populations, predictive modeling can be created by utilizing AI and omics data. In order to forecast disease outbreaks and evaluate the effectiveness of preventive measures, ML and deep learning (DL) models can take into account genetic, molecular, and environmental factors.

Managing Combined Stressors: As a result of changing farming methods and climate change, plants frequently have to deal with several stressors at once. Understanding how plants react to combined biotic and abiotic stress is an important real-world issue that should be the focus of future research. Deciphering the complex relationships between various stressors and how they affect plant defense mechanisms throughout time can be greatly aided by AI-assisted omics approaches. In order to ensure sustainable food production, this information is crucial for creating resilient and comprehensive ways to safeguard crops in complicated stress conditions.

In light of changing agricultural methods and climate change, plants frequently have to deal with several stresses at once. Given that this is a crucial real-world issue, future study should concentrate on understanding how plants react to combined biotic and abiotic stress. In order to understand the complex relationships between various stressors and how they affect plant defense mechanisms over time, AI-assisted omics approaches can be quite helpful. This information is crucial for creating resilient and comprehensive plans to safeguard crops in challenging stress situations, guaranteeing sustainable food production.

Significant opportunities for deciphering complicated biological information have been made possible by the quick developments in ML and biological data creation. Nevertheless, there are significant obstacles to incorporating ML into plant molecular research. Like conventional plant molecular techniques, ML approaches are very context-specific, which emphasizes the significance of careful experimental design. Although ML seeks to develop predictive models, it is important to understand that every ML algorithm has unique advantages and disadvantages that affect prediction effectiveness in particular scenarios. Because of these biological and technical differences, an ML model designed for one dataset may not generalize well to others.

A wealth of information is available due to the quantity of omics datasets. Nonetheless, a significant percentage of these datasets exhibit traits like sparsity and noise. Accurately detecting biological traits is made extremely difficult by this, particularly when integrating different omics data sources (Joyce and Palsson 2006). In ML, the problem of imbalanced datasets where sample quantities differ between categories—occurs frequently. Resampling techniques like oversampling and undersampling are used by researchers to overcome issues (Maimon and Rokach 2005). Li et al. (2021), for example, used the synthetic minority oversampling technique to increase the representation of minority cases, which is an important step in effector protein discovery. These datasets uncertainty and complexity are introduced by noise and sparsity, which may make it more difficult to find significant patterns or features in the biological data. It is difficult for researchers to separate real biological signals from noise, which highlights the necessity of strong analytical techniques to guarantee the validity of results. Furthermore, overfitting is a significant problem, especially in the field of DL. When a model is so heavily customized to the details of a particular dataset that it has trouble generalizing to new, unseen data, this is known as overfitting. This tendency may impair the model's capacity for prediction and make it more difficult to apply to actual situations. Strategies such as dropout have been used to solve this issue (Scholz et al. 2004).

9.6 Conclusion

An innovative method for controlling abiotic stress in plants is provided by the combination of AI and bioinformatics technologies. Scientists can gain a deep understanding of the intricate regulatory networks and molecular mechanisms that underlie stress reactions by combining multi-omics data and utilizing cutting-edge AI algorithms. While AI techniques improve predictive modelling, gene regulatory network interference, and real-time plant monitoring, bioinformatics tools make it easier to identify SRGs, regulatory elements, and molecular markers. The development of stress-resilient plant varieties that can flourish in harsher climatic conditions brought on by anthropogenic activities and global climate change depends on these developments. Real-world uses include discovering salt-tolerant crop types, forecasting gene variations resistant to drought, and using AI-driven phenomics platforms to monitor plant health in real-time under high temperature extremes.

Other agricultural issues, including disease management prediction modelling, resource optimization, genetic improvement, improved crop monitoring, automated farming systems, and climate adaptability, can also be resolved by AI and bioinformatics education. AI can be used, for example, to forecast climate change scenarios to assist farmers in adapting their practices, uncover genetic markers for plant improvement, optimize the use of water and fertilizer, and evaluate big datasets to predict plant disease outbreaks. These technologies also make it easier to automate farming and monitor it in real time, which boosts productivity and lowers labor expenses.

Even with the impressive advancements, a number of obstacles still exist. The successful integration of AI with omics data requires the creation of user-friendly bioinformatics education platforms, the establishment of common data standards, and the resolution of problems including overfitting, data imbalance, and model interpretability. Given that researchers have a wide range of needs, from beginners to professionals, our review highlights how crucial it is to use tools that are appropriate for the user's skill level. To give the upcoming generation of researchers the skills they need to use these cutting-edge tools efficiently, educational programs, interdisciplinary partnerships, and training are crucial. The scientific community can stimulate innovation and more successfully tackle the problems of global food security by creating a collaborative atmosphere.

Acknowledgement

The authors are thankful to the Mohan Babu University and Sam Higginbottom University of Agriculture Technology and Sciences for providing all necessary facilities to prepare this manuscript. We are thankful to Nimbark Chandra Jha for assisting in illustrations.

References

Adil, M.F., Sehar, S., Chen, S. et al. (2022). Stress signaling convergence and nutrient crosstalk determine zinc-mediated amelioration against cadmium toxicity in rice. *Ecotoxicology and Environmental Safety* 230: 113128.

Alipanahi, B., Delong, A., Weirauch, M.T., and Frey, B.J. (2015). Predicting the sequence specificities of DNA- and RNA-binding proteins by deep learning. *Nature Biotechnology* 33: 831–838.

Ambrosino, F., Sabbarese, C., Roca, V. et al. (2020). Analysis of 7-years Radon time series at Campi Flegrei area (Naples, Italy) using artificial neural network method. *Applied Radiation and Isotopes.* 163: 109239.

Araus, J.L., Kefauver, S.C., Vergara-Díaz, O. et al. (2022). Crop phenotyping in a context of global change: what to measure and how to do it. *Journal of Integrative Plant Biology* 64: 592–618.

Banf, M. and Rhee, S.Y. (2017). Enhancing gene regulatory network inference through data integration with markov random fields. *Scientific Reports* 7: 41174. https://doi.org/10.1038/srep41174.

Bateman, A., Martin, M.J., Orchard, S.E. et al. (2023). UniProt: the universal protein knowledgebase in 2023. *Nucleic Acids Research* 51: D523–D531.

Bzdok, D., Altman, N., and Krzywinski, M. (2018). Points of significance: statistics versus machine learning. *Nature Methods* 15: 233–234.

Camacho, D.M., Collins, K.M., Powers, R.K. et al. (2018). Next-Generation machine learning for biological networks. *Cell* 173: 1581–1592. https://doi.org/10.1016/j.cell.2018.05.015.

Cruz, J.A. and Wishart, D.S. (2007). Applications of machine learning in cancer prediction and prognosis. *Cancer Informatics* 2: 59–77.

Dale, J.M., Popescu, L., and Karp, P.D. (2010). Machine learning methods for metabolic pathway prediction. *BMC Bioinformatics* 11: 15.

Datta, T., Kumar, R.S., Sinha, H., and Trivedi, P.K. (2024). Small but mighty: peptides regulating abiotic stress responses in plants. *Plant Cell Environment* 47: 1207–1223.

Deutsch, E.W., Bandeira, N., Sharma, V. et al. (2020). The ProteomeXchange consortium in 2020: enabling 'big data' approaches in proteomics. *Nucleic Acids Research* 48: D1145–D1152.

Faith, J.J., Hayete, B., Thaden, J.T. et al. (2007). Large-scale mapping and validation of *Escherichia coli* transcriptional regulation from a compendium of expression profiles. *PLoS Biology* 5: e8. https://doi.org/10.1371/journal.pbio.0050008.

Foo, S.Q., Tam, W.W., Ho, C.S. et al. (2018). Prevalence of depression among migrants: a systematic review and meta-analysis. *International Journal of Environmental Research and Public Health.* 15: 1986.

Gerstein, M.B., Bruce, C., Rozowsky, J.S. et al. (2007). What is a gene, post-ENCODE? History and updated definition. *Genome Research* 17: 669–681.

Ghosh, S. and Chan, C.K.K. (2016). Analysis of RNA-Seq data using TopHat and Cufflinks. *In Plant Bioinformatics: Methods and Protocols*; Springer: Berlin/Heidelberg, Germany 2: 339–361.

Graur, D., Zheng, Y., Price, N. et al. (2013). On the immortality of television sets: "function" in the human genome according to the evolution-free gospel of ENCODE. *Genome Biology and Evolution* 5: 578–590.

Greenfield, A., Hafemeister, C., and Bonneau, R. (2013). Robust data-driven incorporation of prior knowledge into the inference of dynamic regulatory networks. *Bioinformatics* 29: 1060–1067. https://doi.org/10.1093/bioinformatics/btt099.

Guo, L., Zhao, G., Xu, J.R. et al. (2016). Compartmentalized gene regulatory network of the pathogenic fungus *Fusarium graminearum*. *New Phytology* 211: 527–541. https://doi.org/10.1111/nph.13912.

Haury, A.C., Mordelet, F., Vera-Licona, P., and Vert, J.P. (2012). TIGRESS: trustful inference of gene regulation using stability selection. *BMC Systematic Biology* 6: 145. https://doi.org/10.1186/1752-0509-6-145.

Hruz, T., Laule, O., Szabo, G. et al. (2008). A reference expression database for the meta-analysis of transcriptomes. *Advances in Bioinformatics* 2008: 420747. https://doi.org/10.1155/2008/420747.

Huang, J., Zheng, J., Yuan, H., and McGinnis, K. (2018). Distinct tissue-specific transcriptional regulation revealed by gene regulatory networks in maize. *BMC Plant Biology* 18: 111. https://doi.org/10.1186/s12870-018-1329-y.

Huynh-Thu, V.A., Irrthum, A., Wehenkel, L., and Geurts, P. (2010). Inferring regulatory networks from expression data using tree-based methods. *PLoS One* 5: e12776. https://doi.org/10.1371/journal.pone.0012776.

Islam, S., Reza, M.N., Samsuzzaman, S. et al. (2024). Machine vision and artificial intelligence for plant growth stress detection and monitoring: a review. *Precision Agriculture* 6: 34.

Joshi, S., Kaur, K., Khare, T. et al. (2021). Genome-wide identification, characterization and transcriptional profiling of NHX-type (Na+/H+) antiporters under salinity stress in soybean. *3 Biotech* 11: 1–17.

Joyce, A.R. and Palsson, B.O. (2006). The model organism as a system: integrating'omics' data sets. *Nature Reviews. Molecular Cell Biology* 7 (3): 198–210. https://doi.org/10.1038/nrm1857.

Kamali, S. and Singh, A. (2023). Genomic and transcriptomic approaches to developing abiotic stress-resilient crops. *Agronomy* 13: 2903.

Karp, P.D., Paley, S.M., Krummenacker, M. et al. (2010). Pathway Tools version 13.0: Integrated software for pathway/genome informatics and systems biology. *Briefings in Bioinformatics* 11: 40–79.

Khalid, S., Khalil, T. and Nasreen, S. A. (2014). Survey of feature selection and feature extraction techniques in machine learning. *In Proceedings of the 2014 Science and Information Conference*, London, UK, (27–29 August 2014).

Koda, S., Onda, Y., Matsui, H. et al. (2017). Diurnal transcriptome and gene network represented through sparse modeling in brachypodium distachyon. *Frontiers in Plant Science* 8: 2055. https://doi.org/10.3389/fpls.2017.02055.

Kourou, K., Exarchos, T.P., Exarchos, K.P. et al. (2015). Machine learning applications in cancer prognosis and prediction. *Computational and Structural Biotechnology Journal* 13: 8–17.

Krishna, R., Wang, J., Ahern, W., Sturmfels, P., Venkatesh, P., Kalvet, I., Lee, G.R., Morey-Burrows, F.S., Anishchenko, I. and Humphreys, I.R. (2024). Generalized biomolecular modeling and design with RoseTTAFold All-Atom. *Science* 384: eadl2528. https://doi.org/10.1126/science.adl2528.

Lac, L., Leung, C.K., and Hu, P. (2024). Computational frameworks integrating deep learning and statistical models in mining multimodal omics data. *Journal of Biomedical Informatics* 152: 104629.

Li, J., Wei, L., Guo, F., and Zou, Q. (2021). EP3: an ensemble predictor that accurately identifies type III secreted effectors. *Briefings in Bioinformatics* 22 (2): 1918–1928. https://doi.org/10.1093/bib/bbaa008.

Liu, Z., Xie, M., Yao, Z. et al. (2013). Three meta-analyses define a set of commonly overexpressed genes from microarray datasets on astrocytomas. *Molecular Neurobiology* 47: 325–336.

Lloyd, J.P., Seddon, A.E., Moghe, G.D. et al. (2015). Characteristics of plant essential genes allow for within- and between-species prediction of lethal mutant phenotypes. *The Plant Cell.* 27: 2133–2147.

Lucaciu, R., Pelikan, C., Gerner, S.M. et al. (2019). Bioinformatics guide to plant microbiome analysis. *Frontiers in Plant Science* 10: 474304.

de Luis Balaguer, M.A., Fisher, A.P., Clark, N.M. et al. (2017). Predicting gene regulatory networks by combining spatial and temporal gene expression data in *Arabidopsis* root stem cells. *Proceedings of the National Academy of Sciences* 114: E7632–E7640. https://doi.org/10.1073/pnas.1707566114.

Maimon, O. and Rokach, L. (2005). *Data Mining and Knowledge Discovery Handbook*, vol. 2. New York: Springer.

Marbach, D., Costello, J.C., Kuffner, R. et al. (2012). Wisdom of crowds for robust gene network inference. *Nature Methods* 9: 796–804. https://doi.org/10.1038/nmeth.2016.

Margolin, A.A., Nemenman, I., Basso, K. et al. (2006). ARACNE: an algorithm for the reconstruction of gene regulatory networks in a mammalian cellular context. *BMC Bioinformatics* 7 (Suppl. 1): S7. https://doi.org/10.1186/1471-2105-7-S1-S7.

Mirowski, P. and LeCun, Y. (2009). Dynamic Factor Graphs for Time Series Modeling. In: *Machine Learning and Knowledge Discovery in Databases. ECML PKDD 2009*, Lecture Notes in Computer Science, vol. 5782 (ed. W. Buntine, M. Grobelnik, D. Mladenić, and J. Shawe-Taylor). Berlin, Heidelberg: Springer.

Neeraj, K.N. and Maurya, V. (2020). A review on machine learning (feature selection, classification and clustering) approaches of big data mining in different area of research. *Journal of Critical Reviews* 7: 2610–2626.

Niedziela, G., Niedziela, G., Szabelska-Beręsewicz, A. et al. (2022). Application of edgeR and DESeq2 methods in plant experiments based on RNA-seq technology. *Biometrical Letters* 59: 127–139. https://doi.org/10.2478/bile-2022-0009.

Nyirahabimana, F. and Solmaz, İ. (2024). Cutting-edge of genotyping by sequencing (GBS) for improving drought and salinity stress tolerance in watermelon (*Citrullus lanatus L.*): a review. *Plant Molecular Biology Reports*. 43, 1–10. https://doi.org/10.1007/s11105-024-01465-2.

Omranian, N., Eloundou-Mbebi, J.M., Mueller-Roeber, B., and Nikoloski, Z. (2016). Gene regulatory network inference using fused LASSO on multiple data sets. *Scietific Reports* 6: 20533. https://doi.org/10.1038/srep20533.

Patel, K., Tiwary, G.J., Pandey, K.K., and Asrani, K. (2024). An unsupervised machine learning algorithm: PCA (principal component analysis) comprehensive review. *International Research Journal of Modernization In Engineering Technology and Science* 6: https://doi.org/10.56726/IRJMETS49457.

Redekar, N., Pilot, G., Raboy, V. et al. (2017). Inference of transcription regulatory network in low phytic acid soybean seeds. *Frontiers in Plant Science* 8: 2029. https://doi.org/10.3389/fpls.2017.02029.

Sahoo, J.P., Sahoo, J., Behera, L. et al. (2020). Omics studies and systems biology perspective towards abiotic stress response in plants. *American Journal of Plant Sciecnes* 11: 2172–2194.

Schäfer, J. and Strimmer, K. (2005). A shrinkage approach to large-scale covariance matrix estimation and implications for functional genomics. *Statistical Applications in Genetics and Molecular Biology*. https://doi.org/10.2202/1544-6115.

Schläpfer, P., Zhang, P., Wang, C. et al. (2017). Genome-wide prediction of metabolic enzymes, pathways, and gene clusters in plants. *Plant Physiology* 173: 2041–2059.

Scholz, M., Gatzek, S., Sterling, A. et al. (2004). Metabolite fingerprinting: detecting biological features by independent component analysis. *Bioinformatics* 20 (15): 2447–2454. https://doi.org/10.1093/bioinformatics/bth270.

Sehar, S., Adil, M.F., Askri, S.M.H. et al. (2023). Pan-transcriptomic profiling demarcates *Serendipita indica*-phosphorus mediated tolerance mechanisms in rice exposed to arsenic toxicity. *Rice* 16: 28.

Singh, S., Koyama, H., Bhati, K.K. et al. (2021a). Biotechnological importance of NAC family genes in improvement of food crop. *Journal of Plant Research* 134 (3): 475–495.

Singh, S., Kudapa, H., Garg, V., and Varshney, R.K. (2021b). Comprehensive analysis and identification of drought-responsive candidate NAC genes in three semi-arid tropics (SAT) legume crops. *BMC Genomics* 22: 289.

Sperschneider, J., Dodds, P.N., Gardiner, D.M. et al. (2018). Improved prediction of fungal effector proteins from secretomes with EffectorP 2.0. *Molecular Plant Pathology* 19: 2094–2110.

Toubiana, D., Puzis, R., Wen, L. et al. (2019). Combined network analysis and machine learning allows the prediction of metabolic pathways from tomato metabolomics data. *Communications Biology* 2: 214.

Unamba, C.I., Nag, A., and Sharma, R.K. (2015). Next generation sequencing technologies: The doorway to the unexplored genomics of non-model plants. *Frontiers in Plant Science* 6: 164495.

Varala, K., Marshall-Colón, A., Cirrone, J. et al. (2018). Temporal transcriptional logic of dynamic regulatory networks underlying nitrogen signaling and use in plants. *Proceedings of National Academy of Science, U S A.* 115: 6494–6499.

Wang, X., Lin, P., and Ho, J.W.K. (2018). Discovery of cell-type specific DNA motif grammar in *cis*-regulatory elements using random forest. *BMC Genomics* 19: 929.

Washburn, J.D., Mejia-Guerra, M.K., Ramstein, G. et al. (2019). Evolutionarily informed deep learning methods for predicting relative transcript abundance from DNA sequence. *Proceedings of the National Academy of Sciences USA* 116: 5542–5549.

Webb, S. (2018). Deep learning for biology. *Nature* 554: 555–557. https://doi.org/10.1038/d41586-018-02174-z.

Wilkins, O., Hafemeister, C., Plessis, A. et al. (2016). EGRINs (Environmental Gene Regulatory Influence Networks) in rice that function in the response to water deficit, high temperature, and agricultural environments. *The Plant Cell* 28: 2365–2384. https://doi.org/10.1105/tpc.16.00158.

Willems, D., Javaid, M.K., Pinedo-Villanueva, R. et al. (2022). Importance of time point-specific indirect treatment comparisons of osteoporosis treatments: a systematic literature review and network meta-analyses. *Clinical Therapeutics.* 44: 81–97.

Xiong, W., Wang, C., Zhang, X. et al. (2017). Highly interwoven communities of a gene regulatory network unveil topologically important genes for maize seed development. *The Plant Journal* 92: 1143–1156. https://doi.org/10.1111/tpj.13750.

10

Artificial Intelligence Models for Analysing Plant Transcriptomics

Amaan Arif and Prachi Srivastava

Amity Institute of Biotechnology, Amity University Uttar Pradesh, Lucknow Campus, 226028, India

10.1 Introduction

Plant transcriptomics, the comprehensive study of RNA transcripts within plant cells, plays an important role in decoding the molecular mechanisms underlying gene expression, regulation, and functional dynamics in response to diverse biological and environmental cues (Klepikova and Penin 2019). By profiling messenger RNA (mRNA) levels across various tissues, developmental stages, or stress conditions, transcriptomics provides critical insights into how plants regulate growth, adapt to environmental changes, and respond to biotic and abiotic stresses. This field has become indispensable for advancing plant science, particularly in the context of improving crop traits such as yield, stress tolerance, and disease resistance.

Traditional transcriptomic technologies such as microarrays and next-generation RNA sequencing (RNA-Seq) and others have enabled researchers to generate large amounts of gene expression data (Vikman et al. 2014). Despite their transformative impact, these methods often face challenges when processing massive and heterogeneous datasets typical of high-throughput experiments. Limitations include difficulties in managing data noise, uncovering nonlinear relationships between genes, and integrating transcriptomic information with other omics layers such as proteomics or metabolomics. Additionally, conventional statistical tools may fall short in capturing the intricate, context-dependent patterns embedded within such data.

In recent years, artificial intelligence (AI), particularly through the application of machine learning (ML) and deep learning (DL) frameworks, has emerged as a powerful ally in transcriptomic research (Figure 10.1). These computational approaches excel at identifying complex, multidimensional patterns, making predictions, and automating analytical tasks that were previously labor-intensive. AI-driven models can classify gene expression profiles, infer gene regulatory networks (GRNs), and predict gene function with high precision. Moreover, their capacity to learn from and adapt to new data has opened avenues for real-time decision-making in plant breeding, trait selection, and crop

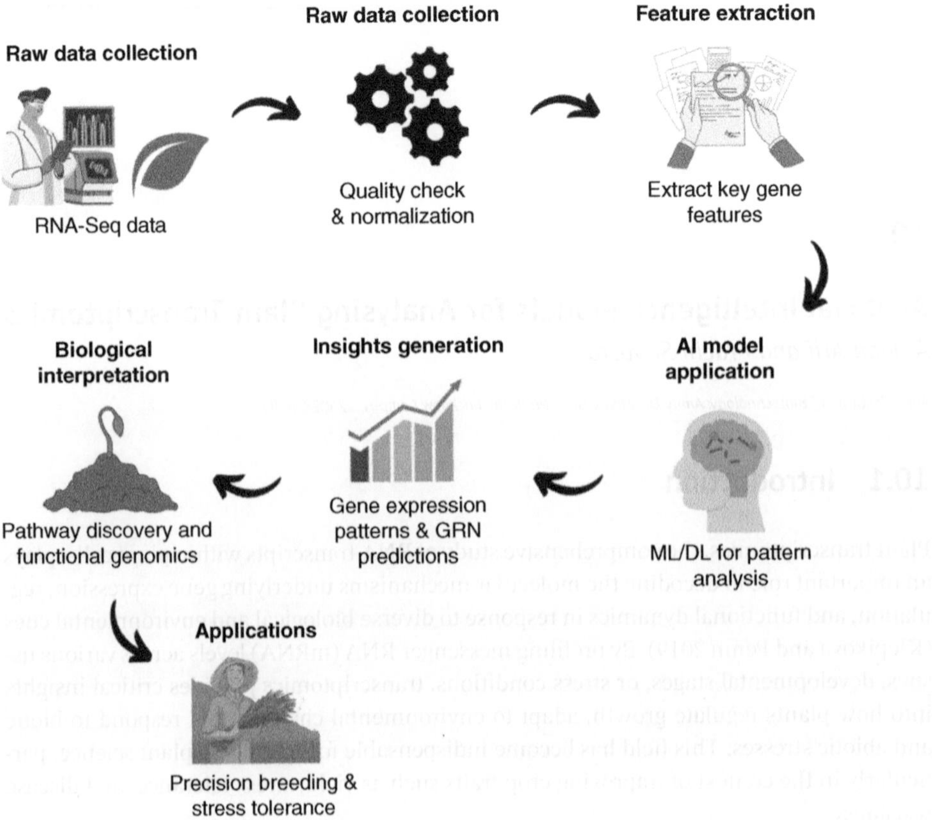

Figure 10.1 AI workflow in plant transcriptomics.

management. As a result, the integration of AI into plant transcriptomics holds tremendous promise for accelerating discoveries and developing resilient, high-performing crops for sustainable agriculture.

10.2 Fundamentals of Transcriptomics in Plants

Transcriptomics is the branch of molecular biology that focuses on the complete set of RNA transcripts produced in a cell or tissue at a given time under specific environmental or developmental conditions. In plants, this field plays a central role in exploring dynamic gene expressions, regulatory mechanisms, and functional responses to both internal and external signals. It allows researchers to observe which genes are active, how strongly they are expressed, and how expression patterns shift during processes such as growth, differentiation, and stress adaptation.

RNA sequencing (RNA-Seq) is a powerful method built upon next-generation sequencing (NGS) platforms. RNA-Seq offers high-resolution, quantitative analysis of transcriptomes. It enables precise measurement of gene expression levels and reveals novel transcripts, isoforms, and splicing variants that may not be detectable using earlier methods. This has significantly enhanced our ability to map transcriptional landscapes across diverse plant species and conditions (Stark et al. 2019).

Next-generation sequencing helps as technological support for transcriptomic study. Its ability to generate large amounts of data rapidly and at relatively low cost has made large-scale transcriptomic profiling feasible. Through NGS, researchers can sequence millions of RNA molecules in parallel, capturing expression data across entire genomes. This is crucial for studying non-model plants, where genomic references may be incomplete or poorly annotated (Satam et al. 2023).

More recently, **spatial transcriptomics** has emerged as a transformative approach. Unlike bulk RNA-Seq, which averages gene expression across tissues, spatial transcriptomics retain the spatial context of transcripts within tissue sections. This enables detailed mapping of gene activity in specific cell types or microenvironments. Although initially developed for animal studies, its application in plant systems is gaining momentum, particularly in developmental biology and stress physiology, where tissue-specific expression patterns are critical (Chen et al. 2023; Rao et al. 2021).

Before the beginning of sequencing-based methods, **microarrays** were widely used for transcriptome analysis. Microarrays rely on hybridization techniques to detect known transcripts using pre-designed probes. While they are cost-effective and suitable for high-throughput studies, their dependence on prior sequence knowledge limits their capacity to identify novel RNA species. In contemporary plant research, microarrays are largely being replaced by more flexible and sensitive sequencing technologies (Irizarry et al. 2005).

Together, these techniques have intensely expanded our understanding of gene expression regulation in plants. They allow researchers to track transcriptional changes across developmental stages, environmental conditions, and biotic interactions. Through transcriptomic profiling, it is possible to dissect molecular responses involved in drought tolerance, nutrient uptake, pathogen resistance, and other complex traits. Such insights are essential for functional genomics, trait discovery, and development of improved crop varieties. As transcriptomic data continue to grow in volume and complexity, new computational frameworks, including AI, are being introduced to manage, analyze, and interpret these datasets. These advances not only deepen our understanding of plant biology but also help in more targeted and sustainable approaches in agriculture.

10.3 Key Challenges in Analyzing Transcriptomic Data

The advancement of transcriptomic technologies has provided plant biologists with tools to explore gene expression at unprecedented resolution. Despite these innovations, the analysis of plant transcriptomic data remains highly complex. Several technical and biological factors continue to pose significant challenges that limit data interpretation, accuracy, and applicability across diverse plant systems:

Challenge	Description	References
Genomic duplication and polyploidy	Polyploidy leads to multiple gene copies (homoeologs), making it difficult to differentiate transcripts during RNA-seq analysis, which results in ambiguous gene expression estimation.	(Ilut et al. 2012)
Absence of standardized procedures	Inconsistencies in sample preparation, sequencing depth, and normalization affect reproducibility, hindering comparison across experiments and labs.	(Grones et al. 2024)
High computational and technical demands	Analysis requires substantial computing power, specialized tools, and trained personnel, which are often limited in many plant research labs.	(Wang et al. 2010)
Limited spatial transcriptomics tools	Difficulty in integrating spatial data with transcriptomic profiles due to technological limitations and the complex structure of plant tissues.	(Chen et al. 2023; Bobrovskikh et al. 2021)
Data storage and sharing issues	Absence of consistent data archiving and sharing frameworks hinders reuse, collaboration, and long-term accessibility of transcriptomic datasets.	(Atta and Fan 2021)
Challenges with non-model species	Lack of reference genomes complicates transcript assembly and annotation, often leading to reliance on related species and potential loss of unique expression data.	(Ward et al. 2012)
Plant-specific cellular features	Structures like plasmodesmata affect transcript distribution, necessitating novel models to account for plant-specific regulatory and spatial dynamics.	(Shaw et al. 2020)

In conclusion, while transcriptomics holds immense promises for advancing plant biology, it faces numerous obstacles that require ongoing attention. Tackling these issues will depend on collaborative efforts to create standardized methodologies, expand genomic resources, and develop advanced analytical tools tailored to the complexities of plant systems. Only then can can full potential of transcriptomic data be harnessed to improve our understanding of plant function, adaptation, and resilience.

10.4 Significance of Gene Expression and Regulatory Networks in Plant Development and Stress Response

In plants, the regulation of gene expression is central to nearly every biological process, from initial development to survival under stress. This regulation is not random but highly

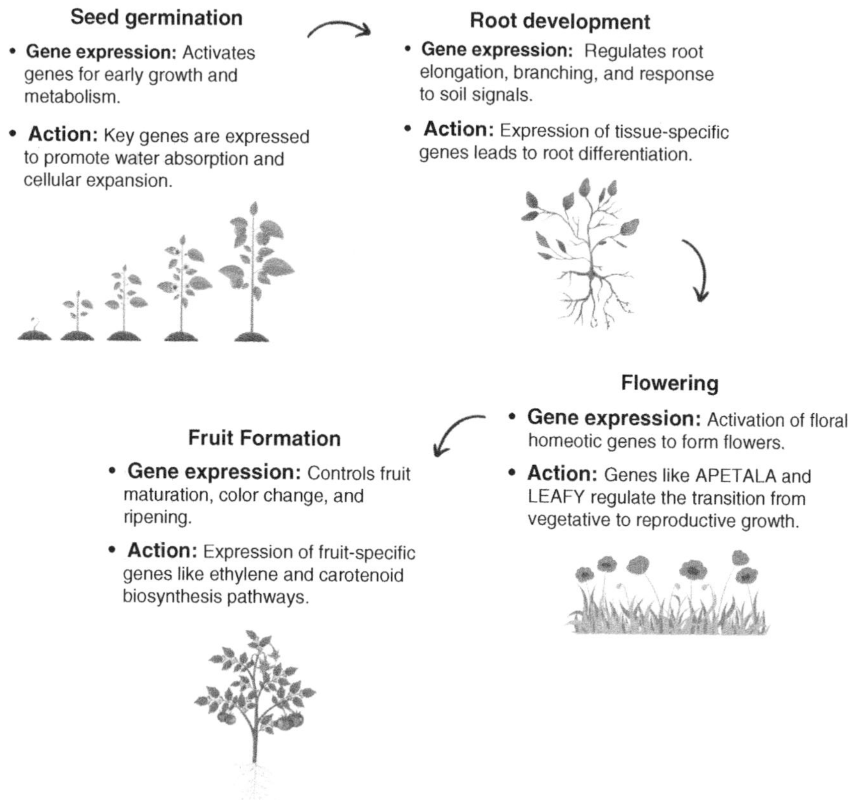

Figure 10.2 Key stages and processes.

coordinated, controlled by complex regulatory systems that enable plants to adapt, grow, and reproduce effectively in changing environments. Transcriptomic approaches have become essential in mapping these processes, offering insight into how plants manage their internal functions in response to both intrinsic signals and external conditions.

Plant development is guided by tightly regulated expression of genes across different tissues and developmental stages (Figure 10.2). The transition from seed to seedling, the formation of roots and shoots, the initiation of flowering, and the maturation of fruits all depend on specific sets of genes being activated or repressed at precise moments. Transcriptomic studies help identify these gene sets and reveal patterns of expression that are unique to certain developmental phases or tissue types. Such analyses provide a molecular framework for understanding how cells differentiate, how organs are formed, and how these events are synchronized to ensure successful plant growth.

Alongside development, plants constantly face a range of environmental challenges. Abiotic stresses such as drought, high temperature, salinity, and nutrient deficiency, as well as biotic stresses like pathogen infection and insect attack, trigger extensive changes in gene expression (Sharma et al., 2020a,b, 2022, 2023; Kaur et al., 2024) (Figure 10.3). Plants respond to these threats by activating regulatory pathways involving stress-responsive

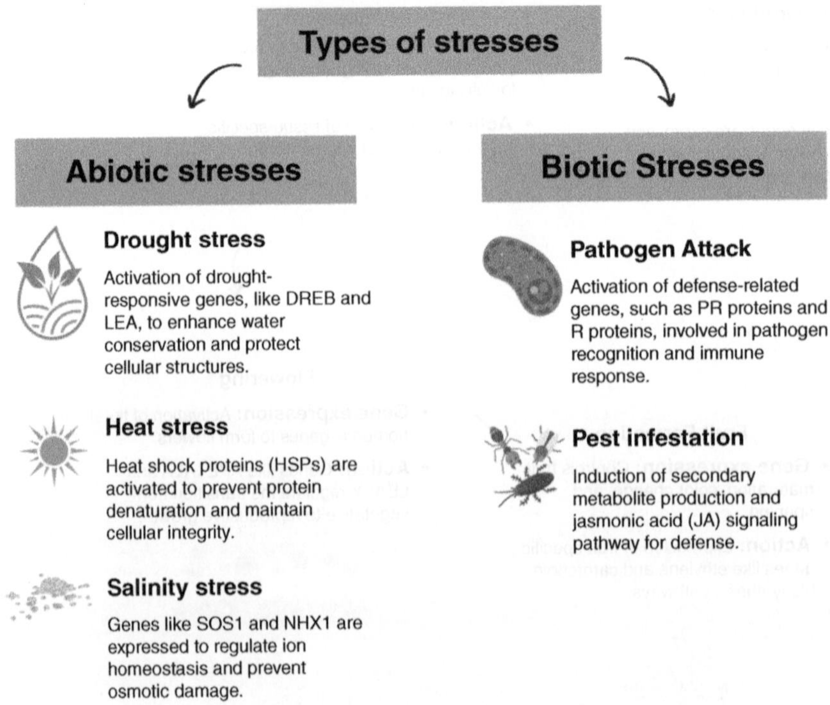

Figure 10.3 Types of stresses and gene expression modulation.

genes, transcription factors, and secondary messengers (Upadhyay 2024; Upadhyay and Shumayla 2022; Srivastava et al., 2022). These components form complex networks that enable rapid adjustments in physiology and metabolism. Through transcriptomics, researchers can identify which genes are turned on or off during stress, revealing mechanisms that confer resilience or susceptibility.

Central to both development and stress response are GRNs. GRNs consist of transcription factors, target genes, and regulatory elements that interact in intricate ways to control gene expression (Figure 10.4). Mapping GRNs allows scientists to discover core drivers of biological functions, such as those involved in hormonal signaling, nutrient transport, or cell cycle regulation. Understanding these networks helps identify candidate genes for genetic engineering or selective breeding, particularly those that contribute to traits like drought tolerance, disease resistance, or improved nutrient use efficiency.

The integration of transcriptomic data with computational modeling and AI has added a new dimension to this research. AI-based algorithms can analyze large expression datasets, detect hidden patterns, and predict gene interactions with high precision. This accelerates the discovery of key regulatory elements and facilitates the development of crops with improved traits. As agriculture faces increasing pressure from climate change and global food demand, such tools are crucial for building more sustainable and productive plant systems (Rao et al. 2021).

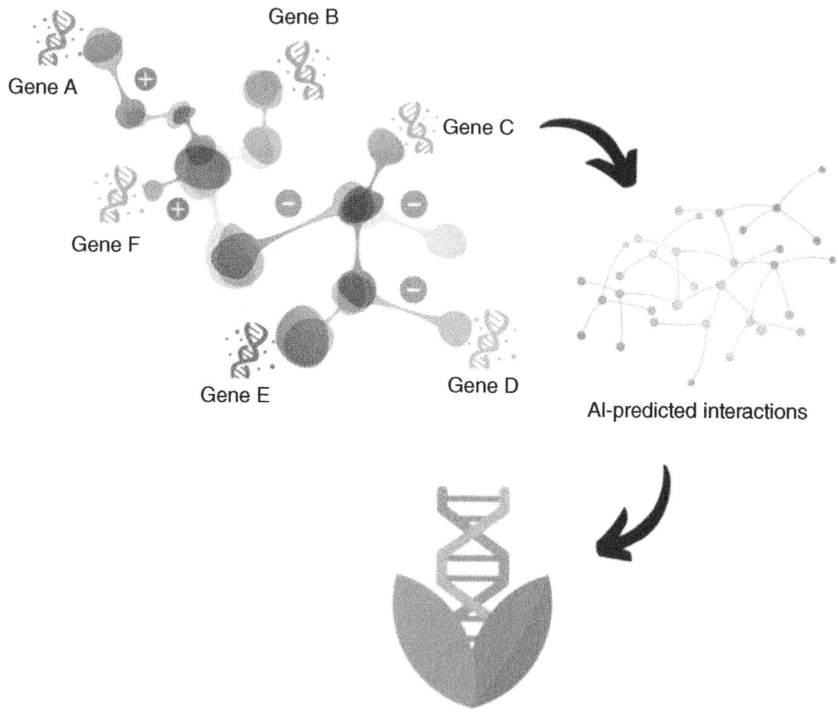

Figure 10.4 Gene regulatory network (GRN) analysis.

10.5 Introduction to AI, ML, and DL in the Context of Transcriptomics

AI encompasses a range of computational approaches designed to replicate aspects of human cognition for the purpose of solving complex tasks. Within the scope of modern biological research, AI has rapidly become indispensable, particularly in the analysis of high-throughput omics data. Its relevance in transcriptomics has grown significantly due to the exponential increase in data volume and complexity arising from NGS technologies (Irizarry et al. 2005).

A critical subset of AI is ML, which focuses on developing algorithms that can learn patterns and relationships directly from data without explicit instruction. These models adapt and improve through experience, making them highly suitable for biological datasets that often contain non-linear patterns and high variability. In the context of plant transcriptomics, ML algorithms have been applied to classify gene expression profiles, identify differentially expressed genes (DEGs), and predict regulatory relationships. Common techniques such as support vector machines (SVM), random forest classifiers, and clustering algorithms like k-means are frequently used to explore gene co-expression, functional annotation, and phenotypic prediction based on expression signatures.

Deep learning (DL) represents a more specialized extension of ML. It involves the use of artificial neural networks composed of multiple layers that allow the system to model complex hierarchical features in large datasets. DL models are particularly effective when working with unstructured or high-dimensional data, as is typical in transcriptomics. In plants, DL frameworks have been employed to detect subtle gene expression changes, recognize stress-specific transcriptomic signatures, and integrate transcriptomics with other omics layers such as proteomics and metabolomics. These models enable a systems-level understanding of biological responses that would be difficult to achieve using traditional statistical approaches.

The integration of AI methods into transcriptomic workflows has substantially improved the efficiency and scalability of gene expression analysis. Traditional analytical tools often struggle with volume, dimensionality, and variability of RNA-Seq data, especially when working with complex plant genomes or multi-omics datasets. In contrast, AI-based approaches offer more robust performance in terms of feature extraction, predictive accuracy, and interpretability. They also support the automation of critical tasks such as data normalization, noise reduction, and discovery of novel transcript isoforms or regulatory interactions.

Overall, the application of AI, ML, and DL in plant transcriptomics is not just a computational enhancement; it represents a paradigm shift in how gene expression data are processed, interpreted, and applied. These technologies bridge the gap between raw data and biological insight, making them essential tools for functional genomics, precision breeding, and the development of resilient plant varieties suited to future environmental challenges.

10.6 Advantages of AI-Driven Approaches Over Conventional Statistical Methods

The analysis of plant transcriptomic data has evolved rapidly with the growing complexity and volume of data generated by modern sequencing technologies. While traditional statistical methods have laid the foundation for interpreting gene expression patterns, they often fall short when applied to large-scale, high-dimensional, and noisy datasets. AI, especially through ML and DL frameworks, offers transformative advantages that address many of these limitations.

One of the most significant benefits of AI-based approaches is their **scalability**. Modern transcriptomic studies often involve thousands of genes measured across a wide range of tissues, developmental stages, or environmental conditions. AI algorithms are designed to process such massive datasets efficiently, enabling simultaneous analysis of multiple variables without compromising accuracy or computational performance.

AI models are also highly effective at **pattern recognition**. Unlike conventional tools that rely on predefined hypotheses or linear models, ML algorithms can detect subtle, non-linear relationships within data. This capacity allows them to uncover hidden gene expression patterns, co-regulation patterns, and complex gene interactions that may be ignored with traditional statistical techniques.

Another critical strength of AI methods lies in their ability to **manage data variability and noise**. Biological datasets, especially those derived from plant tissues exposed to diverse environmental stimuli, are inherently noisy. AI models can filter irrelevant fluctuations and focus on biologically meaningful signals, improving the reliability of the results.

Speed and automation are additional advantages. Tasks such as clustering gene expression profiles, annotating unknown transcripts, or classifying regulatory modules, which once required time-consuming manual processing or multiple statistical tests, can now be completed rapidly through automated AI workflows. This not only accelerates research timelines but also minimizes human error.

Furthermore, AI frameworks excel in **data integration**. Plant systems are complex, and understanding their molecular behavior often requires combining transcriptomic data with other omics layers such as genomics, proteomics, and metabolomics. AI models can integrate these diverse datasets into a unified framework, offering a more complete view of gene function, metabolic pathways, and regulatory dynamics.

Perhaps most importantly, AI approaches support **prediction and generalization**. Once trained, ML models can infer the function of uncharacterized genes, anticipate stress responses, or predict phenotypic outcomes under novel conditions. This predictive capacity has direct implications for crop improvement, helping identify candidate genes for breeding programs or gene editing.

10.7 Key Applications of AI in Analyzing Plant Transcriptomic Data

AI has become an essential benefit in plant transcriptomics, offering analytical capabilities that extend far beyond traditional statistical approaches. The complexity and scale of RNA sequencing data demand tools that can handle high dimensionality, extract meaningful patterns, and deliver predictive insights. AI-based methods, especially ML and DL models, are increasingly used to interpret gene expression data and uncover regulatory mechanisms. Their application has significantly expanded the scope and depth of transcriptomic research in plants.

One major area where AI has proven effective is the **classification and clustering of gene expression data**. Transcriptomic datasets often contain thousands of genes expressed under various environmental or developmental conditions. ML algorithms can classify these genes into functional categories or cluster them based on similar expression patterns. Unsupervised learning techniques, such as k-means and self-organizing maps, group genes with coordinated activity, which helps in identifying co-expression modules associated with specific physiological responses or stress conditions (Haque et al. 2019).

Reconstructing gene regulatory networks is another domain where AI, particularly DL, offers advanced modeling capabilities. These networks represent the complex relationships between transcription factors and their target genes. DL frameworks can infer these interactions by analyzing vast amounts of transcriptomic data, identifying regulatory nodes and their downstream effects with improved precision. Models such as convolutional neural networks (CNNs) or attention-based architectures have shown enhanced accuracy in predicting dynamic regulatory interactions across different tissues or time points (Ghosh and Thirugnanam 2021).

AI is also transforming the **analysis of stress responses**. Plants experience a wide range of environmental challenges, including drought, salinity, heat, and pathogen attack. Transcriptomic data collected under these conditions can be mined using ML algorithms to identify biomarkers, regulatory genes, and key pathways involved in adaptive responses. DL models excel in detecting subtle gene expression signatures that correlate with stress tolerance traits, enabling more targeted crop improvement strategies (Sugiyama and Kotani 2002).

AI also plays a key role in **integrating transcriptomics with other omics datasets**. Systems-level understanding of plant biology requires the combined analysis of genomics, proteomics, metabolomics, and epigenomics. Deep neural networks and other integrative models can merge these complex layers of information to study gene–environment interactions and to understand how molecular networks influence phenotypic traits (Lei et al. 2024).

In parallel, AI methods are used to **identify biomarker genes or gene combinations** that are indicative of a specific trait, disease, or stress condition. These biomarkers can be used for early diagnosis, precision breeding, or monitoring the effectiveness of genetic modifications. ML models can detect DEGs across various conditions and validate them as potential diagnostic or functional markers.

Finally, **phenotype prediction** based on transcriptomic data is a growing area with substantial implications for crop improvement. AI models can predict traits such as yield, flowering time, or drought sensitivity by learning from large-scale expression datasets. This accelerates the selection of promising genotypes and supports decision-making in breeding programs.

10.8 Machine Learning Models in Plant Transcriptomics

ML has emerged as an essential tool in the analysis of plant transcriptomic datasets, offering the ability to extract complex patterns and predictive insights that traditional approaches cannot easily achieve. By modeling relationships within gene expression data, ML facilitates a deeper understanding of gene function, regulation, and response to environmental stimuli. These models are particularly effective at handling large-scale, high-dimensional data generated by RNA sequencing technologies.

ML algorithms fall into two broad categories: supervised and unsupervised learning. In **supervised learning**, models are trained using labeled datasets in which both input features and desired outputs are known. In transcriptomic applications, the input features often consist of gene expression profiles, while the outputs may include phenotypic traits, gene functions, or classification labels related to stress response. Algorithms such as SVM, Random Forests, and Gradient Boosting are commonly employed (Zhang et al. 2024; Sircar and Parekh 2018). These models learn from the patterns present in training data and can then be applied to predict outcomes in new, unseen datasets. For example, a supervised model trained on drought-responsive genes can predict the likelihood of unknown genes being involved in drought tolerance based on their expression behavior under similar conditions.

Unsupervised learning, in contrast, does not rely on predefined labels. Instead, it explores inherent structure within the data to uncover hidden relationships. This type of learning is particularly useful in exploratory transcriptomic studies where underlying biological classes are unknown. Techniques such as k-means clustering, hierarchical clustering, and principal component analysis (PCA) are frequently used (Wu et al. 2023). These methods can group genes based on similar expression profiles across different tissues, developmental stages, or stress conditions. Clusters of co-expressed genes often correspond to shared biological functions or regulatory pathways, providing valuable insight into transcriptional regulation.

10.9 Deep Learning Models in Plant Transcriptomics

DL represents a transformative rise in computational biology, offering enhanced analytical power to interpret complex transcriptomic datasets. As a specialized branch of ML, DL employs multi-layered neural network architectures that can capture complex relationships in high-dimensional data. In plant transcriptomics, these models are increasingly used to analyze gene expression, predict regulatory mechanisms, and integrate multi-omics datasets for a systems-level understanding of plant biology.

CNNs have become essential tools for detecting spatial relationships in gene expression data. Though originally developed for image analysis, CNNs are effective at identifying DEGs and uncovering regulatory motifs within transcriptomic matrices. These networks apply filters across expression profiles to detect subtle patterns, aiding in the discovery of co-regulated genes and functional modules. Their capacity to learn spatial hierarchies makes them particularly suitable for classifying expression profiles and interpreting gene interaction networks (Ho et al. 2021; Abdulhafedh 2021).

Recurrent Neural Networks (RNNs), especially their Long Short-Term Memory (LSTM) variants, are designed to handle temporal data. This architecture is well-suited for analyzing time-series transcriptomic datasets, where gene expression changes over time during plant development or stress exposure. By retaining information from earlier time points, RNNs model dynamic expression behavior and enable researchers to track regulatory shifts across different stages or environmental conditions. These models are invaluable for understanding how plants modulate gene expression in response to fluctuating external stimuli (Gao et al. 2023; Monti et al. 2022).

AI-driven ML and DL models have transformed plant transcriptomics by addressing challenges related to plant transcriptomics, including data dimensionality, noise sensitivity, and integration across data types (Liao et al. 2024; Malashin et al. 2024). These models are accelerating discovery in plant functional genomics and enabling the development of more resilient, productive crop varieties suited to modern agricultural demands (Silva et al. 2019; Wang et al. 2020).

10.10 Integrating AI with RNA-Seq Data

RNA sequencing (RNA-Seq) has revolutionized transcriptomic research by enabling comprehensive and quantitative profiling of gene expression across plant tissues and conditions.

The analysis of RNA-Seq data involves multiple computational steps, each of which plays a critical role in deriving biologically meaningful insights. However, the increasing scale and complexity of transcriptomic datasets demand more efficient, accurate, and scalable analytical approaches. AI offers substantial value in enhancing traditional RNA-Seq workflows by introducing automation, improved pattern recognition, and predictive modeling across all analytical stages.

The first step in RNA-Seq analysis is **data preprocessing**. High-throughput sequencing platforms such as Illumina generate large volumes of raw reads, which often include technical artifacts, low-quality sequences, and adapter contamination. Quality control tools such as FastQC evaluate sequence quality metrics, while tools like Trimmomatic and Cutadapt are employed to trim and clean the data. Cleaned readings are subsequently aligned to a reference genome using alignment tools such as HISAT2, STAR, or Bowtie. Accurate alignment is essential for downstream analyses, including gene quantification and isoform detection (Guerra et al. 2016).

Normalization is an important step to correct for differences in sequencing depth, transcript length, and library composition. Without normalization, comparisons of gene expressions across samples may be biased or misleading. Common normalization metrics include TPM (Transcripts Per Million) and RPKM/FPKM (Reads/Fragments Per Kilobase per Million). Statistical frameworks such as DESeq2 and edgeR apply model-based count normalization to improve accuracy in differential expression analyses (Guerra et al. 2016).

Following normalization, **differential gene expression (DGE) analysis** is performed to identify genes with statistically significant expression changes between experimental conditions. Tools such as DESeq2, Limma-Voom, and edgeR are widely used for this purpose. These tools apply robust statistical models to determine genes that are either upregulated or downregulated under specific treatments, such as exposure to drought or salinity stress (Guerra et al. 2016).

To interpret the biological relevance of DEGs, researchers rely on functional annotation tools. Gene Ontology (GO) analysis categorizes genes into functional classes based on molecular function, biological process, and cellular component. KEGG pathway analysis maps genes to known metabolic and signaling pathways, enabling the identification of disrupted or activated biological processes. Co-expression network analysis, using tools like WGCNA, identifies modules of genes that are co-regulated and may share common biological functions.

AI can be integrated into each of these analytical stages to improve performance, scalability, and depth of insight. In the preprocessing phase, AI models can be trained to identify and correct errors or artifacts in sequence data. During normalization and expression analysis, ML models can uncover hidden biases and improve gene count estimations through adaptive weighting methods. For DGE analysis, AI techniques can enhance sensitivity by modeling complex interactions and nonlinear relationships in expression patterns. DL frameworks, in particular, can detect subtle variations and expression signatures that conventional models may miss. AI also adds value to functional interpretation. Natural language processing models can automate literature mining to link gene sets to published functional annotations. In pathway and network analysis, neural network-based models can infer regulatory interactions and highlight novel gene modules involved in specific stress responses or developmental pathways. These integrative approaches not only reduce manual effort but also increase the interpretability and accuracy of biological conclusions.

10.11 AI Models for Analyzing RNA-Seq Datasets

AI has become a transformative force in RNA-Seq data analysis, offering solutions that overcome the challenges of high dimensionality, data noise, and nonlinear gene regulation. Traditional approaches, while foundational, often lack the scalability and precision needed to interpret the large complexity of transcriptomic landscapes. AI models, particularly ML and DL architectures, provide advanced methods for finding biologically meaningful insights from RNA-Seq datasets and guiding functional genomics studies in plants.

One of the most crucial early applications of AI in RNA-Seq workflows lies in **data preprocessing and denoising**. Raw transcriptomic data frequently contain technical variability and background noise that can obscure weak but biologically significant signals. DL models, such as autoencoders, are highly effective at filtering out irrelevant variation while preserving underlying biological patterns. These models compress gene expression data into lower-dimensional representations and reconstruct cleaner outputs, enabling downstream analyses with higher reliability.

AI also plays a key role in **clustering and classification tasks** within RNA-Seq studies. Unsupervised learning techniques, including k-means and hierarchical clustering, are used to group genes with similar expression patterns across samples or conditions. This facilitates the identification of co-regulated gene modules, which often correspond to shared functions or regulatory controls. In parallel, supervised ML models such as SVM and Random Forests are employed to classify genes or samples based on known labels, such as tissue type or stress condition. These models can also distinguish between DEGs and background noise with high accuracy.

In **reconstructing GRNs**, AI models offer a powerful framework for inferring interactions between transcription factors and their downstream targets. DL architectures such as RNNs are particularly well-suited for modeling temporal or hierarchical gene expression relationships. These networks learn the regulatory architecture embedded in expression data and help identify master regulators that orchestrate critical biological responses in plants.

Functional and pathway analyses also benefit significantly from AI-driven approaches. Integration of transcriptomic data with curated databases such as KEGG and Reactome enables the identification of biological pathways activated under specific environmental or developmental conditions. Graph Neural Networks (GNNs) extend this integration by modeling gene–gene interactions within complex biological networks, offering a topological view of transcriptomic regulation and facilitating the discovery of key pathway components.

AI has become particularly valuable in **predicting plant responses to biotic and abiotic stresses**. Models trained on labeled RNA-Seq datasets can identify stress-responsive genes and regulatory circuits involved in drought, salinity, heat, or pathogen exposure. These predictive capabilities are essential for breeding programs focused on enhancing stress tolerance in crops, as they allow early prioritization of candidate genes and traits.

Table 10.1 AI-powered RNA-seq applications in plant transcriptomics using PISE.

Plant species	No. of RNA-Seq libraries	AI techniques used	Key findings	Biological contexts explored
Arabidopsis thaliana	>25 000	Deep learning	Accurate detection of ~1.65 million intron retention events and splicing dynamics	Developmental stages, tissue-specific expression, stress responses
Zea mays **(Maize)**	>17 000	Machine learning	Identification of splicing shifts related to genetic mutations and environmental variables	Heat stress, drought, nutrient deficiency
Oryza sativa **(Rice)**	>10 000	AI-based isoform detection models	Enhanced identification of splicing isoforms and stress-responsive gene regulation	Abiotic stresses (e.g., drought, salinity), field vs. lab conditions
Glycine max **(Soybean)**	~4000	Supervised learning	Classification of alternative splicing changes linked to signaling and defense	Hormone signaling, pathogen infection, developmental cues

Finally, AI models support the **integration of multi-omics datasets**, which is critical for systems biology approaches in plant science. Combining transcriptomic data with genomic, proteomic, and metabolomic layers offers a comprehensive view of cellular function. DL methods, including variational autoencoders (VAEs), simplify the integration of such diverse data types by learning shared representations that capture essential biological variation. This enables the identification of gene modules and regulatory networks that operate across molecular layers, supporting more holistic interpretations of plant phenotypes.

Together, these case studies in Table 10.1 highlight how AI-powered tools can harness large-scale transcriptomic data to uncover subtle and dynamic post-transcriptional regulatory mechanisms. By facilitating the analysis of intron retention and alternative splicing across model and crop species, resources like PISE contribute to a deeper understanding of gene regulation and stress adaptation. These insights are pivotal for precision breeding and functional genomics, guiding future efforts to engineer climate-resilient and high-yield crops through informed transcriptomic interventions (Zhang et al. 2022).

10.12 Applications for AI in Functional Genomics

AI has redefined how researchers approach functional genomics, particularly in the context of transcriptomic data interpretation. By using advanced ML and DL models, it is now possible to predict gene expression, discover trait-associated genes, and guide precision breeding with remarkable accuracy and efficiency. These approaches allow the exploration of complex biological systems, accelerating discoveries in plant science and crop improvement.

One of the most significant contributions of AI in this field lies in the accurate prediction of gene expression profiles. ML algorithms such as SVM and Random Forests are used to model how gene expression changes in response to diverse environmental stimuli. These models analyze high-dimensional transcriptomic data to uncover regulatory cues that influence gene activation or suppression under specific conditions. For example, the ability to predict which genes are responsive to drought, salinity, or pathogen exposure allows for a deeper understanding of plant stress physiology (Deepika et al., 2024, 2025a,b,c; Sharma et al., 2025a,b; Musich et al. 2021).

DL architectures further enhance this predictive power. CNNs identify spatial patterns within gene expression matrices, offering insights into regulatory motifs and expression zones. In contrast, RNNs, particularly LSTM variants, are suited for modeling time-series transcriptomic data. These models can track dynamic gene expression changes during developmental transitions or environmental stress, revealing critical regulators of plant growth and adaptation (Figure 10.5) (Liu et al. 2021).

Beyond expression profiling, AI also supports the identification of functional genes and the discovery of agriculturally valuable bioactive compounds. Functional gene discovery involves mining transcriptomic data to prioritize genes that are co-expressed or differentially regulated under specific conditions. ML methods cluster genes with shared expression profiles, often uncovering entire pathways associated with important traits such as disease resistance or abiotic stress tolerance. DL techniques add another layer by predicting gene functions from expression signatures and sequence motifs, facilitating the identification of transcription factors like WRKY and DREB known for their roles in stress response (Figure 10.6) (Benny et al. 2020).

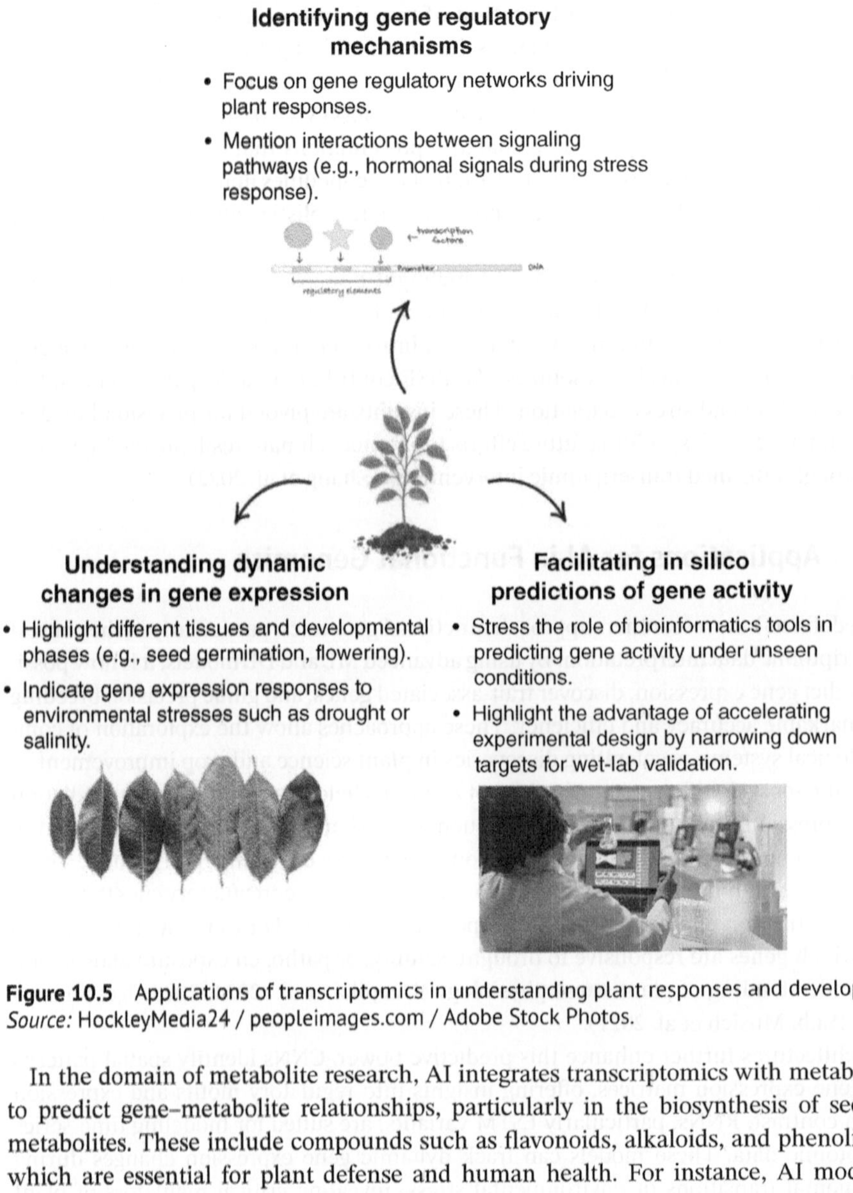

Figure 10.5 Applications of transcriptomics in understanding plant responses and development. *Source:* HockleyMedia24 / peopleimages.com / Adobe Stock Photos.

In the domain of metabolite research, AI integrates transcriptomics with metabolomics to predict gene–metabolite relationships, particularly in the biosynthesis of secondary metabolites. These include compounds such as flavonoids, alkaloids, and phenolic acids, which are essential for plant defense and human health. For instance, AI models can identify key enzymes involved in flavonoid biosynthesis pathways in medicinal plants, enabling metabolic engineering strategies to enhance compound yield (Swindell 2006).

AI-driven tools are also reshaping modern breeding practices. In precision breeding, genomic selection models powered by algorithms such as Deep Neural Networks or ensemble methods are used to forecast trait performance based on genotype and transcriptomic profiles. This allows breeders to identify elite candidates with optimal genetic potential early in the breeding cycle. Marker discovery, a critical component of marker-assisted selection, is also enhanced through AI. ML methods are applied to identify eQTLs and SNPs linked to desirable agronomic traits, streamlining the process of variety development (Hrmova and Hussain 2021; Chen et al. 2016).

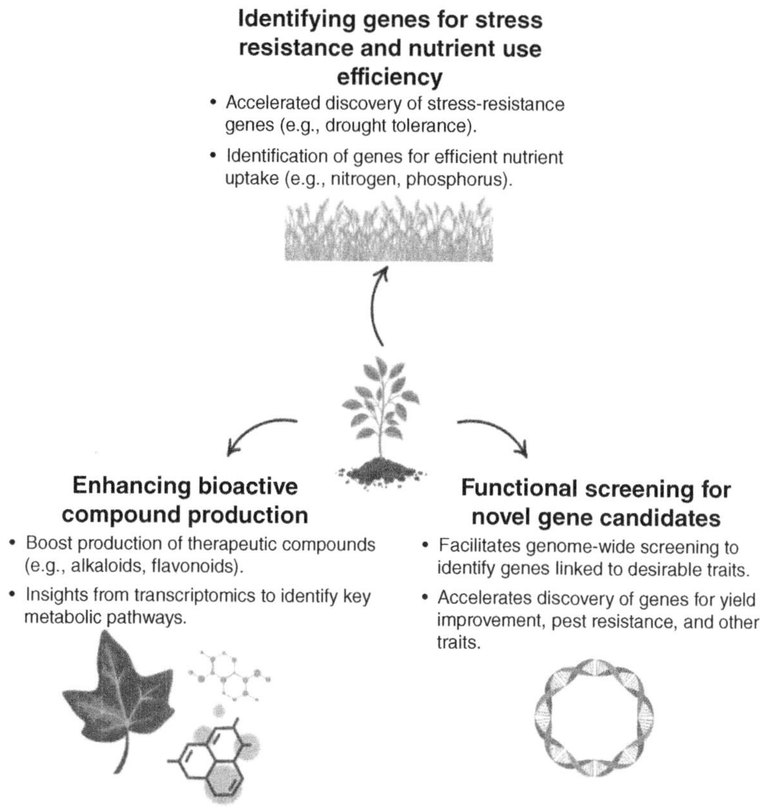

Figure 10.6 Applications of transcriptomics in crop improvement.

Moreover, AI facilitates the breeding of climate-resilient crops by integrating transcriptomic, phenotypic, and environmental data. These integrative models help identify gene networks involved in stress tolerance mechanisms. A relevant example includes the identification of ABA-responsive genes for drought resilience in rice, enabling the breeding of varieties suited for water-limited environments (Figure 10.7) (Wang et al. 2024; Hao et al. 2011).

Finally, AI plays an important role in accelerating breeding timelines. Predictive modeling allows researchers to simulate gene-trait relationships and optimize crossing strategies, thereby reducing the number of field trials and generation cycles. This not only cuts costs but also expedites the delivery of improved cultivars to farmers (Sircar and Parekh 2018; Borah et al. 2017).

In conclusion, AI-enabled functional genomics offers a transformative approach to decoding plant biology. Through predictive modeling, functional gene discovery, and integration with omics data, these tools empower researchers to address pressing challenges in global agriculture. By enhancing our capacity to develop stress-tolerant, high-yielding, and nutritionally superior crops, AI paves the way for a more sustainable and resilient food system.

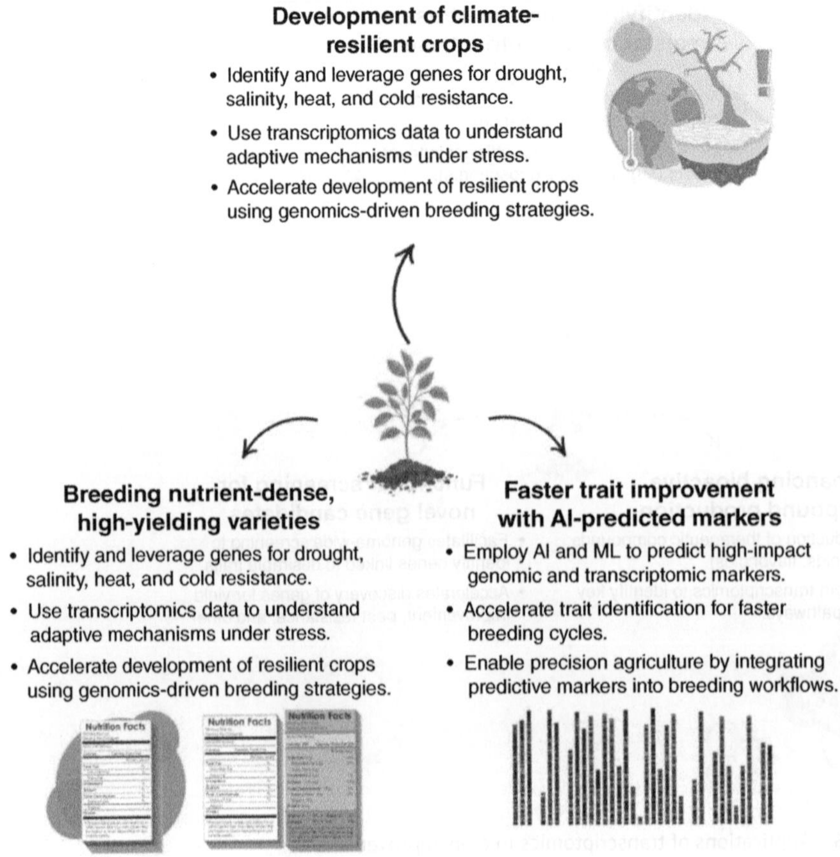

Figure 10.7 Applications of transcriptomics and AI in advancing sustainable agriculture.

10.13 Future Perspectives and Opportunities in AI-Driven Plant Transcriptomics

The integration of AI with transcriptomics is shaping a new frontier in plant science. As datasets become more complex and the demands of global agriculture intensify, AI stands out as a transformative tool in guiding future innovations. Its role in advancing precision breeding and promoting sustainable agricultural practices is particularly compelling.

AI applications in precision breeding offer targeted solutions for crop improvement. Through advanced analysis of transcriptomic datasets, AI algorithms can identify key regulatory genes and molecular markers that govern traits such as drought tolerance, pest resistance, and nutrient efficiency. These insights enable breeders to develop crop varieties tailored to specific environmental or agronomic needs. For instance, predicting gene expression signatures associated with salinity tolerance allows breeders to select resilient genotypes without extensive field trials. This strategic approach accelerates breeding timelines and improves selection accuracy.

Moreover, AI supports the broader goals of sustainable agriculture. By linking transcriptomic data to nutrient uptake pathways, stress responses, and growth regulators, AI can help design crops that require fewer chemical inputs. This not only reduces environmental impact but also ensures more efficient resource use. AI models further assist in simulating plant performance across varying climate scenarios, making it possible to anticipate outcomes and refine breeding strategies before implementation in the field.

In the context of climate change, AI-guided transcriptomic analysis is invaluable. Rising temperatures, erratic rainfall, and increased soil salinity present challenges to global food security. AI models trained on transcriptomic datasets are now able to identify genes involved in adaptive responses to these abiotic stresses. These insights facilitate the development of climate-resilient crops, ensuring food production remains stable even under unpredictable environmental pressures. Equally important is the enhancement of crop nutritional profiles and yield potential through AI-aided selection, directly addressing the needs of a growing population.

Looking ahead, several advancements promise to elevate the synergy between AI and transcriptomics. One area of development is the integration of multi-omics data – bringing together transcriptomic, genomic, proteomic, and metabolomic layers. This holistic perspective provides a more complete view of gene function, regulation, and interaction. As sequencing technologies like Oxford Nanopore continue to evolve, real-time transcriptomics will become a reality. When coupled with AI, these tools will enable rapid, in-field assessments of gene expression in response to immediate environmental changes.

The advent of explainable AI (XAI) is another important milestone. These models provide interpretable outputs, allowing researchers to understand why certain predictions are made. This fosters confidence in AI results and aligns model outputs with known biological pathways. Moreover, edge computing and IoT-enabled sensors will allow AI systems to monitor plant gene expression directly in field conditions, making plant health diagnostics both immediate and precise.

Equally transformative is AI's growing capability to study non-model plant species. Many agriculturally important crops lack comprehensive genomic references. AI models, by learning from available data, can now infer gene functions in these species, reducing dependence on traditional experimental validation. This opens doors to innovation in orphan crops and under-researched plants that hold promise for nutrition, medicine, or climate resilience.

In summary, the integration of AI with plant transcriptomics is poised to redefine the landscape of functional genomics. From enabling data-driven breeding to combating the effects of climate change, AI empowers researchers with tools that are both efficient and insightful. The future of plant science lies in leveraging these advancements to develop resilient, high-yielding, and sustainable crop systems for a rapidly changing world.

10.14 Conclusion

The integration of AI into plant transcriptomics has revolutionized the way researchers analyze and interpret vast and complex gene expression datasets. By overcoming the limitations of traditional methods, AI techniques – such as ML and DL – have enhanced the prediction of gene expression profiles, discovery of functional genes, and identification of regulatory networks critical for plant development and stress responses. AI-driven tools efficiently handle noise, heterogeneity, and scale in RNA-Seq data, enabling accurate identification of DEGs, functional pathways, and stress-responsive regulators. DL models, including CNNs and RNNs, excel in tasks like GRN reconstruction, gene function prediction, and multi-omics integration. AI has enabled groundbreaking advancements in precision breeding, facilitating the development of climate-resilient crops, improved yields, and enhanced nutritional quality – essential for sustainable agriculture and global food security. AI's transformative role lies in its ability to automate complex analyses, extract biologically meaningful insights, and accelerate crop improvement programs. By integrating AI with functional genomics, researchers can optimize plant traits, predict stress responses, and develop solutions to address climate change challenges.

In the future, AI-powered innovations, such as real-time transcriptomics, explainable AI (XAI), and multi-omics integration, will further drive advancements in plant biology. These tools will play a critical role in enabling precision agriculture, fostering sustainability, and ensuring food security for a growing global population. By harnessing the power of AI, plant scientists and breeders are better equipped to address pressing agricultural challenges, paving the way for a more resilient and productive agricultural ecosystem.

References

Abdulhafedh, A. (2021). Incorporating K-means. *Hierarchical Clustering and PCA in Customer Segmentation.* 3: 12–30. https://doi.org/10.12691/JCD-3-1-3.

Atta, L. and Fan, J. (2021). Computational challenges and opportunities in spatially resolved transcriptomic data analysis. *Nature Communications* 12: https://doi.org/10.1038/s41467-021-25557-9.

Benny, J., Marchese, A., Giovino, A. et al. (2020). Gaining insight into exclusive and common transcriptomic features linked to drought and salinity responses across fruit tree crops. *Plants* 9: https://doi.org/10.3390/plants9091059.

Bobrovskikh, A., Doroshkov, A., Mazzoleni, S. et al. (2021). A sight on single-cell transcriptomics in plants through the prism of cell-based computational modeling approaches: benefits and challenges for data analysis. *Frontiers in Genetics* 12: https://doi.org/10.3389/fgene.2021.652974.

Borah, P., Sharma, E., Kaur, A. et al. (2017). Analysis of drought-responsive signalling network in two contrasting rice cultivars using transcriptome-based approach. *Scientific Reports* 7: https://doi.org/10.1038/srep42131.

Chen, C., Ge, Y., and Lu, L. (2023). Opportunities and challenges in the application of single-cell and spatial transcriptomics in plants. *Frontiers in Plant Science* 14: https://doi.org/10.3389/fpls.2023.1185377.

Chen, T., You, L., Hardillo, J., and Chien, M. (2023). Spatial transcriptomic technologies. *Cells* 12: https://doi.org/10.3390/cells12162042.

Chen, X., Tian, L., Wang, Y.W.H. et al. (2016). A R2R3-MYB transcription factor regulates the flavonol biosynthetic pathway in a traditional Chinese medicinal plant. *Epimedium sagittatum. Frontiers in Plant Science* 7: https://doi.org/10.3389/fpls.2016.01089.

Deepika, Madhu, Shekhawat, J. et al. (2025a) Pre-mRNA processing factor 4 kinases (PRP4Ks): Exploration of molecular features, interaction network and expression profiling in bread wheat. J Plant Growth Regul 44, 806–820. https://doi.org/10.1007/s00344-024-11489-3

Deepika, Madhu, Ishu, and Upadhyay, S. K. (2025c). A LysM domain-containing protein TaLysM1-A confers salinity and drought tolerance in transgenic Arabidopsis. *Plant Physiology and Biochemistry*, 229, 110629. https://doi.org/10.1016/j.plaphy.2025.110629

Deepika, Madhu, and Upadhyay, S. K. (2024). Deciphering the features and functions of serine/arginine protein kinases in bread wheat. *Plant Gene*, 38, 100451. https://doi.org/10.1016/j.plgene.2024.100451

Deepika, Shekhawat, J., Madhu et al. (2025b) *Genome*-Wide Analysis Unveiled the Characteristic Features and Functions of AFC Genes of Bread Wheat (Triticum aestivum L.). J Plant Growth Regul 44, 2466–2484. https://doi.org/10.1007/s00344-024-11559-6

Gao, Z., Tang, J., Xia, J. et al. (2023). CNNGRN: a convolutional neural network-based method for gene regulatory network inference from bulk time-series expression data. *IEEE/ACM Transactions on Computational Biology and Bioinformatics* 20: 2853–2861. https://doi.org/10.1109/TCBB.2023.3282212.

Ghosh, M. and Thirugnanam, A. (2021). Introduction to Artificial Intelligence. *Studies in Big Data*. https://doi.org/10.1007/978-981-16-0415-7_2.

Grones, C., Eekhout, T., Shi, D. et al. (2024). Best practices for the execution, analysis, and data storage of plant single-cell/nucleus transcriptomics. *The Plant Cell* 36: 812–828. https://doi.org/10.1093/plcell/koae003.

Guerra, J., Fernandes, N., Barbosa, J. et al. (2016). *Quality Assessment and Quality Control of Next-generation Sequencing for Leishmania Infantum*. 14: 62–62.

Hao, Z., Li, X., Xie, C., Weng, J., Li, M., Zhang, D., Liang, X., Liu, L., Liu, S., and Zhang, S. (2011). Identification of functional genetic variations underlying drought tolerance in maize using SNP markers. *Journal of Integrative Plant Biology*, 53 8, 641–52. https://doi.org/10.1111/j.1744-7909.2011.01051.x.

Haque, S., Ahmad, J., Clark, N., Williams, C., and Sozzani, R. (2019). Computational prediction of gene regulatory networks in plant growth and development. *Current Opinion in Plant Biology*, 47, 96–105. https://doi.org/10.1016/j.pbi.2018.10.005.

Ho, W., Tang, B., and Wong, S. (2021). Predicting property prices with machine learning algorithms. *Journal of Property Research* 38: 48–70. https://doi.org/10.1080/09599916.2020.1832558.

Hrmova, M. and Hussain, S. (2021). Plant transcription factors involved in drought and associated stresses. *International Journal of Molecular Sciences* 22: https://doi.org/10.3390/ijms22115662.

Ilut, D., Coate, J., Luciano, A., Owens, T., May, G., Farmer, A., and Doyle, J. (2012). A comparative transcriptomic study of an allotetraploid and its diploid progenitors illustrates the unique advantages and challenges of RNA-seq in plant species. *American Journal of Botany*, 99 2, 383–96. https://doi.org/10.3732/ajb.1100312.

Irizarry, R., Warren, D., Spencer, F. et al. (2005). Multiple-laboratory comparison of microarray platforms. *Nature Methods* 2: 345–350. https://doi.org/10.1038/nmeth756.

Kaur, A., Madhu, Sharma, A., et al. (2024). Investigation of two-pore K+ (TPK) channels in Triticum aestivum L. suggests their role in stress response. *Heliyon*, 10(6), e27814. https://doi.org/10.1016/j.heliyon.2024.e27814.

Klepikova, A. and Penin, A. (2019). Gene expression maps in plants: current state and prospects. *Plants* 8: https://doi.org/10.3390/plants8090309.

Lei, Y., Huang, X., Guo, X. et al. (2024). DeepGRNCS: deep learning-based framework for jointly inferring gene regulatory networks across cell subpopulations. *Briefings in Bioinformatics* 25: https://doi.org/10.1093/bib/bbae334.

Liao, Y., Qian, F., Zhang, R., and Kumar, P. (2024). Long short-term memory (LSTM) neural networks for predicting dynamic responses and application in piezoelectric energy harvesting. *Smart Materials and Structures* 33: https://doi.org/10.1088/1361-665X/ad508e.

Liu, S., Wang, Z., Zhu, R. et al. (2021). Three differential expression analysis methods for RNA sequencing: limma, EdgeR, DESeq2. *Journal of Visualized Experiments : JoVE* 175: https://doi.org/10.3791/62528.

Malashin, I., Tynchenko, V., Gantimurov, A. et al. (2024). Applications of long short-term memory (LSTM) networks in polymeric sciences: a review. *Polymers* 16: https://doi.org/10.3390/polym16182607.

Monti, M., Fiorentino, J., Milanetti, E. et al. (2022). Prediction of time series gene expression and structural analysis of gene regulatory networks using recurrent neural networks. *Entropy* 24: https://doi.org/10.3390/e24020141.

Musich, R., Cadle-Davidson, L., and Osier, M. (2021). Comparison of short-read sequence aligners indicates strengths and weaknesses for biologists to consider. *Frontiers in Plant Science* 12: https://doi.org/10.3389/fpls.2021.657240.

Rao, A., Barkley, D., França, G., and Yanai, I. (2021). Exploring tissue architecture using spatial transcriptomics. *Nature* 596: 211–220. https://doi.org/10.1038/s41586-021-03634-9.

Satam, H., Joshi, K., Mangrolia, U. et al. (2023). Next-generation sequencing technology: current trends and advancements. *Biology* 12: https://doi.org/10.3390/biology12070997.

Sharma, A., Madhu, Kaur, A., et al. (2023). Thaumatin-domain containing receptor-like kinases in plants. *Elsevier*. https://doi.org/10.1016/b978-0-323-90594-7.00011-9.

Sharma, A., Sharma, H., Rajput, R., et al. (2022). Molecular Characterization Revealed the Role of Thaumatin-Like Proteins of Bread Wheat in Stress Response. *Frontiers in Plant Science*, 12. https://doi.org/10.3389/fpls.2021.807448.

Sharma, A., Shumayla, Tyagi, S., Alok, A., et al. (2020b). Thaumatin-like protein kinases: Molecular characterization and transcriptional profiling in five cereal crops. *Plant Science*, 290, 110317. https://doi.org/10.1016/j.plantsci.2019.110317.

Sharma, H., Taneja, M., and Upadhyay, S. K. (2020a). Identification, characterization and expression profiling of cation-proton antiporter superfamily in Triticum aestivum L. and functional analysis of TaNHX4-B. Genomics, 112(1), 356–370. https://doi.org/10.1016/j.ygeno.2019.02.015.

Sharma, Y., Deepika, Kaur, A., and Upadhyay, S. K. (2025a). Exploration of mitochondrial calcium uniporters suggested their role in abiotic stress response in bread wheat. *Plant Physiology and Biochemistry*, 227, 110180. https://doi.org/10.1016/j.plaphy.2025.110180.

Sharma, Y., Dixit, S., Singh, K. et al. (2025b) Vascular plant one-zinc finger transcription factors: exploration of characteristic features, expression, coexpression and interaction suggested their diverse role in bread wheat (Triticum aestivum L.). Plant Growth Regul 105, *71–87*. https://doi.org/10.1007/s10725-024-01264-5.

Shaw, R., Tian, X., and Xu, J. (2020). Single-cell transcriptome analysis in plants: advances and challenges. *Molecular Plant*. https://doi.org/10.1016/j.molp.2020.10.012.

Silva, J., Teixeira, R., Silva, F., Brommonschenkel, S., and Fontes, E. (2019). Machine learning approaches and their current application in plant molecular biology: a systematic review. *Plant Science : An International Journal of Experimental Plant Biology*, 284, 37–47. https://doi.org/10.1016/J.PLANTSCI.2019.03.020.

Sircar, S. and Parekh, N. (2018). Meta-analysis of drought-tolerant genotypes in Oryza sativa: a network-based approach. *PLoS ONE* 14: https://doi.org/10.1371/journal.pone.0216068.

Srivastava V., Mishra S., Mehrotra S. et al. (2022) Plant Transcription Factors. *Elsevier*. https://doi.org/10.1016/c2020-0-04071-5.

Stark, R., Grzelak, M., and Hadfield, J. (2019). RNA sequencing: the teenage years. *Nature Reviews Genetics* 20: 631–656. https://doi.org/10.1038/s41576-019-0150-2.

Sugiyama, A., and Kotani, M. (2002). Analysis of gene expression data by using self-organizing maps and k-means clustering. Proceedings of the 2002 International Joint Conference on Neural Networks. IJCNN'02 (Cat. No.02CH37290), 2, 1342–1345 vol.2. https://doi.org/10.1109/IJCNN.2002.1007690.

Swindell, W. (2006). The association among gene expression responses to nine abiotic stress treatments in arabidopsis thaliana. *Genetics* 174: 1811–1824. https://doi.org/10.1534/genetics.106.061374.

Upadhyay S.K. (2024) Defense-Related Proteins in Plants. *Elsevier*. https://doi.org/10.1016/c2022-0-02023-7.

Upadhyay S.K. and Shumayla (2022) Plant *Receptor*-Like Kinases.Elsevier. https://doi.org/10.1016/c2020-0-03988-5.

Vikman, P., Fadista, J., and Oskolkov, N. (2014). RNA sequencing: current and prospective uses in metabolic research. *Journal of Molecular Endocrinology*, 53 2, R93–101. https://doi.org/10.1530/JME-14-0170.

Wang, H., Cimen, E., Singh, N., and Buckler, E. (2020). Deep learning for plant genomics and crop improvement. *Current Opinion in Plant Biology*, 54, 34–41. https://doi.org/10.1016/j.pbi.2019.12.010.

Wang, L., Li, P., and Brutnell, T. (2010). Exploring plant transcriptomes using ultra high-throughput sequencing. *Briefings in Functional Genomics*, 9 2, 118–28. https://doi.org/10.1093/bfgp/elp057.

Wang, P., Lehti-Shiu, M., Lotreck, S. et al. (2024). Prediction of plant complex traits via integration of multi-omics data. *Nature Communications* 15: https://doi.org/10.1038/s41467-024-50701-6.

Ward, J., Ponnala, L., and Weber, C. (2012). Strategies for transcriptome analysis in nonmodel plants. *American Journal of Botany*, 99 2, 267–76. https://doi.org/10.3732/ajb.1100334.

Wu, S., Xu, Y., Zhang, Q., and Ma, S. (2023). Gene–environment interaction analysis via deep learning. *Genetic Epidemiology* 47: 261–286. https://doi.org/10.1002/gepi.22518.

Zhang, H., Jia, J., and Zhai, J. (2022). Plant intron-splicing efficiency database (PISE): exploring splicing of ~1,650,000 introns in Arabidopsis, maize, rice, and soybean from ~57,000 public RNA-seq libraries. *Science China Life Sciences* 66: 602–611. https://doi.org/10.1007/s11427-022-2193-3.

Zhang, X., Ibrahim, Z., Khaskheli, M. et al. (2024). Integrative approaches to abiotic stress management in crops: combining bioinformatics educational tools and artificial intelligence applications. *Sustainability*. https://doi.org/10.3390/su16177651.

11

Integrating Artificial Intelligence Technologies with Plant Systems Biology

Anukriti

Bhagwan Mahavir University, Bhagwan Mahavir College of Basic and Applied Sciences, Department of Biotechnology, Vesu, Surat, Gujarat, 395007, India

11.1 Introduction

A plant is a complex biological system. Plants show complex processes like growth and development, interaction with the environment, and other organisms around them. They also present a range of complex process of dealing with the stressful conditions like drought, floods, etc. It is difficult to fully understand all these complex behaviors using the traditional reductionist approach. To understand these complex behaviors of plants, non-reductionist approaches are now being used. Plant systems biology is a non-reductionist approach that depends on the integration of the data from various sources to understand a system. With the help of systems biology, we can easily understand the relation between the molecular characteristics of an organism with its functional and behavioral characteristics (Pazhamala et al. 2021). It helps us in establishing a connecting link between various segments of molecular biology and physiology (Medina 2013). This approach helps in developing and understanding the regulatory networks and the signaling pathways that take place inside a plant while it undergoes various conditions and processes.

Plant systems biology combines data from various biological datasets, such as genomics, transcriptomics, proteomics, metabolomics, and phenomics to create comprehensive models that encompass whole of the plant functions. Such models help us in getting a better understanding of how plants perform their various functions under normal as well as under stressed conditions.However, the extremely large volume and high complexity of the data generated in this field present significant challenges like in the processing of the data or in the integration of the data, which can be in the same or in different formats. It becomes extremely challenging while such huge data is being interpreted by traditional methods. Artificial intelligence (AI) algorithms, like the algorithms of machine learning (ML), deep learning (DL), and natural language processing (NLP), offer powerful tools with the help of which scientists can overcome these challenges (Pengand Rajjou 2024). These techniques help in getting deeper insights into various complex processes of plants at an unmatched speed and are thus helping in improving agricultural practices. It is also helping in understanding the hidden patterns thatscientists might not be able to understand through traditional methods. Though for many years AI has remained as a field which majorly focuses on theory, but now scientists are utilizing the algorithms of AI in the real world for making decisions and predictions. AI is not only helping in plant research but is also making its way toward improving agricultural practices to address the problems of global food security.

Biologists are also collaborating with computer scientists and are using AI for the early identification of diseases, identification of plants, predicting the yield from crop as well as in precision farming. AI-driven algorithms like thoseof MLare playing roles in the prediction of responses of plants against environmental stressors. These algorithms can analyze the gene expression patterns and can thus help in the development of stress-resistant plants and crops (Murmu et al. 2024; Zhang et al. 2024). DL algorithms are specialized in recognizing the extremely complex patterns present in the images and hence they can be used in the detection of plant diseases through the analysis of the images from leaves, stems, roots, etc. DL algorithms can help in the early detection of plant diseases and so can help in reducing the spread of the disease in the crops (Alzahraniand Alsaade 2023; Jung et al. 2023). Another AI approach, namely NLP, can be used to extract information from a vast amount of plant biology literature.

Apart from plant research, AI is also helping to bring changes in the methods of agriculture. AI can help in the identification of the plant and its classification, in predicting the yield of the crops and also in precision farming. It can help farmers in various ways like guiding them when to use which type of fertilizers depending onthe condition of the soil and other environmental conditions with the use of sensors and drones. Summary of the role of AI in agriculture is pictorially shown in Figure 11.1.

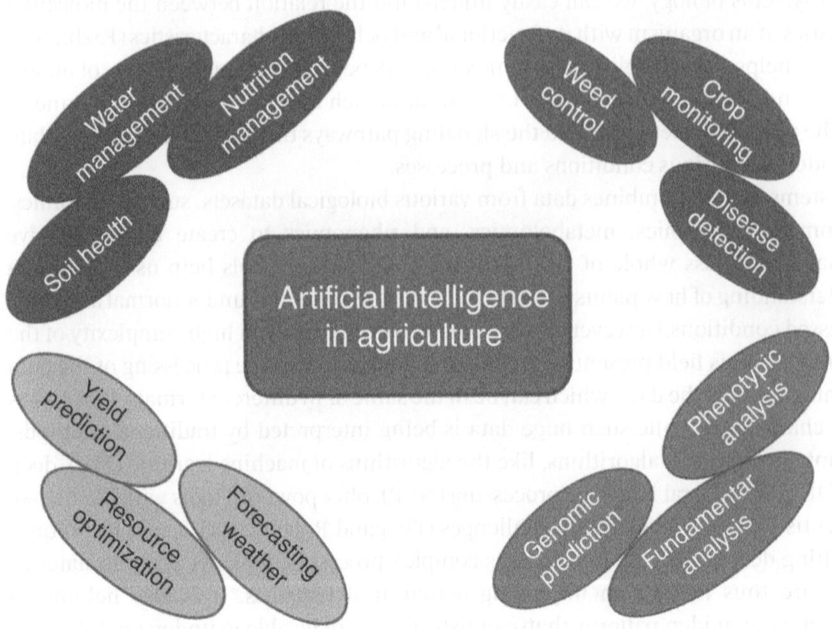

Figure 11.1 Areas where AI is contributing toagriculture.

11.1.1 The Role of AI in Plant Systems Biology

AI technologies have brought a revolution in the ways data is being analyzed in plant systems biology, which is extremely fast in comparison to the traditional statistical methods. AI techniques like ML, DL, etc. have the potential to process and analyze extremely vast datasets with high speed and accuracy that exceeds the traditional statistical methods. The datasets of plant systems are not only vast but are also of different formats. Due to these properties of the datasets, it becomes extremely difficult for traditional methods to integrate the data and interpret the results with high efficiency, which has now become possible due to the algorithms of AI. Key applications of AI in plant systems biology are:

11.1.1.1 Data Integration

Data integration is a process of collecting data from various sources and compiling them into a single unified dataset. This helps in making more precise and fasterdecisions. Plant science also now relies on diverse datasets like image-based datasets, genetic and molecular biology datasets, plant disease datasets, etc., that are usually in different formats. These different formats of the datasets impose a primary and a big challenge in developing and making an accurate understanding ofthe plant systems (Lapatas et al. 2015). Developing models based on such a varied format dataset is difficult and requires a lot of preparatory work to ensure that the model is compatible and can be used extensively.

Analysis of plant data helps scientists in addressing various ecological and environmental issues in agriculture, in developing ways of preventing plants from getting infected withvarious diseases and in mitigating the impacts of environmental stress factors. This also helps in understanding various important features of the plant systems which are earlier unknown to us. AI algorithms, particularly those employing ML, areplaying an important role in this process. Such computational techniques can harmonize and integrate these datasets to generate unified models thatare traceable and shareable among the researchers. These powerful tools can identify patterns, relationships, and features and build a comprehensive system-level understanding of plant biology.

Steps like data collection, data preprocessing, data transformation, data integration, and data validation play important roles in the accurate analysis of the data. Data preprocessing involves cleaning of data and normalization of data, which ensures the reliability of the data. It is an extremely important step. If preprocessing is not performed properly, it can lead to inconsistent and erroneous data, which will result in inaccurate analysis and predictions. Some examples of the use of data integration in plant systems are the integration of the data generated by transcriptomic and metabolomic experiments to identify key genes that are involved in the generation of responses of plants against various environmental stresses like drought (Satrio et al. 2024). Data integration is also being extensively used to increase the yield of the crops and in developing disease-resistant crops (Saiz-Rubioand Rovira-Más 2020).

As extensive data is being generated in plant sciences through transcriptomics, genomics, metabolomics, and proteomics experiments, it is becoming extremely important to develop robust data integration strategies. Challenges of integrating large-scale heterogeneous datasets can be easily overcome by using AI-powered techniques.

11.1.1.2 Predictive Modeling

Predictive modeling is a statistical approach that is used to predict future outcomes based on the data collected from the events that had happened in past and that are currently happening. Models are created to identify the patterns in the data, which helps in the accurate prediction. It is a powerful tool of AIthatfinds its use not only in finance and marketing but also in healthcare and agriculture. It helps in making better data-driven decisions, in optimizingprocesses, in improving efficiency.

In plant systems biology, predictive modeling is the development of mathematicalmodels and various algorithms that help in predicting and analyzing complicated plant processes. The development of these models is based on the data collected from different types of experiments like transcriptomic, genomic, etc., that usually generate data in various formats. This helps in identifying how different genes interact with each other and regulate various processes. It helps in finding the relationship between the genes and their regulatory functions. This further helps the researchers in identifying the key regulatory genes that modulate the plant processes like growth (Jakhmola-Mani et al. 2024). Complex processes like responses of plants against stressed conditions or the development of resistance against diseases can be easily understood by the use of predictive modeling.

With the help of AI, handling massive data indiverse formats becomes extremely easy. With the algorithms of ML and big data analysis, forecasting of disease outbreaks, environmental stresses, and plant growth becomes very easy. Algorithms like Random Forest help in identifying key genes and their interactions with other genes. This is done by analyzing gene expression data. The genes are ranked depending upon the importance of the features, which enable in identifying the genes that are responsible for crucial functions of plants. Analysis of transcriptomic data with the help of a random forest algorithm can help in identifying genes that are responsible for developing resistance against drought (Panahi et al. 2024). When integrated with IoT, satellite imaging, etc., the random forest algorithm can be used to optimize the strategies for farming (Fuentes-Peñailillo et al. 2024). This helps in developing precision farming strategies based on the real-time data of soil moisture, health of the crop, and current weather conditions. When random forest algorithms are used with the data generated by multispectral remote sensing techniques, scientists can do aprecise assessment of the changes in the environmental factors (Mallinis et al. 2020). Understanding can be developed based on how the health of the crop changes with changes in the environment and which genes get impacted byclimate-related stressors.Decision tree algorithms help in predicting the yields of the crops by assessing various environmental factors like humidity, rainfall, condition of the soil, etc. (Malashin et al. 2024). Based on the data, guidance can be provided to the farmers regarding the optimization of the crops and their planting and harvesting schedules. Support vector machine (SVM) algorithms are important tools in distinguishing diseased plants from healthy plants. When the algorithms are trained on vast plant image-based datasets, can be used to distinguish between healthy and diseased plants, which can reduce the risk of spread of any disease in plants (Shoaib et al. 2023).

A continuous advancement can be seen in the field of predictive modeling. Integration of AI, DL, big data analysis, IoT, etc. in predictive modeling is expanding its scope. Involvement of sensors, drones, and satellite imaging techniques helpsin the collection of real-time data, which enhances the role of predictive modeling in the improvement of agricultural practices.

S.No	Algorithm	Name of algorithm	Uses	References
1	Machine learning	Random forest	Used for classification and regression tasks in plant genomics and phenomics.	Esposito (2020); van Dijk et al. (2020)
2		Support vector machine	For classification tasks, such as identifying disease-resistant plant varieties.	Islam (2024)
3		Principal component analysis	PCA can analyze plant growth, physiology, and yield indices.	Ullah et al. (2022)
4		Gradient boosting machine	For high-dimensional data analysis, such as predicting plant phenotypes from genomic data.	Chen et al. (2023)
5	Deep learning	Convolutional neural networks	Image-based phenotyping, such as leaf disease detection or root architecture analysis Enables the automatic extraction of features and the classification of the plant traits.	Li et al. (2021); Murphy et al. (2024)
6		Recurrent neural networks (RNNs)	Environmental response prediction and gene expression analysis over a period of time	Monti et al. (2022); Bhimavarapu et al. (2023) Khaki et al. (2020)
		DeepLab v3+	Image-based approach to identify diseased plants	Yuan et al. (2022)
7		Generative adversarial networks	Used to produce synthetic images of plants. Also helps in developing new training datasets from pre-existing datasets to improve the models.	You et al. (2022); Giuffrida et al. (2017)
8	Explainable AI	SHAP (SHapley Additive exPlanations)	Used to interpret machine learning models and identify key features influencing plant traits.	Lu et al. (2023)
9		LIME (local interpretable model-agnostic explanations)	Interpret predictions of complex models in plant systems biology.	Schallner et al. (2020)

S.No	Algorithm	Name of algorithm	Uses	References
10	Bayesian networks and probabilistic models	Bayesian networks	For modeling gene regulatory networks and predicting plant responses to environmental changes.	Lahiri et al. (2021)
11		Gaussian processes	For modeling complex traits and predicting plant growth under uncertainty.	Ghosh et al. (2022)
12		Reinforcement learning (RL)	Used in controlled environments (e.g., greenhouses) to optimize growth conditions for maximum yield.	Hemming et al. (2019)
13		Graph neural networks (GNNs)	Analyze gene co-expression networks and protein–protein interactions in plants	Lee et al. (2022)

Table with a list of algorithms and their uses in understanding plant systems.

11.1.1.3 Phenotypic Analysis

Phenotypic analysis is the study of the phenotype i.e., the physical characteristicsof an organism, which result from the interaction with the environment. It is done to analyze how genes get perturbed due to the environmental influence and generate certain physical characteristicsin the organism.

In plants, phenotypic analyses are done to understand the physical, structural, and biochemical modifications that happen when they interact with the environment. Though the traditional methods for phenotypic analysis are effective but they are time-consuming and are labor-intensive. Traditional methods are also prone to errors done by humans. Analysis of the plant phenotypes using AI-powered image recognition systems hasbrought a revolution in the field of agriculture (Cembrowska-Lech et al. 2023). Such analyses are highly precise and also reduce the time required. It requires analysis of plant traits like the shape of the leaf, the height of the plant, and the colorof the leaves. Data of such traits is collected using high-resolution images. DL algorithms are being extensively used forimage-based phenotypic analysis like recognizing the disease in a plant by analyzing the leaves and height of the plants. This helps in the early detection of diseases, in the optimization of the health of the crop, and in improving the productivity of the crop. SVMs and random forest algorithms are being used for the extraction and classification of the phenotypic features of the plants from complex data. SVMsuse binary classification to distinguish between healthy and diseased plants. This algorithm analyzes the morphology of the leaves, height of the plants, etc. and categorizes the plants on the basis of height and growth status. Random forest algorithms are used widely to identify the plant traits that are responsible for the growth of the plants under various environmental conditions. This algorithm can be used to identify the content of the chlorophyll in the plant and the changes in the plants against environmental stressors to identify the changes in the phenotypic traits (Shah et al. 2019; Wang et al. 2022).

Convolutional neural networks (CNNs) are also being used to distinguish between diseased plants and plants in stress (Boulent et al. 2019). It can be used to perform automated detection of plant diseases, plant growth monitoring, and even the real-time analysis of stress. Algorithms likerecurrent neural networks (RNNs) and long short-term memory networks (LSTM) are being used for performing time series experiments. Such experiments provide insights into the growth patterns of the plants and in understanding how plants of different ages interact with the environment in different ways (Reuß et al. 2021). It helps in understanding the changes in the plants during different developmental stages in different environmental conditions. Utilizing the information gained by these experiments, better strategies for climatic adaptations can be made.

11.1.1.4 Metabolic Pathway Discovery

Metabolic pathways in plant systems are the chain of chemical reactions that govern the important processes like photosynthesis, respiration, etc. These chains of reactions help in the analysis of how plants respond when they are under stress or when they are being exposed to different environmental conditions. Understanding these pathways helpsus in improving the yield of the crops and in improving their stress tolerance.

AI algorithms can identify and map various pathways with the omics data successfully. It has the potential to make predictions regarding the metabolic pathways that are involved in the plant growth or in the production of secondary metabolites as a response against various environmental stressors (Bao et al. 2023). By the analyzingthe omics data, AI algorithms can predict the metabolic pathways behind various plant processes, uncovering new insights into plant biochemistry and physiology of the plant (Kisiel et al. 2023).

There are supervised and unsupervised algorithms of ML that can predict the unknown metabolic pathways. Using well-labeled datasets, missing enzymes or metabolites and even reactions within a metabolic network can be identified using supervised algorithms of ML (Shah et al. 2021). In unsupervised learning, clusters of similar genes or proteins or enzymes are made, which tryto predict the potential pathway or metabolic interaction (Sampaio et al. 2022). Graph neural networks also identify the patterns within a network and find the missing genes or enzymes within the pathways (Zhang et al. 2021). Probabilistic graphical models are beneficial in the reconstruction of the metabolic pathways by combining the data from various datasets like genomic, metabolic, and transcriptomic datasets (Qi et al. 2014).

AI is transforming the way in which we used to understand the metabolic pathways. AI is not only enhancing the speed of the work being done, but is also helping in understanding how exactly the pathways are working and how they are closely knit with each other. AI is helping in identifying the missing links within and in between the pathways making our understanding more clear. This helps in developing new strategies in the process of crop improvement.

11.1.1.5 Genotype-to-Phenotype Mapping

Genome to Phenotype (G2P) Mapping is a process of finding how the variations in the genes affect the appearance of the organism. This process is extensively used by scientists and breeders in developing new breeds of organisms. Plant scientists use this process to understand how genes control the plant features like height, yield, etc. This is also important to understand how genes get perturbed by the influence of external factors like environment and pests and how plants develop resistance against these factors.

Predicting the phenotypic trait from the genotype is a complex process as the phenotypic traits are not controlled by a single gene. Phenomenon like epistasis (gene–gene interactions), pleiotropy (one gene affecting multiple traits), etc. make it difficult to pinpoint the genetic basis ofa particular phenotypic trait. Moreover, environmental conditions are also responsible for certain phenotypic traits. There are various challengesin G2P mapping like processing large and complex datasets, identifying the hidden patterns, and then making accurate predictions.

AI techniques, such as CNNs, random forest, or SVM, can facilitate understanding the correlation of genetic data with observable traits, enabling the development of more robust crops. DL algorithms can analyze complex high-dimensional data and identify complex relationships that are usually difficult to understand using traditional methods. The predictive potential of AI can predict the phenotypic traits against the genotypes with more accuracy than traditional methods. This potential helps in applications like crop improvements and the development of personalized medicines. CNNs are being used for image-based genotype to phenotype mapping, while RNNs are being extensively used for understanding the growth patterns. Explainatory AI (XAI) is being used to predict the phenotypic traits against the genomic data and predict the crop yield. This is helping in the development of sustainable agriculture (Novielli et al. 2024).

CNNs are extensively used for analyzing visual data. In plant research, CNNs are being used with visual data like leaf morphology, root structure, etc. and analyze the relation between the disease symptoms or stress symptoms and the genetic variations. CNNs have been used to identify the relation between the venation patterns of the leaves in wheat plants and the drought resistance that hasbeen developed (Le Rouxet al. 2024).

Random forests and SVMs are being used to identify Single Nucleotide Polymorphisms (SNPs) that are associated with specific phenotypic traits. They help in classifying the plants based on their phenotypes. Random forests have also been used in finding the content of proteins and other vital elements in rice (Bi et al. 2023).

RNNs like LSTM are trained for time series data analysis to track the plant growth patterns. It is being used to study the growth patterns of the crops under different climatic conditions.

11.2 Case Studies

1) **AI in Crop Breeding:** developing new plant breeds for increased crop yield, resistance to various diseases is an important transformation in agriculture. Researchers have employed AI-based genomic selection techniques to accelerate crop breeding. For example, SVMs and random forest models have been used to predict desirable traits in crops like wheat and rice, reducing the breeding cycle time significantly. Development in various omic techniques like crop phenomics, environmics, etc. are helping in understanding how changes in the environment shape the phenotypic and genotypic characteristics of crops. With the help of this knowledge, scientists are now breeding

climate-resistant smart crops (Khan et al. 2022). Combining various traits of wheat like plant height, root length, branching patterns, etc. and performing pot experiments as well as developing AI models from the data collected helped in developing different wheat varietiesto overcome the food scarcity (Mushtaqet al. 2024).

2) **Stress Response Analysis:** Plants are constantly being exposed to stressful conditions like drought, extremely high temperatures, or heavy rainfall. They are also susceptible toattack from pests. To overcome the unwanted impact of such natural stressed conditions, it is important to understand the processes that plants use to cope with them at the cellular, molecular, and geneticlevels. DL models and ML models have been applied to huge biological data like transcriptomic, metabolomics, and proteomics data to understand how plants respond to abiotic stresses such as drought and salinity (Ali et al. 2024). These insights are vital for the engineering of stress-tolerant plants. Algorithms like random forests or SVMs are being extensively used for the analysis of genomic and transcriptomic data to identify key regulatory genes against stress tolerance (Silva et al. 2019). DL models are used for the analysis of the protein–protein interaction networks. These analyses help in understanding how various genes are interacting with each other and in identifying the hub proteins against various environmental stresses (Ma et al. 2014).
3) **High-Throughput Phenotyping:** High-throughput phenotyping or HTP is access plant traits on a large scale more accurately and quickly. HTP utilizes sensors and thermal cameras to capture high-definition images of plants. It has the ability to capture data from a single cell to the data of the whole plant. This technique is highly useful in practices like sustainable agriculture and in overcoming the issues of food security and global climatic changes. AI-powered image processing tools have revolutionized phenotyping platforms, enabling large-scale, non-destructive analysis of plant traits, such as leaf area, biomass, and chlorophyll content. AI algorithms like ML algorithms and DL algorithms help in automated trait identifications (Niazian and Niedbała 2020), analysis of the images, and extraction of the features (Khan et al. 2021) along with the development of the predictive models.

11.3 Challenges and Opportunities

While the integration of AI into plant systems biology has shown tremendous potential, several challenges remain:

1) **Data Quality and Availability:** Biological systems are multilayered systems, i.e., they have multiple levels of organization. Though algorithms of DL, neural networks, etc. can build multiple-layered systems but not with full efficiency (Batmaz et al. 2019). AI models require large, high-quality datasets. In plant systems biology, scarcity of such high-quality data that represents the diversity in plant species, agricultural practices, and data on various environmental conditions and also the inconsistency of data is a limiting factor for the development of high-performance AI models (Atapattuet al. 2024). There are various studies that highlight the lack of datasets that represent the complete diversity of farmlands,which restricts the training of AI models to a limited set of real-world data. This leads to the development of AI solutions that cannot be generalized. Also,

plants are continuously impacted by the frequent changes in the climatic conditions, composition of the soil, and the attack of pests. Capturing such dynamic data that accurately represents these conditions is still challenging, which reduces the accuracy of the AI models (Gupta et al. 2024).

2) **Interdisciplinary Knowledge Gap:** There is very limited cross-disciplinary training, which leads to inefficient AI models that are not able to capture the complexities of the plant systems. Different disciplines use different terminologies for the same phenomenon, which makes it difficult for an AI model to interpret the data accurately. There are various instances where an AI model has failed to provide optimal results due to the lack of interdisciplinary knowledge. An AI model developed for the integration of the transcriptomic and genomic data from *Arabidopsis thaliana* to study the impact of abiotic stresses failed to provide accurate predictions due to the differences in resolution and annotation of data at different omic layers (Ma et al. 2014). Similarly, due to the absence of standardized data formats and differences in terminologies led to the development of an AI model to predict the interaction between the plants and pathogens (Danchin 2024; Jafar et al. 2024). To overcome such challenges collaboration between specialists of different domains is important. Cross-disciplinary training isnecessary to overcome the barriers and knowledge gaps among different domains.
3) **Interpretability and Explainability:** The terms interpretability and explainability are often used interchangeably but they are different. Interpretability can be defined as the extent to which the human brain can understand the reasons why and how a certain decision is made. It helps in understanding how the inputs and other variables are used to reach the output. Explainability is the ability of an AI model to provide an understandable reasoning for the predictions and decisions made by it. It helps in developing trust and transparency and also helps in debugging and provides an option for improvement of the model (Harfouche et al. 2023). In AI, there is a clear tradeoff between the performance of the AI models and their ability toprovide explainable and interpretable results. Various AI models, especially DL systems, also known as"black box models," provide predictions that are difficult to explain and predict. Along with this, the plant datasets on which the AI models are trained are high dimensional, and the variables are often non-linearly correlated, which further makes the results hard to interpret. On the other hand, there are "white box models" or "glass box models" which provide explainable results but they are not as powerful as black box models (Harfouche et al. 2023). These models are also known as interpretable models or transparent models. Examples of glass box models are linear regression models, decision trees, rule-based systems, etc. In plant systems biology, glass box models are extensively being used in the prediction of gene functions and finding key genetic markers (Watson 2022), in the explainable genotypic and phenotypic predictions, in optimization of the use of fertilizers (Lynch 2011), etc.
4) **Computational Resources:** Training and deploying AI models for large-scale biological data can be computationally intensive, requiring significant resources. Biological systems have multiple and multi-layered feedback loops. To achieve this, AI will require huge computational potential, which is again an intensive process and consumes extremely high energy (Gokcekuyu et al. 2024). To overcome the inefficiencies of AI, specialized types of computing like quantum computing and neuromorphic computing can be used (Shastri et al. 2021; Bhargavi et al. 2024).

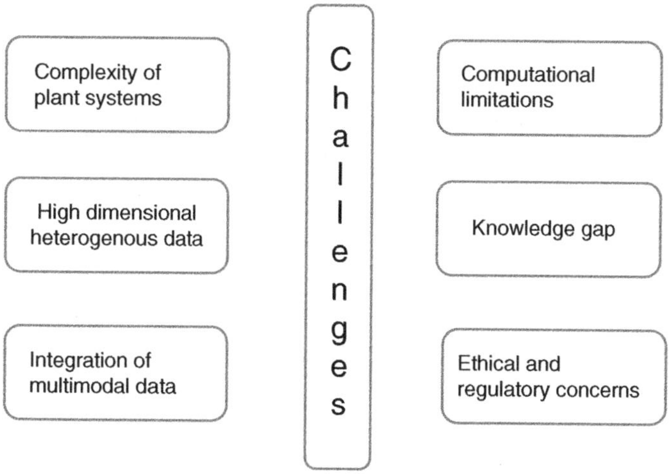

Figure 11.2 Challenges faced by artificial intelligence.

Despite various challenges as discussed in Figure 11.2, AI has immense possibilities in plant systems.

- AI enhances our understanding ofthe fundamental plant processes.
- AI enables the development of precision agriculture by predicting the performance of the crops under varied environmental conditions.
- AI is helping in the development of sustainable agricultural practices.
- AI has also accelerated the discovery of genes and pathways for improved crop traits.

11.4 The Future of AI-Powered Agriculture

As AI is expanding its ways to diverse fields like phylogenetic analysis of plant systems, gene identification, etc., it is easy for us to predict that soon AI will be helpful in designing bioactive compounds that can precisely treat plant diseases. The future of AIdepends on the development ofAI models specificto address biological issues.With the help of DL algorithms, smartphone-based apps can be designed that will provide real-time data of plant health and can predict the changes in the climate and hence forecast the outbreak of pests. AI will not only help in forecasting various climatic and environmental issues but also will help in smart sustainable farming. By Integration of AI with automation, the future of plant-based research and agriculture will shift towarddata-driven models, which will lead to more sustainable agriculture. This will not only increase the food production but will also ensure that less crops will beaffected by the changes in the climatic conditions.

11.5 Conclusion

The involvement of AI in plant systems biology is transforming the way we handle plant systems. This is not only helping in understanding the complexities of the plant system but is also helping in creating new ways of farming and taking prior measures to safeguard our crops and increase their yield.By utilizing the power of AI, researchers can easily understand and analyze the complex biological data, uncover the hidden patterns, and doinnovative farming. Though the future of AI-driven agriculture seems to be encouraging, there are various challenges to overcome like the quality of the data and its availability.

References

Ali, T., Rehman, S.U., Ali, S. et al. (2024). Smart agriculture: utilizing machine learning and deep learning for drought stress identification in crops. *Scientific Reports* 14: 30062. https://doi.org/10.1038/s41598-024-74127-8.

Alzahrani, M.S. and Alsaade, F.W. (2023). Transform and deep learning algorithms for the early detection and recognition of tomato leaf disease. *Agronomy* 13 (5): 1184. https://doi.org/10.3390/agronomy13051184.

Atapattu, A.J., Perera, L.K., Nuwarapaksha, T.D. et al. (2024). Challenges in achieving artificial intelligence in agriculture. In: *Artificial Intelligence Techniques in Smart Agriculture* (ed. S.S. Chouhan, A. Saxena, U.P. Singh, and S. Jain). Singapore: Springer. https://doi.org/10.1007/978-981-97-5878-4_2.

Bao, H., Zhao, J., Zhao, X. et al. (2023). Prediction of plant secondary metabolic pathways using deep transfer learning. *BMC Bioinformatics* 24: 348. https://doi.org/10.1186/s12859-023-05485-9.

Batmaz, Z., Yurekli, A., Bilge, A. et al. (2019). A review on deep learning for recommender systems: challenges and remedies. *Artificial Intelligence Review* 52: 1–37. https://doi.org/10.1007/s10462-018-9654-y.

Bhimavarapu, U., Battineni, G., and Chintalapudi, N. (2023). Improved optimization algorithm in LSTM to predict crop yield. *Computers* 12 (1): 10. https://doi.org/10.3390/computers12010010.

Bi, Y., Yassue, R.M., Paul, P. et al. (2023). Evaluating metabolic and genomic data for predicting grain traits under high night temperature stress in rice. *G3: Genes, Genomes, Genetics* 13 (5): jkad052. https://doi.org/10.1093/g3journal/jkad052.

Boulent, J., Foucher, S., Théau, J., and St-Charles, P.-L. (2019). Convolutional neural networks for the automatic identification of plant diseases. *Frontiers in Plant Science* 10: 941. https://doi.org/10.3389/fpls.2019.00941.

Cembrowska-Lech, D., Krzemińska, A., Miller, T. et al. (2023). An integrated multi-omics and artificial intelligence framework for advance plant phenotyping in horticulture. *Biology* 12 (10): 1298. https://doi.org/10.3390/biology12101298.

Chen, K., Zhao, B., Zhou, L., and Zheng, Y. (2023). Artificial neural network-based seedling phenotypic information acquisition of plant factory. *Agriculture* 13 (4): 888. https://doi.org/10.3390/agriculture13040888.

Ma, C., Xin, M., Feldmann, K.A., and Wang, X. (2014). Machine learning–based differential network analysis: astudy of stress-responsive transcriptomes in *Arabidopsis*. *The Plant Cell* 26 (2): 520–537. https://doi.org/10.1105/tpc.113.121913.

Danchin, A. (2024). Artificial intelligence-based prediction of pathogen emergence and evolution in the world of synthetic biology. *Microbial Biotechnology* 17 (10): e70014. https://doi.org/10.1111/1751-7915.70014.

Esposito, S., Carputo, D., Cardi, T., and Tripodi, P. (2020). Applications and trends of machine learning in genomics and phenomics for next-generation breeding. *Plants* 9 (1): 34. https://doi.org/10.3390/plants9010034.

Fuentes-Peñailillo, F., Gutter, K., Vega, R., and Silva, G.C. (2024). Transformative technologies in digital agriculture: leveraging Internet of Things, remote sensing, and artificial intelligence for smart crop management. *Journal of Sensor and Actuator Networks* 13 (4): 39. https://doi.org/10.3390/jsan13040039.

Bhargavi, G., Das, M., Banumathi, S., Malarvizhi, E. and Surendran, R. (2024). Leveraging NLP and quantum computing for advanced agricultural solutions,*2024 8th International Conference on Electronics, Communication and Aerospace Technology (ICECA)*, Coimbatore, India, pp. 900–906, 10.1109/ICECA63461.2024.10800767.

Ghosh, S., Dey, S., Bhogapurapu, N. et al. (2022). Gaussian process regression model for crop biophysical parameter retrieval from multi-polarized C-band SAR data. *Remote Sensing* 14: 934. https://doi.org/10.3390/rs14040934.

Gokcekuyu, Y., Ekinci, F., Guzel, M.S. et al. (2024). Artificial intelligence in biomaterials: acomprehensive review. *Applied Sciences* 14 (15): 6590. https://doi.org/10.3390/app14156590.

Gupta, D.K., Pagani, A., Zamboni, P., and Singh, A.K. (2024). AI-powered revolution in plant sciences: advancements, applications, and challenges for sustainable agriculture and food security. *Exploration of Foods and Foodomics* 2: 443–459. https://doi.org/10.37349/eff.2024.00045.

Harfouche, A.L., Jacobson, D.A., Kainer, D. et al. (2023). A primer on artificial intelligence in plant digital phenomics: embarking on the data to insights journey. *Trends in Plant Science* 28 (2): 154–184. https://doi.org/10.1016/j.tplants.2023.03.001.

Hemming, S., de Zwart, F., Elings, A. et al. (2019). Remote control of greenhouse vegetable production with artificial intelligence—greenhouse climate, irrigation, and crop production. *Sensors* 19 (8): 1807. https://doi.org/10.3390/s19081807.

Islam, S., Samsuzzaman, Reza, M.N. et al. (2024). Image processing and support vector machine (SVM) for classifying environmental stress symptoms of pepper seedlings grown in a plant factory. *Agronomy* 14 (9): 2043. https://doi.org/10.3390/agronomy14092043.

Jafar, A., Bibi, N., Naqvi, R.A. et al. (2024). Revolutionizing agriculture with artificial intelligence: plant disease detection methods, applications, and their limitations. *Frontiers in Plant Science* 15: 1356260. https://doi.org/10.3389/fpls.2024.1356260.

Jakhmola-Mani, R., Sonali, Pandey, A. et al. (2024). Exploring machine learning algorithms for genefunction prediction in crops. In: *Bioinformatics for Plant Research and Crop Breeding* (ed. J.-T. Chen). Wiley https://doi.org/10.1002/9781394209965.ch6.

Jung, M., Song, J.S., Shin, A.Y. et al. (2023). Construction of deep learning-based disease detection model in plants. *Scientific Reports* 13: 7331. https://doi.org/10.1038/s41598-023-34549-2.

Khaki, S., Wang, L., and Archontoulis, S.V. (2020). A CNN-RNN framework for crop yield prediction. *Frontiers in Plant Science* 10: 1750. https://doi.org/10.3389/fpls.2019.01750.

Khan, M.H.U., Wang, S., Wang, J. et al. (2022). Applications of artificial intelligence in climate-resilient smart-crop breeding. *International Journal of Molecular Sciences* 23 (19): 11156. https://doi.org/10.3390/ijms231911156.

Khan, R.U., Khan, K., Albattah, W., and Qamar, A.M. (2021). Image-based detection of plant diseases: from classical machine learning to deep learning journey. *Wireless Communications and Mobile Computing* 2021: 5541859, 13. https://doi.org/10.1155/2021/5541859.

Kisiel, A., Krzemińska, A., Cembrowska-Lech, D., and Miller, T. (2023). Data science and plant metabolomics. *Metabolites* 13: 454. https://doi.org/10.3390/metabo13030454.

Lahiri, A., Zhou, L., He, P., and Datta, A. (2021). Detecting drought regulators using stochastic inference in Bayesian networks. *PLoS One* 16 (8): e0255486. https://doi.org/10.1371/journal.pone.0255486.

Lapatas, V., Stefanidakis, M., Jimenez, R.C. et al. (2015). Data integration in biological research: an overview. *Journal of Biological Research-Thessaloniki* 22: 9. https://doi.org/10.1186/s40709-015-0032-5.

Le Roux, M.S., Karl, K., Cullis, C., and Botha, A.M. (2024). Unlocking wheat drought tolerance: the synergy of omics data and computational intelligence. *Food and Energy Security* 13: https://doi.org/10.1002/fes3.70024.

Lee, H.J., Chung, Y., Chung, K.Y. et al. (2022). Use of a graph neural network to the weighted gene co-expression network analysis of Korean native cattle. *Scientific Reports* 12: 9854. https://doi.org/10.1038/s41598-022-13796-9.

Li, L., Zhang, S., and Wang, B. (2021). Plant disease detection and classification by deep learning—areview. *IEEE Access* 9: 56683–56698. https://doi.org/10.1109/ACCESS.2021.3069646.

Lu, Y., Fan, X., Zhang, Y. et al. (2023). Machine learning models using SHapley Additive exPlanation for fire risk assessment mode and effects analysis of stadiums. *Sensors* 23 (4): 2151. https://doi.org/10.3390/s2304215.

Lynch, J.P. (2011). Root phenes for enhanced soil exploration and phosphorus acquisition: tools for future crops. *Plant Physiology* 156 (3): 1041–1049. https://doi.org/10.1104/pp.111.175414.

Giuffrida, M.V., Scharr, H. and Tsaftaris, S.A. (2017). ARIGAN: synthetic arabidopsis plants using generative adversarial network, *2017 IEEE International Conference on Computer Vision Workshops (ICCVW)*, Venice, Italy, pp. 2064–2071, 10.1109/ICCVW.2017.242.

Malashin, I., Tynchenko, V., Gantimurov, A. et al. (2024). Predicting sustainable crop yields: deep learning and explainable AI tools. *Sustainability* 16 (21): 9437. https://doi.org/10.3390/su16219437.

Mallinis, G., Chrysafis, I., Korakis, G. et al. (2020). A random forest modelling procedure for a multi-sensorassessment of tree species diversity. *Remote Sensing* 12 (7): 1210. https://doi.org/10.3390/rs12071210.

Medina, M.Á. (2013). Systems biology for molecular life sciences and its impact in biomedicine. *Cellular and Molecular LifeSciences* 70 (6): 1035–1053. https://doi.org/10.1007/s00018-012-1109-z.

Monti, M., Fiorentino, J., Milanetti, E. et al. (2022). Prediction of time series gene expression and structural analysis of gene regulatory networks using recurrent neural networks. *Entropy* 24 (2): 141. https://doi.org/10.3390/e24020141.

Murmu, S., Sinha, D., Chaurasia, H. et al. (2024). A review of artificial intelligence-assisted omics techniques in plant defense: current trends and future directions. *Frontiers in Plant Science* 15: 1292054. https://doi.org/10.3389/fpls.2024.1292054.

Murphy, K.M., Ludwig, E., Gutierrez, J., and Gehan, M.A. (2024). Deep learning in image-based plant phenotyping. *Annual Review of Plant Biology* 75 (1): 771–795. https://doi.org/10.1146/annurev-arplant-070523-042828. PMID: 38382904.

Mushtaq, M.A., Ahmed, H.G.M.-D., and Zeng, Y. (2024). Applications of artificial intelligence in wheat breeding for sustainable food security. *Sustainability* 16: 5688. https://doi.org/10.3390/su16135688.

Niazian, M. and Niedbała, G. (2020). Machine learning for plant breeding and biotechnology. *Agriculture* 10 (10): 436. https://doi.org/10.3390/agriculture10100436.

Novielli, P., Romano, D., Pavan, S. et al. (2024). Explainable artificial intelligence for genotype-to-phenotype prediction in plant breeding: a case study with a dataset from an almond germplasm collection. *Frontiers in Plant Science* 15: 1434229. https://doi.org/10.3389/fpls.2024.1434229.

Panahi, B., Hamid, R., and Jalaly, H.M.Z. (2024). Deciphering plant transcriptomes: leveraging machine learning for deeper insights. *Current Plant Biology* 41: https://doi.org/10.1016/j.cpb.2024.100432.

Pazhamala, L.T., Kudapa, H., Weckwerth, W. et al. (2021). Systems biology for crop improvement. *Plant Genome* 14: e20098. https://doi.org/10.1002/tpg2.20098.

Peng, S. and Rajjou, L. (2024). Advancing plant biology through deep learning-powered natural language processing. *Plant Cell Reports* 43: 208. https://doi.org/10.1007/s00299-024-03294-9.

Qi, Q., Jilong, L., and Cheng, J. (2014). Reconstruction of metabolic pathways by combining probabilistic graphical model-based and knowledge-based methods. *BMC Proceedings* 8: https://doi.org/10.1186/1753-6561-8-S6-S5.

Reuß, F., Greimeister-Pfeil, I., Vreugdenhil, M., and Wagner, W. (2021). Comparison of long short-term memory networks and random forest for sentinel-1 time seriesbased large scale crop classification. *Remote Sensing* 13 (24): 5000. https://doi.org/10.3390/rs13245000.

Saiz-Rubio, V. and Rovira-Más, F. (2020). From smart farming towards agriculture 5.0: areview on crop data management. *Agronomy* 10 (2): 207. https://doi.org/10.3390/agronomy10020207.

Sampaio, M., Rocha, M., and Dias, O. (2022). Exploring synergies between plant metabolic modelling and machine learning. *Computational and Structural Biotechnology Journal* 20: 1885–1900. https://doi.org/10.1016/j.csbj.2022.04.016.

Satrio, R.D., Fendiyanto, M.H., and Miftahudin, M. (2024). Tools and techniques used at global scale through genomics, transcriptomics, proteomics, and metabolomics to investigate plant stress responses at the molecular level. In: *Molecular Dynamics of Plant Stress and its Management* (ed. M. Shahid and R. Gaur). Singapore: Springer. https://doi.org/10.1007/978-981-97-1699-9_25.

Schallner, L., Rabold, J., Scholz, O., and Schmid, U. (2020). Effect of super pixel aggregation on explanations in LIME – acasestudy with biological data. In: *Machine Learning and Knowledge Discovery in Databases. ECML PKDD 2019. Communications in Computer and Information Science*, vol. 1167 (ed. P. Cellier and K. Driessens). Cham: Springer. https://doi.org/10.1007/978-3-030-43823-4_13.

Shah, H.A., Liu, J., Yang, Z., and Feng, J. (2021). Review of machine learning methods for the prediction and reconstruction of metabolic pathways. *Frontiers in Molecular Biosciences* 8: 634141. https://doi.org/10.3389/fmolb.2021.634141.

Shah, S.H., Angel, Y., Houborg, R. et al. (2019). Arandom forest machine learning approach for the retrieval of leaf chlorophyll content in wheat. *Remote Sensing* 11 (8): 920. https://doi.org/10.3390/rs11080920.

Shastri, B.J., Tait, A.N., Ferreira de Lima, T. et al. (2021). Photonics for artificial intelligence and neuromorphic computing. *Nature Photonics* 15: 102–114. https://doi.org/10.1038/s41566-020-00754-y.

Shoaib, M., Shah, B., Ei-Sappagh, S. et al. (2023). An advanced deep learning models-based plant disease detection: a review of recent research. *Frontiers in Plant Science* 14: 1158933. https://doi.org/10.3389/fpls.2023.1158933.

Silva, J.C.F., Teixeira, R.M., Silva, F.F. et al. (2019). Machine learning approaches and their current application in plant molecular biology: a systematic review. *Plant Science* 284: 37–47. https://doi.org/10.1016/j.plantsci.2019.03.020.

Ullah, A., Shakeel, A., Ahmed, H.G.M.-D. et al. (2022). Genetic basis and principal component analysis in cotton (*Gossypiumhirsutum L.*) grown under water deficit condition. *Frontiers in Plant Science* 13: 981369. https://doi.org/10.3389/fpls.2022.981369.

van Dijk, A.D.J., Kootstra, G., Kruijer, W., and de Ridder, D. (2020). Machine learning in plant science and plant breeding. *iScience* 24 (1): 101890. https://doi.org/10.1016/j.isci.2020.101890. PMID: 33364579; PMCID: PMC7750553.

Wang, W., Cheng, Y., Ren, Y. et al. (2022). Prediction of chlorophyll content in multi-temporal winter wheat based on multispectral and machine learning. *Frontiers in Plant Science* 13: 896408. https://doi.org/10.3389/fpls.2022.896408.

Watson, D.S. (2022). Interpretable machine learning for genomics. *Human Genetics* 141 (9): 1499–1513. https://doi.org/10.1007/s00439-021-02387-9.

You, A., Kim, J.K., Ryu, I.H., and Yoo, T.K. (2022). Application of generative adversarial networks (GAN) for ophthalmology image domains: a survey. *Eye and Vision (London, England)* 9 (1): 6. https://doi.org/10.1186/s40662-022-00277-3.

Yuan, H., Zhu, J., Wang, Q. et al. (2022). An improved DeepLab v3+ deep learning network applied to the segmentation of grape leaf black rot spots. *Frontiers in Plant Science* 13: 795410. https://doi.org/10.3389/fpls.2022.795410.

Zhang, X.M., Liang, L., Liu, L., and Tang, M.J. (2021). Graph neural networks and their current applications in bioinformatics. *Frontiers in Genetics* 12: 690049. https://doi.org/10.3389/fgene.2021.690049.

Zhang, X., Ibrahim, Z., Khaskheli, M.B. et al. (2024). Integrative approaches to abiotic stress management in crops: combiningbioinformatics educational tools and artificial intelligence applications. *Sustainability* 16 (17): 7651. https://doi.org/10.3390/su16177651.

12

Disease and Pest Management Using AI Technologies

Acharya Balkrishna, Shalini Bhatt, Rakshit Pathak, and Vedpriya Arya

Patanjali Herbal Research Division, Patanjali Research Foundation Trust, Delhi-Haridwar National Highway, Haridwar, Uttarakhand 249405, India

12.1 Introduction

Artificial Intelligence (AI) refers to the development of systems capable of performing tasks that typically require human intelligence, such as decision-making, learning, problem-solving, and pattern recognition. The integration of AI into agriculture has revolutionized traditional farming methods by automating processes, analyzing large datasets, and optimizing resource utilization (Xu et al., 2021; Talaviya et al., 2020; Subeesh and Mehta, 2021). AI encompasses a wide array of technologies, including machine learning (ML), deep learning (DL), computer vision, natural language processing, robotics, and Internet of Things (IoT) (Soori et al., 2023; Kutyauripo et al., 2023). These technologies work mutually to enhance efficiency, precision, and sustainability in agriculture. Agriculture has always been a data-rich industry, with information derived from weather patterns, soil conditions, crop health, pest infestations, and irrigation needs. However, the challenge lies in interpreting this vast amount of data to make informed decisions promptly. AI addresses this gap by processing data at unprecedented speeds and scales, offering insights that improve productivity while reducing environmental impacts.

Agriculture is at a critical juncture, striving to meet the growing global food demands while contending with limited resources and escalating environmental challenges. By 2050, the global population is projected to surpass 9 billion, necessitating a 60–70% increase in food production (Kutyauripo et al., 2023; Tilman et al., 2011). This formidable task is further exacerbated by several factors. Urbanization and industrial expansion are steadily reducing the availability of arable land. Climate change introduces unpredictable weather patterns, rising temperatures, and extreme events that significantly affect crop yields. Additionally, the emergence of new pests and diseases poses severe threats to crop health, with some infestations capable of devastating entire harvests. Resource scarcity, including water shortages, soil degradation, and declining fertility, underscores the need for efficient resource management. Traditional farming methods, heavily reliant on manual labor and broad, generalized approaches, often fall short of addressing these multifaceted challenges

(Daszkiewicz, 2022; Hossain et al., 2024). However, AI offers a transformative solution through precision agriculture, where farming practices are dynamically adapted to specific conditions in real time, enhancing productivity and sustainability.

One of the most significant applications of AI in agriculture is the management of diseases and pests, which account for up to 40% of global crop production losses annually, resulting in billions of dollars in economic damage (Yuan et al., 2022; Grieve et al., 2019). AI technologies effectively address these challenges through several innovative approaches. AI systems equipped with computer vision and image recognition algorithms can identify symptoms of plant diseases or signs of pest infestations with remarkable accuracy. For instance, drones fitted with cameras and sensors can survey fields, capturing data that AI models analyze to detect abnormalities in plant health. Predictive analytics further enhances this capability by analyzing historical and real-time data, such as weather conditions, soil properties, and pest population dynamics, to forecast potential outbreaks (Grieve et al., 2019). These predictions enable farmers to implement preventive measures, such as targeted treatments, before outbreaks escalate. Additionally, AI-powered decision support systems (DSS) process diverse data sources—including satellite imagery, IoT sensors, and farm records to provide actionable insights, such as optimal timings for planting, harvesting, or applying pesticides. Precision interventions, another hallmark of AI, use algorithms to guide technologies like precision spraying, ensuring that inputs such as pesticides and fertilizers are applied only where needed, thereby reducing waste and environmental impact (Shubhika et al., 2024; Talaviya et al., 2020; Panday et al., 2024). Moreover, the integration of IoT devices, such as soil sensors and weather stations, feeds real-time data into AI systems, further refining the accuracy and efficiency of disease and pest management strategies. These advancements highlight AI's transformative role in safeguarding agricultural productivity and sustainability.

The application of AI in disease and pest management provides numerous advantages, transforming traditional agricultural practices into more efficient and sustainable systems. By automating labour-intensive tasks such as crop monitoring and disease diagnosis, AI significantly boosts efficiency, enabling farmers to focus on strategic decision-making (Vidya Madhuri et al., 2025; Kariyanna and Sowjanya, 2024). This increased efficiency translates directly into higher productivity and profitability. AI also contributes to substantial cost savings by targeting only affected areas for treatment, reducing the indiscriminate use of expensive inputs like pesticides and fertilizers (Balaska et al., 2023). Moreover, timely interventions supported by AI minimize crop losses, further enhancing economic benefits.

The accuracy of AI systems is unparalleled, as they can analyze vast datasets with consistency and precision beyond human capabilities, ensuring effective disease and pest management. This precision is integral to promoting environmental sustainability, as AI-driven precision agriculture minimizes chemical overuse, reducing soil and water pollution (Sow et al., 2024; Mienye et al., 2024). Additionally, AI optimizes resource use, such as irrigation, tailoring it to specific crop requirements and conserving valuable natural resources. One of the most remarkable benefits of AI is its scalability. These technologies can adapt to farms of all sizes, from smallholder plots to large commercial enterprises. Cloud-based AI platforms ensure accessibility, enabling farmers in diverse geographic locations to leverage advanced tools and techniques (Junaid et al., 2021; Javaid et al., 2023). Collectively, these benefits underscore AI's transformative potential in revolutionizing agricultural practices and ensuring sustainable food production for a growing global population. Collectively, AI is revolutionizing agriculture by enhancing productivity, optimizing resource utilization, and minimizing environmental impact. From precision farming to smart supply chain management, AI-driven technologies are transforming traditional agricultural practices, making them more efficient and sustainable. AI applications such as pest and disease detection, autonomous machinery, and weather forecasting help farmers make data-driven decisions, improving crop yields and reducing operational costs. Additionally, AI-powered robotics and automated irrigation systems contribute to labor efficiency and water conservation. Table 12.1 provides an overview of various AI applications in agriculture, highlighting their descriptions and benefits.

Despite its immense potential, the adoption of AI in agriculture faces several challenges. High initial investment costs make AI tools and technologies less accessible, particularly for smallholder farmers in developing countries (Osorio et al., 2024; Mallinger et al., 2024). Many farmers also lack the digital literacy needed to use these systems effectively, necessitating extensive training and capacity-building efforts. The effectiveness of AI relies heavily on accurate and comprehensive data, but gaps in data quality and availability can hinder performance. Additionally, inadequate internet connectivity in rural areas limits the implementation of many AI applications in the agriculture sector (Sharma and Shivandu, 2024). Ethical and legal concerns, including issues of data ownership, privacy, and algorithmic bias, further complicate the widespread adoption of AI in agriculture.

This chapter explores the transformative role of AI in managing agricultural diseases and pests, addressing the significant challenges posed by rising food demands, resource limitations, and environmental threats. It highlights AI's applications in early detection, predictive analytics, decision support, and precision interventions, demonstrating its ability to improve efficiency, accuracy, and sustainability in farming practices. The chapter also examines the benefits of AI, such as cost savings, scalability, and environmental protection, while acknowledging barriers to adoption, including high costs, digital illiteracy, and connectivity issues. Overall, it underscores AI's potential to revolutionize agriculture and enhance global food security.

12.2 AI Technologies in Disease Diagnosis

Automated systems for plant disease detection are crucial in modern agriculture to prevent the frequent outbreaks of plant diseases and the substantial losses they cause. These systems leverage advanced AI methodologies to diagnose diseases effectively. The process begins with collecting plant images using various sensors installed in agricultural fields, followed by image processing and segmentation, where ML algorithms analyze the data to determine whether the plant is healthy or diseased. These frameworks operate in structured phases to ensure efficiency and accuracy (Garg et al., 2024; Jafar et al., 2024).

Table 12.1 AI-driven innovations in agriculture and their impact on productivity and sustainability.

AI application	Description	Benefits
Precision farming	AI-driven sensors and drones collect real-time data on soil health, crop conditions, and weather patterns.	Optimized resource use, improved crop yields, and reduced environmental impact.
Pest and disease detection	AI-powered image recognition and machine learning models detect early signs of pest infestations and plant diseases.	Early intervention reduces crop losses and minimizes pesticide use.
Automated irrigation	AI integrates weather data, soil moisture levels, and crop needs to optimize water usage.	Efficient water management and reduced irrigation costs.
Yield prediction	AI models analyze historical and real-time data to predict crop yields.	Helps farmers plan better and manage supply chains effectively.
Autonomous machinery	AI-powered tractors, harvesters, and drones automate farming operations.	Reduces labor costs and increases efficiency.
Smart supply chain management	AI optimizes logistics, storage, and distribution based on market demand.	Minimizes waste and ensures timely delivery of produce.
Agricultural robotics	AI-powered robots assist in planting, harvesting, and sorting crops.	Increases productivity and reduces dependency on manual labor.
Weather forecasting	AI analyzes climate data to provide accurate weather predictions for farming activities.	Helps farmers make informed decisions on planting and harvesting.
Soil health monitoring	AI-based sensors and imaging techniques assess soil nutrients, pH, and moisture content.	Enables precise fertilization and sustainable soil management.
Crop health monitoring	AI processes satellite images and drone data to track crop growth and detect stress factors.	Ensures timely intervention and higher crop productivity.
Weed detection and management	AI-powered vision systems distinguish between crops and weeds, enabling targeted herbicide application.	Reduces chemical usage and protects the environment.
Smart greenhouses	AI regulates temperature, humidity, and light inside greenhouses for optimal plant growth.	Maximizes yield with controlled growing conditions.
Livestock monitoring	AI-enabled sensors track animal health, movement, and behavior in real-time.	Early disease detection, better breeding management, and increased dairy/meat production.

Table 12.1 (Continued)

AI application	Description	Benefits
Automated crop harvesting	AI-driven robotic systems harvest fruits and vegetables with precision.	Reduces labor dependency and minimizes crop damage.
AI in aquaculture	AI monitors water quality, fish behavior, and feed efficiency in aquaculture systems.	Enhances fish farming productivity and reduces waste.
Agri-FinTech solutions	AI-powered platforms assess credit risk and provide financial services to farmers.	Facilitates access to loans, insurance, and better financial planning.
Blockchain in agriculture	AI integrates with blockchain to track the origin, quality, and safety of agricultural produce.	Increases transparency and prevents fraud in the food supply chain.
AI-Driven market price prediction	AI analyzes market trends, demand, and supply data to forecast crop prices.	Helps farmers make informed decisions about when and where to sell.
Sustainable agriculture optimization	AI models suggest eco-friendly farming techniques based on environmental impact analysis.	Promotes sustainable farming practices and reduces carbon footprint.

A representative framework for plant disease detection illustrates the procedural flow from image acquisition to disease prediction (Figure 12.1).

AI has transformed disease diagnosis in agriculture by enabling automated detection, classification, and monitoring of plant health. By leveraging advanced techniques such as image processing, ML, and DL, AI-driven systems can accurately identify plant diseases at an early stage, reducing crop losses and improving yield (Jackulin and Murugavalli, 2022). These technologies integrate various components, including image acquisition, pre-processing, segmentation, and feature extraction, to enhance diagnostic accuracy. Table 12.2 summarizes key AI technologies and their applications in plant disease diagnosis.

12.2.1 Image Acquisition: The Foundation of Disease Classification

Image acquisition involves capturing high-quality images of plants to facilitate accurate classification. High-resolution cameras, smartphones, and drones are commonly used for this purpose. Smartphones, in particular, have proven useful due to their portability and ability to record images in various supported formats like JPG, PNG, and TIF. These images are then transferred to pre-processing stages for refinement (Dhanya et al., 2022; Li et al., 2025). The efficiency of an automated detection system depends heavily on the quality of the acquired images. Numerous agricultural datasets, such as PlantVillage, New Plant Diseases, Integrated Pest Management (IPM) Images, APS Images, Plant Doc, and PLD, are widely used in research (Jafar et al., 2024; Jin et al., 2023). These datasets provide annotated samples of healthy and diseased plant species, enabling researchers to analyze various infections and train ML models effectively.

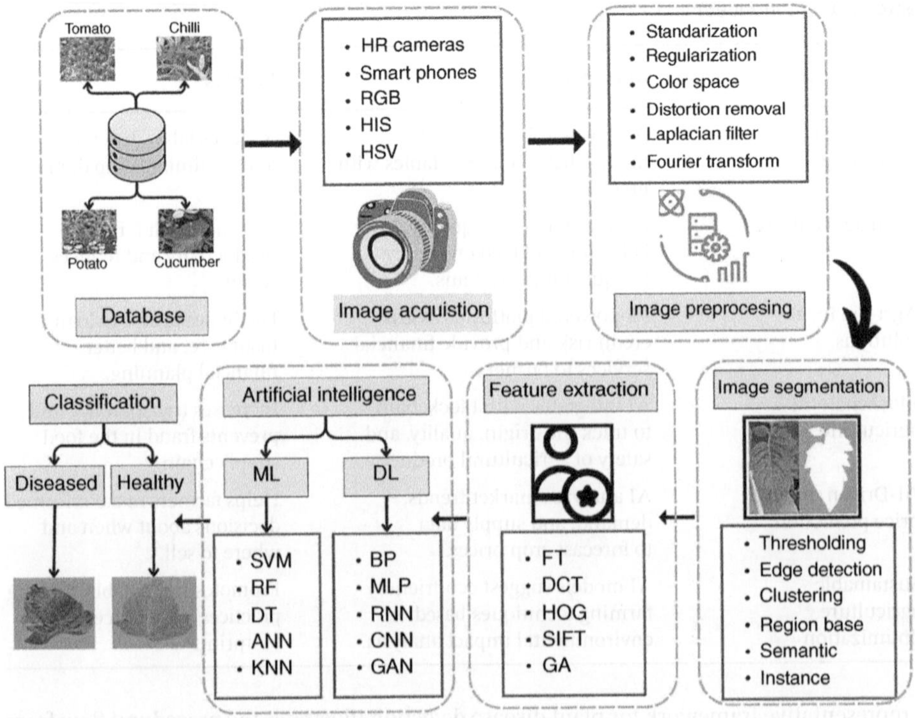

Figure 12.1 Comprehensive plant disease prediction system with key steps. Jafar et al. 2024 / Frontiers Media S.A / CC BY 4.0.

12.2.2 Image Pre-Processing: Refining Raw Data

Image pre-processing is a critical step that ensures raw data quality for optimal disease detection. Captured images often contain noise, distortions, poor illumination, or distracting backgrounds. Pre-processing techniques, such as resizing, noise reduction, and distortion removal, address these issues. High-frequency noise is minimized using low-pass filters, while techniques like Laplacian filtering enhance image clarity. Researchers employ advanced methods such as the Fourier transform (FT) and Gaussian filters to improve the quality of plant disease images. Adjustments, such as using a standardized white background, further enhance interpretability (Zhang et al., 2020; Ziyadinov and Tereshonok, 2023; Fan et al., 2019). Hyperspectral imaging (HSI) is commonly used, with its H (Hue) component being the most relevant for detailed analysis. Histograms are frequently utilized to enhance image power distribution, improving the diagnostic value of processed images. For instance, Nagasubramanian et al. also explored the potential of HSI combined with DL for plant disease identification, specifically targeting charcoal rot in soybean crops (Nagasubramanian et al., 2019). HSI involves capturing detailed spectral information across multiple wavelengths to analyze plant health (Figure 12.2).

In this study, healthy and infected soybean stem samples were collected at 3, 6, 9, 12, and 15 days after charcoal rot infection. Hyperspectral data cubes of inoculated and

Table 12.2 AI technologies in disease diagnosis for agriculture.

AI technology	Function	Application in disease diagnosis
Image acquisition	Captures high-quality images of plants using cameras, drones, and smartphones	Enables real-time monitoring and data collection for disease identification
Image pre-processing	Enhances image quality by reducing noise, adjusting contrast, and refining details	Improves clarity for better disease detection and classification
Image segmentation	Divides images into regions to isolate diseased areas from healthy plant parts	Essential for highlighting infected regions and pinpointing disease spread
Feature extraction	Identifies crucial image characteristics like color, texture, and morphology	Enables accurate disease classification using machine learning models
Machine Learning (ML) algorithms	Uses decision trees, random forests, and SVM for disease classification	Classifies plant diseases based on extracted features and datasets
Deep Learning (DL) Models	CNNs, ResNet, and MobileNet for advanced image recognition	Enhances disease detection accuracy in complex plant images
Hyperspectral Imaging (HSI)	Captures spectral data to analyze plant health beyond visible wavelengths	Differentiates between healthy and diseased plants using spectral signatures
Drones & IoT Sensors	Provides real-time field monitoring using AI-integrated UAVs and sensors	Facilitates early detection and precision agriculture strategies
Predictive Analytics	Analyzes historical and real-time data to forecast disease outbreaks	Helps in proactive disease management and preventive actions
AI Technology	Function	Application in Disease Diagnosis

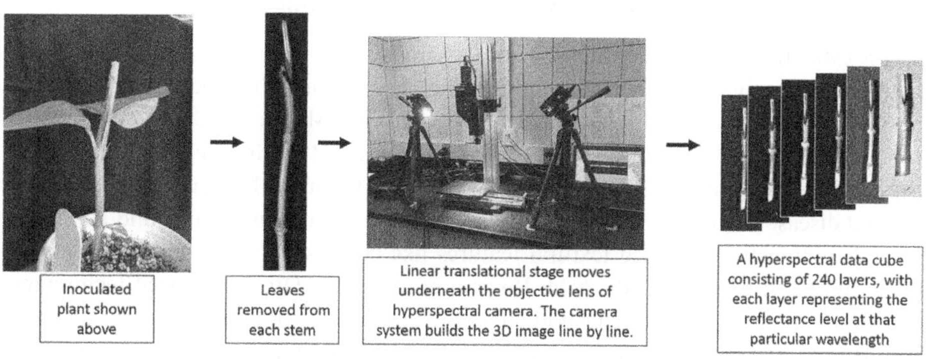

Figure 12.2 Illustration of the hyperspectral data generation procedure. Nagasubramanian et al. 2019 / Springer Nature / CC BY 4.0.

mock-inoculated stems were captured using a Pika XC hyperspectral line imaging scanner, consisting of an imager, a translational stage, a laptop with Spectronon-Pro software, and stable illumination from quartz-tungsten-halogen lamps. The imaging system covered a spectral range of 400–1000 nm with a resolution of 2.5 nm, allowing precise identification of disease symptoms. The authors developed a novel 3D deep convolutional neural network designed to process hyperspectral data, achieving a classification accuracy of 95.73% and an infected class F1 score of 0.87. Using saliency maps, the study visualizes the network's focus on regions with visible disease symptoms, reinforcing the model's physiological relevance. Notably, the near-infrared spectrum was identified as crucial for classification, aligning with established plant health assessment techniques. The study is just an example of the impact of HSI in disease management.

12.2.3 Image Segmentation: Isolating Disease Features

Segmentation divides an image into smaller, meaningful regions for detailed analysis, enabling the distinction between healthy and diseased plant areas. This step is essential for pinpointing disease-affected regions and extracting critical features. Traditional segmentation techniques, such as thresholding, edge detection, and clustering, rely on mathematical and image processing principles (Gill et al., 2022; Rajak et al., 2023). Thresholding, for example, segments images based on pixel intensity values, while edge detection identifies image boundaries. Clustering techniques like K-means and Fuzzy C-means group pixels with similar properties. Despite the utility of traditional methods, they often struggle with complex images (Yang et al., 2024; Chuang et al., 2006). Modern DL-based approaches, such as semantic segmentation and instance segmentation, overcome these limitations by leveraging neural networks. Semantic segmentation assigns a category label to each pixel, while instance segmentation identifies distinct objects in the image (Lei et al., 2024; Emek Soylu et al., 2023). These methods enhance accuracy and are especially useful in challenging scenarios.

12.2.4 Feature Extraction: Unlocking Key Data for Classification

Feature extraction involves identifying critical characteristics of plant images, such as color, texture, and morphology, which are then used for classification. Techniques like color histograms, Gabor filters, and the Grey-Level Co-occurrence Matrix (GLCM) are commonly employed to isolate relevant features (Zhang et al., 2017; Mostajer Kheirkhah and Asghari, 2019; Javidan et al., 2024). Texture analysis, in particular, has proven effective for plant disease diagnosis. Methods such as wavelet packet decomposition and FT enable researchers to extract detailed textural features like contrast, homogeneity, and entropy (Lei et al., 2024; Njoroge et al., 2024; Vishnoi et al., 2022). Other advanced techniques, such as Histogram of Oriented Gradients and Speeded-Up Robust Features, enhance the precision of feature extraction (Routray et al., 2017). Morphological traits, such as leaf shape and size, are also vital for identifying disease patterns, especially in localized infections.

12.2.5 Artificial Intelligence: Transforming Disease Detection

AI has revolutionized plant disease detection by introducing automation and scalability. ML algorithms and DL models are central to these advancements. ML techniques, such as decision trees (DTs), random forests (RFs), and support vector machines (SVMs), are widely used for disease classification. These algorithms classify leaves into categories of healthy or diseased based on labeled datasets (Natarajan et al., 2024; Singla et al., 2024). Supervised learning approaches rely on annotated datasets, while unsupervised learning identifies patterns without explicit labels. Semi-supervised learning, which combines elements of both, is particularly useful in scenarios where labeled data is limited. DL has further enhanced disease detection through its ability to process complex image data. Convolutional neural networks (CNNs), for example, have demonstrated remarkable accuracy in recognizing disease symptoms. In a recent study, Magesh et al. (2025) explained a ML-based approach to optimizing agricultural tillage and fertilizer application. Given the environmental concerns surrounding excessive tillage—such as soil erosion, carbon release, and nutrient runoff—the study employs a CNN to assess tillage intensity from camera-captured field images. This data, combined with soil sensor readings and external environmental inputs, feeds into a 10-parameter algorithm that recommends optimal tilling and fertilizer levels. An illustration of end-to-end solution with inputs, outputs, and core algorithm for real-time data capturing and decision-making process is shown in Figure 12.3.

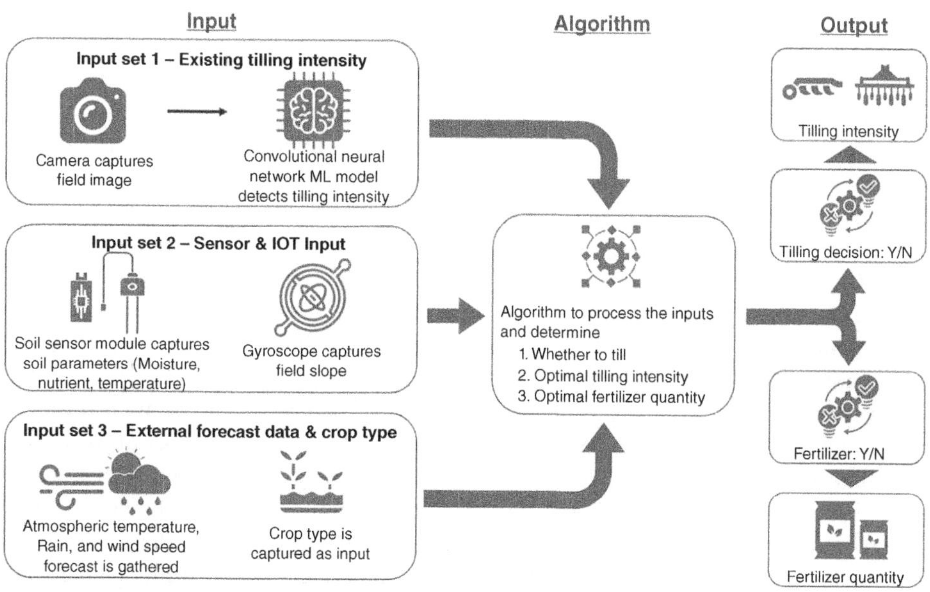

Figure 12.3 Illustration representing the end-to-end solution with inputs, outputs, and core algorithm for real-time data capturing and decision-making process. Magesh, S. (2025) / with permission of Springer Nature.

Advanced architectures like ResNet, DenseNet, and MobileNet outperform traditional methods in identifying subtle disease indicators (Ozdemir and Pacal, 2025; El Sakka et al., 2025). Pre-trained models tailored to agricultural research, such as VGG-16 and Inception, streamline the process by leveraging existing knowledge.

12.2.6 Future Innovations in AI-Driven Disease Diagnosis

The integration of AI with emerging technologies like drones, HSI, and the IoT promises transformative potential in agriculture. Drones equipped with advanced sensors can monitor large fields, capturing real-time data for disease detection. IoT-enabled systems provide continuous monitoring, enabling farmers to take timely preventive measures (Rejeb et al., 2022; Kondo et al., 2024). Additionally, AI-driven frameworks are expected to incorporate predictive analytics, offering insights into disease trends and facilitating proactive interventions. These innovations will ensure sustainable agricultural practices and enhance global food security.

12.2.7 Approaches in AI-Based Disease Detection

Plant pathology is an essential field that categorizes plant diseases into two broad groups: biotic and abiotic diseases. Biotic diseases are caused by living organisms, such as fungi, bacteria, insects, and viruses, which can significantly harm plant health. Conversely, abiotic diseases arise from non-living factors, including environmental stresses like nutritional deficiencies, chemical imbalances, metal toxicity, or physical damage. Plants often exhibit symptoms of abiotic stress under unfavourable conditions such as high temperatures, excessive moisture, inadequate light, a lack of essential nutrients, acidic soil pH, or exposure to greenhouse gases. Recognizing these conditions and their impact on plants is crucial for effective disease management. Plant diseases are often challenging to detect and classify due to the complex nature of symptoms, which may overlap between different diseases. Common symptoms like irregular leaf development, distorted pigmentation, stunted growth, and weakened or reduced pods can be difficult to interpret, making accurate diagnosis challenging. Such visible signs, however, remain crucial for identifying specific diseases and devising control measures. Maintaining plant health is vital not only for sustaining the ecosystem but also for maximizing vegetable production and ensuring the economic viability of agriculture.

The integration of AI-based technologies into plant pathology represents a transformative approach to managing vegetable diseases. Automated frameworks utilizing ML and DL techniques offer unparalleled accuracy and efficiency in detecting and classifying plant diseases, paving the way for sustainable agricultural practices (Rane et al., 2024; Guerri et al., 2024). By addressing the challenges associated with early diagnosis, these innovations contribute significantly to maximizing crop yield and ensuring global food security. Continued research and technological advancements hold the promise of even greater accuracy and applicability, empowering farmers and researchers to combat the growing threat of plant diseases effectively. This section explores cutting-edge approaches for automated disease detection and classification in vegetables using AI, particularly ML and DL. A growing body of research has demonstrated the effectiveness of these technologies in identifying and classifying diseases affecting commonly cultivated vegetables such as tomatoes, chilies, potatoes, and cucumbers.

12.2.8 Automated Disease Detection in Grains

AI-based detection of grain diseases representsa transformative approach in agriculture, using advanced computational methods to enhance disease management and crop productivity (Tirkey et al., 2023). The adoption of image processing and ML techniques has significantly improved the ability to diagnose diseases at an early stage, ensuring timely intervention and reducing crop losses. The process begins with image acquisition using cameras or sensors, which capture high-quality visuals of grain plants. These images undergo pre-processing to remove noise and improve clarity. Techniques like histogram equalization and median filtering are commonly used to enhance contrast and sharpness, ensuring the disease-affected regions are prominently visible (Kong et al., 2024). For instance, applying histogram-based segmentation to rice or maize images has proven effective in isolating diseased areas for further analysis. Segmentation plays a pivotal role in delineating healthy and diseased regions within plant images. Advanced methods like K-means clustering, fuzzy C-means, and edge detection algorithms are widely employed (Ghazal et al., 2024). These algorithms group similar pixel values, making it easier to identify disease-specific patterns such as rust spots or fungal infections. K-means clustering, for instance, has achieved high segmentation accuracy in wheat diseases like yellow rust and powdery mildew. Following segmentation, feature extraction identifies critical attributes such as color, texture, and shape, which are indicative of specific diseases. Methods like Scale-Invariant Feature Transform and GLCM extract these features, converting visual data into numerical parameters (Salehin et al., 2020; Iqbal et al., 2021). These features are then used to train classifiers like SVMs or artificial neural networks (ANNs), which distinguish between healthy and infected plants with remarkable accuracy (Ngugi et al., 2024; Ahmed et al., 2012). For example, ANN-based classifiers have reported up to 94% accuracy in detecting maize diseases such as leaf blight and common rust. To further enhance the precision of disease detection, feature selection and fusion are employed. Techniques such as principal component analysis (PCA) and Genetic Algorithms reduce redundant data, retaining only the most relevant features. This dimensionality reduction improves the performance of classification models, as seen in studies where PCA optimized classifiers achieved up to 95% accuracy in detecting rice blast (Xiao et al., 2018). The integration of DL architectures like CNN has elevated the effectiveness of AI-based detection systems. CNNs automatically learn hierarchical features from images, eliminating the need for manual feature extraction. For instance, CNN-based models have demonstrated over 97% accuracy in identifying fungal and bacterial infections in rice and soybean leaves (Ahila Priyadharshini et al., 2019; Darwish et al., 2020; Ramesh and Vydeki, 2020). Moreover, hybrid models combining CNN with other classifiers, such as SVM, have shown exceptional performance in multi-disease classification tasks (Hasan et al., 2019). Thus, AI-based detection of grain diseases not only ensures early diagnosis but also supports precision agriculture by minimizing manual interventions and resource wastage. These systems, driven by image processing, ML, and DL, are pivotal in tackling global challenges of food security and sustainable crop management. Continued advancements in computational algorithms and integration with IoT devices hold promise for even more robust and scalable solutions in the future.

12.2.9 Automated Tomato Disease Detection

Tomatoes (Solanum Lycopersicon), a widely cultivated crop with high nutritional value, are susceptible to several diseases caused by bacterial infections, fungal pathogens, and pest infestations. These diseases can severely impact production, necessitating efficient detection and management strategies. Several studies have proposed AI-based frameworks for tomato disease detection. For instance, researchers have utilized CNNs to differentiate between healthy and diseased tomato leaves with high accuracy. One such model, consisting of four convolutional layers followed by pooling layers and dense layers, demonstrated an accuracy of 87% when trained on 3663 images from the PlantVillage dataset (Francis and Deisy, 2019). Another study employed ML algorithms such as k-nearest neighbors, logistic regression, DTs, RFs, and SVMs to classify tomato leaf diseases, achieving a remarkable 94% accuracy with RF (Basavaiah and Arlene Anthony, 2020). Further advancements include probabilistic neural networks, which effectively classified diseases such as Verticillium wilt, powdery mildew, Septoria leaf spot, and spider mite infestations, achieving an accuracy of 91.88% (Balakrishna and Rao, 2019). Optimized models combining feature extraction techniques like K-means clustering with backpropagation neural networks have reached up to 99.4% accuracy (Vadivel and Suguna, 2022). Similarly, fine-tuned architectures like ResNet and Xception have achieved classification accuracies as high as 99.95% for distinguishing early blight in tomatoes (Chakravarthy and Raman, 2020).

12.2.10 Automated Chili Disease Detection

Chilies, known for their diverse uses as spices, oils, and medicinal ingredients, are prone to various diseases caused by bacterial, viral, and fungal pathogens. Approximately 45 different viruses are known to infect chili plants, significantly impacting their health and yield. Automated detection methods have demonstrated promise in addressing this challenge. Researchers have employed DL models such as VGG19, ResNet, and DarkNet53 for chili disease classification (Naik et al., 2022). ML techniques such as SVM, RF, and ANNs have also been effective in identifying diseases, with combined DL and SVM approaches yielding accuracies of over 92%. Advanced techniques like Bayesian learning integrated into residual networks have enhanced classification performance, achieving nearly 99% accuracy on datasets including PlantVillage images (Sachdeva et al., 2021). Data augmentation and pre-processing strategies have further improved model performance, enabling reliable disease classification in chili plants.

12.2.11 Automated Potato Disease Detection

Potatoes, the fourth most cultivated crop globally, are highly susceptible to diseases such as black scurf, common scab, and pink rot, which can cause substantial losses. AI-driven frameworks have proven effective in diagnosing potato diseases and improving crop health management. ML models such as RF, SVM, and ANN have been utilized to classify potato leaf diseases, with ANN achieving an accuracy of 92% (Patil et al., 2017). Another approach integrated fuzzy c-means clustering with ML algorithms, yielding 91.67% accuracy in detecting late blight and early blight (Suttapakti and Bunpeng, 2019). DL techniques have also been extensively explored; CNN-based architectures such as AlexNet have demonstrated accuracies exceeding 98% in distinguishing healthy and diseased potato leaves (Arya and Singh, 2019). Innovative methods like reweighted cross-entropy loss functions in DenseNet architectures have addressed issues of data imbalance, achieving classification accuracies of 97.2% (Mahum et al., 2023). Additionally, custom-built CNN models trained on augmented datasets have reached accuracies as high as 99.62%, showcasing their potential for reliable disease diagnosis (Singh and Yogi, 2023).

12.2.12 Automated Cucumber Disease Detection

Cucumbers, valued for their refreshing qualities and high-water content, are vulnerable to diseases like powdery mildew, anthracnose, and angular leaf spots. These diseases can significantly affect leaf health and crop yield, highlighting the need for effective detection strategies. AI-based models have shown promise in cucumber disease detection. For instance, CNN-based U-Net architectures have achieved pixel-level segmentation accuracies exceeding 96% for powdery mildew (Lin et al., 2019). Other approaches involve using pre-trained models like VGG19 and VGG-M for feature extraction, combined with SVM for classification, achieving an overall accuracy of 98.08% (Khan et al., 2020). Advanced techniques such as the global pooling dilated CNN have enhanced disease recognition by expanding convolutional receptive fields and utilizing dilated convolution algorithms, resulting in accuracies of over 95% (Zhang et al., 2019). Sparse representation classifiers and hybrid frameworks integrating data augmentation and feature selection have further improved classification performance, achieving recognition rates of 85% to 93% across various cucumber diseases (Zhang et al., 2017; Kianat et al., 2021).

12.3 AI in Pest Identification and Control

AI is revolutionizing pest identification and control, providing solutions to the limitations of traditional methods. Pest infestations account for significant global agricultural losses, and traditional pest management often involves excessive pesticide use, harming ecosystems and beneficial organisms (Aktar et al., 2009). AI addresses these challenges by offering advanced tools for accurate pest detection, monitoring, and sustainable control strategies. It also offers unparalleled potential to revolutionize pest identification and control. By combining innovative technologies like image recognition, IoT devices, and predictive modeling, AI facilitates accurate pest management while minimizing environmental impacts. In a nutshell, AI is transforming pest management by providing accurate, efficient, and sustainable solutions to pest infestations. Traditional pest control methods often involve excessive pesticide use, leading to environmental damage, pesticide resistance, and harm to beneficial organisms. AI-driven technologies, including image recognition, IoT devices, unmanned aerial vehicles (UAVs), and predictive analytics, have revolutionized

Table 12.3 Key applications of AI in pest identification and control.

AI application	Technology used	Functionality	Examples of applications
Pest Identification and Classification	Image recognition, Deep Learning (CNNs)	Identifies pest species and assesses crop damage using high-resolution images	Plant Village, Plantix, CottonAce, PestID
Monitoring and Surveillance	Smart traps, IoT sensors, Cloud-based analytics	Real-time pest detection, automatic species identification, and continuous monitoring	AI-based pheromone traps, DeepAgro, PestNet
Precision Pest Control	UAVs (drones), ML-based decision support systems	Targeted pesticide application, reducing overuse and optimizing pest management	AI-driven UAVs for precision spraying
Sensor Networks & IoT Integration	AI-powered IoT devices, Predictive analytics	Predicts pest outbreaks, automates monitoring, and improves intervention strategies	Trapview, Spornado

pest identification, monitoring, and precision control (Kariyanna and Sowjanya, 2024). These technologies enable early detection, reduce manual efforts, and optimize pesticide application, ultimately enhancing crop yield and sustainability. Table 12.3 key AI applications in pest identification and control, followed by a detailed description of each approach.

12.3.1 Pest Identification and Classification

AI-driven image recognition technologies are transforming pest management by enabling precise identification of pest species and assessment of crop damage. High-resolution images captured through drones, smartphones, or field cameras serve as the primary data input for these systems. ML algorithms, particularly DL models like CNNs, analyze these images to classify pests with remarkable accuracy. This rapid identification capability is critical in mitigating infestations and preserving crop yields. Applications such as Plant Village, Plantix, CottonAce, and PestID utilize AI-powered algorithms to streamline pest identification. Farmers and agricultural workers can upload images of affected crops or visible pests, which are analyzed against vast datasets of pest species and damage scenarios. These systems can differentiate pests with subtle morphological differences and assess the severity of damage, providing detailed recommendations for targeted interventions. For instance, CottonAce specializes in predicting and addressing infestations like the pink bollworm in cotton fields, allowing farmers to act swiftly.

DL models extend the scope of pest identification by analyzing diverse data types, including acoustic signals and environmental data. Platforms like DeepAgro and PestNet demonstrate the versatility of DL by offering real-time identification in field conditions (Kariyanna and Sowjanya, 2024). These tools enhance detection accuracy by integrating multi-modal data, making them invaluable for identifying hidden or hard-to-detect pests.

The efficiency of AI-driven technologies in pest identification contributes significantly to sustainable agriculture. By enabling targeted pest control measures, these tools reduce the need for broad-spectrum pesticide applications, minimizing environmental damage, and preventing pesticide resistance. Moreover, their ability to deliver timely, accurate results helps farmers save time and resources, ensuring healthier crops and higher yields. Thus, the integration of AI-powered image recognition and DL models into pest management represents a significant leap forward in agricultural practices (Leybourne et al., 2025). These advancements empower farmers with the tools to address pest infestations more effectively, fostering sustainable farming and contributing to global food security.

12.3.2 Monitoring and Surveillance

AI-powered systems are transforming pest control through real-time monitoring and surveillance, offering innovative solutions that reduce manual labor and improve precision. Smart traps integrated with AI algorithms and sensors, such as those developed for managing red palm weevils, automatically detect and count pests (Ghate and Kulkarni, 2024). These systems identify species based on visual or behavioral cues and relay the collected data to cloud networks. Farmers can access this data via user-friendly platforms, enabling them to take timely and informed actions against pest infestations. Unlike traditional methods that rely on manual counting and periodic field inspections, AI-based tools provide continuous monitoring with greater accuracy and efficiency. This is particularly valuable in pheromone-based pest management systems, where the ability to track pest activity in real-time ensures that interventions occur at the optimal time, improving the overall effectiveness of control measures. One of the key advantages of AI-enabled systems is their ability to analyze historical and real-time data to identify patterns and predict potential pest outbreaks (Ali et al., 2023). These predictive insights empower farmers to adopt preventive measures, such as deploying biological controls or adjusting irrigation schedules, before infestations escalate. Additionally, these tools reduce the overuse of pesticides by enabling targeted applications, preserving beneficial organisms and minimizing environmental harm. The scalability of AI-powered monitoring systems makes them suitable for various agricultural setups, from smallholder farms to large-scale operations. By automating labour-intensive tasks, these systems lower costs and make advanced pest management technologies accessible to a broader range of farmers. Therefore, AI-driven monitoring and surveillance tools offer a game-changing approach to pest control. By providing real-time, accurate data and actionable insights, they enhance the efficiency of pest management practices, mitigate crop losses, and support sustainable agriculture. These innovations represent a significant step forward in ensuring global food security while reducing the environmental footprint of agricultural activities.

12.3.3 Precision Pest Control

AI has revolutionized precision agriculture, offering advanced tools for targeted pest control. By harnessing AI's ability to analyze large datasets and deliver actionable insights, farmers can address pest issues with greater accuracy, efficiency, and sustainability. This approach not only minimizes the environmental impact of pest management but also reduces operational costs and optimizes resource use. One of the most impactful innovations in precision pest control is the integration of UAVs equipped with AI systems (Gao et al., 2019). These drones use ML algorithms to identify affected areas by analyzing crop health, pest activity, and damage patterns. By distinguishing between spray and non-spray zones, UAVs guide the precise application of pesticides, significantly reducing waste and the risk of over-application. This targeted method ensures that only the necessary areas are treated, preserving beneficial organisms and maintaining ecological balance.

AI-driven tools also enhance IPM strategies, which combine biological, cultural, and chemical practices to control pests (Prabha et al., 2022). Smart traps, equipped with sensors and AI algorithms, automatically detect and monitor pest populations, providing real-time data to farmers. These traps help pinpoint infestation hotspots, enabling timely interventions. Automated robots further complement this system by navigating farm fields and selectively applying control measures where they are most needed. Moreover, AI analyzes environmental data, weather patterns, and historical pest activity to predict outbreaks. These predictions allow farmers to act proactively, deploying preventive measures rather than reactive solutions. For instance, AI-powered DSS recommend the most effective pest control methods, taking into account factors such as pest species, crop type, and field conditions. The integration of AI in precision agriculture significantly enhances productivity and sustainability. By reducing reliance on broad-spectrum chemical pesticides, it minimizes environmental contamination and supports the long-term health of agricultural ecosystems. This innovative approach ensures higher crop yields, improved food quality, and reduced costs for farmers, while also contributing to global efforts in sustainable farming and food security.

12.3.4 Sensor Network and Integration with IoT and Predictive Analysis

The integration of AI with IoT devices has revolutionized pest management by creating an intelligent, interconnected system capable of real-time monitoring and data-driven decision-making (Sharma and Shivandu, 2024). By combining sensor networks with advanced AI algorithms, farmers can proactively manage pest populations and minimize crop damage, fostering more efficient and sustainable agricultural practices. IoT devices equipped with sensors and cameras are deployed across agricultural fields to continuously monitor pest activity and environmental conditions such as temperature, humidity, and soil moisture. These devices collect vast amounts of data in real time and transmit it to centralized databases via cloud networks. AI algorithms then analyze this data to identify pest population trends, predict potential outbreaks, and generate actionable insights. Additionally, integration of AI and IoT enhances the precision of pest control strategies. By analyzing environmental data alongside pest activity, AI can recommend optimal intervention times and methods. For instance, sensors detecting a rise in pest populations might trigger an alert, prompting localized pesticide application only where necessary. This targeted approach conserves resources, reduces pesticide use, and minimizes environmental impact.

Furthermore, innovative platforms like Trapview and Spornado exemplify the effectiveness of integrating AI and IoT for pest management (Kariyanna and Sowjanya, 2024;

Badia-Melis et al., 2016). Trapview uses IoT-enabled traps with cameras to automatically capture images of pests, which are analyzed by AI to identify species and estimate their population size. Similarly, Spornado employs spore detection systems to monitor fungal pathogens in the air, alerting farmers to disease risks early enough to take preventive action. These systems automate pest monitoring, eliminating the need for labor-intensive manual inspections while ensuring accuracy and efficiency.

12.4 AI for Predictive Modeling in Disease and Pest Management

AI is helping farmers predict pest attacks and plant diseases before they happen. By studying past data on pest outbreaks, weather conditions, soil health, and crop growth, AI can forecast when and where pests or diseases might appear. This allows farmers to take action early, preventing damage and reducing the need for excessive pesticide use. AI works by analyzing patterns in large amounts of data that humans might miss. For example, AI can track how changes in temperature, humidity, and rainfall affect pests like aphids or locusts (Kapetas et al., 2025). If conditions are right for an outbreak, AI can send alerts so farmers can prepare in advance by using traps, adjusting irrigation, or applying natural pest control methods. Some AI-powered systems, like **PlantVillage Nuru** and **IBM Watson Decision Platform for Agriculture**, provide early warnings by using satellite images, sensors, and field reports (Mmbando, 2025). These systems help farmers decide the best time to spray pesticides, apply fertilizers, or change their farming methods to protect crops. Instead of spraying chemicals everywhere, AI helps target only the affected areas, which saves money and is better for the environment. AI also helps in managing plant diseases caused by bacteria, fungi, and viruses (Garg et al., 2024). By using special cameras and AI algorithms, early signs of disease can be detected before they are visible to the naked eye. This means farmers can take quick action, such as changing plant nutrition, isolating infected crops, or choosing disease-resistant seeds for the next planting season. In short, AI makes pest and disease management smarter and more efficient. Instead of waiting for problems to appear, farmers can use AI to predict and prevent them, leading to healthier crops, better yields, and a more sustainable way of farming.

12.5 Economic and Environmental Impacts of AI in Pest and Disease Management

The integration of AI in pest and disease management is bringing both economic and environmental benefits to agriculture. By enhancing pest detection, optimizing pesticide use, and improving overall efficiency, AI is helping farmers reduce costs while ensuring sustainable farming practices (Mana et al., 2024). These advancements not only boost productivity but also contribute to environmental conservation by minimizing the negative impacts of traditional pest control methods. The adoption of AI in pest and disease management is transforming agriculture by offering cost-effective and environmentally sustainable solutions. AI-driven technologies enhance pest detection, optimize pesticide usage, and improve overall efficiency, leading to significant economic benefits for farmers

Table 12.4 Economic and environmental impacts of AI in pest and disease management.

Category	AI application	Key benefits	Examples
Economic benefits	AI-driven pest detection and targeted control	Reduces pesticide costs, minimizes crop losses, increases productivity	Smart traps, AI-based monitoring systems, UAV sprayers
	Predictive modeling for pest outbreaks	Early warning system, optimized intervention timing	AI-powered forecasting models, IoT-based data analytics
	Automation in pest management	Reduces labor costs, improves farm efficiency	AI-assisted robots, autonomous pest removal drones
Environmental benefits	Precision pesticide application	Minimizes chemical usage, protects beneficial organisms	AI-guided UAV spraying, Smart pest traps
	AI-driven Integrated Pest Management (IPM)	Promotes sustainable practices, reduces pesticide resistance	AI-supported biological control strategies, habitat modification
	Climate-smart AI solutions	Optimizes resource use, reduces carbon footprint	AI-enabled irrigation and fertilization systems

while reducing the environmental footprint of traditional pest control methods. Table 12.4 summarizes the key economic and environmental impacts of AI in pest and disease management, followed by a detailed discussion on these aspects.

12.5.1 Economic Benefits

AI-driven pest management reduces farming costs by ensuring more precise pest detection and targeted interventions. Traditional methods often rely on excessive pesticide applications, which can be expensive and inefficient (Brunelle et al., 2024). AI-powered tools, such as smart traps, drones, and automated monitoring systems, identify pest infestations early and recommend precise treatment areas (Kariyanna and Sowjanya, 2024). This reduces the overuse of pesticides, lowering costs for farmers while maintaining crop health. Additionally, AI helps improve crop yields by predicting and preventing pest and disease outbreaks before they become severe. By using predictive modeling, AI analyzes factors such as weather patterns, soil conditions, and past infestation trends to alert farmers about potential risks (Elbasi et al., 2023). This early warning system allows for timely action, reducing crop losses and increasing overall farm productivity. With better yields and lower input costs, farmers can maximize profits and improve their financial stability. AI also reduces labor costs by automating tasks such as pest monitoring, data collection, and decision-making. Smart surveillance systems eliminate the need for frequent manual field inspections, allowing farmers to focus on other essential aspects of farm management. In large-scale agriculture, autonomous robots equipped with AI can identify and remove pests without human intervention, further reducing labor expenses.

12.5.2 Environmental Benefits

One of the most significant environmental advantages of AI in pest management is the reduction of chemical pesticide use. Excessive pesticide application not only affects soil and water quality but also harms beneficial insects, pollinators, and other wildlife. AI-powered precision farming ensures that pesticides are used only in necessary areas, minimizing their environmental impact. Furthermore, AI-driven IPM strategies promote sustainable farming by integrating biological and cultural pest control methods with AI-based decision-making (Elbasi et al., 2023). Instead of relying solely on chemical treatments, AI helps farmers adopt alternative solutions such as crop rotation, habitat modification, and the use of natural predators. These environmentally friendly approaches maintain ecological balance and reduce the risks associated with pesticide resistance (Barathi et al., 2024). AI also contributes to climate change mitigation by optimizing resource use. For example, AI-enabled irrigation and fertilization systems ensure that water and nutrients are applied efficiently, preventing wastage and soil degradation (Aijaz et al., 2025). By improving overall farm sustainability, AI helps reduce the carbon footprint of agricultural activities. In conclusion, the economic and environmental impacts of AI in pest and disease management are profound. AI not only enhances farm profitability by reducing costs and increasing yields but also supports sustainable agriculture by minimizing pesticide use and conserving natural resources. As AI technology continues to evolve, its role in fostering eco-friendly and economically viable farming practices will become even more significant, paving the way for a smarter and more sustainable future in agriculture.

12.6 Challenges

The effectiveness of AI in pest and disease management heavily depends on the availability and quality of data. However, many regions, especially in developing countries, lack well-documented and diverse datasets (Purushotham et al., 2024). AI models are often trained on region-specific data, limiting their ability to function accurately in different climatic and geographical conditions. Additionally, inconsistencies and biases in data can lead to incorrect predictions, reducing the reliability of AI-powered pest management tools. To improve accuracy, it is essential to standardize and continuously update datasets using IoT devices, remote sensing technologies, and contributions from farmers (Ennouri et al., 2021). Beyond data limitations, ethical and legal concerns also pose challenges to AI adoption in agriculture. AI-driven pest monitoring systems collect extensive data through cloud-based platforms, raising questions about privacy, ownership, and control. Farmers may hesitate to adopt AI if they fear that their data could be exploited by corporations or used without their consent. Additionally, AI-driven decision-making creates uncertainty regarding accountability—if an AI system provides incorrect pest control recommendations leading to crop failure, it is unclear whether responsibility falls on the farmer, the technology provider, or the AI system itself. To build trust and encourage AI adoption, clear legal frameworks and ethical guidelines must be established to protect farmers' interests.

Another major challenge is the high cost of implementing AI-powered pest management systems. While these technologies can reduce costs in the long run by improving efficiency, the initial investment in drones, smart traps, and IoT-based monitoring systems can be too expensive for small-scale farmers. Additionally, many AI applications require reliable internet access, electricity, and technical knowledge, which may not be available in remote or underdeveloped farming regions. Bridging this digital divide requires targeted government initiatives, subsidies, and the development of more affordable, user-friendly AI solutions tailored to smallholder farmers. Furthermore, AI must continuously adapt to the complexities of agricultural environments. Pest populations and disease patterns are influenced by multiple environmental factors such as climate change, soil health, and biodiversity. AI models must evolve to recognize new pest behaviors, emerging plant diseases, and shifts in environmental conditions (Atapattu et al., 2024). However, many current AI systems struggle with real-world complexities and require frequent updates and human intervention to maintain accuracy. Developing self-learning AI models that can automatically refine their predictions over time will be crucial for enhancing the precision and reliability of AI-driven pest and disease management (Dhanya et al., 2022).

12.7 Future Directions

The future of AI in pest and disease management will be driven by advancements in robotics, predictive analytics, genetic research, and accessibility. One of the most promising developments is the integration of AI-powered robotics and automation (Sutikno, 2024). Autonomous robots and drones will be equipped with AI to detect, analyze, and respond to pest infestations without human intervention. These intelligent systems will be capable of identifying pest hotspots, applying targeted treatments, and even physically removing pests, reducing dependence on chemical pesticides and promoting sustainable pest control. Another key advancement will be in predictive analytics for early pest detection (Gupta et al., 2024). AI models will become even more precise by incorporating real-time data from weather stations, soil sensors, and satellite imagery. By analyzing environmental conditions and historical pest trends, these models will generate accurate forecasts, allowing farmers to take proactive measures before infestations become severe. This predictive capability will help minimize crop losses and reduce the need for excessive pesticide use, leading to more efficient and environmentally friendly pest management practices. In addition to monitoring and control, AI will revolutionize crop breeding for pest resistance (Zhang et al., 2025). By analyzing vast genomic datasets, AI will help scientists identify genes responsible for natural pest and disease resistance in crops. This will accelerate the development of genetically improved crop varieties that require fewer chemical treatments while offering higher resilience against pests. AI-driven genetic research will play a crucial role in reducing farmers' reliance on pesticides and ensuring more sustainable agricultural practices. To make AI-driven pest management more accessible, the future will also focus on user-friendly AI solutions for small-scale farmers. Many smallholder farmers face challenges in adopting AI due to a lack of technical expertise and resources. Future AI tools will be designed with simplified interfaces, multilingual support, and mobile compatibility, ensuring that farmers of all backgrounds can use them effectively. AI-powered virtual assistants will provide step-by-step guidance on pest identification, disease management, and treatment recommendations, making AI a practical and inclusive tool for all agricultural settings (Assimakopoulos et al., 2024). As AI continues to evolve, its role in pest and disease management will become even more impactful, enabling farmers to adopt smarter, more efficient, and environmentally sustainable practices. By integrating robotics, predictive analytics, genetic advancements, and user-friendly applications, AI will help transform agriculture, ensuring higher crop yields, reduced losses, and a more secure global food supply.

12.8 Conclusion

AI has significantly transformed pest and disease management by enabling early detection, predictive modeling, and precision control. AI-driven technologies, such as smart traps, drones, IoT sensors, and ML algorithms, have improved the accuracy and efficiency of pest monitoring while reducing reliance on chemical pesticides. These innovations have not only enhanced agricultural productivity but also promoted environmentally sustainable practices by minimizing resource wastage and ecological damage. Looking ahead, AI's role in sustainable agriculture will continue to expand with advancements in robotics, automation, predictive analytics, and genetic research. The integration of AI with IoT, remote sensing, and user-friendly mobile applications will make precision pest management more accessible to farmers worldwide. As AI evolves, it will help build more resilient agricultural systems, ensuring higher yields, reduced crop losses, and a balanced approach to pest control that supports both food security and environmental conservation. By embracing AI-driven solutions, the future of agriculture will be smarter, more sustainable, and better equipped to address global challenges in food production.

References

Ahila Priyadharshini, R., Arivazhagan, S., Arun, M., and Mirnalini, A. (2019). Maize leaf disease classification using deep convolutional neural networks. *Neural Comput Appl* 31: 8887–8895. https://doi.org/10.1007/s00521-019-04228-3.

Ahmed, F., Al-Mamun, H.A., Bari, A.S.M.H. et al. (2012). Classification of crops and weeds from digital images: a support vector machine approach. *Crop Protection* 40: 98–104. https://doi.org/10.1016/j.cropro.2012.04.024.

Aijaz, N., Lan, H., Raza, T. et al. (2025). Artificial intelligence in agriculture: advancing crop productivity and sustainability. *J Agric Food Res* 20: 101762. https://doi.org/10.1016/j.jafr.2025.101762.

Aktar, W., Sengupta, D., and Chowdhury, A. (2009). Impact of pesticides use in agriculture: their benefits and hazards. *Interdiscip Toxicol* 2: 1–12. https://doi.org/10.2478/v10102-009-0001-7.

Ali, M.A., Dhanaraj, R.K., and Nayyar, A. (2023). A high performance-oriented AI-enabled IoT-based pest detection system using sound analytics in large agricultural field. *Microprocess Microsyst* 103: 104946. https://doi.org/10.1016/j.micpro.2023.104946.

Arya S, Singh R (2019) A Comparative Study of CNN and AlexNet for Detection of Disease in Potato and Mango leaf. In: 2019 International Conference on Issues and Challenges in Intelligent Computing Techniques (ICICT). IEEE, pp 1–6

Assimakopoulos, F., Vassilakis, C., Margaris, D. et al. (2024). Artificial intelligence tools for the agriculture value chain: status and prospects. *Electronics (Basel)* 13: 4362. https://doi.org/10.3390/electronics13224362.

Atapattu, A.J., Perera, L.K., Nuwarapaksha, T.D. et al. (2024). Challenges in achieving artificial intelligence in agriculture. In: *Artificial Intelligence Techniques in Smart Agriculture*, 7–34. Singapore: Springer Nature Singapore.

Badia-Melis, R., Qian, J.P., Fan, B.L. et al. (2016). Artificial neural networks and thermal image for temperature prediction in apples. *Food Bioproc Tech* 9: 1089–1099. https://doi.org/10.1007/s11947-016-1700-7.

Balakrishna, K. and Rao, M. (2019). Tomato plant leaves disease classification using KNN and PNN. *International Journal of Computer Vision and Image Processing* 9: 51–63. https://doi.org/10.4018/IJCVIP.2019010104.

Balaska, V., Adamidou, Z., Vryzas, Z., and Gasteratos, A. (2023). Sustainable crop protection via robotics and artificial intelligence solutions. *Machines* 11: 774. https://doi.org/10.3390/machines11080774.

Barathi, S., Sabapathi, N., Kandasamy, S., and Lee, J. (2024). Present status of insecticide impacts and eco-friendly approaches for remediation-a review. *Environ Res* 240: 117432. https://doi.org/10.1016/j.envres.2023.117432.

Basavaiah, J. and Arlene Anthony, A. (2020). Tomato leaf disease classification using multiple feature extraction techniques. *Wirel Pers Commun* 115: 633–651. https://doi.org/10.1007/s11277-020-07590-x.

Brunelle, T., Chakir, R., Carpentier, A. et al. (2024). Reducing chemical inputs in agriculture requires a system change. *Commun Earth Environ* 5: 369. https://doi.org/10.1038/s43247-024-01533-1.

Chakravarthy AS, Raman S (2020) Early blight identification in tomato leaves using deep learning. In: 2020 International Conference on Contemporary Computing and Applications (IC3A). IEEE, pp 154–158

Chuang, K.-S., Tzeng, H.-L., Chen, S. et al. (2006). Fuzzy c-means clustering with spatial information for image segmentation. *Computerized Medical Imaging and Graphics* 30: 9–15. https://doi.org/10.1016/j.compmedimag.2005.10.001.

Darwish, A., Ezzat, D., and Hassanien, A.E. (2020). An optimized model based on convolutional neural networks and orthogonal learning particle swarm optimization algorithm for plant diseases diagnosis. *Swarm Evol Comput* 52: 100616. https://doi.org/10.1016/j.swevo.2019.100616.

Daszkiewicz, T. (2022). Food production in the context of global developmental challenges. *Agriculture* 12: 832. https://doi.org/10.3390/agriculture12060832.

Dhanya, V.G., Subeesh, A., Kushwaha, N.L. et al. (2022). Deep learning based computer vision approaches for smart agricultural applications. *Artificial Intelligence in Agriculture* 6: 211–229. https://doi.org/10.1016/j.aiia.2022.09.007.

El Sakka, M., Ivanovici, M., Chaari, L., and Mothe, J. (2025). A review of CNN applications in smart agriculture using multimodal data. *Sensors* 25: 472. https://doi.org/10.3390/s25020472.

Elbasi, E., Zaki, C., Topcu, A.E. et al. (2023). Crop prediction model using machine learning algorithms. *Applied Sciences* 13: 9288. https://doi.org/10.3390/app13169288.

Emek Soylu, B., Guzel, M.S., Bostanci, G.E. et al. (2023). Deep-learning-based approaches for semantic segmentation of natural scene images: a review. *Electronics (Basel)* 12: 2730. https://doi.org/10.3390/electronics12122730.

Ennouri, K., Smaoui, S., Gharbi, Y. et al. (2021). Usage of artificial intelligence and remote sensing as efficient devices to increase agricultural system yields. *J Food Qual* 2021: 1–17. https://doi.org/10.1155/2021/6242288.

Fan, L., Zhang, F., Fan, H., and Zhang, C. (2019). Brief review of image denoising techniques. *Vis Comput Ind Biomed Art* 2: 7. https://doi.org/10.1186/s42492-019-0016-7.

Francis M, Deisy C (2019) Disease detection and classification in agricultural plants using convolutional neural networks — a visual understanding. In: 2019 6th International Conference on Signal Processing and Integrated Networks (SPIN). IEEE, pp 1063–1068.

Gao, P., Zhang, Y., Zhang, L. et al. (2019). Development of a recognition system for spraying areas from unmanned aerial vehicles using a machine learning approach. *Sensors* 19: 313. https://doi.org/10.3390/s19020313.

Garg, T., Dwivedi, P., Mishra, M.K. et al. (2024). Artificial intelligence in plant disease identification. *Empowering agriculture* 55: 179–193. https://doi.org/10.1016/bs.mim.2024.05.007.

Ghate, U. and Kulkarni, H. (2024). Artificial intelligence & nature-based solutions in agriculture: a BT cotton pest management case study in India. *Qeios*. https://doi.org/10.32388/NFGP2F.

Ghazal, S., Munir, A., and Qureshi, W.S. (2024). Computer vision in smart agriculture and precision farming: techniques and applications. *Artificial Intelligence in Agriculture* 13: 64–83. https://doi.org/10.1016/j.aiia.2024.06.004.

Gill, T., Gill, S.K., Saini, D.K. et al. (2022). A comprehensive review of high throughput phenotyping and machine learning for plant stress phenotyping. *Phenomics* 2: 156–183. https://doi.org/10.1007/s43657-022-00048-z.

Grieve, B.D., Duckett, T., Collison, M. et al. (2019). The challenges posed by global broadacre crops in delivering smart agri-robotic solutions: a fundamental rethink is required. *Glob Food Sec* 23: 116–124. https://doi.org/10.1016/j.gfs.2019.04.011.

Guerri, M.F., Distante, C., Spagnolo, P. et al. (2024). Deep learning techniques for hyperspectral image analysis in agriculture: a review. *ISPRS Open Journal of Photogrammetry and Remote Sensing* 12: 100062. https://doi.org/10.1016/j.ophoto.2024.100062.

Gupta, D.K., Pagani, A., Zamboni, P., and Singh, A.K. (2024). AI-powered revolution in plant sciences: advancements, applications, and challenges for sustainable agriculture and food security. *Exploration of Foods and Foodomics* 2: 443–459. https://doi.org/10.37349/eff.2024.00045.

Hasan, M.J., Mahbub, S., Alom, M.S., and Md, A.N. (2019). Rice disease identification and classification by integrating support vector machine with deep convolutional neural network. In: *2019 1st International Conference on Advances in Science*, 1–6. IEEE: Engineering and Robotics Technology (ICASERT).

Hossain, M.M., Sultana, F., Mostafa, M. et al. (2024). Plant disease dynamics in a changing climate: impacts, molecular mechanisms, and climate-informed strategies for sustainable management. *Discover Agriculture* 2: 132. https://doi.org/10.1007/s44279-024-00144-w.

Iqbal, N., Mumtaz, R., Shafi, U., and Zaidi, S.M.H. (2021). Gray level co-occurrence matrix (GLCM) texture based crop classification using low altitude remote sensing platforms. *PeerJ Comput Sci* 7: e536. https://doi.org/10.7717/peerj-cs.536.

Jackulin, C. and Murugavalli, S. (2022). A comprehensive review on detection of plant disease using machine learning and deep learning approaches. *Measurement: Sensors* 24: 100441. https://doi.org/10.1016/j.measen.2022.100441.

Jafar, A., Bibi, N., Naqvi, R.A. et al. (2024). Revolutionizing agriculture with artificial intelligence: plant disease detection methods, applications, and their limitations. *Front Plant Sci* 15: 0–20. https://doi.org/10.3389/fpls.2024.1356260.

Javaid, M., Haleem, A., Khan, I.H., and Suman, R. (2023). Understanding the potential applications of artificial intelligence in agriculture sector. *Advanced Agrochem* 2: 15–30. https://doi.org/10.1016/j.aac.2022.10.001.

Javidan, S.M., Banakar, A., Rahnama, K. et al. (2024). Feature engineering to identify plant diseases using image processing and artificial intelligence: a comprehensive review. *Smart Agricultural Technology* 8: 100480. https://doi.org/10.1016/j.atech.2024.100480.

Jin, D., Yin, H., Zheng, R. et al. (2023). PlantInfoCMS: scalable plant disease information collection and management system for training AI models. *Sensors* 23: 5032. https://doi.org/10.3390/s23115032.

Junaid, M., Shaikh, A., Hassan, M.U. et al. (2021). Smart agriculture cloud using AI based techniques. *Energies (Basel)* 14: 5129. https://doi.org/10.3390/en14165129.

Kapetas, D., Christakakis, P., Faliagka, S. et al. (2025). AI-driven insect detection, real-time monitoring, and population forecasting in greenhouses. *AgriEngineering* 7: 29. https://doi.org/10.3390/agriengineering7020029.

Kariyanna, B. and Sowjanya, M. (2024). Unravelling the use of artificial intelligence in management of insect pests. *Smart Agricultural Technology* 8: 100517. https://doi.org/10.1016/j.atech.2024.100517.

Khan, M.A., Akram, T., Sharif, M. et al. (2020). An automated system for cucumber leaf diseased spot detection and classification using improved saliency method and deep features selection. *Multimed Tools Appl* 79: 18627–18656. https://doi.org/10.1007/s11042-020-08726-8.

Kianat, J., Khan, M.A., Sharif, M. et al. (2021). A joint framework of feature reduction and robust feature selection for cucumber leaf diseases recognition. *Optik (Stuttg)* 240: 166566. https://doi.org/10.1016/j.ijleo.2021.166566.

Kondo, S., Yoshimoto, N., and Nakayama, Y. (2024). Farm monitoring system with drones and optical camera communication. *Sensors* 24: 6146. https://doi.org/10.3390/s24186146.

Kong, L., Huang, M., Zhang, L., and Chan, L.W.C. (2024). Enhancing diagnostic images to improve the performance of the segment anything model in medical image segmentation. *Bioengineering* 11: 270. https://doi.org/10.3390/bioengineering11030270.

Kutyauripo, I., Rushambwa, M., and Chiwazi, L. (2023). Artificial intelligence applications in the agrifood sectors. *J Agric Food Res* 11: 100502. https://doi.org/10.1016/j.jafr.2023.100502.

Lei, L., Yang, Q., Yang, L. et al. (2024). Deep learning implementation of image segmentation in agricultural applications: a comprehensive review. *Artif Intell Rev* 57: 149. https://doi.org/10.1007/s10462-024-10775-6.

Leybourne, D.J., Musa, N., and Yang, P. (2025). Can artificial intelligence be integrated into pest monitoring schemes to help achieve sustainable agriculture? An entomological, management and computational perspective. *Agric For Entomol* 27: 8–17. https://doi.org/10.1111/afe.12630.

Li, X., Niu, Z., Morales-Ona, A.G. et al. (2025). A portable high-resolution snapshot multispectral imaging device leveraging spatial and spectral features for non-invasive corn nitrogen treatment classification. *Sensors* 25: 1320. https://doi.org/10.3390/s25051320.

Lin, K., Gong, L., Huang, Y. et al. (2019). Deep learning-based segmentation and quantification of cucumber powdery mildew using convolutional neural network. *Front Plant Sci* 10: 1–10. https://doi.org/10.3389/fpls.2019.00155.

Magesh, S. (2025). A convolutional neural network model and algorithm driven prototype for sustainable tilling and fertilizer optimization. *npj Sustainable Agriculture* 3: 5. https://doi.org/10.1038/s44264-024-00046-w.

Mahum, R., Munir, H., Mughal, Z.-U.-N. et al. (2023). A novel framework for potato leaf disease detection using an efficient deep learning model. *Human and Ecological Risk Assessment: An International Journal* 29: 303–326. https://doi.org/10.1080/10807039.2022.2064814.

Mallinger, K., Corpaci, L., Neubauer, T. et al. (2024). Breaking the barriers of technology adoption: explainable AI for requirement analysis and technology design in smart farming. *Smart Agricultural Technology* 9: 100658. https://doi.org/10.1016/j.atech.2024.100658.

Mana, A.A., Allouhi, A., Hamrani, A. et al. (2024). Sustainable AI-based production agriculture: exploring AI applications and implications in agricultural practices. *Smart Agricultural Technology* 7: 100416. https://doi.org/10.1016/j.atech.2024.100416.

Mienye, I.D., Sun, Y., and Ileberi, E. (2024). Artificial intelligence and sustainable development in Africa: a comprehensive review. *Machine Learning with Applications* 18: 100591. https://doi.org/10.1016/j.mlwa.2024.100591.

Mmbando, G.S. (2025). Harnessing artificial intelligence and remote sensing in climate-smart agriculture: the current strategies needed for enhancing global food security. *Cogent Food Agric* 11: https://doi.org/10.1080/23311932.2025.2454354.

Mostajer Kheirkhah, F. and Asghari, H. (2019). Plant leaf classification using GIST texture features. *IET Computer Vision* 13: 369–375. https://doi.org/10.1049/iet-cvi.2018.5028.

Nagasubramanian, K., Jones, S., Singh, A.K. et al. (2019). Plant disease identification using explainable 3D deep learning on hyperspectral images. *Plant Methods* 15: 98. https://doi.org/10.1186/s13007-019-0479-8.

Naik, B.N., Malmathanraj, R., and Palanisamy, P. (2022). Detection and classification of chilli leaf disease using a squeeze-and-excitation-based CNN model. *Ecol Inform* 69: 101663. https://doi.org/10.1016/j.ecoinf.2022.101663.

Natarajan, S., Chakrabarti, P., and Margala, M. (2024). Robust diagnosis and meta visualizations of plant diseases through deep neural architecture with explainable AI. *Sci Rep* 14: 13695. https://doi.org/10.1038/s41598-024-64601-8.

Ngugi, H.N., Ezugwu, A.E., Akinyelu, A.A., and Abualigah, L. (2024). Revolutionizing crop disease detection with computational deep learning: a comprehensive review. *Environ Monit Assess* 196: 302. https://doi.org/10.1007/s10661-024-12454-z.

Njoroge, T.K., Sindu, K.M., and Rachael, K. (2024). Feature extraction and segmentation methods in plant disease detection: a multimodal approach. *Andalasian International Journal of Applied Science, Engineering and Technology* 4: 206–223. https://doi.org/10.25077/aijaset.v4i3.182.

Osorio, C.P., Leucci, F., and Porrini, D. (2024). Analyzing the relationship between agricultural AI adoption and government-subsidized insurance. *Agriculture* 14: 1804. https://doi.org/10.3390/agriculture14101804.

Ozdemir, B. and Pacal, I. (2025). An innovative deep learning framework for skin cancer detection employing ConvNeXtV2 and focal self-attention mechanisms. *Results in Engineering* 25: 103692. https://doi.org/10.1016/j.rineng.2024.103692.

Panday, D., Bhusal, N., Das, S., and Ghalehgolabbehbahani, A. (2024). Rooted in nature: the rise, challenges, and potential of organic farming and fertilizers in agroecosystems. *Sustainability* 16: 1530. https://doi.org/10.3390/su16041530.

Patil, V.U., Vanishree, G., Pattanayak, D. et al. (2017). Complete mitogenome mapping of potato late blight pathogen, *Phytophthora infestans* A_2 mating type. *Mitochondrial DNA Part B* 2: 90–91. https://doi.org/10.1080/23802359.2017.1280699.

Prabha, R., Kennedy, J.S., Vanitha, G. et al. (2022). Android application development for identifying maize infested with fall armyworms with Tamil Nadu agricultural university integrated proposed pest management (TNAU IPM) capsules. *Journal of Applied and Natural Science* 14: 138–144. https://doi.org/10.31018/jans.v14iSI.3599.

Purushotham P, Rakhonde G, Reddy NK, et al (2024) Harvesting the future: navigating the challenges and limitations of AI in agriculture. pp 547–564.

Rajak, P., Ganguly, A., Adhikary, S., and Bhattacharya, S. (2023). Internet of Things and smart sensors in agriculture: scopes and challenges. *J Agric Food Res* 14: 100776. https://doi.org/10.1016/j.jafr.2023.100776.

Ramesh, S. and Vydeki, D. (2020). Recognition and classification of paddy leaf diseases using optimized deep neural network with jaya algorithm. *Information Processing in Agriculture* 7: 249–260. https://doi.org/10.1016/j.inpa.2019.09.002.

Rane J, Kaya Ö, Mallick SK, Rane NL (2024) Smart farming using artificial intelligence, machine learning, deep learning, and ChatGPT: Applications, opportunities, challenges, and future directions. In: Generative Artificial Intelligence in Agriculture, Education, and Business. Deep Science Publishing

Rejeb, A., Abdollahi, A., Rejeb, K., and Treiblmaier, H. (2022). Drones in agriculture: a review and bibliometric analysis. *Comput Electron Agric* 198: 107017. https://doi.org/10.1016/j.compag.2022.107017.

Routray, S., Ray, A.K., and Mishra, C. (2017). Analysis of various image feature extraction methods against noisy image: SIFT, SURF and HOG. In: *2017 Second International Conference on Electrical*, 1–5. IEEE: Computer and Communication Technologies (ICECCT).

Sachdeva, G., Singh, P., and Kaur, P. (2021). Plant leaf disease classification using deep convolutional neural network with Bayesian learning. *Mater Today Proc* 45: 5584–5590. https://doi.org/10.1016/j.matpr.2021.02.312.

Salehin I, Talha IM, Saifuzzaman Mohd, et al (2020) An Advanced Method of Treating Agricultural Crops Using Image Processing Algorithms and Image Data Processing Systems. In: 2020 IEEE 5th International Conference on Computing Communication and Automation (ICCCA). IEEE, pp 720–724

Sharma, K. and Shivandu, S.K. (2024). Integrating artificial intelligence and Internet of Things (IoT) for enhanced crop monitoring and management in precision agriculture. *Sensors International* 5: 100292. https://doi.org/10.1016/j.sintl.2024.100292.

Shubhika, S., Patel, P., Singh, R. et al. (2024). Application of artificial intelligence techniques to addressing and mitigating biotic stress in paddy crop: a review. *Plant Stress* 14: 100592. https://doi.org/10.1016/j.stress.2024.100592.

Singh, G. and Yogi, K.K. (2023). Performance evaluation of plant leaf disease detection using deep learning models. *Archives of Phytopathology and Plant Protection* 56: 209–233. https://doi.org/10.1080/03235408.2023.2183792.

Singla, A., Nehra, A., Joshi, K. et al. (2024). Exploration of machine learning approaches for automated crop disease detection. *Curr Plant Biol* 40: 100382. https://doi.org/10.1016/j.cpb.2024.100382.

Soori, M., Arezoo, B., and Dastres, R. (2023). Artificial intelligence, machine learning and deep learning in advanced robotics, a review. *Cognitive Robotics* 3: 54–70. https://doi.org/10.1016/j.cogr.2023.04.001.

Sow, S., Ranjan, S., Seleiman, M.F. et al. (2024). Artificial intelligence for maximizing agricultural input use efficiency: exploring nutrient, water and weed management strategies. *Phyton (B Aires)* 93: 1569–1598. https://doi.org/10.32604/phyton.2024.052241.

Subeesh, A. and Mehta, C.R. (2021). Automation and digitization of agriculture using artificial intelligence and internet of things. *Artificial Intelligence in Agriculture* 5: 278–291. https://doi.org/10.1016/j.aiia.2021.11.004.

Sutikno, T. (2024). The future of artificial intelligence-driven robotics: applications and implications. *IAES International Journal of Robotics and Automation (IJRA)* 13: 361. https://doi.org/10.11591/ijra.v13i4.pp361-372.

Suttapakti U, Bunpeng A (2019) Potato Leaf Disease Classification Based on Distinct Color and Texture Feature Extraction. In: 2019 19th International Symposium on Communications and Information Technologies (ISCIT). IEEE, pp 82–85

Talaviya, T., Shah, D., Patel, N. et al. (2020). Implementation of artificial intelligence in agriculture for optimisation of irrigation and application of pesticides and herbicides. *Artificial Intelligence in Agriculture* 4: 58–73. https://doi.org/10.1016/j.aiia.2020.04.002.

Tilman, D., Balzer, C., Hill, J., and Befort, B.L. (2011). Global food demand and the sustainable intensification of agriculture. *Proceedings of the National Academy of Sciences* 108: 20260–20264. https://doi.org/10.1073/pnas.1116437108.

Tirkey, D., Singh, K.K., and Tripathi, S. (2023). Performance analysis of AI-based solutions for crop disease identification, detection, and classification. *Smart Agricultural Technology* 5: 100238. https://doi.org/10.1016/j.atech.2023.100238.

Vadivel, T. and Suguna, R. (2022). RETRACTED ARTICLE: Automatic recognition of tomato leaf disease using fast enhanced learning with image processing. *Acta Agric Scand B Soil Plant Sci* 72: 312–324. https://doi.org/10.1080/09064710.2021.1976266.

Vidya Madhuri, E., Rupali, J.S., Sharan, S.P. et al. (2025). Transforming pest management with artificial intelligence technologies: the future of crop protection. *Journal of Crop Health* 77: 48. https://doi.org/10.1007/s10343-025-01109-9.

Vishnoi, V.K., Kumar, K., and Kumar, B. (2022). A comprehensive study of feature extraction techniques for plant leaf disease detection. *Multimed Tools Appl* 81: 367–419. https://doi.org/10.1007/s11042-021-11375-0.

Xiao, M., Ma, Y., Feng, Z. et al. (2018). Rice blast recognition based on principal component analysis and neural network. *Comput Electron Agric* 154: 482–490. https://doi.org/10.1016/j.compag.2018.08.028.

Xu, Y., Liu, X., Cao, X. et al. (2021). Artificial intelligence: A powerful paradigm for scientific research. *The Innovation* 2: 100179. https://doi.org/10.1016/j.xinn.2021.100179.

Yang, Z., Niu, H., Wang, X., and Fan, L. (2024). A segmentation method based on the deep fuzzy segmentation model in combined with SCANDLE clustering. *Pattern Recognit* 146: 110027. https://doi.org/10.1016/j.patcog.2023.110027.

Yuan, G.N., Marquez, G.P.B., Deng, H. et al. (2022). A review on urban agriculture: technology, socio-economy, and policy. *Heliyon* 8: e11583. https://doi.org/10.1016/j.heliyon.2022.e11583.

Zhang, S., Wu, X., You, Z., and Zhang, L. (2017). Leaf image based cucumber disease recognition using sparse representation classification. *Comput Electron Agric* 134: 135–141. https://doi.org/10.1016/j.compag.2017.01.014.

Zhang, X., Cui, J., Wang, W., and Lin, C. (2017). A study for texture feature extraction of high-resolution satellite images based on a direction measure and gray level co-occurrence matrix fusion algorithm. *Sensors* 17: 1474. https://doi.org/10.3390/s17071474.

Zhang, S., Zhang, S., Zhang, C. et al. (2019). Cucumber leaf disease identification with global pooling dilated convolutional neural network. *Comput Electron Agric* 162: 422–430. https://doi.org/10.1016/j.compag.2019.03.012.

Zhang, N., Yang, G., Pan, Y. et al. (2020). A review of advanced technologies and development for hyperspectral-based plant disease detection in the past three decades. *Remote Sens (Basel)* 12: 3188. https://doi.org/10.3390/rs12193188.

Zhang, Y., Huang, G., Zhao, Y. et al. (2025). Revolutionizing crop breeding: next-generation artificial intelligence and big data-driven intelligent design. *Engineering* 44: 245–255. https://doi.org/10.1016/j.eng.2024.11.034.

Ziyadinov, V. and Tereshonok, M. (2023). Low-pass image filtering to achieve adversarial robustness. *Sensors* 23: 9032. https://doi.org/10.3390/s23229032.

13

Climate-Resilient Crop Improvement Through AI Technologies

Akhouri Nishant Bhanu[1], Bangar Vaibhav[1], Mohammad Yasin[1], and Sadhana Singh[2]

[1]*Rajmata Vijayaraje Scindia Krishi Vishwa Vidyalaya, Department of Genetics and Plant Breeding, R.A.K. College of Agriculture, Ichhawar Road, Sehore, 466001, Madhya Pradesh, India*
[2]*School of Sciences, U.P. Rajarshi Tandon Open University, Phaphamau, Prayagraj, 211021, Uttar Pradesh, India*

13.1 Introduction

A key tactic for tackling the mounting problems brought on by climate change, water scarcity, environmental degradation, and an excessive dependence on traditional energy sources is climate-resilient agriculture. There has never been a greater need for creative solutions to guarantee food security and sustainable development as global agricultural systems struggle with rising temperatures, changing precipitation patterns, and an increase in the frequency of extreme weather events (Kurgat et al. 2020). The development of climate-resilient crops—varieties that can sustain or improve yields and quality under a range of environmental stresses—is at the heart of this endeavor. Millions of farmers depend on these products to support their lives, especially in rural areas where agriculture continues to be the main source of income (Delcheva et al. 2021).

Crop improvement plays a fundamental role in fostering climate resilience by developing varieties that can withstand both biotic and abiotic stresses. Since the dawn of civilization, plant breeding has been at the heart of agricultural progress, driving advancements that enhance crop productivity, adaptability, and nutritional value. This evolutionary journey (Table 13.1) began with the domestication of cereal grains and has since undergone a series of transformative technological revolutions, improving the efficiency, accuracy, and speed of crop development. The earliest form of plant breeding, referred to as Breeding 1.0, began 10–12 thousand years ago with basic selection methods that eventually led to the cultivation of nearly 7000 food plant varieties worldwide (Eli-Chukwu 2019). The late 19th and early 20th centuries ushered in Breeding 2.0, characterized by the advent of scientific breeding techniques such as controlled crossings, replicated field trials, and statistical analysis (Kaur and Sidhu 2020). Around three decades ago, Breeding 3.0 emerged with the integration of molecular markers and genomic data into traditional phenotypic selection, revolutionizing plant breeding by significantly improving precision and scope (Khan et al. 2022).

Table 13.1 Evolution of plant breeding: Phases, descriptions, and technological innovations.

Breeding phase	Description	Technological innovations
Breeding 1.0	Domestication and rudimentary selection (10–12K years ago)	Selection of superior plants
Breeding 2.0	Scientific breeding (late 19th - early 20th century)	Replicated field trials, controlled crossings, statistical analyses
Breeding 3.0	Integration of molecular markers (last 30 years)	Genomic selection, marker-assisted selection
Breeding 4.0	AI and omics data integration (current era)	Machine learning, big data analytics, high-throughput phenotyping

Today, agriculture is transitioning into Breeding 4.0, an era driven by the integration of omics data and advanced informatic technologies. This transition is necessitated by the urgent need to address global challenges, including rapid population growth, declining arable land, and increasing climate variability. The Green Revolution brought substantial gains through the development of semi-dwarf, nutrient-responsive, and hybrid cultivars; however, the current climate crisis demands even more sophisticated and data-driven approaches (Matyushenko et al. 2016). A paradigm shift in agriculture is underway, moving away from conventional methods towards modern, automated approaches that reimagine farms as highly optimized production units, treating plants and animals as integral components of a precision-driven system (Gryshova et al. 2024).

Recent years have seen the emergence of artificial intelligence (AI) as a revolutionary force in plant breeding, with enormous promise to hasten the creation of crops resistant to climate change. According to Jarrahi (2018), artificial intelligence (AI) is a collection of technologies that mimic human intellect in analyzing large datasets, identifying hidden patterns, and reaching well-informed conclusions. These technologies include machine learning, data mining, and predictive analytics. In agriculture, AI systematically integrates and analyzes phenotypic, genotypic, and envirotypic data to optimize breeding decisions, enhance resource use efficiency, and improve crop performance under diverse climatic conditions (Bao and Xie 2022).

AI technologies solve major bottlenecks in contemporary breeding and genotyping by facilitating high-throughput phenotyping, gene functional analysis, and effective data processing (Khan et al. 2022). Breeders can create high-yielding crop varieties that are resistant to biotic and abiotic challenges and support environmentally friendly and sustainable food production by using AI-driven techniques. The concept of smart agriculture, which incorporates AI, machine learning, and data-driven technologies, is a defining feature of the Agriculture 4.0 revolution. AI now plays a critical role in optimizing every aspect of the agricultural cycle, from soil preparation to harvesting, through the use of automation, robotics, and computer-assisted decision-making (Khan et al. 2024).

This chapter explores the impactful role of AI in advancing climate-resilient crop improvement. It highlights how AI technologies enable high-throughput phenotyping, improve gene functional analysis, and address the complexities of large-scale data processing in contemporary breeding programs. By integrating AI into breeding pipelines, researchers can significantly accelerate the development of robust crop varieties, ensuring global food security in the face of climate change while fostering sustainable agricultural practices. Ultimately, the convergence of AI and plant breeding represents a pivotal step toward addressing the interconnected challenges of food security, environmental sustainability, and climate resilience.

13.2 Artificial Intelligence: Transforming the Future of Agriculture

Today, agriculture, the backbone of human civilization, faces hitherto unheard-of difficulties, including resource scarcity, climate change, environmental degradation, and an ever-increasing global population. The sector is at a critical juncture where traditional practices must evolve to meet the demands of sustainability, productivity, and resilience. In this transformative era, **Artificial Intelligence (AI)** has emerged as a powerful catalyst, offering groundbreaking solutions that reimagine agricultural practices from farm to fork.

13.2.1 Defining AI and Its Role in Modern Agriculture

Artificial Intelligence (AI) involves creating computer systems capable of carrying out functions typically requiring human intellect, such as analyzing patterns, making decisions, solving problems, and understanding language. In modern agriculture, these capabilities are increasingly vital for enhancing efficiency and innovation in farming methods (Zhang et al. 2025).

AI technologies encompass a broad spectrum of methodologies and tools, each offering unique contributions to various facets of agriculture. Among the most prominent AI technologies (Table 13.2) are Machine Learning (ML), Deep Learning (DL), Natural Language Processing (NLP), and Computer Vision.

13.2.1.1 Machine Learning

A branch of artificial intelligence called machine learning is concerned with creating algorithms that let computers analyze data, learn from it, and make predictions or

Table 13.2 Key AI technologies and their applications in agriculture.

Technology	Application
Machine Learning (ML)	Genomic prediction, crop yield forecasting, disease detection
Deep Learning (DL)	Image-based phenotyping, variant calling, genomic selection
Computer Vision	Automated weeding, pest detection, crop health monitoring
Natural Language Processing (NLP)	Data interpretation, automated reporting, knowledge extraction
Internet of Things (IoT)	Real-time environmental monitoring, precision irrigation
Big Data Analytics	Data integration from multiple sources for informed decision-making

judgments (Helm et al. 2020). By examining both historical and current data, machine learning algorithms in agriculture can forecast crop yields, identify illnesses, and maximize resource utilization (Bao and Xie 2022). As ML models are exposed to more data, they get better over time, allowing for more precise and effective decision-making. Random forests (RF) and support vector machines (SVM), two popular machine learning algorithms in agriculture, aid in the classification of crop illnesses and the prediction of environmental effects on agricultural productivity (Elbasi et al. 2023; Attri et al. 2024).

13.2.1.2 Deep Learning (DL)

A sophisticated form of machine learning called "deep learning" uses multi-layered neural networks—thus the term "deep"—to examine intricate data patterns (Sarker 2021). For high-throughput phenotyping and remote sensing applications in precision agriculture, deep learning algorithms—like deep neural networks (DNN) and convolutional neural networks (CNNs)—are indispensable because they excel at picture identification tasks (Shrestha and Mahmood 2019). For example, drone-captured photos can be analyzed using DL to detect crop nutrient deficits or insect infestations (Mohd Robi et al. 2024).

13.2.1.3 Natural Language Processing (NLP)

Computers can now comprehend, interpret, and produce human language thanks to natural language processing (Chowdhary and Chowdhary 2020). In agriculture, NLP can be utilized to analyze textual data from research articles, weather reports, or farmer feedback to inform breeding decisions and crop management strategies (Yang et al. 2024). It also facilitates the development of conversational AI systems, such as virtual agricultural assistants, that provide real-time advice to farmers.

13.2.1.4 Computer Vision

A branch of artificial intelligence called computer vision teaches machines to recognize and analyze visual data from their environment (Szeliski 2022). In agriculture, computer vision technologies are applied in tasks such as automated weeding, crop health monitoring, and yield estimation (Tian et al. 2020). By leveraging cameras and sensors, computer vision systems can detect plant diseases, assess soil quality, and monitor crop growth, contributing significantly to precision farming practices (Zhongzhi 2019).

Together, these AI technologies are revolutionizing modern agriculture by enhancing productivity, optimizing resource use, and ensuring the sustainability of farming practices in the face of climate change and other global challenges.

13.2.1.5 The Digital Revolution in Agriculture

The integration of AI into agriculture is often referred to as the advent of Agriculture 4.0, a digital revolution characterized by smart farming practices, precision agriculture, and data-driven decision-making (Javaid et al. 2022; Batra et al. 2024). Smart agriculture employs AI to optimize every step of the agricultural process, from soil preparation and seeding to harvesting and distribution. IoT-based solutions further enhance this by connecting sensors and devices to provide real-time monitoring of crops, soil conditions, and environmental factors (Ayaz et al. 2019; Javaid et al. 2023).

According to Farooq et al. (2024), investments in AI technologies in the agricultural sector were valued at $518.7 million in 2017 and are expected to grow at an annual rate of 16.2% to reach $2.6 billion by 2025. This surge reflects the increasing recognition of AI's potential to revolutionize agriculture, particularly in areas like crop management, livestock monitoring, and precision farming.

13.2.2 AI Applications in Agriculture

AI's applications in agriculture are diverse, addressing both micro and macro-level challenges. Below are key areas where AI is making a significant impact:

1) **High-Throughput Genomics and Phenomics:** AI accelerates plant breeding through high-throughput sequencing and phenotyping, allowing for rapid identification of desirable traits and genetic markers. Machine learning algorithms are pivotal in genomic prediction, genomic selection, and marker-assisted selection.
2) **Precision Farming:** AI-driven precision farming optimizes resource usage by analyzing soil conditions, weather patterns, and crop health. This leads to efficient irrigation, targeted fertilization, and pest control, reducing waste and enhancing yield.
3) **Automated Agricultural Machinery:** The advent of autonomous tractors, harvesters, and drones powered by AI has revolutionized labor-intensive tasks. These machines perform harvesting, plowing, pruning, and weeding with minimal human intervention, increasing efficiency and reducing labor costs.
4) **Crop and Soil Health Monitoring:** AI technologies enable real-time monitoring of crop health and soil conditions. Computer vision and sensor data help detect diseases, nutrient deficiencies, and environmental stressors early, allowing for timely interventions.
5) **Climate-Resilient Crop Development:** AI plays a crucial role in developing climate-resilient crops by analyzing vast datasets related to genomic, phenotypic, and envirotypic information. This facilitates the breeding of crops that can withstand biotic and abiotic stresses, ensuring food security in the face of climate change (Khan et al. 2022).
6) **Supply Chain Optimization:** AI enhances supply chain management by predicting demand, optimizing logistics, and reducing post-harvest losses. This ensures that agricultural products reach markets efficiently, minimizing waste and maximizing profits.

13.2.3 The Future of AI in Agriculture

AI in agriculture has a bright future ahead of it, since continued developments are expected to completely transform the industry. By using sophisticated computation and analysis methods, AI-enabled breeding platforms are speeding up genetic gains (Farooq et al. 2024). By improving agricultural productivity and resource management, the combination of AI and nanotechnology opens up new possibilities in precision farming (Yadav et al. 2023; Ashique et al. 2025).

But there are drawbacks to the broad use of AI as well, including as ethical issues, data privacy issues, and the requirement for qualified staff to handle and analyze complicated data (Rane et al. 2024). To effectively utilize AI's promise in agriculture, these problems must be resolved.

13.3 Integration of Artificial Intelligence in Plant Breeding for Crop Improvement

The incorporation of Artificial Intelligence (AI) into plant breeding has fundamentally reshaped the development of new crop varieties (Figure 13.1). By leveraging high-throughput genomics and phenomics, AI technologies have accelerated the breeding process, enabling breeders to efficiently analyze large volumes of genetic and phenotypic data (Farooq et al. 2024). This advancement has enhanced the precision, speed, and success rate of breeding programs, leading to more targeted crop improvements.

This diagram illustrates how AI and big data are transforming crop breeding. It begins with high-throughput phenotype acquisition using advanced imaging tools for data collection. Next, AI-driven bioinformatics automates genomic analysis. The AI-based breeding process then predicts and selects optimal plant traits using genomic data. Finally, personalized intelligent breeding leverages AI to develop high-yield, climate-resilient crops with multiple elite genes, ensuring sustainable agriculture.

AI is poised to drive plant breeding in four fundamental ways:

1) **AI-Assisted Breeding Systems:** AI will significantly contribute to theoretical studies, evaluation processes, selection strategies, breeding procedure development, and field management. These systems will streamline the breeding pipeline, making it more efficient and data-driven (Fang 2024).
2) **AI-Equipped Robotics:** Robotics integrated with AI will revolutionize data collection, storage, analysis, sharing, and utilization, thereby upgrading breeding information

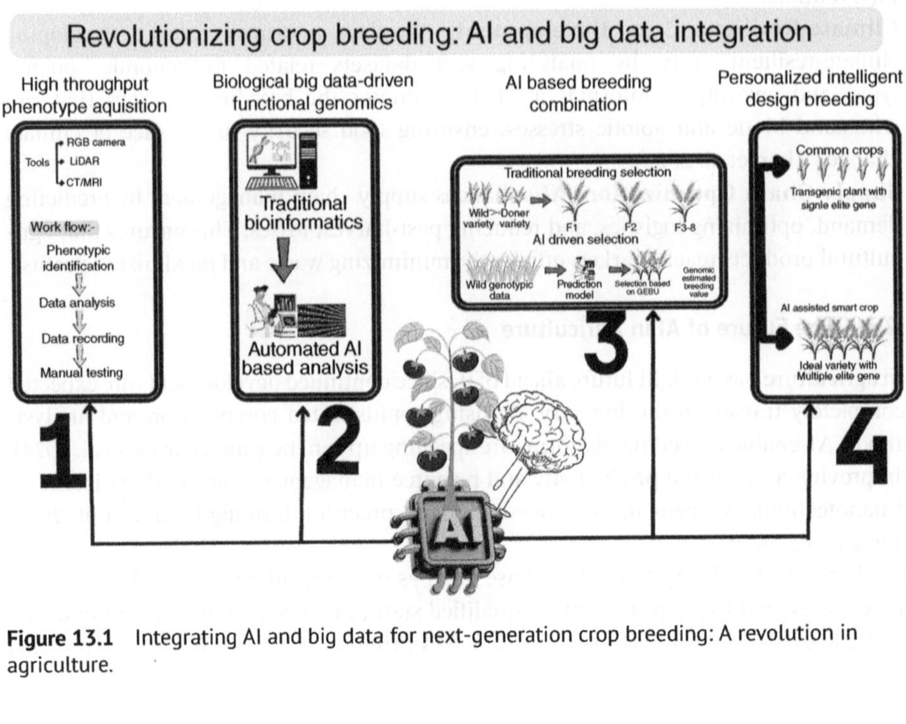

Figure 13.1 Integrating AI and big data for next-generation crop breeding: A revolution in agriculture.

systems. These robots can perform tasks ranging from phenotypic data capture to precision planting and automated harvesting (Riera et al. 2021).

3) **Historical Knowledge Integration:** AI systems will leverage historical breeding data and accumulated knowledge to refine breeding strategies. By learning from past successes and failures, AI can optimize future breeding programs, reducing the time and resources needed for cultivar development (Crossa et al. 2024).
4) **Big Data and Predictive Modeling:** AI-driven breeding systems will utilize big data to enhance design and prediction capabilities through model simulation and optimization. This includes predicting how specific genotypes will perform under varying environmental conditions, thus aiding in the selection of superior plant varieties (Cheng and Wang 2024).

13.4 Characterizing Germplasm Resources with AI to Produce Genomic Big Data

The advent of Artificial Intelligence (AI) has opened new frontiers in the characterization and utilization of germplasm resources, enabling the generation of genomic big data critical for advancing crop breeding programs. The integration of high-throughput remote sensing and plant "omics" technologies has resulted in the creation of extensive multi-dimensional datasets, offering unprecedented opportunities for plant breeders to address complex agricultural challenges (Farooq et al. 2024).

13.4.1 The Role of AI and Machine Learning in Germplasm Characterization

The application of Machine Learning (ML) algorithms in pre-breeding strategies, regional selection, and adaptive marker-assisted selection has proven instrumental in enhancing genetic diversity and accelerating the development of climate-resilient cultivars (Farooq et al. 2024; Yoosefzadeh-Najafabadi et al. 2024). ML models help maintain and restore dynamic genetic processes, which are vital for plant adaptation to changing environmental conditions.

The enormous biological diversity seen in germplasm collections is the source of genetic gain in crops. More than 7 million germplasm accessions, comprising cultivars, landraces, and wild relatives, are stored in more than 1750 genebanks worldwide (Aubry 2023). Even with this abundance of genetic resources, most of their potential has not yet been realized. A potential strategy to better comprehend and utilize these priceless resources is provided by the developing discipline of genebank genomics, which entails genome-wide genotyping of preserved germplasm (Wambugu et al. 2018; Nguyen, and Norton 2020).

Substantial genome-wide genotyping data have been produced for major crops, including over 80,000 wheat accessions (Juliana et al. 2019), 4000 maize accessions, and 20,000 barley accessions (Milner et al. 2019). These datasets provide a foundation for AI-driven predictive genomics, allowing for the targeted selection and optimal testing of accessions in explicit environmental contexts.

13.4.2 Constructing Reference Genomes: The Foundation of Genomic Big Data

A fundamental step in utilizing germplasm resources is the construction of reference genomes. These serve as critical baselines for comparing individuals within a species, facilitating the identification, mapping, and association of allelic mutations with phenotypic traits. The advancement of sequencing technologies has significantly improved the comprehensiveness and precision of reference genome assemblies.

10,000 plant species, covering all known taxonomic groups, are to be sequenced as part of initiatives like the 10KP Project, a crucial part of the larger Earth BioGenome Project (Lewin et al. 2018). This will produce the first full genome-wide assembly for many species, which will be used as a guide for later resequencing attempts to identify allelic variants within species (Varshney et al. 2009).

Crops become more resilient to changing climatic patterns when advantageous features from wild populations are incorporated into elite cultivated types. Exploring genome regulation and variation becomes more feasible as more genome sequences are assembled. In order to maintain and update reference genome annotations and guarantee the distribution of genomic data for researcher-driven analysis, the Ensembl browser (version 93) is essential (Hunt et al. 2018; Farooq et al. 2024).

13.4.3 Addressing Phenotyping Challenges with AI

It is quite difficult to gather phenotypic information for the wide variety of germplasm resources in various settings. However, research has demonstrated that using soil gradient and bioclimatic data from georeferenced sorghum landraces can uncover genetic markers associated with crop adaptability (Lasky et al. 2015).

Breeders can determine the breeding potential of germplasm even in the absence of conventional phenotypic data by combining genomics with passport data, which offers genotypic identity information of accessions. Improved AI-driven genomic prediction models that consider soil gradients, agroclimatic factors, and georeferenced passport data can model the performance of various germplasms, making it easier to choose the best breeding panels. The lack of phenotypic data in target environments is a significant barrier to the use of genebank germplasm (Farooq et al., 2024).

13.4.4 Neural Networks and Deep Learning for Multi-Omics Data

The accuracy and effectiveness of multi-omics data analysis in plant breeding have greatly increased with the use of neural networks (NN) and deep learning (DL). These artificial intelligence techniques use a variety of nonlinear hierarchical techniques to efficiently classify information, simulating brain neurons. A crucial component of contemporary breeding operations, this method is especially helpful for comprehending complex traits and how they interact with environmental factors.

To find trends that might not be visible using conventional techniques, AI models, for example, can concurrently assess environmental, phenotypic, and genotypic data (Harfouche et al. 2019). Breeders can use this to forecast how desired qualities, like disease resistance or drought tolerance, would appear in various environmental settings and to find hidden genetic markers linked to these traits.

13.4.5 Future Directions in AI-Driven Germplasm Characterization

The future of Artificial Intelligence (AI) in germplasm characterization is poised to revolutionize plant breeding by seamlessly integrating big data analytics, high-throughput phenotyping, and genomic sequencing. These advancements will enable the development of comprehensive breeding databases that facilitate precise and efficient crop improvement strategies. As agricultural research advances, the next generation of AI-driven systems will be essential in bridging the phenotype–genotype gap, unlocking the full potential of diverse germplasm resources. By harnessing AI's ability to analyze vast and complex datasets, breeders can accelerate the discovery of key genetic traits that contribute to crop resilience, productivity, and adaptability to changing environmental conditions.

Improving prediction models to better take into consideration the complex relationships between genetic information and environmental influences is one of the key areas in AI-based germplasm characterisation. The development of advanced AI systems that can precisely capture and model these intricate interactions must be the main goal of future research. AI will make it possible to identify superior crop varieties with improved nutritional profiles, increased yield stability across a range of agro-climatic zones, and increased resistance to biotic and abiotic stresses by improving the predictive power of genomic selection and marker-assisted breeding programs.

A paradigm shift in the creation, analysis, and application of genomic big data for plant breeding is represented by AI-enabled characterisation of germplasm resources. AI presents a revolutionary approach to climate-resilient agriculture by expediting the identification of significant genetic characteristics and improving breeding techniques. The use of AI technology into plant breeding will be essential in tackling the issues of global food security and advancing effective and sustainable farming methods as these technologies develop further. In addition to improving crop improvement precision, the ongoing development of AI-driven techniques will open the door to a more robust and effective global food system.

13.5 AI for Overcoming Phenomics Bottlenecks Through Digitalization and Phenotyping Data Collection

In crop breeding, phenotypic data are essential because they provide the basis for comprehending plant characteristics and how they react to environmental factors. However, because of their labor-intensive nature and limited capacity for data capture, traditional phenotyping techniques have long been a bottleneck. By providing high-throughput, accurate, and efficient phenotyping, the recent rise of plant phenomics and AI-driven digitalization offers solutions to these limitations (Dhondt et al. 2013; Tardieu et al. 2017; Ahmad et al. 2024).

13.5.1 The Evolution of Plant Phenomics and AI Integration

Plant phenomics systematically studies plant phenotypes, utilizing advanced technologies to capture detailed trait data across various environmental conditions. Phenomic platforms equipped with sophisticated imaging sensors have revolutionized large-scale phenotyping. These platforms employ both stationary and mobile sensors to monitor diverse plant traits, enabling comprehensive data collection across growth phases.

For instance, rice growth, nitrogen content, leaf area index, and insect prevalence have all been tracked using fixed phenotyping towers fitted with digital cameras (Chawade et al. 2019). These towers are limited to specific areas even though they are simple to maintain. To fully describe crop canopy development, more sophisticated systems, such as the rail-based field scanalyzer at Rothamsted Research, combine a visible camera, 3D laser scanner, thermal infrared (TIR) camera, chlorophyll fluorescence sensor, and hyperspectral camera (Jayasuriya et al. 2024).

13.5.2 Mobile Platforms and UAVs for Enhanced Data Collection

Mobile phenotyping platforms were created to get over spatial restrictions. For example, the high-throughput phenotyping system Crop3D measures 3D plant architecture, leaf temperature, and other important characteristics using several image sensors housed on a moving gantry (Fu and Jiang 2022). Furthermore, canopy characteristics like plant height, normalized difference vegetation index (NDVI), temperature, and RGB imagery have been recorded in crops like soybean and wheat using manually operated carts and self-propelled tractors fitted with a variety of sensors (Bai et al. 2016; Rui et al. 2024).

Environmental factors, such as varying light conditions, can impact imaging accuracy in open fields. Technologies like BreedVision (tractor-pulled multi-sensor phenotyping platform) address these challenges by using a movable dark chamber that blocks ambient light, allowing for consistent imaging. BreedVision utilizes a combination of 3D time-of-flight cameras, laser distance sensors, hyperspectral imaging, and RGB sensors to non-destructively facilitate assessment of traits like plant height, tiller density, grain yield, and moisture content (Busemeyer et al. 2013).

Multispectral and hyperspectral cameras on unmanned aerial vehicles (UAVs), such as the DJI Matrice 300 RTK, Parrot Sequoia+, and SenseFly eBee X, have become dynamic instruments that can quickly gather high-resolution data across large areas. Because UAV photography offers fine-grained canopy color and texture details with remarkable spatial and temporal precision, it makes phenotyping chores easier. Tasks like plant height, canopy leaf area index (LAI), one lodging resistance characteristic (lodging area) in maize (Su et al. 2019), and wheat ear identification and senescence quantification (Yue et al. 2019) have all been successfully completed with this technique.

13.5.3 AI-Driven Data Processing and Feature Extraction

High-throughput phenotyping platforms generate large volumes of complex data, necessitating advanced analytical approaches to uncover relevant patterns and insights. Deep learning (DL) techniques excel in deciphering intricate structures within these high-dimensional datasets (Chen et al. 2020).

To predict important agricultural features, a variety of machine learning (ML) and deep learning (DL) methods have been used extensively. These include random forest (RF), neural networks (NN), partial least squares (PLS), k-nearest neighbor (KNN), and support vector machines (SVM). Plant trait analysis commonly uses advanced DL architectures like as ResNet-50, AlexNet, DenseNet, GoogLeNet, VGG-16, and YOLOv5 (Rodríguez-Lira et al. 2024). For example, convolutional neural networks (CNNs) are used in RGB image processing to support tasks like leaf counting and biotic and abiotic stress detection, as well as to segment plant roots in species like wheat, chicory, and rapeseed (Yasrab et al. 2019; Smith et al. 2020; Gaggion et al. 2021).

Furthermore, models like Faster R-CNN have significantly improved the precision of spike detection in cereal crops like wheat and barley (Li et al. 2022). In soybean, CNNs combined with saliency map analysis of multispectral and hyperspectral imagery have proven useful for identifying biotic stress factors (Nagasubramanian et al. 2019; Shu et al. 2021). In drought-stress assessment for sorghum, linear discriminant analysis (LDA) and PLS techniques have also demonstrated effectiveness (Tauro et al. 2022).

13.5.4 Integrating Multi-Source Data for Enhanced Predictions

Results from combining data from several sources are frequently more accurate than those obtained from using only one dataset. For instance, using a deep neural network (DNN) to integrate RGB, TIR, and multispectral data greatly enhanced the forecast of soybean production. Strong correlations between hyperspectral vegetation indices (HVIs) and soybean seed yield were shown by Yoosefzadeh-Najafabadi et al. et al. (2021), underscoring the possibility of combining genomic and phenomic data for crop development.

Additionally, parameters like soybean nitrogen concentration, leaf area index, above-ground biomass, and chlorophyll content have been estimated using extreme learning machines (Maimaitijiang et al. 2017). Designing crop ideotypes that are suited for particular habitats, especially in light of climate change, requires integrating genotypic, phenotypic, and environmental data. Additionally, *Alternaria alternata*, Anthracnose, Bacterial blight, and Cercospora leaf spot are four common plant diseases that have been automatically detected and classified using ELM (Aqel et al. 2022).

13.5.5 The Role of AI in Field Phenotyping and Data Management

The development of algorithms to transform sensory data into phenotypic information, the construction of models to comprehend genotype–phenotype-environment (G × E) relationships, and the creation of databases to support resource management and data sharing are the three main areas of phenomic data management where artificial intelligence (AI) is essential (Nabwire et al. 2021).

To ensure robust phenotypic data, experiments require repeated trials under diverse conditions to accurately screen for desirable traits, such as grain size, abiotic stress tolerance, and yield potential. While controlled environments provide valuable data, they do not fully represent open-field conditions, leading to a lab-to-field performance gap. Therefore, field-based phenotyping remains a critical focus area for crop breeding.

13.5.6 Future Prospects: Smarter, Faster, and More Accessible Phenotyping

The integration of Artificial Intelligence (AI) into phenotyping technologies is driving a new era of precision agriculture, offering smarter, faster, and more cost-effective solutions for crop improvement. While controlled indoor phenotyping facilities currently surpass field-based phenotyping in throughput and accuracy, efforts are underway to develop high-throughput field phenotyping platforms to bridge this gap, ensuring that AI-driven insights translate effectively to real-world agricultural conditions (Crossa et al. 2024). These advancements are critical for accelerating breeding cycles and enhancing the resilience of crops to biotic and abiotic stresses.

AI-driven phenotyping relies on large, diverse datasets to improve prediction accuracy and optimize breeding strategies. The increasing availability of smartphones and citizen science initiatives presents a promising opportunity to expand the scale and diversity of agricultural datasets. Equipped with advanced sensors, smartphones provide a cost-effective and accessible solution for real-time field data collection, allowing farmers and researchers to contribute to large-scale AI-driven phenotyping networks (Aarif et al. 2025). Despite challenges such as limited battery life and processing power, their widespread adoption makes them valuable tools in modern agricultural research and breeding programs.

Recent research underscores the growing success of AI-based phenotyping approaches in enhancing predictions of crop traits and improving breeding efficiency. For example, Selvaraj et al. (2020) demonstrated that machine learning models such as k-nearest neighbors (kNN), random forest (RF), and support vector machines (SVM) delivered strong predictive performance for cassava root yield, with R^2 values of 0.67, 0.66, and 0.64, respectively. In a similar vein, AI-enabled high-throughput phenotyping has proven effective in major crops like wheat, maize, and oilseeds, supporting tasks such as growth stage monitoring, disease resistance assessment, and yield forecasting (Kundu et al. 2024). These advancements illustrate AI's transformative impact on accelerating phenotypic data processing and driving genetic improvements in crop breeding. As AI tools continue to advance, their integration into phenotyping workflows is expected to play a pivotal role in promoting sustainable agriculture and strengthening global food systems.

13.6 Examining AI's Potential for Genomic Predictions and Gene Function Analysis

The rapid evolution of high-throughput technologies in biological sciences has resulted in an unprecedented accumulation of data, particularly in fields collectively known as "omics" sciences. These disciplines, which encompass genomics, transcriptomics, proteomics, and metabolomics, generate vast, complex datasets that are difficult to analyze using traditional statistical or visual method. The study of such datasets has been completely transformed by the development of Artificial Intelligence (AI) and Machine Learning (ML), which offer tools that can handle massive amounts of data, reveal hidden patterns, and gradually increase forecast accuracy.

13.6.1 AI Applications in Plant Genomics

AI has enormous promise for future development and is already making notable progress in plant genomics. Protein-coding gene identification, cis-regulatory element detection, gene expression profiling, subcellular localization, protein–protein interactions, gene ontology, metabolic pathway analysis, phenotypic predictions, and genomic selection are just a few of the machine learning tools and algorithms used in bioinformatics analyses. Using deep learning approaches, tools like as DeepBind and DeepSEA have been developed to study gene functions and predict genetic traits (Sokolova et al. 2024).

For example, AI algorithms can be used in comparative genomics to transfer knowledge from model plants like *Arabidopsis thaliana* to crops of interest, facilitating the discovery of functional genes and regulatory elements in less-studied species (Wang et al. 2023; MacNish et al. 2025). Moreover, AI-driven models can predict gene functions and assess the differential impacts of gene expression on phenotypic traits, offering new avenues for trait improvement and crop breeding.

13.6.2 AI for Genomic Data Interpretation and Variant Calling

The exponential growth of genomic data due to advances in high-throughput sequencing has created challenges in data storage, processing, and interpretation. AI technologies enable the simultaneous measurement of gene expressions across diverse conditions, providing biologists with more comprehensive representations of their data and facilitating integrative analyses (Li et al. 2018; Hamamoto et al. 2019).

In variant calling, AI has shown remarkable capabilities. Tools like DeepVariant, developed by Poplin et al. (2018), use sophisticated algorithms to predict single-nucleotide polymorphisms (SNPs) and small insertion/deletion (indel) variants. DeepVariant employs a convolutional neural network (CNN) to analyze images of sequencing reads and accurately distinguish between different allelic combinations (Ramachandran et al. 2021). Similarly, Clairvoyante, proposed by Luo et al. (2019), predicts SNPs, indels, and zygosity from long-read alignments, enhancing variant detection across multiple sequencing platforms such as Illumina, PacBio, and Oxford Nanopore.

Long-read sequencing technologies, while offering more detailed structural variation and haplotype information, introduce higher error rates that complicate variant identification (Zhao et al. 2021). AI-driven models employing haplotype-aware modeling, image-based representation, and local realignment have demonstrated superior performance in reducing errors and discovering novel variants in challenging genomic regions (Farooq et al. 2024).

13.6.3 AI in Plant Stress Biology and Metabolic Networks

Artificial Intelligence (AI) is increasingly instrumental in deepening our understanding of plant stress physiology and metabolic regulation. For example, AI has been used to predict genomic crossover events in maize, facilitating the identification of mutation-prone genomic regions—an essential step in accelerating breeding strategies. AI-driven analyses have also been applied to examine DNA methylation patterns under stress conditions in maize, distinguishing between functional genes and pseudogenes (Farooq et al. 2024).

In *A. thaliana* and maize, AI algorithms have been used to identify gene promoters and cis-regulatory elements, as demonstrated by Uygun et al. (2019). These elements are crucial for understanding gene expression regulation under environmental stress. Other studies have leveraged AI to decode complex biosynthetic gene networks involved in nitrogen use efficiency, starch biosynthesis, and secondary metabolite production, particularly in rice and *A. thaliana* (Varala et al. 2018; Li et al. 2020).

Moreover, AI has shown promise in the bioenergy sector. For instance, Meena et al. (2021) highlighted the application of AI in optimizing biomass accumulation in various plant species and algal systems, demonstrating its potential to improve biofuel production efficiency. These advancements underscore AI's expanding role not only in crop improvement but also in sustainable energy solutions.

13.6.4 AI for Enhancing Genomic Selection and Marker-Assisted Breeding

The precision of marker-assisted breeding (MAB) and genomic selection (GS) is directly improved by the increased accuracy of AI in variant calling (Farooq et al. 2024; Sun et al. 2024). This makes it possible for breeders to decide on crop development initiatives with greater knowledge. Recombination rates in genomes have been successfully examined using supervised machine learning algorithms. For instance, Schrider and Kern (2018) distinguished between distinct recombination rate levels in *Drosophila melanogaster* based on DNA patterns using a random forest (RF) classifier.

In population genetics, AI models facilitate the analysis of complex genetic interactions, improving our ability to predict agricultural diseases and streamline bioinformatics pipelines. Companies like Trace Genomics use patented ML algorithms to analyze soil health and identify key factors influencing crop performance, while Sequentia Biotech offers AI RNA-seq (AIR) platforms to simplify transcriptomic data generation and interpretation (Rosignoli et al. 2024).

13.6.5 Future Directions: AI in Comparative Genomics and Multi-Species Analysis

The future of Artificial Intelligence (AI) in genomics lies in its ability to advance multi-species analyses and comparative genomics, paving the way for unprecedented discoveries in plant breeding. Deep learning methodologies are expected to play a critical role in transferring knowledge from well-characterized model plants to less-explored target crops. By leveraging AI-driven comparative genomics, researchers can identify conserved genetic elements, regulatory networks, and key adaptive traits that influence crop performance and resilience. These insights will accelerate breeding efforts and facilitate the development of improved cultivars capable of thriving under diverse environmental conditions.

A key frontier in AI-driven genomics is the integration of phenomic, genomic, and environmental datasets to gain a deeper understanding of genotype-environment interactions (G × E). By analyzing these complex relationships, AI models can enhance predictions of how genetic traits express themselves across varying climatic and agronomic conditions. This will support the development of climate-resilient crop varieties, ensuring higher productivity and stability in the face of climate change. Moreover, AI-enabled modeling will contribute to the design of crop ideotypes—customized plant architectures optimized for specific environmental and management conditions. Such precision breeding approaches will play a crucial role in sustainable agriculture by enhancing resource-use efficiency and mitigating the impacts of environmental stressors.

As AI continues to evolve, its application in gene function analysis and genomic predictions is set to transform plant biology and breeding practices. By uncovering novel gene-trait associations and improving predictive accuracy, AI will revolutionize the way breeders develop improved crop varieties. The integration of AI into comparative genomics holds immense potential for addressing global challenges in food security, climate resilience, and sustainable development. By harnessing these technological advancements, researchers and breeders can drive the next generation of agricultural innovation, ensuring a more secure and sustainable food system.

13.7 Multi-Omic Big Data Integration in Plant Breeding

Multi-omic big data integration has become a game-changer in contemporary plant breeding, offering previously unheard-of insights into the molecular processes underlying phenotypic diversity. The term "omics" refers to a broad category of molecular data sources, such as transcriptomics, proteomics, metabolomics, protein-DNA interactions, genomics, and epigenomics, all of which provide vital information for comprehending intricate biological processes. Large omic datasets have been produced over the last ten years as a result of developments in high-throughput technologies, providing the scientific community with important insights into proteomics, metabolomics, transcriptomics, and genomes (Dai and Shen 2022).

13.7.1 The Role of "Omics" in Information-Intensive Biology

Sub-molecular biological studies that produce enormous datasets in numerous domains have propelled biology's quick transformation into an information-intensive science. In bioinformatics, thorough inventories at the transcriptome, proteome, metabolome, and genome levels have become indispensable tools that provide in-depth understanding of systems biology, ranging from individual genetic components to intricate regulatory networks. These databases offer a once-in-a-lifetime chance to better understand plant growth and stress responses, as well as to reveal the biochemical underpinnings of phenotypic features (Kanaya et al. 2014; Rudakiya et al. 2024).

13.7.2 Machine Learning for Multi-Omic Data Analysis

Due to the vastness and intricacy of omics data, machine learning (ML) has become a vital tool for comprehensive data analysis and interpretation. ML approaches are routinely employed to manage large-scale, multidimensional datasets, identify complex relationships, and integrate insights across various omic domains. Each omic platform features well-established profiling methods (Johnson et al. 2007; Boyle et al. 2008; Buenrostro et al. 2013; Sherwood et al. 2014; Raj et al. 2015), including:

- **RNA-seq**, which enables genome-wide quantification of gene expression
- **ChIP-seq (Chromatin immunoprecipitation with deep sequencing)**, used to map protein–DNA interactions and understand regulatory networks

- **DAP-seq (DNA affinity purification sequencing)**, a method for detecting transcription factor binding sites
- **ATAC-seq (Assay for transposase-accessible chromatin sequencing)**, which facilitates the study of chromatin accessibility

These advanced technologies provide in-depth insights into gene regulation, epigenetic changes, and metabolic pathways, which are critical for understanding the biological mechanisms underlying phenotypic traits. Nevertheless, the integration of such diverse datasets into a unified model remains a major hurdle especially when applied to the improvement of crop varieties.

13.7.3 AI Applications in Genomic and Population Genetics

ML has demonstrated significant utility in identifying various genomic regions and functional elements within the genome. For example, in maize, ML models trained on DNA methylation features have been used to classify functional genes and pseudogenes, offering insights into the functional landscape of the genome (Sartor et al. 2019). Additionally, ML-based methods have been developed to predict crossover probabilities along chromosomes, enhancing our understanding of genetic recombination and inheritance patterns.

ML is also finding applications in population genetics, though much of this work has focused on human genetics. Nevertheless, plant studies have begun to leverage ML to predict the fixation of mutations retained by natural selection, demonstrating its potential in evolutionary biology and crop improvement.

13.7.4 The 9 Versus of Modern Breeding Big Data

Nine essential dimensions viz., volume, value, velocity, variety, veracity, variability, validity, visibility, and vexing complexity are being used to characterize breeding-related big data because of the emergence of multi-omics and spatiotemporal omic technologies (Zhang et al. 2025). These factors highlight how important artificial intelligence (AI) is to the handling and evaluation of this kind of data. Novel approaches are being developed to unravel the intricacy of crop attributes controlled by multilayer genetic and regulatory networks because of the ongoing development of AI and the increasing volume of omics data.

13.7.5 Breakthroughs in Multi-Omic Integration

A landmark study conducted in 2023 by researchers at Huazhong Agricultural University resulted in the development of the first-generation integrated multi-omics network for maize (Zhang et al. 2025). This comprehensive model incorporates more than two million network interactions across different molecular layers, including genomics, transcriptomics, translatomics, and proteomics. Employing sophisticated machine learning algorithms, the research team successfully predicted key functional genes and mapped critical regulatory pathways influencing traits such as flowering time in maize.

Impressively, the integrated network accurately pinpointed 62 of the 63 previously cloned genes related to kernel development, validating its predictive strength (Han et al. 2023).

Among the newly identified genes, a pentatricopeptide repeat (PPR) protein was experimentally confirmed to have a significant role in kernel formation. Furthermore, genes associated with flowering time were categorized into eight functionally similar subnetworks, and 20 candidate genes underwent successful functional validation. This study introduces a transformative approach for dissecting gene function and unraveling the molecular architecture of complex traits, paving the way for next-generation plant genetics research.

13.7.6 Challenges and Future Directions in Multi-Omic Data Integration

The integration of multi-omic data in plant breeding presents a groundbreaking opportunity to enhance crop improvement strategies. However, significant challenges persist, primarily in harmonizing diverse data layers and linking them to environmental stress responses. Capturing the complexity of genotype-environment interactions (G × E) requires advanced AI algorithms capable of modeling entire biological systems with precision.

The creation of AI-driven models that incorporate transcriptomic, proteomic, metabolomic, genomic, and epigenomic data must be a top priority for future studies. By improving phenotypic trait predictions, these models will make it possible to pinpoint the genetic factors that contribute to resistance to both biotic and abiotic stressors. Furthermore, systems biology techniques driven by AI will speed up the identification of important targets for crop development by facilitating the discovery of new regulatory mechanisms (Sheikh et al. 2024).

By leveraging AI and machine learning (ML), the integration of multi-omic big data has the potential to revolutionize agriculture. AI-driven multi-omics analysis enhances gene function discovery and enables the construction of precise regulatory network models, offering unprecedented insights into plant biology.

13.8 AI-Driven Integration to Bridge the Genotype–Phenotype Gap in Modern Crop Breeding

The ability to close the genotype–phenotype gap, a crucial problem in modern agricultural science, has been transformed by the use of Artificial Intelligence (AI) into crop breeding.By leveraging high-throughput phenotyping, genomic data analysis, and advanced machine learning (ML) models, researchers can now make more accurate predictions about how genetic variations manifest as observable traits. This capability is essential for developing climate-resilient, high-yielding crop varieties.

13.8.1 AI for Non-Destructive Crop Trait Estimation and Genetic Studies

Artificial intelligence (AI) has shown considerable promise in enabling non-invasive trait estimation and facilitating genetic analysis in crops. AI-powered techniques have been effectively applied to both hyperspectral and RGB imaging data to forecast early-stage wheat yield and to discover novel alleles through quantitative genomics. For example, Li et al. (2022) developed an automated spike-counting system using ground-based RGB imagery, significantly improving the performance of the Faster R-CNN model by expanding the dataset and refining image annotations.

Another area where AI has proven valuable is in disease diagnostics. A notable case involved the use of deep learning (DL) in a genome-wide association study (GWAS) for detecting sudden death syndrome (SDS) in soybean. This approach enabled the identification of key single nucleotide polymorphisms (SNPs) located near SDS-related candidate genes (Rairdin et al. 2022).

Furthermore, multimodal DL frameworks and machine learning models—including phenotype–genotype multiple instance learning (PheGeMIL), leave-one-environment-out (LOEO) cross-validation, support vector regression (SVR), and gradient boosting machines (GBM) have been applied to improve genomic predictions of wheat yield. These models enhance prediction accuracy by integrating both genotypic and phenotypic data from multiple environments (Montesinos-López et al. 2023, 2024; Togninalli et al. 2023).

13.8.2 Integrating Phenomics, Genomics, and AI for Crop Improvement

The convergence of genomics, phenomics, and artificial intelligence (AI) presents a comprehensive strategy for enhancing the monitoring of crop performance, evaluating stress responses, and discovering new genes and quantitative trait loci (QTLs) (Zhang et al. 2025). This integrative approach is especially valuable for accelerating the development of crops with improved resilience to climate change.

A persistent issue in genetic research is the problem of *missing heritability*, where identified genetic markers do not fully explain the heritable variation in complex traits. To address this, incorporating elements like transposable elements and epigenetic modifications such as DNA methylation—offers a more complete understanding of phenotypic expression. Recent breakthroughs in transposon classification have been achieved through deep learning (DL) algorithms, including tools like Inpactor2 and TEsorter, which facilitate deeper insights into genome structure and variability (Zhang et al. 2022; Orozco-Arias et al. 2023).

AI's influence also extends into the field of functional genomics. For instance, generative adversarial networks (GANs) have been employed to design synthetic promoters by leveraging the characteristics of naturally occurring promoter sequences in *Escherichia coli* (Wang et al. 2020). Further advancements include feedback GANs (FBGANs), which integrate a feedback mechanism to iteratively enhance synthetic DNA sequences based on performance evaluations from functional analyzers ultimately aiding in protein function optimization (Gupta and Zou 2019; Yasmeen et al. 2023).

13.8.3 Using AI to Connect Crop Genome to Phenome

Modern breeding approaches increasingly focus on accurately linking genomic information to crop phenotypes. However, connecting whole-genome data to high-throughput phenotypic traits remains a significant challenge, impeding the full potential of field phenotyping and omics integration.

AI facilitates the differentiation of phenomics and genomics data from germplasm collections and mapping populations (Figure 13.2). Techniques like SNP detection, QTL analysis,

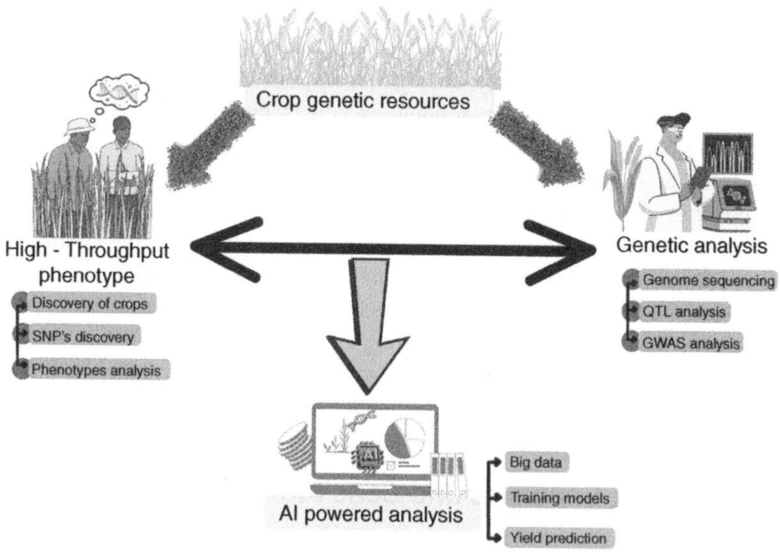

Figure 13.2 Leveraging artificial intelligence for high-throughput crop phenotyping and gene function analysis in modern breeding.

GWAS, and genomic selection generate vast datasets. AI can efficiently evaluate and link these data, improving breeding strategies by predicting gene functions and crop traits.

AI platforms can analyze large and heterogeneous datasets, including genome sequencing and photo imaging, outperforming conventional analytical methods. Applications include genome assembly analysis, multi-omics integration (metabolomics, proteomics, genomics, transcriptomics), and the interpretation of gene expression cascades (Quazi 2022). AI models also identify significant SNPs in polyploid plants and enhance high-throughput crop stress phenotyping (Khan et al. 2022).

This diagram illustrates how Artificial Intelligence (AI) enhances modern crop breeding by integrating high-throughput phenotypic and genotypic data from diverse crop germplasm and breeding populations. AI processes large datasets, including phenotypic diversity, SNP polymorphisms, QTL analysis, GWAS analysis, and genome sequencing, to predict crop traits and improve breeding efficiency. Through big data analysis, training models, and yield prediction, AI enables whole-genome selection and novel breeding strategies for crop improvement.

13.8.4 Applications in Next-Generation Sequencing and Crop Improvement

NGS-based genotyping has transformed crop improvement by increasing gene-mapping precision and enabling the discovery of important genetic loci. Genome-wide association studies (GWAS) utilizing NGS have been extensively applied in crops such as soybean to uncover loci linked to critical agronomic traits, including seed weight, protein content, pod dehiscence, nitrogen fixation, plant height, and disease resistance (Yoosefzadeh-Najafabadi et al. 2021). These advancements provide a deeper understanding of the genetic architecture underlying complex traits, enabling more precise breeding strategies.

Bulk segregant analysis (BSA) and its modified approaches have also become integral in identifying candidate genes associated with various agronomic and stress-related traits. In soybean, BSA has been successfully used to map genes responsible for resistance to soybean mosaic virus, charcoal rot, and powdery mildew, as well as those regulating flowering time. The recent development of DeepBSA, a deep-learning algorithm for QTL mapping and functional gene cloning in maize, exemplifies the transformative impact of AI in accelerating genetic discoveries and refining breeding methodologies (Wang et al. 2023).

AI-driven approaches are closing the genotype–phenotype gap by combining genomic, phenotypic, and environmental data, representing a major advancement in modern crop breeding. Utilizing AI in NGS-based research improves the accuracy and efficiency of gene discovery, supporting the development of high-yielding and climate-resilient crops. As AI technology advances, its impact on genomic research will be crucial in driving sustainable agriculture and ensuring global food security.

13.9 AI for Functional Genomics and Gene Mining

The emergence of Artificial Intelligence (AI) and Machine Learning (ML) has significantly advanced functional genomics and gene mining, providing advanced tools to identify and rank genes linked to important agronomic traits. These computational techniques allow researchers to investigate gene functions, protein interactions, and sequence variations with exceptional accuracy, expediting crop improvement initiatives.

13.9.1 Machine Learning Methodologies for Gene Prioritization

Numerous machine learning techniques have been created to rank the genes that are important for agronomic qualities. These include methods that consider gene annotation and sequence variation, examine protein interactions, and study gene functions. For example, Lin et al. (2019) found that transcription factors are essential for ranking genes linked to characteristics in rice and *Arabidopsis thaliana*. Both shared and species-specific genomic patterns were highlighted by Demirci et al. (2018), who showed that DNA shape traits might predict crossover occurrences across a variety of plant species, including *A. thaliana*, tomato, maize, and rice.

13.9.2 AI in Abiotic Stress Gene Identification and Single-Cell Analysis

AI models have proven instrumental in identifying genes responsible for abiotic stress responses. Techniques like support vector machines (SVMs) and deep learning (DL) have been applied to single-cell RNA sequencing data, facilitating the exploration of developmental processes and environmental responses within heterogeneous tissues (Vadapalli et al. 2022; Adeyanju and Ogunjobi 2024). These extensive datasets, encompassing thousands of cells and tens of thousands of genes, necessitate unsupervised ML techniques to identify patterns and organize data without predefined labels.

13.9.3 Metabolomics and Multi-Omic Integration

Metabolomics presents unique challenges due to the unknown characteristics of many components. ML methodologies have been developed to integrate and analyze metabolomic data, predicting metabolic pathways in crops such as tomatoes. Multi-omic analyses, integrating transcriptomic, proteomic, and metabolomic data, have also benefited from ML. For example, McLoughlin et al. utilized extensive multi-omic analysis to examine maize autophagy mutants, revealing cellular dependencies on autophagy under varying nitrogen conditions (McLoughlin et al. 2018).

13.9.4 AI in Gene Regulation and GWAS

The mlDNA R package, utilizing a random forest (RF) algorithm, was developed to identify salt stress-related genes in *A. thaliana* (Ma et al. 2014). Additionally, integrating gene regulation networks (GRNs) with prioritization algorithms has led to the discovery of important drought-related transcription factors such as *OsbHLH148* in rice (Gupta et al. 2021). While genome-wide association studies (GWAS) are commonly used to detect genes associated with traits, linkage disequilibrium (LD) often results in large quantitative trait loci (QTLs) encompassing numerous genes. ML-based methods, including penalized regression, gradient boosting machines (GBMs), Bayesian approaches, and DL, have been developed to prioritize genes within these QTLs (Yan and Wang 2023; Farooq et al. 2024).

13.9.5 AI for Abiotic Stress and Salt Tolerance Gene Discovery

The importance of ML and DL in locating salt-tolerance genes in plants has been highlighted by recent research. To find genes that are differentially expressed under salt stress, a large language model (LLM)-based gene function prediction model was created, revealing important pathways involved in stress reactions. Yang et al. (2023) identified 16 *Spartina alterniflora* HKT genes with varying expression patterns by using DeepGOPlus, a DL-based gene annotation tool, to find high-affinity K+ transporters (HKTs) associated with salt tolerance. Additionally, Gao et al. (2023) highlighted the enrichment of genes linked to ion transport and detoxification pathways by combining machine learning (ML) with coevolutionary gene information to discover genes relevant to salt stress.

13.9.6 Deep Learning for Large-Scale Gene Expression Analysis

A novel DL-based method was developed to detect influential genes in the genome, enabling large-scale expression profiling at reduced costs. This method integrates AutoEncoder and DeepLIFT to assess the impact of each gene, offering a data-driven, unbiased approach over traditional methods (Hussein et al. 2024). The computational framework incorporates nonlinear feature extraction and feature scoring, demonstrating superior accuracy and robustness in predicting target genes.

13.9.7 Applications in Natural Product Discovery and Molecular Mining

In natural product (NP) discovery, molecular representations like SMILES (simplified input line entry system), SMARTS (SMILES arbitrary target specification) and InChI (international chemical identifier) have evolved into formats like DeepSMILES and SELFIES, integrated into ML tools for efficient substructure searching and data storage. AI tools have been developed to predict target proteins, including protein kinase B (PKBβ), cytochrome P450 (CYP450), and estrogen receptor a (ERa), enhancing our understanding of protein functions and interactions (Dhakal et al. 2022).

In Hannigan et al. (2019), developed DeepBGC, a DL model employing BiLSTM and word embedding techniques for detecting biosynthetic gene clusters (BGCs). The subsequent version, Deep-BGCpred, improved prediction accuracy and reduced false positives by incorporating multi-source domain encoding (Yang et al. 2021b).

13.9.8 Applications in Functional Genomics and Bioinformatics

The integration of Artificial Intelligence (AI) with bioinformatics and genome sequencing has revolutionized the interpretation of complex molecular networks, enabling deeper insights into genetic regulation and trait associations. AI-driven approaches facilitate the analysis of transcription factor binding sites, long non-coding RNAs (lncRNAs), microRNAs (miRNAs), epistatic interactions, coding genes, targeted polyadenylation sites, and cis-regulatory elements (CREs) (Nagarajan et al. 2024). By leveraging these advanced computational models, researchers can systematically decode genetic mechanisms, enhancing the precision of gene discovery and functional characterization.

Crop breeding has greatly benefited from AI-powered analysis of vast phenotypic and genotypic datasets stored in genomic databases. These datasets provide breeders with critical insights into candidate genes and trait-associated loci, streamlining the development of superior crop varieties. Beyond genetic analysis, AI also plays a pivotal role in interpreting variations in crop yield, climatic assessments, and stress phenotyping. Additionally, AI models can assess environmental factors such as UV radiation, wind patterns, and hail damage, offering a more comprehensive understanding of how crops respond to diverse agro-climatic conditions (Khan et al. 2022).

By integrating multi-omic data with AI-driven computational models, researchers can accelerate the identification of key genetic determinants and metabolic pathways underlying complex traits. This convergence of AI and functional genomics is driving a new era of precision breeding, enabling the development of climate-resilient and high-yielding crops.

13.10 Using AI to Discover Exceptional Alleles and Causal Variants in Omic Data

The identification of excellent alleles and causal variants from omic datasets represents a crucial advancement in modern plant breeding and genetics. By harnessing the power of Artificial Intelligence (AI) and Machine Learning (ML), researchers are now able to delve deeper into complex biological systems, uncovering relationships that traditional methods often miss.

13.10.1 Resolving Omic Data's Dimensionality Curse

The integration of multi-omic data—encompassing genomics, transcriptomics, proteomics, and metabolomics—offers a comprehensive view of biological systems. However, the ultra-high dimensionality of these datasets poses significant challenges, often referred to as the "curse of dimensionality" (Ramstein et al. 2019; Zaman et al. 2022). This phenomenon complicates both single-omic and multi-omic analyses, making it essential to reduce dimensionality before meaningful insights can be extracted (Ramstein and Buckler (2022).

To address this, researchers employ advanced dimensionality reduction techniques. For instance, Kang et al. (2022) proposed a roadmap for multi-omic data integration, utilizing autoencoders for data denoising and Convolutional Neural Networks (CNNs) for feature extraction. Although initially designed for biomedical applications, this approach serves as a valuable framework for plant studies, enabling the analysis of large datasets with improved accuracy and efficiency.

13.10.2 Multi-Omic Integration for Enhanced Predictions

Predictions from the integration of several omic layers are frequently more accurate than those from single-omic analysis. The main reason for this is that multi-omic approaches provide a comprehensive knowledge of biological systems by taking into consideration intricate relationships and correlations among many biological data types (Chakraborty et al. 2024). AI models can more accurately forecast phenotypic outcomes and pinpoint causal variations linked to features by capturing these complex interactions.

13.10.3 Functional Characterization of Genes Using AI

To breed crops via genetic modification and genome editing, functional studies of plant genes—especially those underpinning agronomic traits—provide crucial information. Finding genes participating biological processes is based on characterizing gene functions using multi-omic data (Yang et al. 2021a; Mahmood et al. 2022; Roychowdhury et al. 2023).

Yang et al. (2021a) effectively used a Deep Learning (DL)-based model to evaluate multi-omic data and discovered several genes linked to salt tolerance in non-model halophytic plants, then verified their activities. These uses demonstrate how DL may be used to identify important genetic elements that support crops' resistance to stress.

13.10.4 AI in Identifying Causal Phenotypic Variants

In plant genetics, directly defining causative phenotypic variations is still a difficult but crucial objective. In this area, AI models have shown a great deal of promise. To infer functional changes brought about by mutations in maize, for example, the machine learning-based technique PICNC (Prediction of Mutation Impact by Calibrated Nucleotide Conservation) was created (Ramstein and Buckler 2022; Farooq et al. 2024).

PICNC incorporates mutation significance data from UniRep, a technique for assessing protein structure and mutations based on LSTM (Long Short-Term Memory). Researchers can uncover causative variants that contribute to phenotypic variety by using this method to forecast the functional effects of mutations (Ramstein and Buckler 2022).

13.10.5 Building Advanced Tools for Biological Interpretation

Foundational AI tools such as UniRep and PICNC pave the way for developing more advanced systems for biological interpretation (Ramstein and Bucker (2022).

By combining ML algorithms with biological knowledge, researchers can create robust models capable of predicting allelic effects, identifying regulatory elements, and uncovering the genetic basis of complex traits.

13.11 AI-Enabled Genomic Selection for Practical Plant Breeding by Phenotype Prediction

The rise of Artificial Intelligence (AI) and Machine Learning (ML) has greatly reshaped plant breeding, especially through genomic selection (GS). By utilizing genome-wide markers to predict phenotypic traits, genomic selection has become a valuable tool for speeding up crop improvement. This method outperforms traditional marker-assisted selection (MAS) by overcoming its challenges in identifying genes with minor effects on complex traits.

13.11.1 Limitations of Traditional Approaches and the Rise of Genomic Selection

MAS has been instrumental in plant breeding; however, its effectiveness diminishes when dealing with complex traits influenced by multiple genes of small effect. This shortfall arises because MAS often captures only a fraction of the genetic variation associated with these traits, leading to lower prediction accuracy compared to phenotypic selection. To overcome this, genomic selection (GS) offers a more comprehensive strategy by utilizing genome-wide markers, thus enabling the prediction of complex traits with higher precision.

Based on the use of prior knowledge and parameter settings, genomic selection techniques can be generally divided into parametric and nonparametric approaches. Regularized linear regression (RLR) models, like ridge regression (RR) and least absolute shrinkage and selection operator (LASSO), are parametric techniques that help address the overparameterization problems that come with simple linear models. However, nonparametric approaches, especially those based on machine learning (ML) algorithms such as random forests (RF), support vector machines (SVM), and artificial neural networks (ANN), provide flexibility in modeling complex, nonlinear relationships between genotypes and phenotypes (Meuwissen et al. 2001); Gianola et al. 2011; Long et al. 2011; Holliday et al. 2012; Budhlakoti et al. 2022).

13.11.2 Machine Learning in Genomic Selection

Machine learning enhances genomic selection by enabling the development of models that can learn from existing data and predict outcomes for new, unseen data. Unlike traditional statistical models, ML can accommodate the complexity and scale of modern genomic datasets, uncovering hidden patterns and relationships that would otherwise remain undetected.

Supervised learning models are predominantly used in GS, where the goal is to predict target phenotypic traits based on input genomic data. SVMs, for instance, are highly effective in both classification and regression tasks due to their ability to detect subtle patterns within complex datasets. By utilizing kernel functions, SVMs can model nonlinear relationships between genotypes and phenotypes, enhancing prediction accuracy (Noble 2006).

ANNs have also shown great potential in GS applications. These networks can identify intricate patterns in data and predict complex functions, making them particularly useful for modeling epistatic interactions and dominance effects in genomic markers (Rosado et al. 2020). Moreover, ANNs do not rely on assumptions about phenotypic distributions, allowing for flexible and accurate estimation of genetic effects (Gianola et al. 2011).

13.11.3 Deep Learning: The Future of Genomic Selection

Deep learning (DL), an advanced subset of ML, builds upon ANNs by incorporating multiple layers of interconnected nodes, forming deep neural networks (DNNs). These architectures, including multilayer perceptrons (MLPs), convolutional neural networks (CNNs), and recurrent neural networks (RNNs), have demonstrated superior performance in genomic prediction tasks.

MLPs, characterized by at least one hidden layer, are widely used in GS due to their simplicity and effectiveness in prediction tasks. However, they may suffer from overfitting, which can reduce prediction accuracy when applied to real-world datasets. CNNs, primarily used in computer vision tasks, have been adapted for GS to efficiently process genomic data. By reducing input size and sharing parameters, CNNs optimize computational efficiency while maintaining high prediction accuracy (Pook et al. 2020).

RNNs, which incorporate feedback loops to process sequential data, offer unique advantages in modeling time-series genomic data. However, their training requires significant computational resources, making them less practical for some GS applications (Montesinos-López et al. 2021).

A notable DL-based approach in GS is DNN genomic prediction (DNNGP), which integrates multi-omic data to predict plant phenotypes (Wang et al. 2023). This method addresses overfitting and enhances convergence speed, outperforming traditional statistical models in prediction accuracy, particularly in large populations.

13.11.4 Integrating Prior Knowledge and Multi-Omic Data

Incorporating prior biological knowledge such as gene functions, expression profiles, and molecular interactions substantially improves the performance of AI-based genomic selection (GS) models. This integration not only simplifies the high-dimensional nature of genomic datasets but also strengthens the interpretability and reliability of trait predictions from a biological standpoint (Farooq et al. 2021). Common strategies for embedding such knowledge into GS models include the use of customized kernel functions, partitioning genetic variance, and constructing neural network architectures informed by sources like gene ontology (GO), transcriptomic datasets, and findings from genome-wide association studies (GWAS) (Li et al. 2024; Wu et al. 2024).

Moreover, multi-omic data integration, encompassing genomics, transcriptomics, proteomics, and metabolomics, provides a comprehensive view of the biological system. AI models that leverage these diverse data sources can more accurately predict phenotypic traits and uncover the underlying genetic mechanisms.

13.11.5 Challenges and Future Prospects

Despite the significant promise of AI-enabled genomic selection (GS), several challenges must be addressed to fully realize its potential in plant breeding. One major limitation is the issue of overfitting, particularly in deep learning (DL) models, which can lead to reduced generalizability and inaccurate predictions. This challenge can be mitigated through the application of Bayesian methods and careful optimization of model architectures to ensure robustness and reliability in genomic predictions. Additionally, the successful implementation of DL models requires specialized expertise in hyperparameter selection, model tuning, and adjustment processes, making accessibility and scalability key concerns.

To harness AI's full potential in GS, further research is needed to refine model architectures and integrate diverse data sources. Combining phenotypic, climatic, and breeder experience data with genomic information will enhance the accuracy and applicability of AI-driven predictions. Expanding the scale and diversity of training datasets, improving computational frameworks, and optimizing AI algorithms will further contribute to the development of precise and reliable genomic selection models.

13.12 Enhancing Gene Editing with Generative AI

Mutagenesis, hybridization, and genetic engineering are examples of traditional plant breeding techniques that have historically made a substantial contribution to raising crop output and quality. Nevertheless, these techniques have intrinsic drawbacks, including lengthy breeding cycles, high unpredictability, poor precision, partial loss of gene function, and time-consuming screening procedures (Dixit et al. 2024). A new age of accurate and effective molecular breeding techniques has been brought about by the development of genome sequencing technologies. Because it allows for precise and effective genomic changes, CRISPR/Cas9 technology has transformed agricultural quality enhancement (Dixit et al. 2024).

13.12.1 The Role of AI in Advancing Genome Editing

Artificial Intelligence (AI) is increasingly embedded in genome editing pipelines, where it significantly enhances the design and operational efficiency of tools such as CRISPR/Cas9. AI-based platforms—including DeepCRISPR, CRISTA, and DeepHF—are pivotal in guide RNA (gRNA) design, enabling precise identification of optimal target sequences (Dixit et al. 2024). These models factor in key variables such as genomic context, the specific Cas enzyme in use, the intended type of edit, and the likelihood of off-target activity, thus improving both specificity and success rates in genome editing workflows.

In addition to conventional genome editing, AI plays an essential role in refining more advanced techniques such as base editing, prime editing, and epigenome editing. These technologies allow for highly targeted DNA modifications without relying on homology-directed repair mechanisms. By forecasting editing outcomes and optimizing strategies, AI supports the application of these tools in plant breeding programs, promoting genetic improvements for enhanced stress tolerance and agronomic performance (Dixit et al. 2024).

13.12.2 Generative AI in Protein Engineering and Gene Editing

Generative AI models have significantly advanced gene editing capabilities, particularly in protein structure prediction and optimization. AlphaFold2, a prominent AI tool, predicts protein structures with remarkable accuracy, providing a foundational resource for designing more efficient genome-editing proteins. This tool has been pivotal in identifying novel functional clusters of deaminases, which are essential in developing efficient base editors (Dixit et al. 2024).

The creation of OpenCRISPR-1, the first de novo generated gene editor created utilizing a large language model (LLM) trained on more than one million CRISPR operons, is a significant advancement in this field (Ruffolo et al. 2024). With AI models directly producing functional genome-editing proteins suited for agricultural uses, this invention ushers in a new era in protein creation (Dixit et al. 2024).

13.12.3 Compact and Efficient Genome Editing Tools

Designing compact genome-editing tools is crucial for enhancing precision and efficiency in genetic manipulation. AI technologies are at the forefront of this effort, enabling the creation of streamlined, multifunctional genome editors. By leveraging AI-driven protein structure prediction and optimization, researchers can develop genome-editing tools that are both effective and adaptable to various crop improvement programs (Dixit et al. 2024; Zhang et al. 2025).

However, optimizing these proteins for practical applications requires extensive training datasets and multidisciplinary collaboration. Ensuring the functionality and safety of redesigned proteins in vivo necessitates comprehensive validation processes and an integrative approach that combines bioinformatics, molecular biology, and plant sciences (Dixit et al. 2024).

13.12.4 Epigenome Editing and the Role of AI

Unlike conventional genome editing, which directly modifies the DNA sequence, epigenome editing (epi-GED) targets the regulatory mechanisms that control gene expression—without making permanent changes to the genetic code (Dixit et al. 2024; Vidanagamachchi and Waidyarathna 2024). This technique involves the precise manipulation of DNA methylation, histone modifications, and RNA editing, offering significant potential for applications in functional genomics, regenerative medicine, and disease modulation. Through epi-GED, researchers can selectively activate or repress specific genes or pathways implicated in various physiological and pathological processes.

One of the most promising platforms for epigenome editing is CRISPR/Cas-based epi-GED, which operates by coupling catalytically inactive CRISPR/Cas systems with epigenetic effector domains to control gene expression without altering DNA sequences (Goell and Hilton 2021). The integration of machine learning (ML) and deep learning (DL) has further advanced the field by enabling sophisticated analyses of complex epigenomic landscapes and supporting the reconstruction of regulatory networks (Machnicka and Wilczynski 2020).

A notable advancement in this area is EpiCas-DL, developed by Yang et al. (2023), a DL-based model designed to predict the efficiency of single guide RNAs (sgRNAs) in CRISPR-mediated epi-GED. The model leverages four key epigenetic inputs: gene expression, DNA methylation, chromatin accessibility, and proximity to transcription start sites (TSS), resulting in an improved prediction accuracy with an AUC of 0.87. EpiCas-DL also identifies critical determinants influencing sgRNA performance in both gene activation and repression scenarios.

Despite these advances, the application of AI in epigenome editing remains relatively nascent compared to its use in base and prime editing. This slower adoption is likely due to the inherent complexity and dynamic variability of epigenomic data, including temporal changes in chromatin state, methylation patterns, and histone modification landscapes.

13.12.5 Ethical Concerns of AI-Assisted Gene Editing

The integration of Artificial Intelligence (AI) in gene editing, while offering unprecedented precision and efficiency, raises important ethical concerns related to informed consent, autonomy, and the potential for undue influence in decision-making. AI-guided genome editing especially using tools like CRISPR/Cas9 has opened up transformative possibilities in disease treatment and genetic research. However, these advancements must be carefully aligned with robust ethical standards and regulatory oversight to prevent misuse and ensure equitable access.

AI significantly improves the accuracy, speed, and cost-effectiveness of gene editing, making it especially valuable in tackling inherited disorders such as sickle cell anemia and thalassemia. Models like DeepCRISPR, CRISTA, and DeepHF help design guide RNAs (gRNAs) with high specificity by incorporating variables such as the genomic context, type of Cas enzyme, on- and off-target scores, and expected editing outcomes (Dixit et al. 2024). This level of precision reduces the risk of unintended genetic alterations, which is critical in clinical applications.

Beyond conventional CRISPR-based editing, AI also facilitates the development and optimization of advanced genome editing technologies, including base editing (BED), prime editing (PED), and epigenome editing (epi-GED). AI tools such as BE-Hive and PE-Design are employed to select the most appropriate editing systems based on target sequence characteristics, mutation types, off-target risk, and the predicted functional consequences of the edits on genes and phenotypes (Dixit et al. 2024).

Notwithstanding these developments, concerns regarding algorithmic bias, data privacy, and the possibility of growing health disparities are raised by the application of AI in gene editing. Addressing these issues and preserving public confidence require ensuring algorithm design transparency, encouraging responsible use, and include stakeholders in regulatory deliberations.

13.12.6 AI-Driven Genome Editing: Future Directions in Plant Breeding

AI-driven genome editing is transforming plant breeding by enabling precise genetic modifications that enhance crop resilience, yield, and nutritional quality. The continued advancement of AI technologies is expected to further refine genome-editing tools, improving their efficiency and specificity. Future research should focus on optimizing AI models through expanded training datasets, enhanced model interpretability, and the integration of multi-omic data. These improvements will provide deeper insights into gene function and regulatory mechanisms, facilitating more targeted and effective genetic modifications (Dixit et al. 2024).

As AI-driven genome editing continues to evolve, its application will lead to more robust, sustainable, and efficient agricultural practices. The ability to precisely manipulate genetic traits will not only accelerate crop improvement but also contribute to global food security and environmental sustainability by developing cultivars that are more resilient to biotic and abiotic stresses.

13.13 AI Providing Access to Envirotyping Data for Crop Breeding

13.13.1 Introduction to Envirotyping in Crop Breeding

Climate change has significantly impacted global agricultural systems, affecting both crop productivity and food security. To counter these challenges, modern breeding programs have evolved to incorporate environmental data alongside traditional genotyping and phenotyping. This has led to the emergence of envirotyping, a systematic method for characterizing environmental influences on crop growth and performance, forming a crucial third pillar in contemporary plant breeding (Xu 2016). Envirotyping involves the collection, analysis, and integration of diverse environmental variables, such as biotic and abiotic stresses, soil composition, climatic conditions, and agricultural management practices, into predictive breeding models (Xu 2016).

13.13.2 The Role of Envirotyping in Predictive Breeding

Envirotyping is essential for understanding genotype-by-environment interactions (GEI), which influence how genetic traits are expressed under varying environmental conditions. By capturing environmental signals and their interactions with responsive genes, breeders can model and predict phenotypic outcomes across diverse ecological contexts (Cortés et al. 2022). The integration of envirotypic data with genomic and phenomic datasets facilitates the development of climate-smart crops that are better adapted to environmental changes (Xu et al. 2022).

To improve crop adaptation techniques, Cortés et al. (2022) created genome-environment associations (GEAs), which combine evolutionary genomics with ecological and climatic data. This method supports the use of multi-dimensional machine learning (ML) models to account for polygenic evolutionary adaptations in breeding by allowing the computation of genomic estimated adaptive values (GEAVs). To increase crop breeding accuracy and efficiency, Xu et al. (2022) have suggested an integrated genomic-enviromics prediction (iGEP) scheme that makes use of big data technologies, artificial intelligence (AI), and multi-omics data.

13.13.3 AI-Driven Integration of Envirotyping Data

The rapid accumulation of environmental data, alongside advancements in high-throughput phenotyping and genomics, necessitates robust computational tools for effective data management and interpretation. AI, particularly ML and deep learning (DL), has emerged as a transformative technology in this field, enabling the integration and analysis of envirotyping data within breeding programs (Cortés et al. 2022).

AI algorithms are instrumental in conducting association studies that identify genomic targets linked to adaptive, climate-resilient traits. For example, ML models have successfully identified novel abiotic stress tolerance variations from wild crop relatives, aiding the development of drought-resistant cultivars (Cortés et al. 2022). By combining genotypic, phenotypic, and envirotypic data, AI-driven models can predict crop performance across diverse environmental conditions, thereby guiding breeders in selecting climate-adapted varieties.

13.13.4 Framework for AI-Enabled Envirotyping in Breeding

A breeding framework has been projected that integrates genotypic, phenotypic, and envirotypic data to enhance the predictive accuracy of AI-driven models (Figure 13.3). High-throughput phenotyping systems operating in both controlled environments and open fields using robotic platforms collect extensive data, which is then transmitted to centralized servers through the Internet for advanced computational analysis (Xu 2016). This integrated data-driven approach enables a comprehensive assessment of genotype-environment (G × E) interactions, providing deeper insights into how environmental factors influence genetic expression and trait performance.

Machine learning (ML) and deep learning (DL) technologies play a crucial role in analyzing these complex datasets, facilitating the identification of crop cultivars optimized for specific or broad environmental conditions. By leveraging AI-powered predictive analytics, breeders can make informed decisions on cultivar selection, optimizing crop adaptation across diverse agro-climatic zones. This precision-oriented breeding paradigm not only improves agricultural productivity but also aligns with sustainability goals by promoting robust performance under variable climatic conditions.

This diagram illustrates how AI and big data optimize crop breeding by integrating phenomics, genomics, and environmental data. High-throughput phenotyping and genotyping data are processed through internet servers and computer systems. These datasets are analyzed using ontologies, which help AI predict breeding outcomes. The results guide cultivar selection, crop improvement strategies, and adaptation to specific environments, ultimately enhancing breeding efficiency and sustainability.

13.13.5 Envirotyping and Environmental Classification

Envirotyping involves classifying and characterizing environmental variables based on their predictability, repeatability, and manageability. Environmental factors can be grouped into mega-environmental variables (e.g., temperature, light, precipitation, soil properties), which are largely predictable over time, and minor environmental variables, which exhibit greater variability and are less predictable (Cooper and Hammer 1996; Lobell et al. 2011).

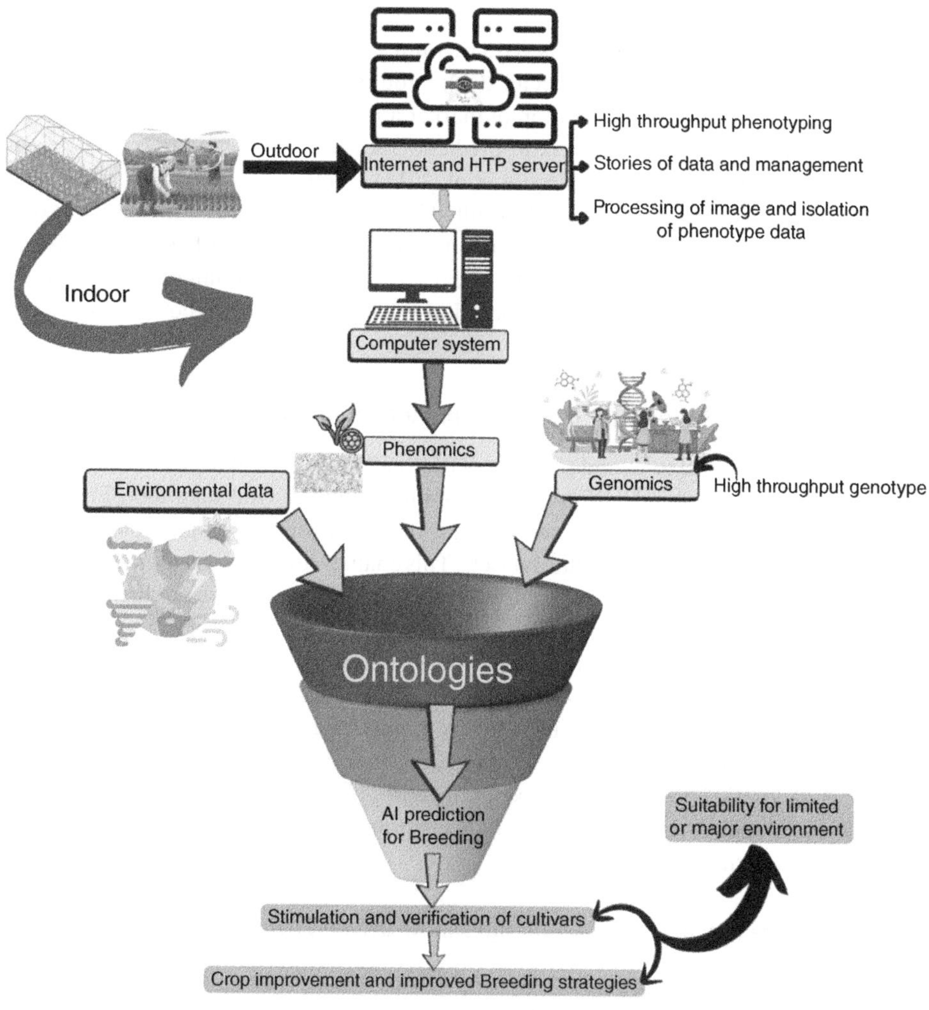

Figure 13.3 AI-driven crop breeding: Integrating phenomics, genomics, and environmental data for precision agriculture.

The concept of heritability extends to envirotypes, with genotypes exhibiting 100% heritability across environments, while envirotypes display varying heritability based on the proportion of environmental variance that is consistently explained across spatial and temporal scales. This classification enables breeders to identify repeatable GEI patterns and develop environment-specific cultivars, thereby enhancing selection reliability and genetic gains (Yan et al. 2022).

13.13.6 AI Applications in Genomic Prediction Across Environments

Incorporating envirotyping data into genomic prediction (GP) models has emerged as a powerful strategy for improving the prediction of complex agronomic traits such as yield, disease resistance, and tolerance to abiotic stress. Studies have shown that accounting for genotype-by-environment interactions (GEI) within predictive frameworks substantially increases model accuracy (Burgueño et al. 2012; Jarquín et al. 2014).

For example, in wheat, the integration of marker and pedigree information to explicitly model GEI led to a 34% improvement in prediction accuracy compared to models based solely on additive genetic effects (Jarquín et al. 2014). Similarly, in maize, incorporating environmental covariates into GP models resulted in a 12% to 22% increase in predictive performance across multiple growing environments (Acosta-Pech et al. 2017). These findings underscore the critical role of envirotyping in capturing the nuanced interactions between genotypic variation and environmental conditions, thereby enhancing the robustness and utility of AI-powered breeding tools.

13.13.7 Advanced AI Tools for Envirotyping Data Integration

A number of AI-powered tools and software platforms have been developed to facilitate the integration of envirotyping data into plant breeding pipelines. One such tool, EnvRtype, provides functionalities for remote sensing and the processing of ecophysiological variables derived from raw environmental datasets. This enables comprehensive environmental profiling and the characterization of environmental quality for genomic prediction models (Costa-Neto et al. 2021).

Another example is CERIS-JGRA, an analytical framework that improves genomic prediction (GP) by incorporating environmental indices (EIs) into model structures. This tool has demonstrated success in predicting complex traits such as flowering time, plant height, and grain yield in key crops like maize, wheat, and oats across multiple growing conditions (Li et al. 2021). These platforms exemplify cost-effective, AI-enhanced envirotyping solutions capable of leveraging high-resolution enviromic data to improve the accuracy and efficiency of modern breeding programs.

13.13.8 Future Prospects and Challenges in AI-Driven Envirotyping

The integration of envirotyping data into AI-driven breeding programs holds immense potential for advancing precision agriculture and climate-resilient crop development. However, several challenges must be addressed to fully harness its benefits. A major obstacle is the need for high-dimensional spatiotemporal enviromic data, which captures environmental influences across diverse locations and time scales. Advanced modeling frameworks, such as Space × Time × Genome × Mega-genome × Metagenome, are essential to effectively analyze the intricate interactions between genetic and environmental factors. These comprehensive models will allow breeders to refine selection strategies and optimize crop performance in variable agro-climatic conditions.

The successful application of AI-enabled envirotyping also requires multidisciplinary collaboration and the establishment of extensive databases integrating genomic, phenomic, and enviromic data. By overcoming these challenges, AI-driven envirotyping has the potential to revolutionize crop breeding, enabling the development of high-yielding, climate-resilient cultivars that contribute to global food security.

The integration of AI technologies into envirotyping represents a transformative step in modern crop improvement. By systematically analyzing environmental influences on crop performance, AI-driven models will enhance the precision and efficiency of breeding programs. This advancement will facilitate the development of climate-smart crops that are better adapted to evolving environmental conditions, ensuring sustainable agricultural practices and strengthening global food security.

While machine learning (ML) provides a powerful framework for analyzing genotype–phenotype-environment (G × P × E) interactions, no single model is universally optimal. Assumptions that perform well in one domain may not be effective in another, highlighting the need for adaptable AI approaches (Wolpert 1996). Therefore, when incorporating envirotypic data into integrated genomic-enviromic prediction (iGEP), a range of models, algorithms, and hyperparameter sets should be tested to determine the best fit for specific datasets. Multivariate models, such as principal component analysis (PCA), along with advanced AI methodologies, will play a crucial role in refining enviromic variables and enhancing environmental indices (EI). These innovations will drive the next generation of AI-driven breeding programs, ensuring improved crop adaptability and resilience in response to global climate challenges.

13.14 Conclusion

The advancement of agricultural systems has been shaped by continuous innovations aimed at improving crop productivity, quality, and sustainability. However, the limitations of conventional breeding techniques like mutagenesis, hybridization, and genetic engineering have been made clear by the growing constraints brought on by climate change and the world's growing food demands. These conventional approaches often involve extended breeding cycles, limited precision, and high resource demands, making them less adaptable to the dynamic agricultural landscape (Dixit et al. 2024). As a result, artificial intelligence (AI) technologies have become game-changing instruments with the potential to completely alter crop management and breeding methods.

By combining genomic, phenomic, and envirotyping data, AI-driven methods in plant breeding have enabled a paradigm change by enabling breeders to model intricate genotype-environment interactions (GEI) and forecast crop performance under a variety of circumstances. When compared to conventional genomic selection (GS) methods, machine learning (ML) and deep learning (DL) algorithms have shown better predictive powers, making it possible to identify desired features and alleles for crop improvement with more accuracy (Cortés et al. 2022). Furthermore, by utilizing sophisticated computer models and extensive datasets to maximize genetic advantages, AI-powered breeding platforms are speeding up the creation of agricultural varieties that are adaptable to climate change.

The accuracy and effectiveness of agricultural techniques have been significantly improved by the integration of AI with sensor technologies, cloud-based platforms, and Internet of Things (IoT) devices. Precision nutrient management systems minimize fertilizer waste and lessen environmental contamination, while AI-powered irrigation systems with soil moisture sensors maximize water use. By increasing resource efficiency and lowering the environmental impact of crop production, these technologies support sustainable agriculture practices (Xu et al. 2022).

Furthermore, AI is playing a pivotal role in advancing genomics and gene-editing tools, such as CRISPR/Cas9. AI-driven models enhance the design of guide RNAs, optimize protein functionality, and improve the precision of genome-editing techniques. Generative AI models, such as AlphaFold2, have revolutionized protein structure prediction, enabling the development of more precise and efficient genome-editing tools tailored for agricultural applications (Dixit et al. 2024). AI's potential in designing future crop cultivars adapted to evolving environmental conditions underscores its critical role in climate-resilient agriculture.

However, realizing the full potential of AI in crop improvement requires addressing several challenges. These include the need for comprehensive, high-quality datasets, fostering multidisciplinary collaboration, and strengthening capacity building in the agricultural sector. Public sector breeding programs, such as those under the One CGIAR initiative, must prioritize the development of unified AI-driven infrastructures to avoid data silos and enhance the efficiency of predictive breeding models (Cortés et al. 2022). Additionally, ensuring responsible AI deployment—including the integration of ethical considerations and gender-inclusive approaches—is essential for equitable access to technological advancements and the promotion of sustainable agricultural development.

In conclusion, AI technologies hold immense potential to revolutionize crop breeding and management practices, addressing the pressing challenges of climate change and global food security. By harnessing AI's capabilities, the agricultural sector can accelerate the development of climate-resilient, high-yielding crop varieties, optimize resource efficiency, and promote sustainable farming practices. As AI continues to evolve, its integration into agriculture will play a pivotal role in driving the next agricultural revolution, fostering resilience, sustainability, and prosperity for future generations.

References

Aarif, K.O.M., Alam, A., and Hotak, Y. (2025). Smart sensor technologies shaping the future of precision agriculture: recent advances and future outlooks. *Journal of Sensors* 2025 (1): 2460098.

Acosta-Pech, R., Crossa, J., de Los Campos, G. et al. (2017). Genomic models with genotype× environment interaction for predicting hybrid performance: an application in maize hybrids. *Theoretical and Applied Genetics* 130: 1431–1440.

Adeyanju, S.A. and Ogunjobi, T.T. (2024). *Machine Learning in Genomics: Applications in Whole Genome Sequencing, Whole Exome Sequencing, Single-Cell Genomics, and Spatial Transcriptomics*. Medinformatics.

Ahmad, A., Liew, A.X., Venturini, F. et al. (2024). AI can empower agriculture for global food security: challenges and prospects in developing nations. *Frontiers in Artificial Intelligence* 7: 1328530.

Aqel, D., Al-Zubi, S., Mughaid, A., and Jararweh, Y. (2022). Extreme learning machine for plant diseases classification: a sustainable approach for smart agriculture. *Cluster Computing* 25 (3): 2007–2020.

Ashique, S., Raikar, A., Jamil, S. et al. (2025). Artificial intelligence integration with nanotechnology: a new frontier for sustainable and precision agriculture. *Current Nanoscience* 21 (2): 242–273.

Attri, I., Awasthi, L.K., and Sharma, T.P. (2024). Machine learning in agriculture: a review of crop management applications. *Multimedia Tools and Applications* 83 (5): 12875–12915.

Aubry, S. (2023). Genebanking plant genetic resources in the postgenomic era. *Agriculture and Human Values* 40 (3): 961–971.

Ayaz, M., Ammad-Uddin, M., Sharif, Z. et al. (2019). Internet-of-Things (IoT)-based smart agriculture: toward making the fields talk. *IEEE Access* 7: 129551–129583.

Bai, G., Ge, Y., Hussain, W. et al. (2016). A multi-sensor system for high throughput field phenotyping in soybean and wheat breeding. *Computers and Electronics in Agriculture* 128: 181–192.

Bao, X. and Xie, Y. (2022). AI-enhanced agricultural practices. *Journal of Agricultural Informatics* 45 (3): 212–220.

Batra, I., Sharma, C., Malik, A. et al. (2024). Industrial revolution and smart farming: a critical analysis of research components in Industry 4.0. *The TQM Journal.* 37: 1–29.

Boyle, A.P., Davis, S., Shulha, H.P. et al. (2008). High-resolution mapping and characterization of open chromatin across the genome. *Cell* 132 (2): 311–322.

Budhlakoti, N., Kushwaha, A.K., Rai, A. et al. (2022). Genomic selection: a tool for accelerating the efficiency of molecular breeding for development of climate-resilient crops. *Frontiers in Genetics* 13: 832153.

Buenrostro, J.D., Giresi, P.G., Zaba, L.C. et al. (2013). Transposition of native chromatin for fast and sensitive epigenomic profiling of open chromatin, DNA-binding proteins and nucleosome position. *Nature Methods* 10 (12): 1213–1218.

Burgueño, J., de los Campos, G., Weigel, K., and Crossa, J. (2012). Genomic prediction of breeding values when modeling genotype × environment interaction using pedigree and dense molecular markers. *Crop Science* 52 (2): 707–719.

Busemeyer, L., Mentrup, D., Möller, K. et al. (2013). BreedVision—a multi-sensor platform for non-destructive field-based phenotyping in plant breeding. *Sensors* 13 (3): 2830–2847.

Chakraborty, S., Sharma, G., Karmakar, S., and Banerjee, S. (2024). Multi-OMICS approaches in cancer biology: new era in cancer therapy. *Biochimica et Biophysica Acta (BBA) - Molecular Basis of Disease* 1870 (5): 167120.

Chawade, A., Van Ham, J., Blomquist, H. et al. (2019). High-throughput field-phenotyping tools for plant breeding and precision agriculture. *Agronomy* 9 (5): 258.

Chen, R., Yang, L., Goodison, S., and Sun, Y. (2020). Deep-learning approach to identifying cancer subtypes using high-dimensional genomic data. *Bioinformatics* 36 (5): 1476–1483.

Cheng, Q. and Wang, X. (2024). Machine learning for AI breeding in plants. *Genomics, Proteomics and Bioinformatics* 22 (4): qzae051.

Chowdhary, K. and Chowdhary, K.R. (2020). Natural language processing. In: *Fundamentals of Artificial Intelligence*, 603–649. Springer.

Cooper, M. and Hammer, G.L. (1996). Plant adaptation and crop improvement. In: *CAB International*. IRRI.

Cortés, A.J., López-Hernández, F., and Blair, M.W. (2022). Genome–environment associations, an innovative tool for studying heritable evolutionary adaptation in orphan crops and wild relatives. *Frontiers in Genetics* 13: 910386.

Costa-Neto, G., Crossa, J., and Fritsche-Neto, R. (2021). Enviromic assembly increases accuracy and reduces costs of the genomic prediction for yield plasticity in maize. *Frontiers in Plant Science* 12: 717552.

Crossa, J., Montesinos-Lopez, O.A., Costa-Neto, G. et al. (2024). Machine learning algorithms translate big data into predictive breeding accuracy. *Trends in Plant Science* 30 (2): 167–184.

Dai, X. and Shen, L. (2022). Advances and trends in omics technology development. *Frontiers in Medicine* 9: 911861.

Delcheva, E., Nencheva, I., and Penev, N. (2021). Regulatory challenges for the marketing of agricultural products in Bulgaria and the European Union. In *Agribusiness and Rural Areas-Economy, Innovation and Growth 2021 Conference Proceedings* (No. 1, pp. 348–356). University publishing house "Science and Economics", University of Economics-Varna.

Demirci, S., Peters, S.A., de Ridder, D., and van Dijk, A.D. (2018). DNA sequence and shape are predictive for meiotic crossovers throughout the plant kingdom. *The Plant Journal* 95 (4): 686–699.

Dhakal, A., McKay, C., Tanner, J.J., and Cheng, J. (2022). Artificial intelligence in the prediction of protein–ligand interactions: recent advances and future directions. *Briefings in Bioinformatics* 23 (1): bbab476.

Dhondt, S., Wuyts, N., and Inzé, D. (2013). Cell to whole-plant phenotyping: the best is yet to come. *Trends in Plant Science* 18 (8): 428–439.

Dixit, S., Kumar, A., Srinivasan, K. et al. (2024). Advancing genome editing with artificial intelligence: opportunities, challenges, and future directions. *Frontiers in Bioengineering and Biotechnology* 11: 1335901.

Elbasi, E., Zaki, C., Topcu, A.E. et al. (2023). Crop prediction model using machine learning algorithms. *Applied Sciences* 13 (16): 9288.

Eli-Chukwu, N.C. (2019). Plant breeding evolution and its impact. *Agriculture and Biology Journal* 12 (3): 87–95.

Fang, J. (2024). Breeding 5.0: AI-driven revolution in designed plant breeding. *Molecular Plant Breeding* 15: 27–33.

Farooq, M., Van Dijk, A.D., Nijveen, H. et al. (2021). Prior biological knowledge improves genomic prediction of growth-related traits in Arabidopsis thaliana. *Frontiers in Genetics* 11: 609117.

Farooq, M.A., Gao, S., Hassan, M.A. et al. (2024). Artificial intelligence in plant breeding. *Trends in Genetics* 40 (10): 891–908. https://doi.org/10.1016/j.tig.2024.07.001.

Fu, X. and Jiang, D. (2022). High-throughput phenotyping: the latest research tool for sustainable crop production under global climate change scenarios. In: *Sustainable Crop Productivity and Quality Under Climate Change*, 313–381. Academic Press.

Gaggion, N., Ariel, F., Daric, V. et al. (2021). ChronoRoot: high-throughput phenotyping by deep segmentation networks reveals novel temporal parameters of plant root system architecture. *GigaScience* 10 (7): giab052.

Gao, S., Chen, S., Yang, M. et al. (2023). Mining salt stress-related genes in Spartina alterniflora via analyzing co-evolution signal across 365 plant species using phylogenetic profiling. *aBIOTECH* 4 (4): 291–302.

Gianola, D., Okut, H., Weigel, K.A., and Rosa, G.J. (2011). Predicting complex quantitative traits with Bayesian neural networks: a case study with Jersey cows and wheat. *BMC Genetics* 12: 1–14.

Goell, J.H. and Hilton, I.B. (2021). CRISPR/Cas-based epigenome editing: advances, applications, and clinical utility. *Trends in Biotechnology* 39 (7): 678–691.

Gryshova, I., Balian, A., Antonik, I. et al. (2024). Artificial intelligence in climate smart in agricultural: toward a sustainable farming future. *Access Journal* 5 (1): 125–140.

Gupta, A. and Zou, J. (2019). Feedback GAN for DNA optimizes protein functions. *Nature Machine Intelligence* 1 (2): 105–111.

Gupta, C., Ramegowda, V., Basu, S., and Pereira, A. (2021). Using network-based machine learning to predict transcription factors involved in drought resistance. *Frontiers in Genetics* 12: 652189.

Hamamoto, R., Komatsu, M., Takasawa, K. et al. (2019). Epigenetics analysis and integrated analysis of multiomics data, including epigenetic data, using artificial intelligence in the era of precision medicine. *Biomolecules* 10 (1): 62.

Han, L., Zhong, W., Qian, J. et al. (2023). A multi-omics integrative network map of maize. *Nature Genetics* 55 (1): 144–153.

Hannigan, G.D., Prihoda, D., Palicka, A. et al. (2019). A deep learning genome-mining strategy for biosynthetic gene cluster prediction. *Nucleic Acids Research* 47 (18): e110–e110.

Harfouche, A.L., Jacobson, D.A., Kainer, D. et al. (2019). Accelerating climate resilient plant breeding by applying next-generation artificial intelligence. *Trends in Biotechnology* 37 (11): 1217–1235.

Helm, J.M., Swiergosz, A.M., Haeberle, H.S. et al. (2020). Machine learning and artificial intelligence: definitions, applications, and future directions. *Current Reviews in Musculoskeletal Medicine* 13: 69–76.

Holliday, J.A., Wang, T., and Aitken, S. (2012). Predicting adaptive phenotypes from multilocus genotypes in Sitka spruce (Picea sitchensis) using random forest. *G3: Genes, Genomes, Genetics* 2 (9): 1085–1093.

Hunt, S.E., McLaren, W., Gil, L. et al. (2018). Ensembl variation resources. *Database* 2018: bay119.

Hussein, A., Prasad, M., and Braytee, A. (2024). Explainable AI methods for multi-omics analysis: a survey. *arXiv preprint* arXiv:2410.11910.

Jarquín, D., Crossa, J., Lacaze, X. et al. (2014). A reaction norm model for genomic selection using high-dimensional genomic and environmental data. *Theoretical and Applied Genetics* 127: 595–607.

Jarrahi, M.H. (2018). Artificial intelligence and the future of work: human-AI symbiosis in organizational decision making. *Business Horizons* 61 (4): 577–586.

Javaid, M., Haleem, A., Singh, R.P., and Suman, R. (2022). Enhancing smart farming through the applications of Agriculture 4.0 technologies. *International Journal of Intelligent Networks* 3: 150–164.

Javaid, M., Haleem, A., Khan, I.H., and Suman, R. (2023). Understanding the potential applications of artificial intelligence in agriculture sector. *Advanced Agrochem* 2 (1): 15–30.

Jayasuriya, N., Guo, Y., Hu, W., and Ghannoum, O. (2024). Image based crop monitoring technologies in protected horticulture: a review. *arXiv preprint* arXiv:2401.13928.

Johnson, D.S., Mortazavi, A., Myers, R.M., and Wold, B. (2007). Genome-wide mapping of in vivo protein-DNA interactions. *Science* 316 (5830): 1497–1502.

Juliana, P., Poland, J., Huerta-Espino, J. et al. (2019). Improving grain yield, stress resilience and quality of bread wheat using large-scale genomics. *Nature Genetics* 51 (10): 1530–1539.

Kanaya, S., Altaf-Ul-Amin, M., Kiboi, S.K., and Afendi, F.M. (2014). Big data and network biology. *BioMed Research International* 2014: 836708.

Kang, M., Ko, E., and Mersha, T.B. (2022). A roadmap for multi-omics data integration using deep learning. *Briefings in Bioinformatics* 23 (1): bbab454.

Kaur, P. and Sidhu, R. (2020). The progression of plant breeding phases. *Journal of Crop Science* 18 (2): 45–59.

Khan, M.H.U., Wang, S., Wang, J. et al. (2022). Applications of artificial intelligence in climate-resilient smart-crop breeding. *International Journal of Molecular Sciences* 23 (19): 11156.

Khan, I.U., Taherdoost, H., Madanchian, M. et al. (2024). *Future Tech Startups and Innovation in the Age of AI* (ed. H. Rahman). CRC Press.

Kundu, S., Saini, D.K., Meena, R.K. et al. (2024). High-throughput phenotyping and AI technologies for deciphering crop resilience to heat stress. *Plant Physiology Reports* 1–17.

Kurgat, B.K., Lamanna, C., Kimaro, A. et al. (2020). Adoption of climate-smart agriculture technologies in Tanzania. *Frontiers in Sustainable Food Systems* 4: 55.

Lasky, J.R., Upadhyaya, H.D., Ramu, P. et al. (2015). Genome-environment associations in sorghum landraces predict adaptive traits. *Science Advances* 1 (6): e1400218.

Lewin, H.A., Robinson, G.E., Kress, W.J. et al. (2018). Earth BioGenome project: sequencing life for the future of life. *Proceedings of the National Academy of Sciences* 115 (17): 4325–4333.

Li, Y., Wu, F.X., and Ngom, A. (2018). A review on machine learning principles for multi-view biological data integration. *Briefings in Bioinformatics* 19 (2): 325–340.

Li, Z., Cao, L., Zhao, L. et al. (2020). Identification and biotechnical potential of a Gcn5-related N-acetyltransferase gene in enhancing microalgal biomass and starch production. *Frontiers in Plant Science* 11: 544827.

Li, X., Guo, T., Wang, J. et al. (2021). An integrated framework reinstating the environmental dimension for GWAS and genomic selection in crops. *Molecular Plant* 14 (6): 874–887.

Li, L., Hassan, M.A., Yang, S. et al. (2022). Development of image-based wheat spike counter through a Faster R-CNN algorithm and application for genetic studies. *The Crop Journal* 10 (5): 1303–1311.

Li, H., Li, X., Zhang, P. et al. (2024). Smart breeding platform: a web-based tool for high-throughput population genetics, phenomics, and genomic selection. *Molecular Plant* 17 (5): 677–681.

Lin, F., Fan, J., and Rhee, S.Y. (2019). QTG-Finder: a machine-learning based algorithm to prioritize causal genes of quantitative trait loci in Arabidopsis and rice. *G3: Genes, Genomes, Genetics* 9 (10): 3129–3138.

Lobell, D.B., Schlenker, W., and Costa-Roberts, J. (2011). Climate trends and global crop production since 1980. *Science* 333 (6042): 616–620.

Long, N., Gianola, D., Rosa, G.J., and Weigel, K.A. (2011). Application of support vector regression to genome-assisted prediction of quantitative traits. *Theoretical and Applied Genetics* 123: 1065–1074.

Luo, R., Sedlazeck, F.J., Lam, T.W., and Schatz, M.C. (2019). A multi-task convolutional deep neural network for variant calling in single molecule sequencing. *Nature Communications* 10 (1): 998.

Ma, C., Xin, M., Feldmann, K.A., and Wang, X. (2014). Machine learning–based differential network analysis: a study of stress-responsive transcriptomes in Arabidopsis. *The Plant Cell* 26 (2): 520–537.

Machnicka, M.A. and Wilczynski, B. (2020). Machine learning and deep learning for the advancement of epigenomics. In: *Epigenetics of the Immune System* (ed. D. Kabelitz and J. Bhat), 217–237. Academic Press.

MacNish, T.R., Danilevicz, M.F., Bayer, P.E. et al. (2025). Application of machine learning and genomics for orphan crop improvement. *Nature Communications* 16 (1): 982.

Mahmood, U., Li, X., Fan, Y. et al. (2022). Multi-omics revolution to promote plant breeding efficiency. *Frontiers in Plant Science* 13: 1062952.

Maimaitijiang, M., Ghulam, A., Sidike, P. et al. (2017). Unmanned Aerial System (UAS)-based phenotyping of soybean using multi-sensor data fusion and extreme learning machine. *ISPRS Journal of Photogrammetry and Remote Sensing* 134: 43–58.

Matyushenko, I., Sviatukha, I., and Grigorova-Berenda, L. (2016). Modern approaches to classification of biotechnology as a part of NBIC-technologies for bioeconomy. *British Journal of Economics, Management and Trade* 14 (4): 1–14.

McLoughlin, F., Augustine, R.C., Marshall, R.S. et al. (2018). Maize multi-omics reveal roles for autophagic recycling in proteome remodelling and lipid turnover. *Nature Plants* 4 (12): 1056–1070.

Meena, M., Shubham, S., Paritosh, K. et al. (2021). Production of biofuels from biomass: Predicting the energy employing artificial intelligence modelling. *Bioresource Technology* 340: 125642.

Meuwissen, T.H., Hayes, B.J., and Goddard, M. (2001). Prediction of total genetic value using genome-wide dense marker maps. *Genetics* 157 (4): 1819–1829.

Mohd Robi, S.N.A., Ahmad, N., Izhar, M.A.M. et al. (2024). Utilizing UAV data for neural network-based classification of melon leaf diseases in smart agriculture. *International Journal of Advanced Computer Science and Applications* 15 (1): 1212–1219.

Montesinos-López, O.A., Montesinos-López, A., Pérez-Rodríguez, P. et al. (2021). A review of deep learning applications for genomic selection. *BMC Genomics* 22: 1–23.

Montesinos-López, A., Rivera, C., Pinto, F. et al. (2023). Multimodal deep learning methods enhance genomic prediction of wheat breeding. *G3: Genes, Genomes, Genetics* 13 (5): jkad045.

Montesinos-López, A., Crespo-Herrera, L., Dreisigacker, S. et al. (2024). Deep learning methods improve genomic prediction of wheat breeding. *Frontiers in Plant Science* 15: 1324090.

Nabwire, S., Suh, H.K., Kim, M.S. et al. (2021). Application of artificial intelligence in phenomics. *Sensors* 21 (13): 4363.

Nagarajan, R., Wang, C., Walton, D., and Walton, N. (2024). Artificial intelligence applications in genomics. *Advances in Molecular Pathology* 7 (1): 145–154.

Nagasubramanian, K., Jones, S., Singh, A.K. et al. (2019). Plant disease identification using explainable 3D deep learning on hyperspectral images. *Plant Methods* 15: 1–10.

Nguyen, G.N. and Norton, S.L. (2020). Genebank phenomics: a strategic approach to enhance value and utilization of crop germplasm. *Plants* 9 (7): 817.

Noble, W.S. (2006). What is a support vector machine? *Nature Biotechnology* 24 (12): 1565–1567.

Orozco-Arias, S., Humberto Lopez-Murillo, L., Candamil-Cortés, M.S. et al. (2023). Inpactor2: a software based on deep learning to identify and classify LTR-retrotransposons in plant genomes. *Briefings in Bioinformatics* 24 (1): bbac511.

Pook, T., Freudenthal, J., Korte, A., and Simianer, H. (2020). Using local convolutional neural networks for genomic prediction. *Frontiers in Genetics* 11: 561497.

Poplin, R., Chang, P.C., Alexander, D. et al. (2018). A universal SNP and small-indel variant caller using deep neural networks. *Nature Biotechnology* 36 (10): 983–987.

Quazi, S. (2022). Artificial intelligence and machine learning in precision and genomic medicine. *Medical Oncology* 39 (8): 120.

Rairdin, A., Fotouhi, F., Zhang, J. et al. (2022). Deep learning-based phenotyping for genome wide association studies of sudden death syndrome in soybean. *Frontiers in Plant Science* 13: 966244.

Raj, A., Shim, H., Gilad, Y. et al. (2015). msCentipede: modeling heterogeneity across genomic sites and replicates improves accuracy in the inference of transcription factor binding. *PLoS One* 10 (9): e0138030.

Ramachandran, A., Lumetta, S.S., Klee, E.W., and Chen, D. (2021). HELLO: improved neural network architectures and methodologies for small variant calling. *BMC Bioinformatics* 22: 1–31.

Ramstein, G.P. and Buckler, E.S. (2022). Prediction of evolutionary constraint by genomic annotations improves functional prioritization of genomic variants in maize. *Genome Biology* 23 (1): 183.

Ramstein, G.P., Jensen, S.E., and Buckler, E.S. (2019). Breaking the curse of dimensionality to identify causal variants in Breeding 4. *Theoretical and Applied Genetics* 132 (3): 559–567.

Rane, N., Choudhary, S.P., and Rane, J. (2024). Acceptance of artificial intelligence: key factors, challenges, and implementation strategies. *Journal of Applied Artificial Intelligence* 5 (2): 50–70.

Riera, L.G., Carroll, M.E., Zhang, Z. et al. (2021). Deep multiview image fusion for soybean yield estimation in breeding applications. *Plant Phenomics* 2021: 9846470.

Rodríguez-Lira, D.C., Córdova-Esparza, D.M., Álvarez-Alvarado, J.M. et al. (2024). Trends in machine and deep learning techniques for plant disease identification: a systematic review. *Agriculture* 14 (12): 2188.

Rosado, R.D.S., Cruz, C.D., Barili, L.D. et al. (2020). Artificial neural networks in the prediction of genetic merit to flowering traits in bean cultivars. *Agriculture* 10 (12): 638.

Rosignoli, S., Boi, D., Rubini, E., Asteriti, I. A., Guarguaglini, G., and Paiardini, A. (2024). Developing tools for structural bioinformatics: from python to bedside. *In Bioinformatics and Computational Biology Conference*, Naples, Italy, 27–29 November 2024.

Roychowdhury, R., Das, S.P., Gupta, A. et al. (2023). Multi-omics pipeline and omics-integration approach to decipher plant's abiotic stress tolerance responses. *Genes* 14 (6): 1281.

Rudakiya, D.M., Patel, R., and Patel, H. (2024). Omics technologies: an advanced approach to understand the systems using bioinformatics tools. In: *Current Omics Advancement in Plant Abiotic Stress Biology* (ed. D. Bhatt, M. Nath, S. Badoni, and R. Joshi), 301–322. Academic Press.

Ruffolo, J.A., Nayfach, S., Gallagher, J. et al. (2024). Design of highly functional genome editors by modeling the universe of CRISPR-Cas sequences. *BioRxiv* 2024-04: https://doi.org/10.1101/2024.04.22.590591.

Rui, Z., Zhang, Z., and Chen, T. (2024). A review of field plant phenotyping platform, sensing and navigation technology. In: *Transforming Technologies in Smart Agriculture: New Applications During the Past Decade* (ed. Z. Zhang, Y. Jiang, C. Wen, et al.), 1–46. Springer.

Sarker, I.H. (2021). Deep learning: a comprehensive overview on techniques, taxonomy, applications and research directions. *SN Computer Science* 2 (6): 1–20.

Sartor, R.C., Noshay, J., Springer, N.M., and Briggs, S.P. (2019). Identification of the expressome by machine learning on omics data. *Proceedings of the National Academy of Sciences* 116 (36): 18119–18125.

Schrider, D.R. and Kern, A.D. (2018). Supervised machine learning for population genetics: a new paradigm. *Trends in Genetics* 34 (4): 301–312.

Selvaraj, M.G., Valderrama, M., Guzman, D. et al. (2020). Machine learning for high-throughput field phenotyping and image processing provides insight into the association of above and below-ground traits in cassava (Manihot esculenta Crantz). *Plant Methods* 16: 1–19.

Sheikh, M., Iqra, F., Ambreen, H. et al. (2024). Integrating artificial intelligence and high-throughput phenotyping for crop improvement. *Journal of Integrative Agriculture* 23 (6): 1787–1802.

Sherwood, R.I., Hashimoto, T., O'donnell, C.W. et al. (2014). Discovery of directional and nondirectional pioneer transcription factors by modeling DNase profile magnitude and shape. *Nature Biotechnology* 32 (2): 171–178.

Shrestha, A. and Mahmood, A. (2019). Review of deep learning algorithms and architectures. *IEEE Access* 7: 53040–53065.

Shu, M., Shen, M., Zuo, J. et al. (2021). The application of UAV-based hyperspectral imaging to estimate crop traits in maize inbred lines. *Plant Phenomics* 2021: 9890745.

Smith, A.G., Petersen, J., Selvan, R., and Rasmussen, C.R. (2020). Segmentation of roots in soil with U-Net. *Plant Methods* 16 (1): 13.

Sokolova, K., Chen, K.M., Hao, Y. et al. (2024). Deep learning sequence models for transcriptional regulation. *Annual Review of Genomics and Human Genetics* 25.

Su, W., Zhang, M., Bian, D. et al. (2019). Phenotyping of corn plants using unmanned aerial vehicle (UAV) images. *Remote Sensing* 11 (17): 2021.

Sun, L., Lai, M., Ghouri, F. et al. (2024). Modern plant breeding techniques in crop improvement and genetic diversity: from molecular markers and gene editing to artificial intelligence—a critical review. *Plants* 13 (19): 2676.

Szeliski, R. (2022). *Computer Vision: Algorithms and Applications*. Springer Nature.

Tardieu, F., Cabrera-Bosquet, L., Pridmore, T., and Bennett, M. (2017). Plant phenomics, from sensors to knowledge. *Current Biology* 27 (15): R770–R783.

Tauro, F., Maltese, A., Giannini, R., and Harfouche, A. (2022). Latent heat flux variability and response to drought stress of black poplar: a multi-platform multi-sensor remote and proximal sensing approach to relieve the data scarcity bottleneck. *Remote Sensing of Environment* 268: 112771.

Tian, H., Wang, T., Liu, Y. et al. (2020). Computer vision technology in agricultural automation—a review. *Information Processing in Agriculture* 7 (1): 1–19.

Togninalli, M., Wang, X., Kucera, T. et al. (2023). Multi-modal deep learning improves grain yield prediction in wheat breeding by fusing genomics and phenomics. *Bioinformatics* 39 (6): btad336.

Uygun, S., Azodi, C.B., and Shiu, S.H. (2019). Cis-regulatory code for predicting plant cell-type transcriptional response to high salinity. *Plant Physiology* 181 (4): 1739–1751.

Vadapalli, S., Abdelhalim, H., Zeeshan, S., and Ahmed, Z. (2022). Artificial intelligence and machine learning approaches using gene expression and variant data for personalized medicine. *Briefings in Bioinformatics* 23 (5): bbac191.

Varala, K., Marshall-Colón, A., Cirrone, J. et al. (2018). Temporal transcriptional logic of dynamic regulatory networks underlying nitrogen signaling and use in plants. *Proceedings of the National Academy of Sciences* 115 (25): 6494–6499.

Varshney, R.K., Nayak, S.N., May, G.D., and Jackson, S.A. (2009). Next-generation sequencing technologies and their implications for crop genetics and breeding. *Trends in Biotechnology* 27 (9): 522–530.

Vidanagamachchi, S.M. and Waidyarathna, K.M.G.T.R. (2024). Opportunities, challenges and future perspectives of using bioinformatics and artificial intelligence techniques on tropical disease identification using omics data. *Frontiers in Digital Health* 6: 1471200.

Wambugu, P.W., Ndjiondjop, M.N., and Henry, R.J. (2018). Role of genomics in promoting the utilization of plant genetic resources in genebanks. *Briefings in Functional Genomics* 17 (3): 198–206.

Wang, Y., Wang, H., Wei, L. et al. (2020). Synthetic promoter design in Escherichia coli based on a deep generative network. *Nucleic Acids Research* 48 (12): 6403–6412.

Wang, K., Abid, M.A., Rasheed, A. et al. (2023). DNNGP, a deep neural network-based method for genomic prediction using multi-omics data in plants. *Molecular Plant* 16 (1): 279–293.

Wolpert, D.H. (1996). The lack of a priori distinctions between learning algorithms. *Neural Computation* 8 (7): 1341–1390.

Xu, Y. (2016). Envirotyping for deciphering environmental impacts on crop plants. *Theoretical and Applied Genetics* 129: 653–673.

Xu, Y., Zhang, X., Li, H. et al. (2022). Smart breeding driven by big data, artificial intelligence, and integrated genomic-enviromic prediction. *Molecular Plant* 15 (11): 1664–1695.

Yadav, A., Yadav, K., Ahmad, R., and Abd-Elsalam, K.A. (2023). Emerging frontiers in nanotechnology for precision agriculture: advancements, hurdles and prospects. *Agrochemicals* 2 (2): 220–256.

Yan, J. and Wang, X. (2023). Machine learning bridges omics sciences and plant breeding. *Trends in Plant Science* 28 (2): 199–210.

Yang, Y., Saand, M.A., Huang, L. et al. (2021a). Applications of multi-omics technologies for crop improvement. *Frontiers in Plant Science* 12: 563953.

Yang, Z., Liao, B., Hsieh, C. et al. (2021b). Deep-BGCpred: a unified deep learning genome-mining framework for biosynthetic gene cluster prediction. *BioRxiv* 2021–2011. https://doi.org/10.1101/2021.11.15.468547.

Yang, M., Chen, S., Huang, Z. et al. (2023). Deep learning-enabled discovery and characterization of HKT genes in Spartina alterniflora. *The Plant Journal* 116 (3): 690–705.

Yang, T., Mei, Y., Xu, L. et al. (2024). Application of question answering systems for intelligent agriculture production and sustainable management: a review. *Resources, Conservation and Recycling* 204: 107497.

Yasmeen, E., Wang, J., Riaz, M. et al. (2023). Designing artificial synthetic promoters for accurate, smart, and versatile gene expression in plants. *Plant Communications* 4 (4): 100558.

Yasrab, R., Atkinson, J.A., Wells, D.M. et al. (2019). RootNav 2.0: deep learning for automatic navigation of complex plant root architectures. *GigaScience* 8 (11): giz123.

Yoosefzadeh-Najafabadi, M., Tulpan, D., and Eskandari, M. (2021). Using hybrid artificial intelligence and evolutionary optimization algorithms for estimating soybean yield and fresh biomass using hyperspectral vegetation indices. *Remote Sensing* 13 (13): 2555.

Yoosefzadeh-Najafabadi, M., Hesami, M., and Eskandari, M. (2024). Machine learning-enhanced utilization of plant genetic resources. In: *Sustainable Utilization and Conservation of Plant Genetic Diversity* (ed. J.M. AI-Khayri, S.M. Jain, and S. Penna), 619–639. Singapore: Springer Nature Singapore.

Yue, J., Yang, G., Tian, Q. et al. (2019). Estimate of winter-wheat above-ground biomass based on UAV ultrahigh-ground-resolution image textures and vegetation indices. *ISPRS Journal of Photogrammetry and Remote Sensing* 150: 226–244.

Zaman, E.A.K., Mohamed, A., and Ahmad, A. (2022). Feature selection for online streaming high-dimensional data: a state-of-the-art review. *Applied Soft Computing* 127: 109355.

Zhang, R.G., Li, G.Y., Wang, X.L. et al. (2022). TEsorter: an accurate and fast method to classify LTR-retrotransposons in plant genomes. *Horticulture Research* 9: uhac017.

Zhang, Y., Huang, G., Zhao, Y. et al. (2025). Revolutionizing crop breeding: next-generation artificial intelligence and big data-driven intelligent design. *Engineering* 44: 245–255.

Zhao, X., Collins, R.L., Lee, W.P. et al. (2021). Expectations and blind spots for structural variation detection from long-read assemblies and short-read genome sequencing technologies. *The American Journal of Human Genetics* 108 (5): 919–928.

Zhongzhi, H. (2019). *Computer Vision-Based Agriculture Engineering*. CRC Press.

14

Integration of AI with Precision Agriculture Technologies

Tanisha Anand[1], Sumita Mishra[1], Sachin Kumar[1], Rajesh K. Tiwari[2], and Mala Trivedi[2]

[1]*Amity University, Amity School of Engineering & Technology, Lucknow, Uttar Pradesh*
[2]*Amity University, Amity Institute of Biotechnology, Lucknow, Uttar Pradesh*

14.1 Introduction

Agricultural practices specially in India combined with advanced plant breeding and modern technologies are providing food security to millions of people across the globe. Nevertheless, as the world encounters increasing difficulties in the twenty-first century, the agricultural industry must adjust to tackle substantial challenges. With the global population projected to reach 9.7 billion by 2050 (Sharma and Shivandu 2024), food production must significantly escalate to satisfy demand. Still, the industry struggles with resource constraints such as water shortages, diminishing arable land, and ecological degradation. These obstacles are magnified by climate change, which disrupts agricultural cycles and causes additional unpredictability in climate patterns. Traditional farming techniques are no longer adequate to fulfill these escalating requirements.

In reaction to these hardships, PA has surfaced as a groundbreaking, data-oriented method of farming. This contemporary approach allows farmers to enhance their resource utilization, boost crop yields, and lessen environmental effects. Central to this shift is artificial intelligence (AI), which plays an essential role in leading the PA initiative. By utilizing AI, farmers can gather data from multiple sources to make knowledgeable choices at every phase of the farming process. From monitoring crop health to managing resources and forecasting yields, AI equips farmers with actionable information.

The incorporation of AI into agriculture signifies a major transition from traditional techniques, where planting, watering, and harvesting decisions were mainly based on knowledge and observation. In contrast, AI-driven PA depends on real-time data and sophisticated analytics, enabling farmers to anticipate results, fine resource distribution, and swiftly react to shifts in weather or pest infestations.

However, AI does not function independently. It is bolstered by various other enabling technologies that further improve its functionalities. Global positioning systems (GPS),

Remote Sensing, and unmanned aerial vehicles (UAVs) are vital in offering real-time, location-specific information that AI systems interpret. GPS facilitates accurate mapping and monitoring of agricultural equipment, permitting Variable Rate Technology (VRT), where fertilizers, pesticides, and water are utilized in correct amounts and at appropriate locations. Remote sensing, either through satellites or UAVs, grants farmers crucial insights into soil health, crop conditions, and potential dangers like drought or disease (Sharma and Shivandu 2024).

UAVs (Linaza et al. 2021; Senoo et al. 2024), commonly referred to as drones, have transformed how farmers oversee their fields. These aerial devices, equipped with high-resolution cameras and sensors, supply real-time images that assist farmers in evaluating crop health, tracking pest outbreaks, and even administering fertilizers and pesticides in a precise and targeted way. UAVs diminish the necessity for labor-intensive field assessments, enabling farmers to oversee extensive areas effectively (Senoo et al. 2024).

In addition to AI, GPS, and UAVs, the Internet of Things (IoT) is crucial in PA. IoT connects different farm sensors and devices to a central system, facilitating real-time data gathering and oversight. IoT sensors monitor soil moisture, nutrient levels, temperature, humidity, and other essential environmental factors. When combined with AI, IoT systems can streamline irrigation, fertilization, and pest control, guaranteeing that crops obtain precisely what they require when they need it.

The combination of AI, GPS, UAVs, and IoT in PA signifies a transformative change in farming practices. Farmers no longer have to depend solely on experience or instinct; rather, they can make decisions based on data that enhance every component of their operations. AI-driven systems can evaluate extensive datasets to forecast crop yields, assess soil health, identify early signs of pest problems, and adjust to shifting climate conditions (Figure 14.1). This technology empowers farmers to take proactive measures that boost productivity, conserve resources, and lessen environmental impact (Feng et al. 2019; Haneklaus et al. 2016; Khardia et al. 2022).

In this chapter, we investigate how AI and these supporting technologies have revolutionized contemporary farming. We explore the applications of AI in monitoring crop health, predicting yields, managing resources (such as optimizing fertilizers, overseeing irrigation, automating weed control, and precisely spraying fertilizers and pesticides),

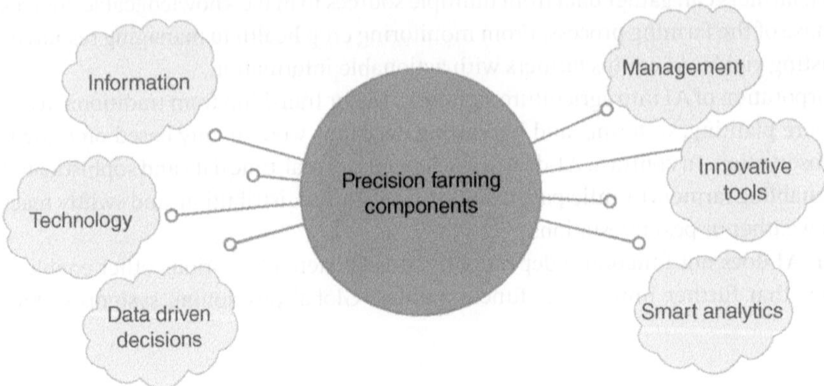

Figure 14.1 Components of precision farming.

adapting to climate, and optimizing supply chains (Jain et al. 2023; Lange and Peake 2020). By looking into these applications, we emphasize how PA is transforming farming for a more sustainable and efficient future. As AI continues to advance, its integration into agriculture will only strengthen, assisting farmers worldwide in tackling the increasing challenges of food production, environmental sustainability, and climate resilience (Sinha et al. 2022).

14.2 Enabling Technologies for Precision Agriculture

AI assisted PA leverages cutting edge technologies like sensors, drones, and data analytics to optimize farming practices. By using soil and weather sensors to collect real-time data, satellite imagery to monitor crops, and machine learning to perform predictive analytics, farmers can make informed decisions that improve productivity and sustainability (Figure 14.2). This integration enables precise resource management, increasing yields, reducing costs (Miranda et al. 2022; Jimenez Lopez et al. 2015).

14.2.1 Artificial Intelligence

AI is swiftly changing the landscape of agriculture. It enables farmers to handle extensive datasets in real-time, fostering accuracy in every stage from planting to harvesting. Machine learning models and neural networks assist in identifying patterns and delivering actionable insights for enhancing crop yields and lowering resource wastage (Lakhiar et al. 2024). Some of AI's notable functions in agriculture encompass forecasting crop diseases,

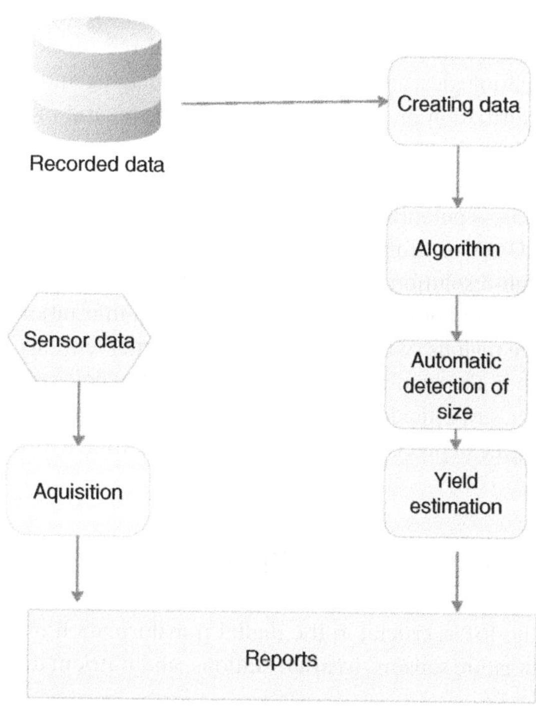

Figure 14.2 Data collection and extraction to be applied in precision farming.

automating irrigation schedules, and analyzing satellite data for weather predictions. AI algorithms are capable of assimilating data from diverse sensors, farming equipment, and remote imagery to offer farmers a comprehensive perspective of their operations facilitating more precise and timely decision-making (Shaikh et al. 2022; Oliveira and Silva 2023).

AI also bolsters robotics in agriculture, with autonomous machines able to carry out tasks such as weeding, planting, and even harvesting. These innovations have significantly lowered labor expenses and increased the efficiency of farming operations. One of the primary advantages of employing AI in PA is its capability to amalgamate data from various sources and provide predictive insights. For instance, AI models can evaluate data from soil sensors, weather stations, and satellite imagery to forecast the optimal times for planting, irrigating, and applying fertilizers. These forecasts assist farmers in maximizing their resource utilization, lessening costs, and curtailing waste (Karunathilake et al. 2023; Lakhiar et al. 2024).

Another utilization of AI is in monitoring crop health. By employing computer vision and machine learning algorithms, AI systems can scrutinize crop images to identify early indications of diseases or nutrient shortages (Lassoued et al. 2021; Masongsong 2024). This empowers farmers to undertake preventive measures before issues escalate, enhancing crop yields and diminishing the necessity for chemical interventions.

14.2.2 GPS, Remote Sensing, and UAV (Unmanned Aerial Vehicle)

GPS play a vital role in PA as they enable accurate mapping and monitoring of farming activities. By supplying precise location information, GPS facilitates VRT (Nath 2024), which modifies the quantities of fertilizers, pesticides, and water applied according to specific soil and crop requirements at various spots in a field (Leong et al. 2023; Liu et al. 2023).

This guarantees the effective use of resources and minimizes waste. Remote sensing technologies, encompassing satellite imaging and aerial reconnaissance, provide real-time information on crop health, soil moisture, and meteorological trends. These technologies utilize electromagnetic radiation to gauge crop, evaluate plant health, and identify early indications of stress. This information supports farmers in making proactive choices to address potential issues.

UAVs or drones have emerged as vital instruments for data gathering. Outfitted with high-resolution cameras and multispectral sensors, UAVs can obtain detailed images of crops and pinpoint areas of stress or pest infestations. They allow farmers to survey expansive regions swiftly, enabling targeted interventions and reducing the necessity for manual labor. For example, drones can scan entire fields to locate zones that need irrigation, fertilizer, or pesticide application, and they can even carry out these tasks autonomously. This degree of precision surpasses that of traditional farming practices (Liu et al. 2023).

14.3 Internet of Things (IoT)

The IoT is crucial in the digital transformation of agriculture. IoT devices, including soil moisture sensors, weather stations, and nutrient detectors, constantly gather data from the

Table 14.1 Different types of sensors.

Sensor type	Applications	Advantages	Parameters	References
Ultrasound Sensors	Used for distance measurement, obstacle detection, and liquid level detection	Affordable, small size, suitable for indoor use	Measurement Range: Typically 2 cm to 5 meters. Accuracy: ±1–2 cm (depending on the distance). Operating frequency: Typically between 20 kHz to 40 kHz	(Sharma and Shivandu, 2024; Haneklaus et al., 2016; Oliveira and Silva, 2023; Lassoued et al., 2021)
LiDAR (Light Detection and Ranging)	Applied in autonomous vehicles, forestry	Highly accurate, long distance capability, effective	Range: 10 m to 300 m, depending on the model and application. Accuracy: Typically ±2 to 3 cm. Laser Wavelength: 850 nm to 1550 nm.	(Linaza et al., 2021; Senoo et al., 2024; Shafi et al., 2019; Feng et al., 2019 Shaikh et al., 2022)
Color Cameras	Used in plant health monitoring, fruit detection, and precision agriculture.	Captures rich data, widely available, cost-effective	Resolution: Typically 1 MP to 20 MP or higher. Frame Rate: 15 fps to 60 fps (frames per second). Field of View (FoV): 30° to 120° horizontal. Image processing: RGB color space.	(Sharma and Shivandu, 2024; Shafi et al., 2019; Lange and Peake, 2020; Sinha et al., 2022)
			Spectral Range: 400 nm to 2500 nm (visible to near-infrared). Spectral Resolution: Typically 5 nm to 10 nm. Spatial Resolution: Varies from a few millimeters to meters per pixel, depending on the application. Field of View (FoV): 20° to 40° horizontal.	(Senoo et al., 2024; Miranda, 2022; Lakhiar et al., 2024; Shaikh et al., 2022)
Laser Sensor-based Spraying	Used for precision spraying, crop monitoring, and optimizing yield	Provides high precision, real-time canopy parameter measurements	Measurement Range: 0.5 m to 50 m. Accuracy: ±1-2 cm. Laser Wavelength: 850 nm to 1550 nm. Field of View (FoV): 360° horizontal, 10 to 30 vertical.	(Linaza et al., 2021; Jain et al., 2023; Lakhiar et al., 2024)

field. These sensors transmit information wirelessly to central systems, delivering real-time updates on environmental conditions (Nemade et al. 2023).

When combined with AI, IoT facilitates autonomous reactions. For instance, if soil moisture levels drop below a specific threshold, an AI system can automatically trigger an irrigation system based on real-time weather predictions, ensuring crops are given optimal water without waste.

Farmers are utilizing IoT devices to oversee livestock health, enhance greenhouse conditions, and track the status of autonomous farm vehicles. The integration of AI and IoT has resulted in more intelligent farming methods, decreasing labor expenses and boosting crop yields while reducing environmental impact (Nemade et al. 2023).

By employing a network of connected devices, IoT can collect and analyze extensive amounts of data that would otherwise be challenging for farmers to gather manually. This encompasses monitoring soil health, weather patterns, pest activity, and equipment functionality in real time. AI algorithms then analyze this information to provide suggestions or initiate automated actions, such as modifying irrigation schedules or deploying drones for pesticide application (Table 14.1).

IoT's significance in agriculture extends beyond crop surveillance (Ontiri and Amuhaya 2022). It is also applicable to precision livestock farming, where sensors are affixed to animals to observe their health, behavior, and reproductive patterns.

14.4 Role/Applications of AI in Precision Agriculture

AI significantly impacts PA by facilitating data-driven choices that enhance farming techniques, boost yields, and promote sustainability. By utilizing AI-driven image analysis, farmers can track crop health in real-time, identifying problems such as nutrient shortages, pest outbreaks, and diseases promptly. AI-powered predictive models examine ecological data and past trends to predict harvest outputs, enabling improved planning and distribution of resources (Figure 14.3). In precision irrigation, AI algorithms utilize information from soil moisture sensors, weather predictions, and crop water needs to enhance water efficiency, minimizing waste and guaranteeing that crops get the appropriate amount of water at the optimal time.

14.4.1 Crop Health Monitoring and Assessment

AI-powered systems are essential for monitoring crop health. Utilizing satellite images, drone observations, and sensors in the field, farmers can constantly track the condition of their crops. AI algorithms evaluate this data to identify diseases, nutrient shortages, and pest problems. By recognizing issues early, farmers can implement preventive measures to lessen yield reductions.

Sophisticated AI models employ spectral imaging to assess plant health by examining reflectance at particular light wavelengths. For instance, the Normalized Difference Vegetation Index (NDVI) data is extensively utilized to measure plant robustness. AI interprets these metrics to ascertain which sections of a field require additional focus (Pandey and Srivastava 2021).

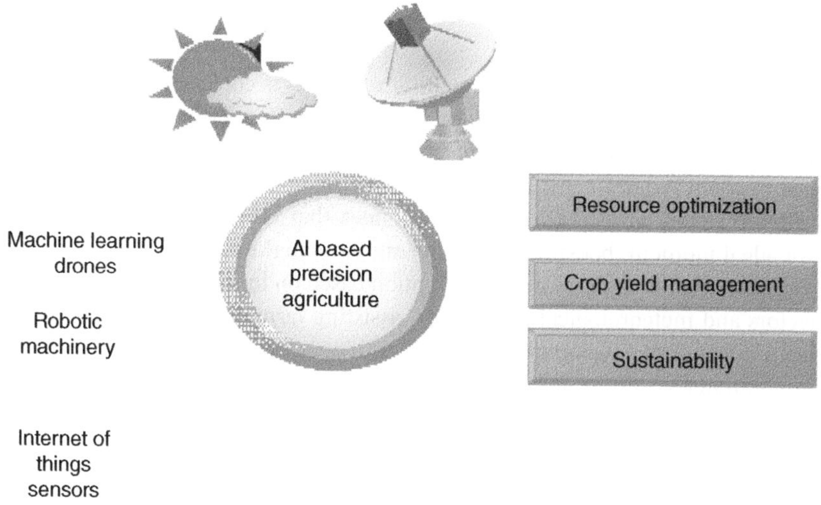

Figure 14.3 AI in agriculture.

AI-driven crop health monitoring systems are adept at processing information from various sources, including multispectral cameras mounted on drones and soil sensors. These systems utilize computer vision methods to spot irregularities in plant growth, such as discolored leaves, which may suggest nutrient shortage or disease incidences. Early identification allows for focused interventions, like applying fertilizers or pesticides solely where necessary, lowering expenses and environmental consequences.

14.4.2 Predictive Analytics for Crop Yield Estimation

AI-enhanced predictive analytics utilize historical crop information, climatic trends, and soil health assessments to project future crop yields. These forecasts enable farmers to enhance their planning and distribute resources more effectively (Piccoli et al. 2023).

Predictive models apply machine learning methods to evaluate factors such as soil moisture, temperature, planting times, and pest pressures. These models furnish farmers with comprehensive insights on anticipated yields, assisting them in fine-tuning harvests and handling market demand.

One instance of AI's forecasting capabilities is its capacity to estimate yields based on weather trends (Piccoli et al. 2023). By scrutinizing historical weather records along with present conditions, AI can determine the required water or fertilizer for crops in the upcoming weeks. This aids farmers in preventing either excessive or insufficient resource applications, thus ensuring optimal crop development.

Predictive models can also project pest invasions or disease outbreaks. For instance, AI algorithms have the ability to evaluate weather information, including temperature and humidity levels, to foresee when conditions are favorable for pests like aphids or fungi to flourish. Consequently, farmers can adopt preventive measures, such as applying pesticides or modifying irrigation schedules to mitigate the risk of crop loss.

14.5 Resource Management

In agriculture that integrates AI, the management of resources is considerably improved. through solutions driven by data that enhance the utilization of fertilizers, water, and pesticides. AI plays a role in fertilizer enhancement by examining soil information to identify specific nutrient requirements, preventing excessive application and diminishing detrimental runoff (Piccoli et al. 2023). This guarantees that crops acquire the correct nutrients at the ideal moment, boosting efficiency and reducing expenses. In water management AI-enabled systems adjust water consumption based on immediate data from soil moisture detectors and meteorological predictions, assisting farmers in conserving water while ensuring optimal crop development. Automated weed control employs AI-enabled machines and computer vision to detect and eliminate weeds without damaging crops, lessening dependence on herbicides and encouraging sustainable methods. Additionally, AI facilitates targeted spraying of fertilizers and pesticides, utilizing drones and automated sprayers to deliver chemicals precisely where needed, minimizing environmental consequences and cutting down on input expenses. In summary, AI advances resource management in agriculture by enhancing efficiency minimizing waste, and promoting sustainability.

14.5.1 Fertilizer Optimization

AI assists farmers in refining their fertilizer usage by evaluating soil information to establish the precise nutrient requirements of crops. Conventional farming practices frequently lead to excessive application, which not only squanders resources but also contributes to environmental harm. AI guarantees that fertilizers are administered in accurate amounts, lowering expenses and minimizing detrimental runoff into aquatic environments.

By examining data from soil sensors, AI systems can furnish farmers with instantaneous advice on the quantity and timing of fertilizer application. These suggestions are grounded in soil conditions, crop requirements, and climate predictions, ensuring that farmers utilize fertilizers effectively and lessen their ecological impact.

AI-driven precision fertilization systems are created to modify the application of nutrients according to the distinct needs of every plant. These systems employ machine learning algorithms to evaluate data from sensors embedded in the soil, drones, and satellite imagery to anticipate nutrient deficiencies and suggest the most suitable time and amount for fertilizer application.

14.5.2 Irrigation Management

Water scarcity represents oneof the largest challenges in agriculture, particularly in arid areas. AI-powered irrigation systems guarantee that crops are provided with the appropriate quantity of water at the correct time. These systems assess data from soil moisture sensors, weather predictions, and crop water needs to supply water effectively, minimizing waste and enhancing crop yields.

In conjunction with IoT devices, AI has the capability to automate irrigation schedules based on real-time information, ensuring that fields are neither excessively irrigated nor insufficiently irrigated. Irrigation management is especially crucial in areas where water resources are limited. Through the use of AI, farmers can apply precision irrigation practices that supply crops with exactly the right amount of water required for optimal growth. This not only preserves water but also guarantees that plants do not experience overwatering, which can cause root damage and reduced yields.

AI-based irrigation management also enables farmers to foresee water needs ahead of time by examining weather data, soil moisture content, and crop growth phases. This assists farmers in preparing for periods of drought or heavy rainfall and taking essential measures to shield their crops from water stress.

14.5.3 Automated Weed Management

Weed control is another field where AI is making noteworthy advancements. AI-driven robots can discern between crops and weeds by utilizing computer vision, facilitating precise weed elimination without harming the crop. These robots diminish the necessity for herbicides, lower production expenses, and reduce environmental impact.

Historically, weed management has been labor-intensive and demanded extensive chemical use. AI-powered machines, including autonomous weeders, utilize advanced image recognition technologies to spot weeds and eradicate them mechanically. This decreases the dependence on chemical herbicides, encouraging more sustainable agricultural practices. Moreover, these machines can operate autonomously, enabling farmers to concentrate on other vital tasks.

14.5.4 Precision Spraying of Fertilizer and Pesticide

AI-controlled drones and automated sprayers guarantee that fertilizers and pesticides are applied only where they are necessary. These technologies rely on real-time data from sensors and imaging systems to identify the specific zones that need treatment. Precision spraying minimizes the overapplication of chemicals, reduces production costs, and diminishes environmental damage.

Precision spraying is particularly efficient in dealing with pest outbreaks and supplying nutrients during critical growth phases. By focusing on specific areas of a field, AI-guided systems assist in decreasing the total volume of chemicals utilized, thus minimizing their environmental repercussions. In addition to boosting crop health, this strategy can substantially lower input expenses for farmers.

14.6 Climate Adaptation

The unpredictable nature of climate change presents a considerable risk to agriculture. AI models are capable of aiding farmers in adapting by predicting changes in weather and recommending suitable farming practices. Whether it's altering planting schedules, adjusting irrigation systems, or choosing crop varieties that can withstand drought or flooding, AI delivers the insights necessary to risks.

For instance, AI models can evaluate historical weather patterns, soil conditions, and plant genetics to suggest ideal planting periods or determine which crops are best suited for enduring extreme temperatures. These predictive insights empower farmers to make knowledgeable choices that boost the resilience of their crops against climate variability.

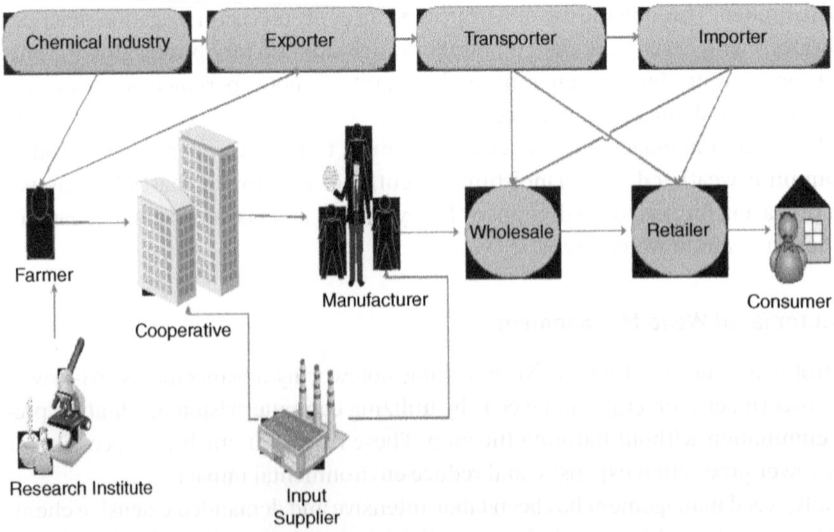

Figure 14.4 Process of supply chain in agriculture.

Moreover, AI-driven climate adaptation tools can assist farmers in anticipating extreme weather occurrences, such as storms or heatwaves, and taking preventative actions to safeguard their crops. By offering early alerts, AI allows farmers to lessen potential losses and enhance their overall farming management practices.

14.7 Supply Chain Optimization

In addition to the farm, AI significantly contributes to enhancing the agricultural supply chain. By forecasting demand and improving inventory management, AI can aid in minimizing waste and ensuring that fresh produce is delivered to markets promptly (Pierre et al. 2023).

AI-enabled platforms can monitor the transport of crops from farm to market, facilitating better logistics and decreasing post-harvest losses. This degree of transparency and efficiency enhances profitability for farmers and guarantees consumers receive fresh products (Figure 14.4).

Optimizing the supply chain is especially crucial in minimizing food waste and ensuring that agricultural goods reach consumers at peak freshness. AI technologies are capable of analyzing market trends, transportation schedules, and storage conditions to enhance the efficiency of the entire supply chain.

14.8 Result and Discussion

14.8.1 5G Technology and Real-Time Data

The implementation of 5G networks in agriculture will significantly enhance data transmission speeds and facilitate real-time decision-making. With quicker connections, farmers can instantly access data from IoT sensors, enabling immediate modifications to irrigation, fertilization, or pest control systems.

5G will also boost the functionality of autonomous farming equipment, allowing for more accurate operations like planting and spraying, in addition to improving coordination among multiple machines in the field.

As 5G technology becomes increasingly prevalent, it will support real-time surveillance and regulation of farming activities. Farmers will have the ability to make data-driven decisions more swiftly and accurately, leading to more efficient resource utilization and greater crop yields (Pierre et al. 2023).

14.8.2 Edge Computing for Faster Processing

Edge computing permits data processing to take place nearer to the source, thereby decreasing latency and enhancing the speed of decision-making in the field. This is especially valuable for applications that necessitate immediate responses, such as autonomous machines and drone-based crop monitoring (Pierre et al. 2023).

By locally processing data on devices like drones, robots, or sensors, edge computing reduces the dependency on continuous data transmission to the cloud, decreasing bandwidth consumption and boosting the efficiency of AI-powered systems. This guarantees that real-time insights can be acted upon without delay, improving the efficacy of PA.

14.8.3 Blockchain for Data Security and Traceability

Blockchain technology provides a secure and transparent method to monitor agricultural products throughout the supply chain. By establishing an unchangeable record of every transaction and procedure, blockchain upholds the integrity of data and enables consumers to verify the origins of their food. This not only aids in ensuring food safety but also facilitates more responsible farming practices, as consumers can trace products back to their origins and confirm whether they were produced sustainably or organically.

Blockchain technology can also be applied to address issues concerning data privacy and security within PA. Farmers can exercise greater control over their data, ensuring that sensitive information is shared solely with trusted parties. This degree of transparency fosters trust between farmers, distributors, and consumers.

Utilizing blockchain for traceability will also be crucial in minimizing food fraud and confirming the authenticity of organic or sustainably produced goods (Prabha and Pathak 2023). In situations of contamination or product recalls, blockchain simplifies tracing the source of the issue and quickly removes affected products from the supply chain, thereby bolstering food safety.

14.8.4 AI-Driven Crop Breeding and Genetic Engineering

AI is progressively being incorporated into crop breeding initiatives to expedite the development of new plant varieties with favorable traits, such as disease resistance, drought tolerance, or enhanced yield. Machine learning algorithms can assess extensive datasets of genetic information to spot the best candidates for breeding. This accelerates the time necessary for traditional breeding methods, which heavily depend on trial and error (Prabha and Pathak 2023).

AI-driven crop breeding can additionally assist farmers in adapting to the shifting climate. For instance, AI models can evaluate weather patterns and soil conditions to suggest the most suitable crops to plant in a specific area, considering elements like drought resistance or temperature tolerance.

Alongside CRISPR (Karunathilake et al. 2023) and other gene-editing technologies, AI can facilitate the creation of crops that are better suited to evolving climates or resistant to particular pests. These innovations hold the promise of transforming the agricultural sector by producing more resilient and fruitful crops.

14.8.5 Autonomous Farm Machinery

Self-driving tractors and harvesters are examples of autonomous farm machinery that are anticipated to become more prevalent in PA. These machines utilize AI to carry out tasks with minimal human involvement, enhancing efficiency and lowering labor expenses.

Autonomous equipment can function continuously, enabling the swift and precise completion of planting, harvesting, and other time-sensitive activities. This technology also minimizes the possibility of human mistakes, ensuring that agricultural operations are conducted with accuracy (Qazi et al. 2022).

Moreover, autonomous machines can come fitted with sensors and cameras to assess crop health and soil conditions in real time. This information is subsequently evaluated by AI algorithms to make prompt adjustments, such as changing planting depth or optimizing fertilizer usage, to guarantee the best possible results for crop yields.

14.8.6 AI and Climate Resilience

AI can be utilized to simulate the long-term effects of climate change on agriculture, permitting farmers to modify their practices to upcoming conditions. By evaluating climate data, AI systems can offer suggestions for crop rotation, soil management, and irrigation methods that will assist farmers in maintaining productivity despite environmental changes (https://doi.org/10.54536/ajaset.v8i3.3843, n.d.).

AI-based climate resilience tools can also aid policymakers and researchers in formulating strategies to alleviate the consequences of climate change on global food security. This will be especially crucial as increasing temperatures, shifting precipitation patterns, and more common extreme weather occurrences continue to pose challenges to agricultural systems worldwide. In areas where climate change is likely to lead to substantial changes in agricultural productivity, AI can support farmers in moving toward more sustainable and resilient farming practices. This might involve planting crops that are better suited to the new climatic conditions, modifying irrigation schedules to save water, or incorporating new technologies to enhance efficiency.

14.9 Conclusion

The incorporation of AI in PA has the ability to transform farming significantly. By enhancing resource utilization, elevating crop health monitoring, and enabling predictive analytics, AI assists farmers in attaining increased productivity while diminishing environmental impacts. The Agriculture 4.0 landscape is described by Araújo et al. (2021). As per him, it is a move towards digitalization and smart farming, the Agriculture 4.0 era is boosted by developments in IoT, sensors, robots, AI, big data, and cloud computing. Also, in Kanagaraj et al. (2023), it introduces a predictive classification model using an Artificial Neural Network (ANN) to forecast crop yield data.

Nonetheless, the extensive adoption of AI encounters obstacles, such as high expenses, concerns regarding data privacy, and the necessity for improved education and infrastructure. Addressing these challenges will be vital to realizing AI's complete potential within the agricultural sector.

The future of agriculture is shaped by AI-driven advancements, as they lay the foundation for sustainable, efficient, and resilient agricultural practices.

Concluding it, the future of PA lies in the continued integration of AI and other emerging technologies. By leveraging AI, farmers can achieve greater efficiency, sustainability, and profitability while addressing the challenges posed by climate change and resource constraints. This step in field of agriculture can be a boon as per Han (2023) which underscores the role of machine learning (ML) and AI in enabling connected, data-driven PA and smart farming systems.

References

Feng, X., Yan, F., and Liu, X. (2019). Study of wireless communication technologies on Internet of Things for precision agriculture. *Wireless Personal Communications* 108 (3): 1785–1802. https://doi.org/10.1007/s11277-019-06496-7.

Haneklaus, S., Lilienthal, H., and Schnug, E. (2016). 25 years precision agriculture in Germany–a retrospective. In: *13th International Conference on Precision Agriculture.*

Jain, S., Bhujel, S., Shrivastava, U. et al. (2023). Advancements in drone technology for fruit crop management: a comprehensive review. *Int. J. Envir. Clim. Change* 13 (11): 4367–4378. https://doi.org/10.9734/ijecc/2023/ v13i113617.

Jimenez Lopez, A.F., Salamanca, J.M., Quiroz Medina, M.J., and erez, O.E.A.Pé. (2015). Crops diagnosis using digital image processing and precision agriculture technologies. *Inge Cuc* 11 (1): https://doi.org/10.17981/ingecuc.11.1.2015.

Karunathilake, E.M.B.M., Le, A.T., Heo, S. et al. (2023). The path to smart farming: Innovations and opportunities in precision agriculture. *Agriculture* 13 (8): 1593. https://doi.org/10.3390/agriculture13081593.

Khardia, N., Sharma, S., and Kumawat, H. (2022). Precision farming. *A Monthly Peer Reviewed Magazine for Agriculture and Allied Sciences* 63: https://doi.org/10.60151/envec/zblx2641.

Lakhiar, I.A., Yan, H., Zhang, C. et al. (2024). A review of precision irrigation water-saving technology under changing climate for enhancing water use efficiency, crop yield, and environmental footprints. *Agriculture* 13 (2): 245–267. https://doi.org/10.3390/agriculture14071141.

Lange, A.F. and Peake, J. (2020). Precision agriculture, position, navigation, and timing technologies in the 21st century: integrated satellite navigation. *Sensor Systems, and Civil Applications* 2: 1735–1747. https://doi.org/10.1002/ 9781119458555.ch56.

Lassoued, R., Macall, D.M., Smyth, S.J., and Phillips, P.W.B. (2021). Expert insights on the impacts of, and potential for, agricultural big data. *Sustainability* 13 (5): 2521. https://doi.org/10.3390/su13052521.

Leong, Y.M., Lim, E.H., Subri, N.F., and Jalil, N. (2023). Transforming agriculture: navigating the challenges and embracing the opportunities of Artificial Intelligence of Things. *2023 IEEE International Conference on Agrosystem Engineering, Technology & Applications (AGRETA), Shah Alam, Malaysia* 2023: 142–147. https://doi.org/10.1109/AGRETA57740.2023.10262747.

Linaza, M.T., Posada, J., Bund, J. et al. (2021). Data-driven artificial intelligence applications for sustainable precision agriculture. *Agronomy* 11 (6): 1227.

Liu, Z., Yu, X., and Choi, A. Y. (2023). Data visualization for designing F2C IoT agricultural system. In Third International Conference on Artificial Intelligence, Virtual Reality, and Visualization (AIVRV 2023), Vol. 12923,pp. 488– 492. 10.1117/12.3011451.

Masongsong, J.R. (2024). Environmental awareness and participation among college students of Mindoro State University – Calapan City Campus. *American Journal of Environment and Climate (AJEC)* 3 (1): 25–29.

Micheni, E., Machii, J., and Murumba, J. (2022). Internet of Things, Big Data Analytics, and Deep Learning for Sustainable Precision Agriculture. *2022 IST-Africa Conference (IST-Africa), Ireland* 2022: 1–12. https://doi.org/10.23919/IST-Africa56635.2022.9845510.

Miranda, J.C., e-Mola, J.G., Arn-o, J., and Gregoria, E. (2022). Fruit data: a dual software application for Azure kinect cameras to acquire and extract informative data in yield tests performed in fruit orchard environments. *SoftwareX* 20: 101231.

Nath, S. (2024). A vision of precision agriculture: balance between agricultural sustainability and environmental stewardship. *Agronomy Journal* 116 (3): 1126–1143.

Nemade, S., Ninama, J., Kumar, S. et al. (2023). Advancements in agronomic practices for sustainable crop production: a review. *International Journal of Plant & Soil Science* 35 (22): 679–689.

Oliveira, R.C. and Silva, R.D.S. (2023). Artificial intelligence in agriculture: benefits, challenges, and trends. *Applied Sciences* 13 (13): 7405. https://doi.org/10.3390/app13137405.

Ontiri, G.K. and Amuhaya, L.L. (2022). Integration of mand automation technology in sustainable farming for achieving food security in Kenya. *European Journal of Electrical Engineering and Computer Science* 6 (1): 66–71. https://doi.org/10.24018/ejece.2022.6.1.413.

Pandey, M.K. and Srivastava, P.K. (2021). A probe into performance analysis of real-time forecasting of endemic infectious diseases using Machine Learning and Deep Learning Algorithms. In: *Advanced Prognostic Predictive Modelling in Healthcare Data Analytics*, Lecture Notes on Data Engineering and Communications Technologies, vol. 64 (ed. S. Roy, L.M. Goyal, and M. Mittal), 241–265. Singapore: Springer.

Piccoli, F., Locatelli, S.G., Schettini, R., and Napoletano, P. (2023). An open-source platform for GIS data management and analytics. *Sensors* 23 (8): 3788. https://doi.org/10.3390/s23083788.

Pierre, N., Viviane, I.V.I., Lambert, U. et al. (2023). AI based real-time weather condition prediction with optimized agricultural resources. *European Journal of Technology* 7 (2): 36–49.

Polwaththa, K.P.G.D.M., Amarasinghe, S.T.C., Amarasinghe, A.A.Y.D., and Amarasinghe, A.A.Y. (2024). Exploring artificial intelligence and machine learning in precision agriculture: a pathway to improved efficiency and economic outcomes in crop production. *American Journal of Agricultural Science, Engineering, and Technology* 8 (3): 2158–8104.

Prabha, C. and Pathak, A. (2023). Enabling technologies in smart agriculture: a way forward towards future fields. In: *International Conference on Advancement in Computation & Computer Technologies (InCACCT), Gharuan, India, 2023*, 821–826. https://doi.org/10.1109/InCACCT57535.2023.10141722.

Qazi, S., Khawaja, B.A., and Farooq, Q.U. (2022). IoT-equipped and AI-enabled next generation smart agriculture: a critical review, current challenges and future trends. *IEEE Access* 10: 21219–21235.

Senoo, E.E.K., Anggraini, L., Kumi, J.A. et al. (2024). IoT solutions with artificial intelligence technologies for precision agriculture: definitions, applications, challenges, and opportunities. *Electronics* 13 (10): 1894. https://doi.org/10.3390/electronics13101894.

Shafi, U., Mumtaz, R., García-Nieto, J. et al. (2019). Precision agriculture techniques and practices: from considerations to applications. *Sensors* 19 (17): 3796. https://doi.org/10.3390/s19173796.

Shaikh, T.A., Mir, W.A., Rasool, T., and Sofi, S. (2022). Machine learning for smart agriculture and precision farming: towards making the fields talk. *Archives of Computational Methods in Engineering* 29 (7): 4557–4597.

Sharma, K. and Shivandu, S.K. (2024). Integrating artificial intelligence and Internet of Things (IoT) for enhanced crop monitoring and management in precision agriculture. *Sensors International* 100292.

Sinha, R., Quir, J.J., Sankaran, S., and Khot, L.R. (2022). High Resolution aerial photogrammetry based 3D mapping of fruit crop canopies for precision inputs management. *Info Process. Agri.* 9 (1): 11–23. https://doi.org/10.1016/j.inpa.2021.01.006.

15

AI-Enabled IoT for Crop Improvement: A Paradigm Shift in Smart Agriculture

Akhouri Nishant Bhanu, Bangar Vaibhav, and Satyam Sanodiya

Rajmata Vijayaraje Scindia Krishi Vishwa Vidyalaya, Department of Genetics and Plant Breeding, R.A.K. College of Agriculture, Ichhawar Road, Sehore, 466001, Madhya Pradesh, India

15.1 Introduction

The global population is projected to reach nearly 10 billion by 2050, posing significant challenges to global food security (World Resources Institute [WRI] 2019). The World Resource Institute estimates that current agricultural practices will be unable to meet this growing demand, leading to a substantial food production shortfall. At present, approximately 30.7% of the world's population is engaged in agriculture, utilizing 2781 million hectares of arable land (FAO 2021). However, agricultural expansion is limited due to constraints such as declining soil fertility, water scarcity, and increasing pest and disease outbreaks (Godfray et al. 2010). Therefore, innovative solutions that optimize agricultural practices without additional resource consumption are imperative.

Artificial intelligence (AI) has emerged as a transformative tool in agriculture, enabling data-driven decision-making to enhance productivity and efficiency (Darwin et al. 2021; Upadhyay 2026). When combined with the Internet of Things (IoT), AI-powered solutions can revolutionize farming practices by leveraging real-time data collection through smart sensors and drones, coupled with advanced computational techniques such as neural networks and image recognition (Liakos et al. 2018). These technologies have the potential to improve agricultural productivity, enhance food security, and minimize resource wastage (Wolfert et al. 2017).

IoT in agriculture refers to the integration of connected devices such as soil moisture sensors, weather stations, drones, and automated irrigation systems that collect and transmit real-time data on soil conditions, temperature, humidity, crop health, and pest infestations (Verdouw et al. 2021). This data-driven approach facilitates precision agriculture, reducing input waste while maximizing yields (Kamilaris and Prenafeta-Boldú 2018).

IoT-enabled smart irrigation systems, for instance, have demonstrated the ability to reduce water consumption by up to 30% while simultaneously improving crop yields (Nawandar and Satpute 2019; Hamdi et al. 2021). By leveraging cloud computing and wireless sensor networks (WSNs), IoT-based agricultural systems can optimize water and fertilizer use, ensuring sustainable resource management (Goedde et al. 2020; Srivastava and Maity 2023). Additionally, IoT applications extend beyond crop cultivation to livestock management, disease prevention, and automated pest control (Rejeb et al. 2022; Cihan 2023).

AI plays a critical role in crop improvement through ML, deep learning (DL), and computer vision techniques that analyze vast amounts of data generated by IoT devices. AI-driven predictive models can forecast weather patterns and recommend optimal sowing times, thereby mitigating the risks associated with climate variability (Liakos et al. 2018; Singh and Ahmed 2021). Additionally, AI-based image recognition systems can identify crop diseases at early stages, allowing for timely interventions and reducing potential yield losses (Chlingaryan et al. 2018; Singh and Bhanot 2020).

In plant breeding and genetic improvement, AI accelerates the identification of desirable traits by analyzing extensive genomic datasets, aiding the development of climate-resilient and high-yielding crop varieties. AI-powered phenotyping, combined with IoT-enabled monitoring, enhances breeding efficiency by enabling high-throughput, non-destructive trait assessments (Singh and Ahmed 2021). These advancements contribute to the development of sustainable agricultural systems that maximize productivity while minimizing environmental impact (Dutta et al. 2025; Torky and Hassanein 2020).

Despite its transformative potential, the integration of IoT and AI in agriculture faces several challenges, including high implementation costs, data security concerns, and the need for robust infrastructure (Kumari and Nafchi 2024). Additionally, interoperability and scalability issues hinder widespread adoption, necessitating collaborative efforts among researchers, policymakers, and industry stakeholders (Bongiovanni and Lowenberg-DeBoer 2004; Monge et al. 2023). Addressing these challenges is critical to ensuring equitable access to IoT-driven agricultural solutions across diverse farming landscapes.

This chapter explores the role of IoT in modern agriculture, focusing on agritech innovations, IoT applications in crop improvement, and the challenges associated with implementing IoT-based solutions. Additionally, it discusses the integration of AI with IoT, highlighting its impact on precision agriculture, plant breeding, and sustainable resource management.

15.2 The Role of IoT in Crop Improvement

The integration of the IoT in agriculture is transforming crop improvement by enabling real-time monitoring, automation, and data-driven decision-making. IoT consists of a network of smart sensors, actuators, and communication devices that work together to optimize various farming operations, improving both productivity and sustainability (Wolfert et al. 2017; Verdouw et al. 2021).

The IoT is a network of interconnected devices and technologies that play a pivotal role in precision agriculture and smart farming. IoT architecture in agriculture integrates agricultural sensors with information and communication technology (ICT) and unmanned aerial vehicles (UAVs), facilitating essential data collection for precision farming (Madakam et al. 2015; Boursianis et al. 2022). The growing presence of IoT and mobile data is crucial in the context of the fourth industrial revolution. Advances in communication technologies and wireless networks, such as 5G, LoRaWAN, NB-IoT, Sigfox, ZigBee, and Wi-Fi, have significantly broadened the application of IoT across multiple sectors. These technologies enable real-time remote management, high-capacity phenotyping, and improvements in coverage, bandwidth, connection density, and end-to-end latency.

15.2.1 IoT Sensors and Data Collection

IoT-enabled sensors play a crucial role in monitoring key agricultural parameters such as soil moisture, temperature, pH, nutrient levels, and environmental conditions. These sensors collect high-resolution data in real time, allowing farmers to make informed decisions regarding irrigation, fertilization, and pest control (Chlingaryan et al. 2018; Kamilaris and Prenafeta-Boldú 2018). The deployment of smart sensors across fields helps identify variations in soil health and crop growth, enabling precision agriculture by optimizing resource application and reducing waste (Dutta et al. 2025).

15.2.2 Smart Irrigation Systems

Water scarcity is a major challenge in global agriculture, and IoT-based smart irrigation systems provide an effective solution. These systems leverage soil moisture sensors and weather data to regulate water supply, ensuring efficient irrigation while minimizing water wastage (Nawandar and Satpute 2019; Hamdi et al. 2021). Research indicates that IoT-driven irrigation management can reduce water consumption by up to 30% while improving crop yields by approximately 20% (Goedde et al. 2020; Srivastava and Maity 2023). AI-driven analytics further enhance efficiency by predicting optimal irrigation schedules based on weather forecasts and soil conditions (Kumari and Nafchi 2024).

15.2.3 IoT in Disease and Pest Management

IoT-enabled pest and disease management systems utilize real-time imaging, temperature, and humidity sensors to detect early signs of crop diseases and pest infestations. Drones equipped with IoT sensors and AI-based image recognition technologies can identify diseased plants, facilitating targeted pesticide application and reducing chemical use (Singh and Bhanot 2020; Zanin et al. 2022). These technologies enhance precision farming by detecting pest outbreaks at an early stage and enabling proactive intervention strategies (Bongiovanni and Lowenberg-DeBoer 2004).

15.2.4 IoT-Enabled Drone and UAV Technology

Drones and UAVs integrated with IoT sensors provide high-resolution imaging and spectral analysis for assessing crop health, detecting nutrient deficiencies, and monitoring field conditions (Liakos et al. 2018; Rahman et al. 2025). These aerial monitoring systems reduce labor costs and increase productivity by offering real-time insights into crop growth and field variability. Additionally, automated drones facilitate precision spraying of fertilizers and pesticides, reducing excessive application and improving overall soil health (Chlingaryan et al. 2018; Rejeb et al. 2022).

15.2.5 IoT-Based Supply Chain and Traceability

Beyond farm operations, IoT technology is transforming agricultural supply chains by ensuring traceability and quality control. Radio frequency identification (RFID) tags and blockchain-integrated IoT systems allow real-time tracking of produce from farm to market, improving food safety and reducing post-harvest losses (Torky and Hassanein 2020; Monge et al. 2023). Smart logistics powered by IoT can optimize transportation routes, monitor storage conditions, and prevent spoilage, leading to a more efficient agricultural value chain (Goedde et al. 2020).

15.2.6 Case Studies and Real-World Applications

- **Smart Farming in the Netherlands:** IoT-driven greenhouses and precision agriculture systems have increased crop yields by 20% while reducing fertilizer use by 30% (Hamdi et al. 2021).
- **IoT-Based Climate-Resilient Agriculture in India:** IoT sensors deployed in drought-prone regions have optimized water use, resulting in a 25% increase in water efficiency and improved soil moisture retention (Srivastava and Maity 2023).
- **Automated Pest Control in the USA:** IoT-enabled pest surveillance systems have reduced crop losses by 15% through real-time detection and targeted interventions (Singh and Bhanot 2020).

15.3 IoT Technologies Used in Plant Breeding, Molecular Breeding, and Gene Editing

The integration of the IoT in plant breeding, molecular breeding, and gene editing has revolutionized the way genetic improvements are achieved in crops. IoT-enabled technologies provide real-time data collection, precision monitoring, and automation, significantly enhancing breeding efficiency and accuracy. These technologies facilitate high-throughput phenotyping (HTP), automated environmental monitoring, and precise genetic modifications, thereby accelerating breeding programs.

15.3.1 IoT in Plant Breeding

Plant breeding is a cornerstone of modern agriculture, aimed at selecting and developing superior crop varieties with enhanced traits such as higher yield, disease resistance, pest tolerance, and resilience to environmental stresses like drought, heat, and salinity. Traditionally, plant breeding relies on labor-intensive field observations and manual phenotypic assessments, often spanning multiple growing seasons. The advent of the IoT has revolutionized this process by integrating advanced sensing technologies, wireless networks, and data analytics into breeding programs. IoT enables breeders to collect real-time, high-resolution data on plant performance and environmental conditions, accelerating the identification of elite genotypes and reducing the time to market for improved varieties. This section delves into three key IoT technologies—HTP Systems,

Smart Sensor Networks, and UAVs and Drones—and their transformative roles in plant breeding, supported by real-world examples and scientific evidence.

15.3.1.1 High-Throughput Phenotyping (HTP) Systems

HTP systems leverage IoT-powered imaging technologies to assess plant traits with remarkable precision and efficiency. These systems utilize multispectral and hyperspectral cameras, LiDAR (Light Detection and Ranging), and thermal sensors to provide real-time data on crucial phenotypic characteristics such as plant height, biomass accumulation, chlorophyll content, canopy temperature, and stress responses. According to Yang et al. (2020), HTP systems allow breeders to evaluate thousands of genotypes simultaneously, overcoming the limitations of traditional visual scoring methods, which are often subjective and time-consuming. The scientific foundation of these technologies lies in their ability to capture plant health indicators using advanced imaging techniques. Multispectral cameras detect light reflectance across specific wavelengths, revealing vital parameters like the Normalized Difference Vegetation Index (NDVI), which correlates with chlorophyll content and photosynthetic activity (Araus and Cairns 2014). Hyperspectral imaging extends this capability by analyzing a broader spectrum, detecting subtle stress signals such as water deficiency or nutrient imbalance. LiDAR technology provides precise three-dimensional structural data, measuring canopy architecture and biomass, while thermal sensors detect variations in leaf temperature, enabling the identification of heat stress (Furbank and Tester 2011). A real-world example of HTP implementation comes from a study conducted at the International Maize and Wheat Improvement Center (CIMMYT) in Mexico. IoT-enabled HTP platforms equipped with multispectral and thermal cameras were deployed to screen wheat lines for heat and drought tolerance, identifying superior genotypes with enhanced canopy cooling and water-use efficiency. This system significantly reduced breeding cycles by approximately 30% (Cossani and Sadras 2018). The impact of these systems lies in their ability to overcome the phenotyping bottleneck, allowing data-driven selection of traits that align with breeding goals, such as yield stability under changing climatic conditions.

15.3.1.2 Smart Sensor Networks

Smart Sensor Networks, often deployed as WSNs integrated with IoT devices, provide continuous, localized monitoring of environmental parameters critical to plant breeding. These networks enable breeders to gather data on soil moisture, temperature, humidity, light intensity, and nutrient levels, generating a comprehensive dataset for evaluating genotype-environment interactions (G × E). According to Cobb et al. (2013), these networks play a crucial role in understanding how genetic potential manifests under varying field conditions, a fundamental aspect of breeding resilient crops. The scientific principle behind these networks is based on spatially distributed sensors connected via IoT communication protocols such as ZigBee and LoRaWAN, transmitting data to cloud platforms for analysis. Soil moisture sensors use capacitance or time-domain reflectometry to quantify water availability, while temperature and humidity sensors track microclimatic fluctuations, helping breeders assess crop responses to environmental stressors like drought and heatwaves (Walter et al. 2017). A notable real-world application of smart sensor networks was observed at the Australian Plant Phenomics Facility (APPF), where

IoT-based WSNs were deployed to monitor barley breeding trials in South Australia. These sensors tracked soil moisture and temperature across multiple plots, enabling researchers to identify barley lines that maintained higher grain fill under water-limited conditions. Published by Findurova et al. (2023), the findings from this study contributed to the development of drought-tolerant barley varieties that are now widely adopted by local farmers. By providing granular, continuous data, smart sensor networks allow breeders to replicate field conditions with high precision, isolate genetic effects, and select varieties that are best suited to specific agroecological zones.

15.3.1.3 Unmanned Aerial Vehicles (UAVs) and Drones

UAVs, commonly known as drones, have emerged as essential tools for large-scale phenotyping and yield estimation in plant breeding. Equipped with IoT-enabled sensors, UAVs collect high-resolution images and data over extensive breeding plots, offering a scalable and efficient alternative to ground-based assessments. As noted by Singh and Ahmed (2021), drones significantly reduce the labor and time required for traditional field evaluations while enhancing the accuracy of trait measurements across diverse genotypes. The scientific foundation of drone-based phenotyping lies in their ability to carry multispectral, hyperspectral, or RGB cameras, capturing detailed data on plant health, vigor, and stress. Advanced UAV models integrate thermal imaging and LiDAR technologies, enabling three-dimensional mapping and heat stress detection. IoT connectivity ensures that the collected data is transmitted in real time to breeders' dashboards, where ML algorithms process and quantify key traits (Shakoor et al. 2017). In India, the Indian Council of Agricultural Research (ICAR) utilized IoT-enabled drones to phenotype rice breeding trials in Punjab. Drones equipped with multispectral cameras mapped NDVI and biomass across 500 plots, helping identify high-yielding rice lines resistant to blast disease. Findings published by Sankaran et al. (2015) demonstrated that UAV-based phenotyping reduced evaluation time by 50% compared to manual methods, accelerating the release of improved rice varieties such as PR126. Similarly, researchers at the University of Illinois used drones to monitor soybean trials, achieving 95% accuracy in detecting pod count and maturity (Miao et al. 2024). The impact of UAV technology in breeding programs is profound, as it enhances scalability, enables rapid screening of large populations, and provides precise yield predictions—critical factors in meeting global food security demands.

15.3.1.4 Integration of AI with IoT in High-Throughput Phenotyping

HTP refers to the rapid, large-scale evaluation of plant traits such as height, leaf area, biomass, and physiological responses, which are critical for understanding plant performance and guiding breeding programs. Traditional phenotyping methods, reliant on manual measurements and visual assessments, are often slow, labor-intensive, and limited in scale, posing a bottleneck in crop improvement efforts. The integration of the IoT with AI has transformed this process by enabling automated, real-time data collection and analysis. IoT devices—such as soil sensors, drones equipped with imaging systems, and automated scanners—continuously monitor plant characteristics across diverse environments, generating vast datasets that capture temporal and spatial variability. AI algorithms, particularly those based on ML and DL, process these data streams, extracting patterns, correlations, and predictive insights that human analysis might overlook (Yang et al. 2020). This synergy enhances the precision, scalability, and speed of phenotypic assessments, allowing breeders to identify superior genotypes more efficiently and accelerate the development of resilient, high-yielding crop varieties.

The combination of IoT and AI addresses key challenges in HTP by automating repetitive tasks and improving data resolution. For instance, IoT sensors can track environmental factors like soil moisture and temperature alongside plant traits, while AI models analyze these multidimensional datasets to reveal genotype-environment interactions (Cobb et al. 2013). This approach not only reduces the time required for trait evaluation but also increases the accuracy of selections in breeding programs, supporting the development of crops tailored to specific climatic or agricultural needs. As a result, the integration of AI with IoT is becoming a cornerstone of modern plant phenomics, driving advancements in precision agriculture and sustainable food production.

15.3.1.5 AI-Based IoT Platforms for Real-Time Plant Trait Data Collection

Several innovative platforms exemplify the power of combining AI and IoT for real-time plant trait assessment, each tailored to specific phenotyping needs. These systems leverage advanced hardware and computational tools to deliver actionable insights, reducing manual effort and enhancing breeding efficiency. Below are three notable examples:

1) **PlantEye F600 H-Portal Scanner**
   Developed by Phenospex in the Netherlands, the PlantEye F600 is a multispectral 3D scanner designed for high-precision plant phenotyping. Deployed at institutions like the University of Southern Queensland (UniSQ) in Australia, this fully automated system uses IoT connectivity to measure traits such as plant height, digital biomass, leaf area, and chlorophyll content with exceptional accuracy. The scanner integrates 3D vision with multispectral imaging, capturing data non-destructively across thousands of plants in growth chambers, greenhouses, or fields (Phenospex 2021). At UniSQ, its deployment has reportedly reduced manual phenotyping workload by approximately 85%, enabling continuous monitoring without disrupting plant growth (Courier Mail 2023). By pairing this IoT-driven data collection with AI algorithms, researchers can analyze growth dynamics and physiological responses in real time, facilitating rapid trait selection for breeding programs.
2) **CropQuant-Air**
   CropQuant-Air is an AI-powered IoT platform designed for field-based phenotypic analysis, utilizing low-cost drones to collect canopy imagery of crops such as wheat. Developed as part of efforts to democratize HTP, this system integrates DL models and image processing algorithms to perform tasks like plot-based spike detection, spectral analysis, and yield classification (Zhou et al. 2017). The platform's IoT component involves drones equipped with RGB and multispectral cameras, which transmit data wirelessly to a central server where AI processes it into actionable insights. A study published in *Frontiers in Plant Science* demonstrated its effectiveness in wheat breeding trials, where it accurately quantified yield-related traits across large field plots, reducing the need for labor-intensive manual counts (Zhou et al. 2017). CropQuant-Air's affordability and scalability make it a valuable tool for researchers and breeders aiming to assess performance traits under real-world conditions, enhancing the precision of selection for improved varieties.

15.3.1.6 Implications and Synergy

The combined use of HTP systems, smart sensor networks, and UAVs creates a synergistic IoT ecosystem for plant breeding. For instance, data from soil sensors can be correlated with drone imagery to validate phenotypic responses, while HTP platforms refine the selection process with detailed physiological insights. A study by Tattaris et al. (2016) demonstrated this integration in wheat breeding, where IoT tools collectively improved the accuracy of yield predictions by 25% under drought conditions.

The PlantEye F600 excels in controlled environments, providing detailed morphological and physiological data; CropQuant-Air scales phenotyping to field conditions with cost-effective aerial imaging; and CropVoice bridges plant biology with technology for early stress detection. Together, they demonstrate how IoT generates high-resolution, real-time data that AI converts into meaningful insights, enhancing the efficiency of breeding programs. Studies like those by Singh et al. (2016) further emphasize that AI-driven analysis of IoT data can improve trait prediction accuracy by up to 20%, accelerating the identification of genotypes suited to challenging environments. As these technologies evolve, their combined use promises to refine phenotyping workflows, supporting the global demand for sustainable crop improvement.

15.3.2 IoT in Molecular Breeding

Molecular breeding represents a sophisticated approach to crop improvement, leveraging genetic information through techniques such as marker-assisted selection (MAS) and genomic selection (GS) to develop varieties with superior traits like yield, disease resistance, and environmental adaptability. Unlike traditional breeding, which relies heavily on phenotypic observation, molecular breeding uses DNA markers and genomic data to guide the selection process, significantly accelerating the development of elite cultivars. The integration of the IoT into molecular breeding introduces a layer of precision, automation, and data connectivity that enhances the efficiency and accuracy of these methods. IoT technologies, including automated genotyping platforms, IoT-integrated bioreactors and growth chambers, and data-driven GS systems, streamline workflows, optimize experimental conditions, and enable real-time data analysis.

15.3.2.1 Automated Genotyping Platforms

Automated genotyping platforms, enhanced by IoT technologies, have revolutionized MAS by enabling high-throughput genotyping with minimal human intervention. MAS involves identifying and selecting plants carrying specific DNA markers linked to desirable traits, a process that requires rapid and accurate genetic analysis. IoT-based systems integrate DNA extraction robots, real-time polymerase chain reaction (PCR) machines, and next-generation sequencing (NGS) platforms, all connected via IoT networks for seamless data transfer and monitoring. Bassi et al. (2016) highlight that such automation reduces the time required for genotyping from weeks to days, making it feasible to screen thousands of plants in a single breeding cycle. These platforms use IoT to monitor equipment performance, calibrate conditions (e.g., temperature, reagent flow), and transmit genotyping results to cloud databases for immediate analysis. A notable real-world example comes from the International Crops Research Institute for the Semi-Arid Tropics (ICRISAT) in India, where IoT-enabled automated genotyping was employed to accelerate MAS in chickpea breeding. According to Varshney et al. (2019), researchers used an IoT-integrated NGS platform to identify drought-tolerance markers in chickpea lines, screening over 10,000 samples in a month and releasing the improved variety international code for variety verification (ICCV) 17,126 ahead of schedule. This efficiency stems from IoT's ability to automate repetitive tasks and ensure data integrity, allowing breeders to focus on interpreting results rather than managing workflows.

The scientific foundation of these platforms lies in their capacity to handle large-scale genomic data with precision. For instance, IoT connectivity ensures that sequencing errors are flagged in real time, and data are synchronized across multiple research sites, facilitating collaborative breeding efforts. A study by Heffner et al. (2010) underscores that high-throughput genotyping, when paired with IoT, increases the throughput of MAS by up to 50%, enabling breeders to target complex traits like polygenic disease resistance that require extensive marker analysis. This integration not only boosts efficiency but also enhances the scalability of molecular breeding programs, making them viable for resource-limited settings where manual genotyping would be impractical.

15.3.2.2 IoT-Integrated Bioreactors and Growth Chambers

IoT-integrated bioreactors and growth chambers provide controlled environments that are critical for molecular breeding experiments, particularly those involving gene expression analysis and transgenic plant development. These systems use IoT sensors to monitor and adjust environmental parameters such as temperature, humidity, light intensity, and CO_2 levels, ensuring optimal conditions for studying gene function and trait expression. Pauli et al. (2020) emphasize that precise environmental control is essential for validating the effects of specific genetic modifications, as even slight variations can skew results. In molecular breeding, where understanding genotype-environment interactions ($G \times E$) is key, IoT-enabled growth chambers allow researchers to replicate field conditions or simulate stress scenarios (e.g., drought, heat) with high fidelity. Data from these sensors is transmitted in real time to cloud platforms, where it can be correlated with genomic data to assess gene performance.

A compelling real-world application is seen in the work of the Donald Danforth Plant Science Center in the United States, where IoT-integrated growth chambers were used to study cassava gene expression under drought stress. Researchers deployed IoT sensors to maintain precise water-deficit conditions, revealing upregulated genes linked to drought tolerance, as reported by Wilson et al. (2021). This experiment, which supported the development of drought-resistant cassava lines, relied on IoT to log environmental data every minute, ensuring consistency across replicates. Similarly, IoT bioreactors have been used in rice molecular breeding at the International Rice Research Institute (IRRI). Here, IoT systems controlled nutrient delivery and temperature in bioreactors to optimize tissue culture conditions for transgenic rice lines carrying submergence-tolerance genes (e.g., SUB1), accelerating their evaluation (Ismail and Mackill 2014). These examples demonstrate how IoT ensures repeatability and precision, critical for translating molecular insights into field-ready varieties.

15.3.2.3 Data-Driven Genomic Selection

GS relies on statistical models to predict the breeding value of plants based on genome-wide marker data, offering a powerful alternative to MAS for complex traits controlled by multiple genes. IoT enhances GS by integrating phenotypic and environmental data collected from field sensors, drones, and growth chambers into cloud-based platforms, enabling real-time access and analysis. Crossa et al. (2017) argue that IoT-driven data integration improves the accuracy of GS prediction models by providing comprehensive datasets that capture G × E interactions. IoT-enabled cloud computing allows breeders to share genomic and phenotypic data across institutions, fostering collaborative research and reducing redundancies. ML algorithms, fed by IoT data streams, refine these models by identifying patterns in trait expression under diverse conditions, making GS more predictive and actionable.

A practical example of this approach is the wheat breeding program at CIMMYT in Mexico, where IoT-enabled data platforms were used to enhance GS for yield improvement. Field sensors and drones collected real-time data on soil moisture, temperature, and plant growth across multiple trial sites, which was integrated with genomic profiles of 5000 wheat lines. According to a study by Juliana et al. (2020), this IoT-supported GS approach increased prediction accuracy by 15% compared to traditional methods, leading to the release of high-yielding varieties like Borlaug 100. Another instance is the soybean breeding program at the University of Arkansas, where IoT cloud platforms processed data from weather stations and genotyping arrays to select lines with enhanced protein content, as documented by Wartha and Lorenz (2024). These cases illustrate how IoT facilitates data-driven decision-making, allowing breeders to prioritize genotypes with the highest potential for commercial success.

15.3.2.4 Integration of AI with IoT in Molecular Breeding

Molecular breeding aims to select plants with specific genetic traits that enhance crop performance and resilience. The integration of AI and the IoT significantly improves the efficiency and precision of this process by enhancing data collection and analysis.

1) **IoT-Based Data Collection:** IoT devices, such as sensors, are deployed throughout the agricultural fields to capture real-time data on environmental factors and plant phenotypic traits like height, leaf area, and biomass. These sensors allow for continuous monitoring of the plants and their growing conditions, reducing the need for manual intervention and ensuring a consistent flow of information (Magar et al. 2024). This data provides a comprehensive overview of the plants' environment and development.
2) **AI-Driven Data Analysis:** The large datasets gathered by IoT devices are processed using AI algorithms, which can detect patterns and relationships between genetic markers and phenotypic traits (Liakos et al. 2018). AI's ability to analyze these complex datasets enhances the accuracy and precision of identifying plants with the most desirable traits. As a result, the breeding process becomes more efficient, leading to improved crop varieties with better performance in terms of yield, disease resistance, and stress tolerance (Mekonnen et al. 2019).

15.3.2.5 Synergistic Impact and Future Directions

The synergy between automated genotyping, IoT-controlled environments, and data-driven GS creates a robust framework for molecular breeding. For instance, real-time phenotypic data from growth chambers can validate marker-trait associations identified through automated genotyping, while cloud-based GS models refine selections based on field sensor inputs. This interconnected system, as noted by Meuwissen et al. (2016), reduces the breeding cycle time by up to 40%, a critical advantage in addressing urgent agricultural challenges like climate change. Looking ahead, the integration of IoT with AI and CRISPR-based gene editing promises even greater precision, enabling breeders to target specific genomic regions with real-time environmental feedback. Thus, Integrating AI with the IoT is transforming molecular breeding by enabling precise, real-time monitoring of plant traits, thereby accelerating the development of superior crop varieties.

15.3.3 IoT in Gene Editing

Gene editing technologies, particularly CRISPR-Cas9, have transformed plant breeding by enabling precise modifications to DNA sequences, allowing researchers to introduce or enhance traits such as disease resistance, drought tolerance, and nutritional quality with unprecedented accuracy (Upadhyay 2021, 2024). While CRISPR-Cas9 offers immense potential, its practical application in plant systems involves complex workflows, including DNA delivery, cell transformation, and regeneration, which can be labor-intensive and prone to variability. The integration of the IoT into gene editing addresses these challenges by introducing automation, real-time monitoring, and data management capabilities that enhance efficiency, reproducibility, and scalability. IoT-enabled systems, such as automated CRISPR workflows, real-time cell culture monitoring, and cloud-based gene data management platforms, streamline the gene editing process, ensuring optimal conditions and facilitating rapid analysis.

15.3.3.1 Automated CRISPR Workflows

Automated CRISPR workflows, supported by IoT-connected robotic systems, have significantly improved the efficiency and precision of genome editing protocols in plant breeding. CRISPR-Cas9 involves multiple steps—designing guide RNAs (gRNAs), delivering editing components into plant cells, and screening for successful edits—all of which benefit from automation to reduce human error and increase throughput (Alok et al., 2021). IoT-enabled robotic liquid handling systems, such as those used for pipetting reagents, preparing CRISPR constructs, and performing transformations, are equipped with sensors and connectivity that monitor performance metrics (e.g., volume accuracy, temperature stability) in real time. Alonge et al. (2022) highlight that such automation can process hundreds of samples simultaneously, a critical advantage when editing multiple plant lines for traits like yield or stress tolerance. These systems transmit data to centralized platforms, allowing researchers to adjust protocols dynamically and ensure consistency across experiments.

A real-world example of this technology in action is the gene editing program at the John Innes Centre in the United Kingdom, where IoT-connected robotic platforms were used to automate CRISPR-Cas9 workflows for wheat improvement. Researchers targeted genes associated with grain size and disease resistance, using an IoT-integrated liquid handler to deliver CRISPR components into wheat protoplasts. According to a study by Borrill (2020), this automation reduced the time for transformation and screening by 40%, enabling the rapid identification of edited lines with enhanced traits, such as resistance to powdery mildew. The IoT component ensured that each step—reagent mixing, incubation, and

DNA extraction—was logged and optimized, minimizing variability. This precision stems from the integration of IoT to coordinate robotic operations with environmental controls, as demonstrated by Elijah et al. (2018), who highlighted that IoT-based automated systems can enhance agricultural efficiency by optimizing resource use and reducing manual intervention by significant margins. Such advancements make gene editing more accessible and scalable, particularly for complex polyploid crops like wheat.

15.3.3.2 Real-Time Cell Culture Monitoring

Real-time cell culture monitoring, facilitated by IoT sensors, is critical for maintaining optimal conditions during the transformation and regeneration phases of gene editing in plants. After CRISPR components are delivered into plant cells (e.g., via Agrobacterium or biolistics), successful editing depends on the survival and proliferation of transformed cells in culture, a process sensitive to environmental factors like pH, oxygen levels, temperature, and nutrient availability. IoT sensors embedded in bioreactors and culture systems continuously track these parameters, providing data that can be analyzed and adjusted in real time to prevent cell stress or death. Nasti and Voytas (2021) emphasize that such monitoring is essential for ensuring high transformation efficiency, as even minor deviations can disrupt gene expression or editing outcomes. IoT connectivity allows these sensors to alert researchers to anomalies and automate corrective actions, such as adjusting pH with buffers or increasing aeration, thereby enhancing reproducibility.

A practical application of this technology is seen in the gene editing efforts at the University of California, Davis, where IoT-enabled bioreactors were used to optimize rice transformation for herbicide resistance. Researchers employed CRISPR-Cas9 to edit the ALS gene, and IoT sensors monitored pH and oxygen levels in callus cultures, ensuring stable conditions during regeneration. According to a report by Zhang et al. (2021), this real-time oversight increased transformation efficiency by 25%, resulting in the successful development of rice lines resistant to imidazolinone herbicides. Similarly, at the Max Planck Institute for Plant Breeding Research in Germany, IoT sensors tracked temperature and CO_2 levels in Arabidopsis cultures edited for drought tolerance (Reddy and Pullaiah 2024). The system maintained precise conditions (e.g., 22°C, 5% CO_2), leading to a higher yield of viable edited plants. These examples underscore how IoT ensures the delicate balance required for plant cell survival post-editing, a process that traditional manual monitoring struggles to achieve consistently.

15.3.3.3 Cloud-Based Gene Data Management

Cloud-based gene data management, powered by IoT platforms, revolutionizes the analysis and storage of genomic data generated during gene editing experiments. CRISPR workflows produce vast datasets, including gene sequences, gRNA designs, off-target predictions, and transformation efficiency metrics, which require robust computational tools for processing. IoT platforms link laboratory instruments (e.g., sequencers, imaging systems) to cloud storage, enabling real-time data upload, sharing, and analysis across research teams. Jeong et al. (2017) note that this connectivity accelerates the design-to-validation pipeline by allowing researchers to access and refine CRISPR strategies on the fly, improving the selection of edited lines with desired traits. ML algorithms integrated into these platforms can predict editing outcomes based on historical data, further enhancing efficiency.

A striking real-world example is the collaboration between the Salk Institute and the Innovative Genomics Institute (IGI) in the United States, where an IoT-enabled cloud platform was used to manage data from CRISPR-edited sorghum lines aimed at improving drought tolerance. IoT-connected sequencers uploaded genomic data to the cloud, where researchers analyzed editing efficiency and trait expression in real time. According to a study by Zhu et al. (2008), this system reduced the analysis timeline from months to weeks, identifying sorghum variants with enhanced water-use efficiency that are now in field trials. Another instance is the tomato breeding program at Wageningen University in the Netherlands, where IoT cloud platforms processed CRISPR data to enhance fruit quality traits like lycopene content. Cloarec-Ung et al. (2024) reported that real-time access to sequencing results enabled rapid iteration of gRNA designs, boosting editing success by 20%. These cases demonstrate how IoT-driven cloud management transforms gene editing into a data-centric process, fostering collaboration and accelerating trait development.

15.3.3.4 Integration of Artificial Intelligence in Gene Editing for Crop Enhancement

Gene editing technologies, such as CRISPR-Cas9, have significantly transformed the potential for modifying plant genomes, enabling the creation of crops with improved characteristics, including higher yields, better disease resistance, and greater adaptability to climate change. Despite their promise, challenges such as off-target effects and the intricate nature of gene interactions persist. AI plays a pivotal role in addressing these challenges by:

1) **Predicting Gene Functions and Interactions:** AI systems analyze vast genomic datasets to predict the functions of specific genes and their interactions within the plant genome. This predictive capability aids in the selection of target genes for editing, thereby increasing the probability of obtaining the desired traits. By leveraging AI, researchers can gain deeper insights into gene behavior, enhancing gene editing outcomes (Smith et al. 2020).
2) **Optimizing Guide RNA Design:** In CRISPR-based gene editing, AI enhances the design of guide RNAs by making them more efficient and specific. This optimization reduces off-target effects and improves the precision of gene edits, ensuring that the changes are confined to the desired genetic loci. AI-driven algorithms help researchers fine-tune guide RNA sequences to maximize editing success rates (Kim et al. 2021).
3) **Analyzing Phenotypic Outcomes:** AI models are used to link genetic modifications with phenotypic changes in plants, providing a prediction of how specific genetic edits will manifest in terms of traits such as growth rate, yield, or resistance to environmental stress. This predictive power allows researchers to focus on the most promising genetic modifications, streamlining the crop improvement process and reducing the time needed for successful breeding (Xu et al. 2022).

15.3.3.5 Synergistic Benefits and Future Potential

The integration of automated workflows, real-time monitoring, and cloud-based data management creates a cohesive IoT ecosystem for gene editing. For example, data from IoT sensors in cell cultures can inform robotic adjustments in CRISPR delivery, while cloud platforms correlate editing outcomes with environmental conditions, refining future experiments. This synergy, as highlighted by Doudna and Charpentier (2014), amplifies the precision and speed of CRISPR applications in plants. Looking forward, IoT's role in gene editing is poised to expand with advancements like AI-driven design tools and miniaturized sensors, potentially enabling field-based editing systems that bypass traditional lab constraints. Integrating AI with gene editing technologies offers transformative potential for crop improvement by enhancing precision, efficiency, and predictability in developing superior plant varieties.

15.3.4 IoT for Controlled Environments in Breeding Trials

15.3.4.1 IoT for Greenhouse and Growth Chamber Monitoring

In breeding trials, where plants are subjected to specific environmental conditions to observe growth, response to stress, and other traits, accurate control and real-time monitoring of environmental factors are crucial. IoT systems help regulate key environmental variables such as temperature, humidity, CO_2 levels, light intensity, and nutrient supply, ensuring optimal conditions for plant growth and facilitating better genetic studies. These advancements allow for increased reproducibility and scalability of experiments, which are fundamental to breeding programs focused on crop improvement.

One prominent example of IoT implementation in controlled environments is Argus Controls, a climate control system that integrates IoT technology to monitor and regulate environmental factors in plant breeding greenhouses. Argus Controls enables real-time monitoring of parameters such as temperature, relative humidity, light intensity, and CO_2 concentration, allowing breeders to adjust these variables remotely through a centralized system. The system can automatically adjust climate settings based on predefined conditions, providing a stable environment that minimizes human error. Additionally, the system offers data logging and analysis, which helps identify patterns in plant growth and development under different environmental conditions, thereby supporting better breeding decisions. By combining real-time data acquisition with automated adjustments, Argus Controls enhances the precision and efficiency of breeding trials (Argus Controls 2025).

Another significant advancement in the IoT-enabled monitoring of breeding trials is the Grodan e-Gro platform. This smart platform is designed to optimize nutrient management in plant breeding trials. Grodan e-Gro utilizes IoT sensors to monitor the water, nutrient, and pH levels in the growth medium, providing real-time insights into plant health and nutrient uptake. The system integrates with a range of environmental control systems, enabling breeders to adjust nutrient concentrations based on the specific needs of the plants under study. Grodan e-Gro also facilitates remote monitoring and intervention, which is especially useful for large-scale breeding operations or when experiments involve a wide variety of plant species with differing nutrient requirements. By optimizing nutrient delivery and preventing over- or under-fertilization, Grodan e-Gro helps create the ideal conditions for plant growth, contributing to the overall success of breeding trials (Grodan 2025).

In addition to environmental and nutrient control, IoT technology also plays a pivotal role in monitoring plant physiology, particularly in relation to photosynthesis and respiration. Li-Cor's Portable Photosynthesis System is an example of an IoT-connected device that measures gas exchange in plants, specifically the rate of photosynthesis and respiration. This system allows researchers to assess how plants are responding to different environmental conditions, which is critical in breeding trials that focus on traits related to stress tolerance, growth rate, and photosynthetic efficiency. The device is capable of measuring CO_2 uptake and release, providing data on plant respiration rates and overall metabolic activity. This information is invaluable for breeding programs aiming to select plants with superior photosynthetic efficiency or resistance to environmental stressors such as drought or high temperatures. By integrating the Li-Cor system into IoT networks, breeders can collect continuous data, track plant performance over time, and make data-driven decisions to optimize breeding strategies (Li-Cor Biosciences 2025).

The impact of IoT in breeding trials extends beyond the control of environmental factors and physiological monitoring. IoT systems also facilitate the collection and analysis of large datasets, which can be used to identify correlations between environmental conditions and plant performance. This data-driven approach enables more precise breeding, as researchers can identify which variables most significantly impact plant traits. For example, in the case of drought-tolerant crop development, IoT systems can help monitor how different levels of soil moisture and atmospheric conditions influence the water usage efficiency of plants. With real-time data on how plants are responding to such stressors, breeders can select the most resilient plants for future propagation.

Eventually, IoT-enabled systems have become integral to modern plant breeding trials. With tools like Argus Controls, Grodan e-Gro, and the Li-Cor Portable Photosynthesis System, researchers can monitor and regulate environmental factors with unprecedented precision, enabling more effective and efficient breeding. By leveraging real-time data collection, automated control systems, and advanced analytics, IoT technologies are enhancing the scalability, accuracy, and reproducibility of plant breeding programs. These innovations are not only transforming how breeding trials are conducted but also paving the way for the development of crops that are better suited to the challenges of a changing climate and growing global population.

15.3.4.2 Integration of AI in Automated Data Collection and Decision Support for Greenhouse and Growth Chamber Monitoring

Greenhouses and growth chambers play a crucial role in controlled agricultural environments, enabling precise manipulation of key factors such as temperature, humidity, light, and CO_2 levels, which are vital for breeding trials. The integration of AI into these environments has greatly enhanced their effectiveness through several innovative mechanisms:

1) **Real-Time Environmental Monitoring and Control:** IoT sensors continuously monitor various environmental parameters within greenhouses and growth chambers. AI algorithms process this data to identify patterns and detect anomalies, facilitating automated adjustments that ensure optimal conditions for plant growth. This dynamic control system maintains environmental factors within the desired range, helping to mitigate stress and prevent disease outbreaks (Harun et al. 2019; Ardiansah et al. 2020).
2) **Predictive Analytics for Crop Management:** By processing both historical and real-time data, AI models predict future environmental conditions and plant responses. This capability allows for proactive management strategies, such as adjusting irrigation schedules in anticipation of temperature shifts, which maximizes resource use and improves crop resilience (Martínez-Chavez et al. 2024).

3) **Automated Phenotyping:** AI-powered image analysis tools enable the automated measurement of plant traits, such as height, leaf area, and biomass. This HTP accelerates the evaluation of genetic variations, streamlining the selection process in breeding trials and minimizing the need for manual labor (Wang et al. 2020).
4) **Early Detection of Pests and Diseases:** AI-driven monitoring systems process data from various sensors to detect early signs of pest infestations or disease outbreaks. Early identification allows for timely interventions, reducing crop damage and minimizing the reliance on chemical treatments (Vidya Madhuri et al. 2025).

15.3.4.3 AI-Based IoT Platforms for Greenhouse and Growth Chamber Monitoring

Several AI-powered IoT platforms have been developed to enhance the monitoring and management of controlled agricultural environments:

1) **AI-Powered Greenhouse Environmental Control System (AI-GECS):** This platform integrates weather forecasts, microclimate predictions, and crop physiological data to automate greenhouse operations. Using a Multi-Model Super Ensemble (MMSE) forecasting framework, AI-GECS ensures precise real-time control of environmental conditions, optimizing plant growth conditions (Morris et al. 2023).
2) **SAgric-IoT:** An IoT-based platform that combines environmental sensor data with convolutional neural networks (CNNs) for precision agriculture. SAgric-IoT monitors key environmental and physical variables, providing early disease detection while automating irrigation and fertilization processes, thereby increasing management efficiency in greenhouses (Contreras-Castillo et al. 2023).
3) **iGrow:** A smart agriculture solution designed for autonomous greenhouse management. By framing the greenhouse control problem as a Markov Decision Process (MDP), iGrow employs a neural network-based simulator to optimize control strategies. Real-world deployments have demonstrated substantial improvements in crop yield and profitability (Cao et al. 2022).
4) **TEKTELIC Greenhouse Monitoring Solution:** A cloud-based IoT platform that facilitates remote monitoring and management of greenhouse environments. By connecting sensors, controllers, and actuators, TEKTELIC enables real-time adjustments and microclimate control, improving overall greenhouse management (Johnson et al. 2020).

15.3.5 IoT in Automated Data Collection and AI-Driven Decision Support in Plant Breeding

The convergence of IoT technology with AI has significantly advanced the precision and efficiency of plant breeding programs. IoT devices are increasingly being used to collect real-time data on various environmental and physiological factors during breeding trials, while AI algorithms analyze this data to support decision-making processes. This synergy between IoT and AI not only improves the accuracy of data collection but also enables breeders to make more informed, data-driven decisions, leading to the identification and development of superior crop varieties. The use of IoT in automated data collection and AI-driven decision support represents a paradigm shift in modern plant breeding, enhancing both the scale and scope of breeding efforts.

15.3.5.1 Teralytic Soil Sensor: IoT-Based Real-Time Soil Analysis

One of the most important aspects of plant breeding is understanding the soil conditions in which plants are cultivated, as soil properties can greatly influence plant growth, nutrient uptake, and overall health. The Teralytic Soil Sensor is a prime example of how IoT can be utilized to provide real-time, continuous soil analysis during breeding trials. This sensor measures several critical soil parameters such as moisture, pH, temperature, and nutrient content, transmitting this data in real-time to a cloud-based platform. By integrating IoT sensors like Teralytic into breeding programs, researchers can monitor how soil conditions vary across different experimental plots, allowing them to adjust irrigation, fertilization, and soil amendments based on the real-time feedback.

For instance, in a breeding trial aimed at developing drought-tolerant crop varieties, the Teralytic sensor can provide continuous soil moisture data, allowing breeders to track soil water levels and determine the precise timing of irrigation. By combining this data with environmental variables such as temperature and humidity, breeders can simulate stress conditions that help identify plants with superior drought resistance. Furthermore, the continuous data stream ensures that breeders can quickly detect soil imbalances or nutrient deficiencies, enabling timely interventions. Teralytic's real-time monitoring capabilities help fine-tune soil conditions, providing the ideal environment for plant growth and accelerating the breeding process (Teralytic 2025).

15.3.5.2 DigiFarm AI: AI-Integrated IoT System for Superior Crop Variety Selection

The development of superior crop varieties is a primary goal of plant breeding programs, and the integration of AI with IoT technology is playing a transformative role in achieving this goal. The DigiFarm AI system is an AI-integrated IoT platform that assists breeders in selecting the best-performing crop varieties based on extensive data analysis. The DigiFarm system collects data from multiple IoT devices deployed across the breeding trial field, including soil sensors, environmental monitors, and plant health detectors. It then applies ML algorithms to this vast dataset, identifying patterns and correlations between environmental factors, plant traits, and yield performance.

For example, in a breeding program focused on improving disease resistance in wheat, DigiFarm AI can analyze data from soil sensors that monitor moisture levels and temperature, combined with data from plant sensors that track growth patterns and detect disease symptoms. Using AI algorithms, DigiFarm can predict how different environmental conditions influence disease susceptibility and identify varieties that show superior resistance traits. This predictive capability allows breeders to make better-informed decisions, accelerating the breeding process by narrowing down the most promising candidates for further development.

Furthermore, DigiFarm AI's integration with IoT devices provides a comprehensive and real-time view of field conditions, ensuring that breeders have access to up-to-date information at every stage of the breeding cycle. This system reduces the guesswork in selecting crop varieties and offers insights that are otherwise difficult to obtain through traditional methods (DigiFarm AI 2025).

15.3.5.3 FieldBook App: IoT-Assisted Mobile App for Plant Breeding Data Management

Efficient data management is essential for successful plant breeding, as large volumes of data are generated during field trials, including plant growth measurements, environmental conditions, and genetic information. The FieldBook App is an IoT-assisted mobile application designed to streamline data collection, organization, and analysis in plant breeding programs. The app integrates with IoT sensors in the field, enabling researchers to collect real-time data on a variety of factors, including soil moisture, temperature, and plant phenotypes, directly into the app. This data is then uploaded to the cloud, where it can be accessed and analyzed by breeders.

The FieldBook App provides breeders with a platform to track all aspects of their breeding trials in one place, offering functionalities such as field mapping, trial design, and data synchronization with IoT devices. This mobile solution improves the accuracy of data collection by reducing the risk of human error and ensures that data is stored in a centralized, easily accessible location. By leveraging the app's real-time data synchronization capabilities, breeders can monitor trial progress and quickly respond to any issues that arise. For instance, if an IoT sensor detects a sudden drop in soil moisture or an increase in temperature, the FieldBook App can alert breeders, enabling them to take immediate action to optimize trial conditions.

Additionally, the app's user-friendly interface and integration with other data analysis tools allow breeders to analyze data on the go, enhancing decision-making in the field. Whether it is tracking genetic traits, monitoring phenotypic changes, or assessing environmental impacts on plant growth, the FieldBook App simplifies the data management process and supports better decision-making in plant breeding (FieldBook 2025).

15.3.5.4 Impact of IoT and AI on Plant Breeding Programs

The integration of IoT in automated data collection, combined with AI-driven decision support, has the potential to significantly transform plant breeding programs. The continuous, real-time data provided by IoT devices allows for more precise monitoring of environmental factors, soil conditions, and plant health. When this data is coupled with AI algorithms, breeders can gain deeper insights into how these factors interact and influence plant traits, leading to more efficient selection and development of superior crop varieties.

By utilizing systems like Teralytic, DigiFarm AI, and the FieldBook App, plant breeders are able to scale up their operations, enhance precision, and make more informed decisions. This integration of IoT and AI in plant breeding not only improves productivity and efficiency but also accelerates the breeding process, contributing to the development of crops that are more resilient, higher yielding, and better suited to the challenges of a changing climate and growing global population.

15.3.5.5 Integration of Artificial Intelligence with Automated Data Collection for Crop Enhancement

Automated data collection in agriculture involves using IoT sensors and remote sensing technologies to continuously monitor critical environmental and crop parameters, including soil moisture, temperature, humidity, and plant health indicators. These sensors provide real-time data, eliminating the need for manual measurements and offering a comprehensive overview of field conditions. AI algorithms process this vast amount of data to detect patterns, predict potential outcomes, and generate actionable insights. This enables farmers to make informed decisions that optimize crop yield and improve resource efficiency (Li et al. 2024).

15.3.5.6 AI-Driven Decision Support Systems for Agriculture

AI-driven decision support systems (DSSs) utilize advanced ML models and data analytics to assist farmers in making better crop management decisions. These systems provide valuable recommendations on areas such as irrigation, fertilization, pest and disease management, and harvest timing. By analyzing both historical and real-time data, AI-based DSS can predict potential challenges, such as pest infestations or disease outbreaks, enabling farmers to take proactive measures to minimize crop loss and improve overall farm management (Joshi et al. 2022).

15.3.5.7 AI-Based IoT Platforms for Real-Time Plant Trait Data Collection

Several AI-powered IoT platforms have been developed to facilitate the real-time collection of plant trait data, which accelerates the selection process in breeding programs and enhances crop management:

1) **Automated IoT Platform by the Alliance of Bioversity International and CIAT:** This platform integrates sensors, actuators, and smart devices for real-time monitoring and control of greenhouse parameters such as irrigation and temperature. It also includes alarms to detect potential crop diseases or stress, reducing human error and improving crop management efficiency (CIAT 2021).
2) **IoT-Based Plant Health Monitoring System:** This AI-powered IoT system autonomously monitors plant conditions and automatically takes corrective actions when necessary. The system remotely sends health reports to farmers, enabling them to track and manage crop conditions proactively (Pavel et al. 2019; Banerjee et al. 2023).
3) **IoT-Enabled Greenhouse Automation Systems:** These systems use IoT sensors to gather accurate, real-time data on greenhouse conditions such as lighting, temperature, soil moisture, and humidity. The system can automatically adjust environmental parameters to optimize conditions for plant growth, significantly reducing the need for manual intervention (Eastern Peak 2023).

15.4 Benefits and Challenges of Integrating AI and IoT in Crop Improvement

15.4.1 Benefits

The integration of AI and the IoT is transforming crop improvement by enhancing precision, efficiency, and sustainability in agricultural practices. This synergy, often referred to as the Artificial Intelligence of Things (AIoT), combines real-time data collection capabilities of IoT with AI-driven advanced analytics to optimize decision-making in farming operations.

15.4.1.1 Enhanced Precision Agriculture

IoT devices, such as soil moisture sensors, weather stations, and drones, continuously monitor environmental and crop-related parameters. AI algorithms process these large datasets to generate actionable insights, allowing farmers to implement precision agriculture strategies. For example, AI can analyze soil properties, weather patterns, and historical yield data to determine optimal planting schedules and locations, thereby improving productivity and resource efficiency (Zhang et al. 2021).

15.4.1.2 Improved Crop Monitoring and Disease Detection

Continuous monitoring enabled by IoT sensors aids in the early detection of plant stress, diseases, and pest infestations. AI-powered image recognition and data analytics can identify subtle changes in crop health, often imperceptible to the human eye. Early interventions based on these insights can mitigate disease outbreaks and reduce crop losses, thus contributing to sustainable agriculture (Mishra and Mishra 2024; Titirmare et al. 2024).

15.4.1.3 Optimized Resource Management

AIoT systems facilitate precise resource allocation by optimizing the use of water, fertilizers, and pesticides. AI models, leveraging IoT data on soil nutrient levels and crop requirements, enable variable-rate applications—ensuring that inputs are delivered only where needed. This targeted approach not only conserves resources but also reduces production costs and minimizes environmental impact (Raza et al. 2023; Qayoom et al. 2024).

15.4.1.4 Predictive Analytics for Yield Forecasting

AI-driven predictive analytics models analyze historical and real-time environmental data to forecast crop yields with high accuracy. These predictions help farmers in harvest planning, storage management, and market strategy formulation. Accurate yield estimates also enhance supply chain efficiency and market stability by reducing uncertainties in production (Titirmare et al. 2024).

15.4.1.5 Automation and Labor Efficiency

The integration of AI with IoT has enabled automation in agriculture, reducing dependence on manual labor. Autonomous machinery, guided by AI, can efficiently perform tasks such as planting, weeding, and harvesting. This mechanization not only addresses labor shortages but also enhances operational efficiency and precision in farming practices (Gupta and Gupta 2024).

15.4.1.6 Data-Driven Decision Support Systems

AIoT platforms aggregate data from multiple sources, including IoT sensors, satellite imagery, and historical farm records, to provide comprehensive decision-support systems. These platforms generate tailored recommendations for crop management based on local environmental conditions and specific crop needs, improving productivity and resilience against climate variability (Joshi et al. 2022).

15.4.2 Challenges

Despite their numerous advantages, the implementation of AI and IoT in crop improvement presents several challenges that must be addressed for widespread adoption (Table 15.1).

Table 15.1 Challenges in implementation of IoT in crop improvement.

S. No.	Challenges	Description	Mitigation strategy	References
1.	Data security and privacy	IoT devices collect large amounts of sensitive agricultural data, including crop health and farm locations, making them vulnerable to cyber threats.	Implement advanced encryption standards and secure network protocols.	Kim et al. (2021); Carvalho et al. (2025)
2.	Lack of technical knowledge	Farmers in rural areas often struggle with using IoT and AI technologies due to limited digital literacy and lack of support.	Provide training programs, troubleshooting services, and peer-support initiatives.	Sinha and Dhanalakshmi (2022)
3.	High costs and investment	The cost of implementing IoT, including setup, maintenance, and data subscriptions, is often high for small and medium-sized farms.	Optimize data use, automate routine processes, and configure cost-effective IoT systems.	Elijah et al. (2018); Farooq et al. (2020)
4.	Infrastructure limitations	IoT systems require stable internet connectivity and reliable power sources, which are often lacking in remote agricultural areas.	Expand rural broadband access, invest in low-power IoT solutions, and develop offline data storage options.	Kritikos (2017)
5.	Reliability issues	IoT devices are prone to environmental factors such as dust, extreme temperatures, and moisture, affecting their durability and accuracy.	Use ruggedized, waterproof enclosures and implement adaptive AI models that compensate for data fluctuations.	Asikainen et al. (2013)
6.	Scalability challenges	Managing a vast network of IoT devices across large farmlands is complex, requiring seamless data integration and processing.	Utilize cloud-based solutions, edge computing, and scalable network architectures.	Al-Fuqaha et al. (2015); Misra et al. (2020)
7.	Interoperability problems	AI and IoT platforms from different manufacturers may not be compatible, making data exchange and automation difficult.	Develop industry-wide IoT integration platforms and promote open-source software for standardization.	Elijah et al. (2018); Patel et al. (2016)
8.	Regulatory and legal barriers	Compliance with data protection laws and evolving regulations related to AI-driven agriculture poses challenges to widespread adoption.	Monitor regulatory changes and align IoT solutions with national and international agricultural policies.	Elijah et al. (2018)

15.4.2.1 Data Security and Privacy

The extensive deployment of IoT devices results in the continuous collection and transmission of vast agricultural datasets, including soil conditions, crop health, and environmental factors. Ensuring data security and privacy is critical, as cyber vulnerabilities could lead to unauthorized access or misuse of sensitive agricultural information. Implementing robust encryption, access control, and cybersecurity protocols is essential to safeguard data integrity (Kim et al. 2021; Garg et al. 2022; Carvalho et al. 2025).

15.4.2.2 Technical Expertise and Farmer Training

Effective utilization of AI and IoT in agriculture requires specialized technical knowledge, which many farmers—especially in developing regions—may lack. Limited digital literacy can hinder the adoption of these advanced technologies. Training programs and extension services are necessary to equip farmers with the necessary skills to operate and interpret AIoT systems, ensuring widespread and effective usage (Sinha and Dhanalakshmi 2022).

15.4.2.3 Infrastructure Limitations

Reliable internet connectivity is crucial for real-time data transmission from IoT devices. However, many rural and agricultural regions lack adequate internet infrastructure, creating a significant barrier to the implementation of AIoT solutions. Investments in rural broadband expansion and satellite-based internet services can help overcome these connectivity challenges (Kritikos 2017).

15.4.2.4 Economic Constraints

The high initial investment required for AI and IoT technologies poses financial challenges, particularly for small and medium-sized farms. The affordability of hardware, maintenance costs, and uncertainty regarding return on investment can discourage adoption. Financial incentives, subsidies, and cost-sharing programs can facilitate access to these technologies for resource-limited farmers (Elijah et al. 2018; Farooq et al. 2020).

15.4.2.5 Interoperability and Standardization Issues

The integration of diverse IoT devices and AI platforms often faces interoperability challenges due to the lack of standardized communication protocols. Devices from different manufacturers may not seamlessly exchange data, limiting their effectiveness. Establishing industry-wide standards and promoting open-source platforms can enhance compatibility and interoperability among AIoT systems (Patel et al. 2016; Elijah et al. 2018; Nair et al. 2023).

15.4.2.6 Energy Consumption and Sustainability

IoT devices deployed in agricultural settings require a continuous and reliable power source. Many rely on battery-powered operation, which can lead to increased maintenance efforts and operational costs. Developing energy-efficient IoT sensors, optimizing data transmission frequencies, and utilizing renewable energy sources such as solar panels can help mitigate these challenges (Gupta and Singh 2022).

15.4.2.7 Data Management and Quality Assurance

The vast amounts of heterogeneous and unstructured data generated by IoT devices can pose challenges in storage, processing, and quality assurance. Ensuring data accuracy, consistency, and integrity is essential for the reliable functioning of AI models. Implementing robust data preprocessing techniques and adopting standardized protocols for data validation can improve the reliability of AI-driven agricultural insights (Joshi et al. 2022).

15.4.2.8 Social and Cultural Barriers

Resistance to technological change remains a significant barrier to AI and IoT adoption in agriculture. Farmers accustomed to traditional methods may perceive these innovations as complex or unreliable. Demonstration projects, field trials, and participatory approaches that involve farmers in the development and implementation of AIoT technologies can build trust and encourage adoption (Das et al. 2023).

Integrating AI with IoT in crop improvement presents several challenges, stemming from the inherent complexity of AI algorithms, ethical considerations, and the significant computational demands. As AI models require vast amounts of data and processing power, ensuring the system's efficiency and scalability can be difficult. Moreover, the complexity of AI algorithms can make their deployment and maintenance challenging, often requiring specialized expertise. Ethical concerns, such as data privacy and the potential for biased decision-making, also need to be addressed, particularly when sensitive agricultural data is being collected and analyzed.

Additionally, there are several other challenges that must be carefully considered. These include the integration of heterogeneous data sources from IoT devices, ensuring reliable connectivity in remote agricultural areas, and overcoming the high costs associated with implementing these advanced technologies on a large scale. Furthermore, the ability to interpret and act on the vast amounts of data generated by IoT sensors can be overwhelming, necessitating effective data management and analysis frameworks. Overcoming these obstacles is crucial for successfully leveraging AI and IoT in crop improvement. Table 15.2 lists some of the additional challenges that should be considered.

15.5 Future Prospects of IoT and AI Integration in Crop Improvement

The combination of the IoT and AI in agriculture is set to transform crop improvement, driving greater precision, efficiency, and sustainability in farming practices. With advancements in 5G connectivity, edge computing, and AI-driven predictive modeling, these technologies will offer unprecedented opportunities for plant breeding and farm management. As agricultural systems become increasingly data-driven, the future direction of IoT and AI in crop improvement will focus on enhancing real-time data processing, improving predictive capabilities, and ensuring efficient and secure data management. However, to fully realize their potential, several challenges must be addressed, including the need for reliable connectivity, scalable computational frameworks, and user-friendly implementation strategies.

One of the most significant advancements in this field will be the deployment of 5G networks, which will enhance the capabilities of IoT devices by providing faster data transmission rates and lower latency. This will enable real-time monitoring of crop conditions,

Table 15.2 Additional challenges in AI-integrated IoT implementation in crop improvement.

S. No.	Challenges	Description	Mitigation strategy	References
1.	Bias in AI models	AI models trained on limited datasets may introduce bias, leading to inaccurate predictions for diverse agricultural environments.	Develop diverse, high-quality training datasets and implement bias mitigation techniques in AI models.	Zhao et al. (2022); Mehrabi et al. (2021)
2.	Data overload and processing delays	Large-scale IoT networks generate massive amounts of data, requiring extensive computational resources for real-time processing.	Implement efficient data filtering techniques, edge computing, and AI-driven data summarization methods.	Gubbi et al. (2013); Zhang et al. (2020)
3.	Ethical concerns and farmer trust	The use of AI in decision-making may lead to ethical concerns, such as over-reliance on technology and lack of transparency in AI recommendations.	Develop explainable AI (XAI) models that provide clear reasoning for recommendations and involve farmers in the decision-making process.	Binns et al. (2018); Dignum (2018); Dara et al. (2022)
4.	Energy consumption and sustainability	AI-driven IoT systems require high computational power, increasing energy demands, especially in large-scale applications.	Optimize AI models for low-power computing, use solar-powered IoT devices, and implement energy-efficient hardware.	Stojmenovic and Wen (2014); Saleem et al. (2019)
5.	Cybersecurity threats in AI algorithms	AI models can be vulnerable to adversarial attacks and data poisoning, which may lead to incorrect decision-making in agriculture.	Implement robust cybersecurity measures, use federated learning for decentralized AI training, and conduct regular security audits.	Goodfellow et al. (2016); Biggio and Roli (2018)

6.	Lack of standardized AI metrics for agriculture	The evaluation of AI models in agriculture lacks universally accepted performance metrics, making it difficult to compare different solutions.	Develop standardized benchmarks and evaluation criteria for AI models used in agriculture.	Kamilaris and Prenafeta-Boldú (2018); Mohanty et al. (2016)
7.	Limited AI adaptability to dynamic agro-climatic conditions	AI models trained on historical data may struggle to adapt to sudden changes in weather, pests, and diseases.	Develop adaptive AI models that incorporate real-time climate and pest data to improve decision-making.	Schmidhuber and Tubiello (2007); Shafiq and Gu (2022)
8.	Farmer resistance to AI adoption	Traditional farmers may be skeptical of AI-driven decision-making and prefer conventional farming practices.	Conduct awareness campaigns, offer incentives for AI adoption, and provide AI-assisted decision support rather than full automation.	Wolfert et al. (2017); Fountas et al. (2015)

allowing for quicker decision-making processes that can optimize breeding trials and precision agriculture practices (Wang et al. 2024). In conjunction with 5G, edge computing will play a crucial role in processing data closer to the source rather than relying on centralized cloud systems. By enabling on-field AI models to analyze phenotypic and environmental data locally, edge computing will reduce dependence on high-bandwidth internet connections and ensure uninterrupted operations in remote agricultural regions (Shi et al. 2018). This will be particularly beneficial in developing countries where connectivity remains a challenge, as it will facilitate real-time decision-making even in areas with limited infrastructure.

The future of crop improvement will also witness the widespread adoption of AI-driven phenotyping, where DL models will be used to analyze multispectral and hyperspectral imagery, LiDAR scans, and thermal data to assess plant health, growth patterns, and stress responses. AI-powered predictive analytics will allow breeders to identify superior genotypes more efficiently, significantly shortening the breeding cycle and accelerating the development of climate-resilient crops (Singh and Ahmed 2021). The integration of advanced ML techniques, such as CNNs and recurrent neural networks , will further improve the modeling of genotype-by-environment ($G \times E$) interactions, enhancing the selection of high-yielding and stress-tolerant varieties (Montesinos-López et al. 2024). By leveraging these technologies, plant breeders will be able to make data-driven selections with greater precision, thereby improving the efficiency of breeding programs.

Automation and robotics will also play an increasing role in crop management, particularly in tasks such as precision weeding, selective harvesting, and disease monitoring. AI-integrated autonomous robots, equipped with IoT sensors and AI-driven vision systems, will improve labor efficiency and reduce the reliance on chemical inputs by enabling targeted interventions. In particular, UAVs with AI-powered imaging systems will facilitate large-scale, real-time phenotypic assessments, significantly reducing the time and labor required for data collection while improving measurement accuracy (Zhao et al. 2022). These advancements will be instrumental in scaling up breeding trials and improving the efficiency of field-based evaluations.

In addition to improving efficiency and automation, ensuring data security and transparency will be crucial for the future of AI and IoT in agriculture. The integration of blockchain technology with AI-driven IoT systems will enhance data integrity, enabling secure and immutable recording of phenotypic and environmental data. Blockchain-based solutions will facilitate secure data sharing among breeding institutions while addressing concerns related to intellectual property rights and data ownership (Lin et al. 2020; Jain and Sharma 2024). By providing a transparent and tamper-proof platform for recording breeding records, blockchain will improve collaboration between researchers and accelerate the development of new crop varieties.

15.6 Conclusion

The convergence of the IoT and AI is reshaping modern agriculture by enhancing precision, efficiency, and sustainability in crop production. IoT-driven smart farming solutions—such as sensor-based monitoring, automated irrigation, and real-time data analytics—have significantly optimized resource utilization, leading to increased productivity and improved farm management. The integration of AI further amplifies these benefits by enabling predictive analytics, intelligent decision-making, and adaptive automation. Together, these technologies enhance crop yields, reduce input wastage, and strengthen resilience against environmental fluctuations, thereby supporting global food security in the face of climate change.

IoT and AI are also revolutionizing plant breeding and molecular breeding for crop improvement by accelerating genetic advancements and optimizing trait selection. HTP platforms, including sensor-based imaging, drones, and automated growth chambers, facilitate real-time monitoring of key agronomic traits, generating vast datasets with unprecedented precision. AI-driven algorithms analyze these datasets to identify genotype-phenotype correlations, predict superior trait combinations, and enhance marker-assisted and GS strategies. Additionally, AI models leverage IoT-generated environmental and physiological data to refine predictive breeding frameworks, improving stress resilience, yield potential, and crop quality. The integration of these technologies expedites breeding cycles, enhances selection accuracy, and supports the development of climate-resilient crop varieties. Looking ahead, the convergence of IoT, AI, and advanced breeding techniques such as gene editing and speed breeding is expected to further accelerate crop improvement, ensuring agricultural sustainability and food security.

Despite their transformative potential, the widespread adoption of IoT and AI in agriculture faces significant challenges, including issues of interoperability, cybersecurity, scalability, and high implementation costs. Addressing these barriers requires a multidisciplinary approach that integrates technological innovation with economic feasibility and policy support. Future research must focus on enhancing the interoperability of IoT ecosystems, refining AI algorithms for real-time agricultural decision-making, and developing cost-effective, user-friendly solutions tailored to diverse farming systems. Additionally, capacity-building initiatives—such as digital literacy programs and infrastructure development—will be essential to equip farmers, agronomists, and researchers with the necessary skills to fully leverage these technologies.

As agriculture advances towards an era of smart and connected farming, the integration of IoT with emerging technologies such as blockchain, robotics, and edge computing will further drive precision farming, automation, and traceability. A collaborative effort among researchers, policymakers, and industry stakeholders is crucial to overcoming existing challenges and ensuring the seamless deployment of IoT and AI in agricultural systems. By embracing these innovations, the agricultural sector can transition towards a more resilient, resource-efficient, and sustainable food production system—capable of meeting the demands of a rapidly growing global population while minimizing environmental impact.

References

Al-Fuqaha, A., Guizani, M., Mohammadi, M. et al. (2015). Internet of things: a survey on enabling technologies, protocols, and applications. *IEEE Communications Surveys and Tutorials* 17 (4): 2347–2376.

Alonge, M., Lebeigle, L., Kirsche, M. et al. (2022). Automated assembly scaffolding using RagTag elevates a new tomato system for high-throughput genome editing. *Genome Biology* 23 (1): 258.

Alok, A., Chauhan, H., Upadhyay, S. K., Pandey, A., Kumar, J., and Singh, K. (2021). Compendium of Plant-Specific CRISPR Vectors and Their Technical Advantages. *Life*, 11(10), 1021. https://doi.org/10.3390/life11101021.

Araus, J.L. and Cairns, J.E. (2014). Field high-throughput phenotyping: the new crop breeding frontier. *Trends in Plant Science* 19 (1): 52–61.

Ardiansah, I., Bafdal, N., Suryadi, E., and Bono, A. (2020). Greenhouse monitoring and automation using Arduino: a review on precision farming and internet of things (IoT). *International Journal on Advanced Science Engineering and Information Technology* 10 (2): 703–709.

Argus Controls. (2025). Argus control systems for greenhouse management. www.arguscontrols.com (accessed 15 June 2025).

Asikainen, M., Haataja, K., and Toivanen, P. (2013). Wireless indoor tracking of livestock for behavioral analysis. In *2013 9th International Wireless Communications and Mobile Computing Conference (IWCMC)*, (pp. 1833–1838), 1 July–5 July 2013, Sardinia, Italy. IEEE.

Banerjee, A., Lal, E., and Berlin Hency, V. (2023). IoT-based plant health monitoring system using CNN and image processing. *Integrated Green Energy Solutions* 1: 263–290.

Bassi, F.M., Bentley, A.R., Charmet, G. et al. (2016). Breeding schemes for the implementation of genomic selection in wheat (Triticum spp.). *Plant Science* 242: 23–36.

Biggio, B., and Roli, F. (2018). Wild patterns: ten years after the rise of adversarial machine learning. In *Proceedings of the 2018 ACM SIGSAC Conference on Computer and Communications Security* (pp. 2154–2156), 13 October–17 October 2025, Taipei, Taiwan.

Binns, R., Veale, M., and Shadbolt, N. (2018). The ethics of artificial intelligence and robotics. In E. N. Zalta (Ed.), *Stanford Encyclopedia of Philosophy*. Metaphysics Research Lab, Stanford University. https://plato.stanford.edu/archives/win2018.

Bongiovanni, R. and Lowenberg-DeBoer, J. (2004). Precision agriculture and sustainability. *Precision Agriculture* 5: 359–387.

Borrill, P. (2020). Blurring the boundaries between cereal crops and model plants. *New Phytologist* 228 (6): 1721–1727.

Boursianis, A.D., Papadopoulou, M.S., Diamantoulakis, P. et al. (2022). Internet of things (IoT) and agricultural unmanned aerial vehicles (UAVs) in smart farming: a comprehensive review. *Internet of Things* 18: 100187.

Cao, X., Yao, Y., Li, L. et al. (2022, June). igrow: A smart agriculture solution to autonomous greenhouse control. In: *Proceedings of the AAAI Conference on Artificial Intelligence*, vol. 36, No. 11, 11837–11845.

Carvalho, N., Adão, T., Morais, R. et al. (2025). Cybersecurity in precision agriculture: a short review and a practical status assessment over mySense IoT-based platform. *Procedia Computer Science* 256: 255–266.

Chlingaryan, A., Sukkarieh, S., and Whelan, B. (2018). Machine learning approaches for crop yield prediction and nitrogen status estimation in precision agriculture: a review. *Computers and Electronics in Agriculture* 151: 61–69.

CIAT. (2021). Automated IoT platform for greenhouse management. *Alliance of Bioversity International and CIAT*. https://www.ciat.org/ (accessed 15 June 2025).

Cihan, P. (2023). IoT technology in smart agriculture. In *International Conference on Recent Academic Studies* (Vol. 1, pp. 185–192).

Cloarec-Ung, F.M., Beaulieu, J., Suthananthan, A. et al. (2024). Near-perfect precise on-target editing of human hematopoietic stem and progenitor cells. *eLife* 12: RP91288.

Cobb, J.N., DeClerck, G., Greenberg, A. et al. (2013). Next-generation phenotyping: requirements and strategies for enhancing our understanding of genotype–phenotype relationships and its relevance to crop improvement. *Theoretical and Applied Genetics* 126: 867–887.

Contreras-Castillo, J., Guerrero-Ibañez, J.A., Santana-Mancilla, P.C., and Anido-Rifon, L. (2023). SAgric-IoT: an IoT-based platform and deep learning for greenhouse monitoring. *Applied Sciences* 13 (3): 1961.

Cossani, C.M. and Sadras, V.O. (2018). Water–nitrogen colimitation in grain crops. *Advances in Agronomy* 150: 231–274.

Courier Mail. (2023). University of Southern Queensl and adopts PlantEye F600 for advanced plant phenotyping. Courier Mail, Brisbane, Australia.

Crossa, J., Pérez-Rodríguez, P., Cuevas, J. et al. (2017). Genomic selection in plant breeding: methods, models, and perspectives. *Trends in Plant Science* 22 (11): 961–975.

Dara, R., Hazrati Fard, S.M., and Kaur, J. (2022). Recommendations for ethical and responsible use of artificial intelligence in digital agriculture. *Frontiers in Artificial Intelligence* 5: 884192.

Darwin, B.P., Prince, S., Popescu, D.E., and Hemanth, D.J. (2021). Recognition of bloom/yield in crop images using deep learning models for smart agriculture: a review. *Agronomy* 11 (4): 646.

Das, S., Saibabu, N., and Pranaya, D. (2023). Blockchain and intelligent computing framework for sustainable agriculture: theory, methods, and practice. In *Intelligent Engineering Applications and Applied Sciences for Sustainability* (pp. 208–228). IGI Global.

DigiFarm AI. (2025). AI-integrated IoT system for crop variety selection. www.digifarm.com (accessed 15 July 2025).

Dignum, V. (2018). Ethics in artificial intelligence: introduction to the special issue. *Ethics and Information Technology* 20 (1): 1–3.

Doudna, J.A. and Charpentier, E. (2014). The new frontier of genome engineering with CRISPR-Cas9. *Science* 346 (6213): 1258096.

Dutta, M., Gupta, D., Tharewal, S. et al. (2025). Internet of things-based smart precision farming in soilless agriculture: opportunities and challenges for global food security. *IEEE Access* 13: 2169–3536.

Eastern Peak (2023). IoT-enabled greenhouse automation systems for optimal crop management. Eastern Peak Technologies. https://www.easternpeak.com/ (accessed 14 July 2025).

Elijah, O., Rahman, T.A., Orikumhi, I. et al. (2018). An overview of Internet of Things (IoT) and data analytics in agriculture: benefits and challenges. *IEEE Internet of Things Journal* 5 (5): 3758–3773.

FAO (2021). *The State of Food and Agriculture*. Food and Agriculture Organization of the United Nations.

Farooq, M.S., Riaz, S., Abid, A. et al. (2020). Role of IoT technology in agriculture: a systematic literature review. *Electronics* 9 (2): 319.

FieldBook. (2025). FieldBook App: mobile app for plant breeding data management. www.fieldbookapp.com (accessed 14 July 2025).

Findurova, H., Veselá, B., Panzarová, K. et al. (2023). Phenotyping drought tolerance and yield performance of barley using a combination of imaging methods. *Environmental and Experimental Botany* 209: 105314.

Fountas, S., Carli, G., Sørensen, C.G. et al. (2015). Farm management information systems: current situation and future perspectives. *Computers and Electronics in Agriculture* 115: 40–50.

Furbank, R.T. and Tester, M. (2011). Phenomics–technologies to relieve the phenotyping bottleneck. *Trends in Plant Science* 16 (12): 635–644.

Garg, T., Khullar, S., and Kaur, G. (2022). Security issues and challenges for cyber-physical systems. In: *Cyber-Physical Systems* (ed. N. Sharma, L.K. Awasthi, M. Mangla, et al.), 161–176. Chapman and Hall/CRC.

Godfray, H.C.J., Beddington, J.R., Crute, I.R. et al. (2010). Food security: the challenge of feeding 9 billion people. *Science* 327 (5967): 812–818.

Goedde, L., Katz, J., Ménard, A., and Revellat, J. (2020). *Agriculture's Connected Future: How Technology Can Yield New Growth*. McKinsey and Company.

Goodfellow, I., Bengio, Y., and Courville, A. (2016). *Deep Learning*. MIT Press.

Grodan (2025). e-Gro system: nutrient management for greenhouses. www.grodan.com (accessed 14 July 2025).

Gubbi, J., Buyya, R., Marusic, S., and Palaniswami, M. (2013). Internet of Things (IoT): a vision, architectural elements, and future directions. *Future Generation Computer Systems* 29 (7): 1645–1660.

Gupta, N. and Gupta, P.K. (2024). Robotics and artificial intelligence (AI) in agriculture with major emphasis on food crops. In: *Digital Agriculture: A Solution for Sustainable Food and Nutritional Security* (ed. P.M. Priyadarshan, S.M. Jain, S. Penna, and J.M. Al-Khayri), 577–605. Springer.

Gupta, S.K. and Singh, S. (2022). Survey on energy efficient dynamic sink optimum routing for wireless sensor network and communication technologies. *International Journal of Communication Systems* 35 (11): e5194.

Hamdi, M., Rehman, A., Alghamdi, A. et al. (2021). Internet of Things (IoT) Based Water Irrigation System. *International Journal of Online & Biomedical Engineering* 17 (5).

Harun, A.N., Mohamed, N., Ahmad, R. et al. (2019). Improved Internet of Things (IoT) monitoring system for growth optimization of Brassica chinensis. *Computers and Electronics in Agriculture* 164: 104836.

Heffner, E.L., Lorenz, A.J., Jannink, J.L., and Sorrells, M.E. (2010). Plant breeding with genomic selection: gain per unit time and cost. *Crop Science* 50 (5): 1681–1690.

Ismail, A.M. and Mackill, D.J. (2014). Response to flooding: submergence tolerance in rice. In: *Plant Genetic Resources and Climate Change* (ed. M. Jackson, B. Ford-Lloyd, and M. Parry), 251–269. Wallingford UK: CABI.

Jain, N. and Sharma, C. (2024). Blockchain for secure intellectual property rights management. *DME Journal of Management* 5 (2): 1–12.

Jeong, H.H., Kim, S.Y., Rousseaux, M.W. et al. (2017). CRISPRcloud: a secure cloud-based pipeline for CRISPR pooled screen deconvolution. *Bioinformatics* 33 (18): 2963–2965.

Johnson, N., Kumar, M. S., and Dhannia, T. (2020). A study on the significance of smart IoT sensors and data science in digital agriculture. In *2020 Advanced Computing and Communication Technologies for High Performance Applications (ACCTHPA)* (pp. 80–88). IEEE.

Joshi, A., Guevara, D., and Earles, M. (2022). Standardizing and centralizing datasets to enable efficient training of agricultural deep learning models. arXiv preprint arXiv:2208.02707.

Juliana, P., Singh, R.P., Braun, H.J. et al. (2020). Genomic selection for grain yield in the CIMMYT wheat breeding program—status and perspectives. *Frontiers in Plant Science* 11: 564183.

Kamilaris, A. and Prenafeta-Boldú, F. X. (2018). Deep learning in agriculture: a survey.

Kim, Y., Park, H., and Liu, D. (2021). Deep learning for CRISPR guide RNA design and off-target prediction. *Genome Research* 31 (7): 987–999.

Kritikos, M. (2017). Science and technology options assessment. European Parliamentary Research Service. Scientific Foresight Unit (STOA). Brussels: PE, 603.

Kumar, V., Jakhwal, R., and Singh, S. (2023). Decision support system (DSS) for managing fertigation in open field vegetables crops. *Progressive Agriculture* 23 (1): 63–66.

Kumari, K. and Nafchi, A. M. (2024). Sustainable agriculture with AI, machine learning, deep learning, and IoT for future farming. In *2024 ASABE Annual International Meeting* (p. 1), American Society of Agricultural and Biological Engineers.

Li, Y., Xu, Y., and Wang, M. (2024). A review on big data analysis and the application of machine learning and deep learning in crop molecular breeding. *Advances in Resources Research* 4 (4): 703–727.

Liakos, K.G., Busato, P., Moshou, D. et al. (2018). Machine learning in agriculture: a review. *Sensors* 18 (8): 2674.

Li-Cor Biosciences. (2025). Li-Cor portable photosynthesis system: measuring photosynthesis and respiration. www.licor.com (accessed 13 June 2025).

Lin, J., Long, W., Zhang, A., and Chai, Y. (2020). Blockchain and IoT-based architecture design for intellectual property protection. *International Journal of Crowd Science* 4 (3): 283–293.

Madakam, S., Ramaswamy, R., and Tripathi, S. (2015). Internet of Things (IoT): a literature review. *Journal of Computer and Communications* 3 (5): 164–173.

Magar, H. B., Sharma, P., Philip, V., and Ubale, A. S. (2024). Optimizing wireless sensor network performance: TEEN vs. LEACH analysis. In *AIP Conference Proceedings* (Vol. 3156, No. 1), AIP Publishing.

Martínez-Chavez, L.A., Hernández-Ramírez, M.Y., Feregrino-Pérez, A.A., and Esquivel Escalante, K. (2024). Cutting-edge strategies to enhance bioactive compound production in plants: potential value of integration of elicitation, metabolic engineering, and green nanotechnology. *Agronomy* 14 (12): 2822.

Mehrabi, N., Morstatter, F., Saxena, N. et al. (2021). A survey on bias and fairness in machine learning. *ACM Computing Surveys (CSUR)* 54 (6): 1–35.

Mekonnen, Y., Namuduri, S., Burton, L. et al. (2019). Machine learning techniques in wireless sensor network-based precision agriculture. *Journal of the Electrochemical Society* 167 (3): 037522.

Meuwissen, T., Hayes, B., and Goddard, M. (2016). Genomic selection: a paradigm shift in animal breeding. *Animal Frontiers* 6 (1): 6–14.

Miao, Y., Mulla, D.J., and Huang, Y. (2024). Remote sensing for precision agriculture. In: *In Remote Sensing Handbook, Volume III* (ed. P.S. Thenkabail), 229–254. CRC Press.

Mishra, H. and Mishra, D. (2024). AI for data-driven decision-making in smart agriculture: from field to farm management. In: *Artificial Intelligence Techniques in Smart Agriculture* (ed. S.S. Chouhan, U.P. Singh, A. Saxena, and S. Jain), 173–193. Springer Nature Singapore: Singapore.

Misra, N.N., Dixit, Y., Al-Mallahi, A. et al. (2020). IoT, big data, and artificial intelligence in agriculture and food industry. *IEEE Internet of Things Journal* 9 (9): 6305–6324.

Mohanty, S.P., Hughes, D.P., and Salathé, M. (2016). Using deep learning for image-based plant disease detection. *Frontiers in Plant Science* 7: 215232.

Monge, J., Ribeiro, G., Raimundo, A. et al. (2023). AI-based smart sensing and AR for gait rehabilitation assessment. *Information* 14 (7): 355.

Montesinos-Lopez, A., Crespo-Herrera, L., Dreisigacker, S. et al. (2024). Deep learning methods improve genomic prediction of wheat breeding. *Frontiers in Plant Science* 15: 1324090.

Morris, M. R., Sohl-Dickstein, J., Fiedel, N. et al. (2023). Levels of AGI for operationalizing progress on the path to AGI. *arXiv preprint* arXiv:2311.02462.

Nair, R.R., Babu, T., and Ebin, P.M. (2023). Developing edge AI for embedded Systems. In: *Embedded Artificial Intelligence* (ed. A.N. Boruah, M. Goswami, M. Kumar, and O. Loyola-González), 32–54. Chapman and Hall/CRC.

Nasti, R.A. and Voytas, D.F. (2021). Attaining the promise of plant gene editing at scale. *Proceedings of the National Academy of Sciences* 118 (22): e2004846117.

Nawandar, N.K. and Satpute, V.R. (2019). IoT based low cost and intelligent module for smart irrigation system. *Computers and Electronics in Agriculture* 162: 979–990.

Patel, K.K., Patel, S.M., and Scholar, P. (2016). Internet of things-IOT: definition, characteristics, architecture, enabling technologies, application and future challenges. *International Journal of Engineering Science and Computing* 6 (5): 6122–6131.

Pauli, T., Marx, E., and Matzner, M. (2020). Leveraging industrial IoT platform ecosystems: insights from the complementors' perspective. In *ECIS*, 15 June–17 June 2020. Marseille, France (Online).

Pavel, M. I., Kamruzzaman, S. M., Hasan, S. S., and Sabuj, S. R. (2019). An IoT based plant health monitoring system implementing image processing. In *2019 IEEE 4th International Conference on Computer and Communication Systems (ICCCS)* (pp. 299–303), IEEE.

Phenospex. (2021). PlantEye F600 multispectral 3D scanner for plants - PHENOSPEX. https://phenospex.com/products/plant-phenotyping/planteye-f600-multispectral-3d-scanner-for-plants/.

Qayoom, K., Gupta, S., and Kumar, D. (2024). IoT-enabled variable rate technologies for optimal resource use. In: *Agriculture 4.0* (ed. S. Gupta, W. Hasan, S. Singh, et al.), 150–168. CRC Press.

Rahman, M.F.F., Zhang, N., and Abdel-Raheem, E. (2025, May). Intelligent Transportation Systems Utilizing UAVs: Integration with IoT and Machine Learning. In: *2025 International Wireless Communications and Mobile Computing (IWCMC)*, 580–586. IEEE.

Raza, I., Zubair, M., Zaib, M. et al. (2023). Precision nutrient application techniques to improve soil fertility and crop yield: a review with future prospect. *International Research Journal of Educational and Tecnology.* 5 (8): 109–123.

Reddy, R.R. and Pullaiah, T. (2024). *Bio-inspired Technologies for the Modern World: Emerging Trends and Applications*. Apple Academic Press.

Rejeb, K., Abdollahi, A., Al-Turjman, F., and Treiblmaier, H. (2022). The Interplay between the Internet of Things and agriculture: a bibliometric analysis and research agenda. *Internet of Things* 19: 100580.

Saleem, Y., Crespi, N., Rehmani, M.H., and Copeland, R. (2019). Internet of things-aided smart grid: technologies, architectures, applications, prototypes, and future research directions. *IEEE Access* 7: 62962–63003.

Sankaran, S., Khot, L.R., Espinoza, C.Z. et al. (2015). Low-altitude, high-resolution aerial imaging systems for row and field crop phenotyping: a review. *European Journal of Agronomy* 70: 112–123.

Schmidhuber, J. and Tubiello, F.N. (2007). Global food security under climate change. *Proceedings of the National Academy of Sciences* 104 (50): 19703–19708.

Shafiq, M. and Gu, Z. (2022). Deep residual learning for image recognition: a survey. *Applied Sciences* 12 (18): 8972.

Shakoor, N., Lee, S., and Mockler, T.C. (2017). High throughput phenotyping to accelerate crop breeding and monitoring of diseases in the field. *Current Opinion in Plant Biology* 38: 184–192.

Shi, F., Li, Q., Zhu, T., and Ning, H. (2018). A survey of data semantization in internet of things. *Sensors* 18 (1): 313.

Singh, M. and Ahmed, S. (2021). IoT based smart water management systems: a systematic review. *Materials Today Proceedings* 46: 5211–5218.

Singh, R. and Bhanot, N. (2020). An integrated DEMATEL-MMDE-ISM based approach for analysing the barriers of IoT implementation in the manufacturing industry. *International Journal of Production Research* 58 (8): 2454–2476.

Singh, A., Ganapathysubramanian, B., Singh, A.K., and Sarkar, S. (2016). Machine learning for high-throughput stress phenotyping in plants. *Trends in Plant Science* 21 (2): 110–124.

Sinha, B.B. and Dhanalakshmi, R. (2022). Recent advancements and challenges of Internet of Things in smart agriculture: a survey. *Future Generation Computer Systems* 126: 169–184.

Smith, A.M., Walsh, J.R., Long, J. et al. (2020). Standard machine learning approaches outperform deep representation learning on phenotype prediction from transcriptomics data. *BMC Bioinformatics* 21: 1–18.

Srivastava, A. and Maity, R. (2023). Assessing the potential of AI–ML in urban climate change adaptation and sustainable development. *Sustainability* 15 (23): 16461.

Stojmenovic, I. and Wen, S. (2014). The fog computing paradigm: scenarios and security issues. In *2014 Federated Conference on Computer Science and Information Systems* (pp. 1–8). IEEE.

Tattaris, M., Reynolds, M.P., and Chapman, S.C. (2016). A direct comparison of remote sensing approaches for high-throughput phenotyping in plant breeding. *Frontiers in Plant Science* 7: 1131.

Teralytic. (2025). Teralytic soil sensor for real-time soil monitoring in agriculture. www.teralytic.com (accessed 13 June 2025).

Titirmare, S., Margal, P.B., Gupta, S., and Kumar, D. (2024). AI-powered predictive analytics for crop yield optimization. In: *Agriculture 4.0* (ed. S. Gupta, W. Hasan, S. Singh, et al.), 89–110. CRC Press.

Torky, M. and Hassanein, A.E. (2020). Integrating blockchain and the internet of things in precision agriculture: Analysis, opportunities, and challenges. *Computers and Electronics in Agriculture* 178: 105476.

Upadhyay, S. K. (2021). Genome Engineering for Crop Improvement. *Wiley*. https://doi.org/10.1002/9781119672425.

Upadhyay, S. K. (2024). Applications of Genome Engineering in Plants. *Wiley*. https://doi.org/10.1002/9781394183913.

Upadhyay, S. K. (2026). In: *Designer Crops: Implications of Artificial Intelligence and Modern Biotechnological Methods*. In: Elsevier Science.

Varshney, R.K., Pandey, M.K., Bohra, A. et al. (2019). Toward the sequence-based breeding in legumes in the post-genome sequencing era. *Theoretical and Applied Genetics* 132 (3): 797–816.

Verdouw, C., Tekinerdogan, B., Beulens, A., and Wolfert, S. (2021). Digital twins in smart farming. *Agricultural Systems* 189: 103046.

Vidya Madhuri, E., Rupali, J.S., Sharan, S.P. et al. (2025). Transforming pest management with artificial intelligence technologies: the future of crop protection. *Journal of Crop Health* 77 (2): 48.

Walter, A., Finger, R., Huber, R., and Buchmann, N. (2017). Smart farming is key to developing sustainable agriculture. *Proceedings of the National Academy of Sciences* 114 (24): 6148–6150.

Wang, X., Silva, P., Bello, N.M. et al. (2020). Improved accuracy of high-throughput phenotyping from unmanned aerial systems by extracting traits directly from orthorectified images. *Frontiers in Plant Science* 11: 587093.

Wang, J., Wang, Y., Li, G., and Qi, Z. (2024). Integration of remote sensing and machine learning for precision agriculture: a comprehensive perspective on applications. *Agronomy* 14 (9): 1975.

Wartha, C.A. and Lorenz, A.J. (2024). Genomic predictions of genetic variances and correlations among traits for breeding crosses in soybean. *Heredity* 133 (3): 173–185.

Wilson, A.N., Kumar, A., Jha, A., and Cenkeramaddi, L.R. (2021). Embedded sensors, communication technologies, computing platforms and machine learning for UAVs: a review. *IEEE Sensors Journal* 22 (3): 1807–1826.

Wolfert, S., Ge, L., Verdouw, C., and Bogaardt, M.J. (2017). Big data in smart farming–a review. *Agricultural Systems* 153: 69–80.

WRI (2019). Creating a Sustainable Food Future. World Resources Institute Report.

Xu, Y., Zhang, X., Li, H. et al. (2022). Smart breeding driven by big data, artificial intelligence, and integrated genomic-enviromic prediction. *Molecular Plant* 15 (11): 1664–1695.

Yang, W., Feng, H., Zhang, X. et al. (2020). Crop phenomics and high-throughput phenotyping: past decades, current challenges, and future perspectives. *Molecular Plant* 13 (2): 187–214.

Zanin, R., Chen, X., and Kim, H. (2022). Advances in AI-driven agriculture and environmental monitoring. *Computers and Electronics in Agriculture* 190: 106418.

Zhang, X.S., Zhang, X., and Kaparthi, P. (2020). Combat information overload problem in social networks with intelligent information-sharing and response mechanisms. *IEEE Transactions on Computational Social Systems* 7 (4): 924–939.

Zhang, Y., Li, T., Zhao, C. et al. (2021). An integrated smart sensor dressing for real-time wound microenvironment monitoring and promoting angiogenesis and wound healing. *Frontiers in Cell and Developmental Biology* 9: 701525.

Zhao, J., Li, Y., Wang, X., and Zhang, H. (2022). AI-driven IoT in smart agriculture: applications, challenges, and future trends. *Computers and Electronics in Agriculture* 198: 107015. https://doi.org/10.1016/j.compag.2022.107015.

Zhou, J., Reynolds, D., Cornu, T.L. et al. (2017). CropQuant: an automated and scalable field phenotyping platform for crop monitoring and trait measurements to facilitate breeding and digital agriculture. *BioRxiv [Preprints]* 161547.

Zhu, X., Young, D., Watson, B. J. et al. (2008). 1000 islands: integrated capacity and workload management for the next generation data center. In *2008 International Conference on Autonomic Computing* (pp. 172–181). IEEE.

16

Optimizing Crop Management Practices Using AI Technologies

Ambrina Sardar Khan[1] *and Prateek Srivastava*[2]

[1] *Integral University, Department of Environmental Sciences, Lucknow, Uttar Pradesh 226026, India*
[2] *University of Allahabad, Department of Botany, Prayagraj, Uttar Pradesh 211002, India*

16.1 Introduction

Climate change poses a significant challenge to global food security, threatening agricultural productivity and exacerbating micronutrient deficiencies and chronic hunger, especially in the wake of the COVID-19 pandemic. Extreme temperatures, excessive sunlight, elevated CO_2 levels, and disruptions in rainfall patterns are making crops more vulnerable to disease, leading to reduced yields and escalating food insecurity (Hasegawa et al. 2018; Ray et al. 2019). This situation is compounded by a rise in the global hunger index and a rapidly growing population (Ukhurebor et al. 2021). Conventional breeding methods, which are time-consuming and labor-intensive are proving inefficient in developing climate-resilient crops.

To address this, there is an urgent need to accelerate the plant breeding cycle using artificial intelligence (AI) to analyze plant responses to environmental changes in real-time. AI-driven insights can aid in the development of climate-smart crops for biomass production and ecosystem services, offering a more effective solution to mitigate the impacts of climate change on food security. Additionally, the close connection between food security and economic growth, which is also influenced by climate change, further complicates the challenge, making AI integration in plant breeding essential for both immediate and long-term solutions (Rai 2022).

To address the pressing challenges of climate change and food security, next-generation plant breeding technologies are crucial for developing climate-smart crops with enhanced nutritional value (Watson et al. 2018; Ghosh et al. 2018). While conventional breeding has played a key role in ensuring food and nutritional security by strengthening genetic diversity and identifying novel genes (Ahmar et al. 2020), it is a slow process, typically taking 6–7 years due to the lengthy breeding cycle, which includes parent selection, hybridization, data recording, analysis, and field evaluation (Al-Khayri et al. 2015; Qaim 2020). This delays the release of improved cultivars to farmers and limits the rate of genetic gain (Al-Khayri et al. 2015).

To overcome this bottleneck, modern breeding techniques, such as speed breeding powered by AI, are now being widely adopted (Harfouche et al. 2019; Razzaq et al. 2021). Speed breeding accelerates crop improvement by manipulating the growing environment to optimize light intensity and duration, stimulating photoreceptors that regulate circadian rhythms and trigger rapid reproductive development (Ghosh et al. 2018). Researchers categorize plants into short-day, long-day, and day-neutral groups, adjusting light exposure to enhance generation advancement. Longer light durations accelerate growth for short-day and day-neutral plants, while shorter exposure optimizes development for long-day plants (Ghosh et al. 2018; Leal et al. 2022). This innovative approach significantly speeds up the breeding cycle, facilitating the faster development of climate-resilient and high-yielding crop varieties (Godwin et al. 2019).

The ultimate goal of plant breeders is to enhance yield and resilience by accurately predicting which cultivars will produce superior hybrids. Integrating classical breeding with OMICS technologies such as phenomics, genomics, transcriptomics, proteomics, and metabolomics has revolutionized breeding by providing vast amounts of high-quality data (Sartor et al. 2019). However, managing this complex, large-scale multi-omics data poses a significant challenge, potentially hindering cultivar selection and breeding efficiency (Sartor et al. 2019; Raja Sundaram and Selbig 2016). AI in agriculture offers a transformative solution by rapidly processing big data and linking it to underlying biological processes under varying environmental conditions.

By integrating speed breeding with AI-driven data analysis, breeding programs can be accelerated, addressing the limitations of conventional methods and enhancing genetic gains. This combination enables the development of stress-tolerant, high-yielding cultivars tailored to withstand climate change challenges (Crews and Cattani 2018). Moving toward digital agriculture, this approach explores how AI can be seamlessly integrated with speed breeding to optimize the processing of big OMICS data, paving the way for next-generation breeding technologies that will revolutionize crop improvement and strengthen global food and nutritional security.

16.2 Traditional Breeding: The Foundation of Agricultural Advancement

Ancient farmers initially practiced plant breeding to domesticate crops within their environment. Over time, this practice evolved into a widely recognized approach for improving crop yield and disease resistance (Ahmar et al. 2020). The earliest form of plant breeding involved selecting wild cultivars with desirable traits and crossing them with local varieties, followed by rigorous selection over multiple generations (Al-Khayri et al. 2015; Thudi et al., 2021). Traditionally, breeders identified variations among progenies based on morphological traits such as plant height, branch number, and yield per plant, analyzing them through statistical methods (Gupta et al. 2010). With advancements in molecular biology, plant breeders began utilizing genetic markers, which provided a more precise and efficient means of assessing genetic variation compared to morphological and biochemical markers (Al-Khayri et al. 2015; Gupta et al. 2010). These molecular tools have played a crucial role in accelerating breeding programs and enhancing the conservation of genetic resources (Kushwah et al. 2020). Integrating molecular techniques with classical breeding has unlocked the hidden genetic potential of landraces, wild relatives, and cultivated varieties by facilitating the identification of quantitative trait loci (QTLs) and novel genes absent in local cultivars (Gupta et al. 2010; Dormatey et al. 2020).

Gene pyramiding, a technique used to accumulate multiple beneficial genes into elite cultivars, has expanded the genetic diversity available for crop improvement (Kushwah et al. 2020). Researchers have developed various genetic, linkage, and QTL maps for essential crops, enabling breeders to identify favorable genetic variations using molecular markers (Gupta et al. 2010). Whole-genome and RNA sequencing have further refined genome-wide analyses by detecting single nucleotide polymorphisms (SNPs), though their accuracy depends on sample size and marker efficiency (Kushwah et al. 2020; Gupta et al. 2010). High-resolution QTL mapping requires a robust sample size and comprehensive genomic coverage, as gaps in genetic maps can hinder marker-trait associations and gene identification (Dormatey et al. 2020).

Conventional breeding has historically played a significant role in developing new perennial crops by domesticating superior wild cultivars or hybridizing them with cultivated varieties (Crews and Cattani 2018). Domestication involves establishing a crop in a new environment and selecting superior genotypes based on phenotypic traits (Ahmar et al. 2020). However, hybridization is often more practical, as crossing genetically contrasting cultivars can produce hybrids with enhanced agronomic traits (Crews and Cattani 2018). This method has successfully improved several perennial crops, including sorghum (*Sorghum bicolor* × *S. halepense*), wheat (Triticum spp. × Thinopyrum spp.), rice (*Oryza sativa* × *Oryza longistaminata*), and buckwheat (Fagopyrum spp. × Fagopyrum spp.) (Crews and Cattani 2018). Despite their effectiveness, conventional breeding techniques are time-consuming and require extensive data collection. To overcome these challenges, researchers have integrated molecular breeding approaches, significantly accelerating the development of improved crop cultivars.

16.3 Advancements in Molecular Breeding Techniques

Genomics-assisted tools have revolutionized plant breeding, offering a powerful way to enhance crop growth and productivity in response to changing environmental conditions. By using DNA-based markers, breeders can efficiently select the best crossbreeds through marker-assisted selection (MAS) (Al-Khayri et al. 2015). When combined with traditional breeding methods, plant genomics provides deep insights into genetic diversity at both phenotypic and molecular levels, significantly accelerating crop improvement programs (Ahmar et al. 2020). The process of genomics-assisted breeding (GAB) begins with selecting cultivars or wild species that possess desirable traits for biotic or abiotic stress tolerance (Ashkani et al. 2015). These selections are then linked to their genetic profiles through specific molecular markers, followed by rigorous selection to develop climate-resilient crops. High-density DNA fingerprinting of economically significant crops further connects plant physiology with phenological traits, helping identify ideal genotypes with unique characteristics (Gupta et al. 2010). Researchers have confirmed that precise genotyping, followed by phenotyping and statistical analysis, can uncover valuable information, leading to the identification of QTLs. High-density genetic linkage maps developed using biparental and double haploid mapping populations have been instrumental in identifying QTLs responsible for disease resistance and agronomic traits in cereals and legumes (Wang et al. 2016; Bhadauria et al. 2017; Nzuki et al. 2017).

Advancements in next-generation sequencing (NGS) have accelerated the discovery of QTLs linked to yield, stress tolerance, and other critical traits. These technologies have enabled the development of robust genetic markers such as SNPs and insertions/deletions (InDels), which facilitate the identification of novel genes. By integrating genotyping-by-sequencing (GBS) with genome-wide association studies (GWAS) (Wang et al. 2016), researchers have been able to pinpoint alleles that contribute to important agronomic traits. As a result, GAB has significantly improved crop characterization and accelerated breeding processes by identifying genetic variations linked to essential traits (Bhadauria et al. 2017; Bhatta et al. 2018; Varshney et al. 2019).

In recent years, GAB has driven remarkable progress in over 100 agriculturally significant crops, enhancing their yield, productivity, and resilience to extreme environments (Hickey et al. 2019). It has been extensively used to identify key QTLs associated with disease resistance in crops such as rice (bacterial blight, blast disease), barley (yellow mosaic virus, powdery mildew), and legumes (cyst nematode in soybean, rust resistance in peanuts) (Varshney et al. 2021a). GAB has also played a crucial role in identifying QTLs related to abiotic stress tolerance and nutritional quality, such as salt and drought resistance, grain protein content, amylose levels, and oleic acid content in wheat, rice, and groundnut (Hickey et al. 2019).

Despite its effectiveness, conventional GAB is time-consuming, as it involves extensive crossing, allele mining, and rigorous phenotyping. To address these challenges, researchers have developed GAB 2.0 an advanced version of the original approach. GAB 2.0 integrates MAS, GWAS, and genome editing tools like CRISPR-Cas9, alongside speed breeding techniques that accelerate the creation of novel genetic variations. This approach enables more precise and cost-effective breeding for stress-tolerant and nutritionally enhanced crops (Upadhyay, 2021, 2024; Varshney et al. 2021b). Mutation breeding has also been widely used to create genetic variations for crop improvement (Raina et al. 2016). This technique relies on chemical agents or high-energy radiation to induce mutations in specific genome regions, increasing genetic diversity (Chaudhary et al. 2019). Target-induced local lesions in genomes (TILLING) provide a more controlled and precise way to introduce mutations compared to traditional chemical mutagens like ethyl methane sulfonate (EMS) or methyl methane sulfonate (MMS).

Mutation breeding has been successfully applied to enhance growth and stress tolerance in crops such as wheat, rice, tomatoes, and legumes. However, the process is highly labor-intensive and requires extensive screening to identify desired traits (Chaudhary et al. 2019; Bado et al. 2015). Recently, Meta-QTL (MQTL) analysis has been introduced to refine QTL identification and mapping by leveraging data from multiple studies (Khahani et al. 2021). For example, researchers used the BioMercator program to analyze complex abiotic and biotic stress traits in durum wheat, pinpointing genes responsible for quality and disease resistance (Soriano et al. 2021). Similarly, a GWAS study on rice identified 1052 QTLs and 144 MQTLs across 122 rice populations, successfully linking them to key agronomic traits for future breeding programs (Khahani et al. 2021).

While these advanced breeding techniques have significantly improved crop development, they are not without challenges. Genomics-assisted approaches such as GBS, GWAS, and MQTL analysis can be limited by genetic drag, gene erosion, hybridization incompatibilities, and complex selection processes. To overcome these hurdles, researchers have incorporated functional genomic tools and gene cloning technologies, enabling the development of genetically modified (GM) crops that carry essential genes to meet the growing global food demand. By integrating classical and modern genetic approaches, breeders can now shuffle and recombine alleles to create superior crop varieties tailored for higher yields, stress resistance, and improved nutritional value ensuring food security in an ever-changing climate.

16.4 Speed Breeding: A New Era in Agriculture

Speed breeding is a groundbreaking technique that has revolutionized plant breeding by significantly accelerating crop growth, development, and commercialization (Watson et al. 2018; Ghosh et al. 2018; Varshney et al. 2021b). This method enhances yield potential, nutritional value, and resilience to environmental stresses, making it a game-changer in modern agriculture. Unlike conventional breeding, which can take years to develop improved cultivars, speed breeding shortens the crop cycle by one to two months, allowing for faster advancements in breeding programs (Watson et al. 2018). Originally inspired by NASA's research on growing wheat under artificial lights, speed breeding has since been adapted for various crop species to enhance agronomic traits (Ghosh et al. 2018). The process mimics natural day and night conditions using artificial light with specific wavelengths and controlled temperatures for up to 22 hours a day (Ghosh et al. 2018; Ahmar et al. 2020). This extended exposure to light boosts photosynthesis and accelerates key physiological and metabolic processes, leading to faster flowering, fruiting, and seed development (Godwin et al. 2019).

In recent years, speed breeding has transformed global agriculture by streamlining critical breeding processes such as crossing, backcrossing, gene pyramiding, MAS, and the development of GM crops (Watson et al. 2018; Ghosh et al. 2018; Ahmar et al. 2020). Researchers have successfully achieved up to four to six generations of crops like *Brassica napus*, *Pisum sativum*, *Triticum aestivum*, *Hordeum vulgare*, and *Cicer arietinum* within a single year, a significant leap from traditional breeding methods, which typically allow only two generations per year. Moreover, speed breeding integrates seamlessly with MAS and GWAS, enabling plant breeders to develop stable, homozygous genotypes faster and facilitate the release of improved cultivars (Ghosh et al. 2018; Jähne et al. 2020). The effectiveness of speed breeding largely depends on the intensity and quality of light, as well as optimal temperature and CO_2 levels (Jähne et al. 2020). High-intensity lighting and controlled environmental conditions dramatically enhance photosynthesis, reducing the time needed for plants to reach the flowering stage.

However, implementing speed breeding requires careful consideration of energy costs and temperature fluctuations, which can influence plant growth. Advances in technology have helped overcome these challenges, with modern light systems and sensors automatically adjusting to maintain optimal conditions at reduced costs (Jähne et al. 2020).

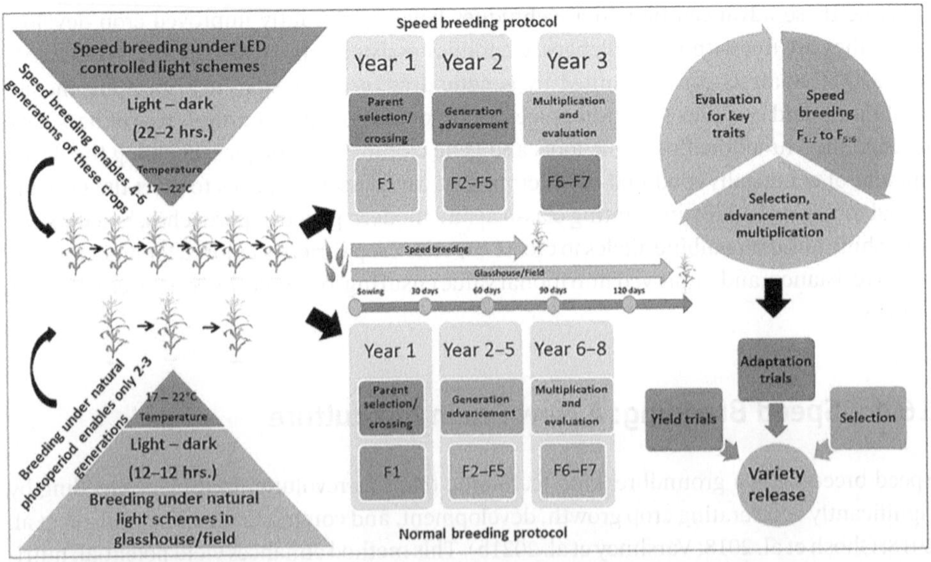

Figure 16.1 Speed breeding protocol: Accelerating breeding cycles for enhanced growth and yield compared to conventional approaches under standard photoperiod (Rai 2022).

The introduction of LED-based lighting systems has further refined speed breeding protocols, allowing precise control over light duration and intensity. This innovation has successfully optimized flowering times and even suppressed unwanted traits in crops like lupins and soybeans (Chiurugwi et al. 2019). As speed breeding continues to evolve, it promises to play a vital role in meeting the growing global demand for high-yield, stress-tolerant crops in a cost-effective and time-efficient manner (Figure 16.1).

16.5 Traditional Meets Speed Breeding: Shaping Agricultural Progress

Integrating speed breeding with traditional breeding methods requires careful planning and the selection of cultivars with high genetic potential (Watson et al. 2018; Ghosh et al. 2018; Jähne et al. 2020). This strategic combination accelerates breeding programs, enabling the development of high-yield and stress-tolerant crop varieties. Choosing inbred lines with strong genetic growth allows breeders to fast-track crop improvement, making it possible to identify superior phenotypes at an early stage (Watson et al. 2018; Ghosh et al. 2018).

Genomic selection further enhances this process by incorporating MAS and GWAS to predict the most promising individuals for breeding (González-Barrios et al. 2021; Bhat et al. 2016). Until the 1990s, phenotyping and genotyping were expensive and inefficient, significantly slowing down crop improvement efforts. However, the advent of NGS transformed plant breeding by making genomic selection more accessible and cost-effective (Bhat et al. 2016). This breakthrough enabled the identification of crucial QTLs and genes, paving the way for targeted crop improvements through forward breeding approaches (González-Barrios et al. 2021; Bhat et al. 2016).

Despite these advancements, time remains a critical constraint in generating superior genetic combinations through hybridization and recombination (Delfina et al. 2016). To overcome this, researchers are increasingly integrating molecular breeding techniques with speed breeding technology. This fusion significantly accelerates the development of new crop varieties by allowing multiple generations to be cultivated in a short span (Bhat et al. 2016; Delfina et al. 2016).

One of the most effective applications of speed breeding is the single seed descent (SSD) method, which enables the rapid development of genetically stable populations at a fraction of the cost of traditional double haploid techniques (Varshney et al. 2019). SSD-generated populations offer higher genetic gains, ultimately leading to the creation of improved cultivars when incorporated into breeding programs (Ghosh et al. 2018; Ahmar et al. 2020). Studies confirm that speed breeding, when combined with MAS and GWAS, facilitates the rapid introgression of desirable traits into elite cultivars (González-Barrios et al. 2021; Bhat et al. 2016). Additionally, this approach accelerates the development of recombinant inbred lines (RILs) and near-isogenic lines (NILs), expediting the identification of QTLs for specific traits (Delfina et al. 2016; Varshney et al. 2019).

Speed breeding also enhances the efficiency of genome editing technologies like CRISPR-Cas9 by drastically reducing the time required for new generations to develop after successful gene editing (Wolter et al. 2019; Sushmita et al., 2021). This integrated approach has been successfully tested in crops such as chickpeas, peas, lentils, faba beans, and pigeon peas. Researchers have demonstrated that by maintaining extended light exposure (20–22 hours per day) and optimal temperature conditions, they can achieve five to six generations per year. These generations are then analyzed for high-yield and disease-resistant traits, significantly advancing breeding programs. However, for this integration to be truly effective, it requires specialized training, pre-breeding research, a well-planned breeding strategy, and access to diverse germplasm suited to the desired traits. When implemented correctly, speed breeding in combination with classical breeding methods has the potential to revolutionize agricultural innovation, making crop improvement faster, more efficient, and more sustainable (Singh et al., 2024; Saxena et al. 2019; Cazzola et al. 2020; Samineni et al. 2020; Idrissi 2020; Mobini et al. 2020; Kumar et al., 2023).

16.6 Artificial Intelligence (AI) in Plant Breeding

Advancements in plant OMICS research have led to an explosion of complex datasets, making it crucial to decode their significance for precise crop trait identification (Harfouche et al. 2019). NGS has further accelerated genome sequencing, generating vast genetic information (Godwin et al. 2019). When combined with transcriptomics and proteomics, techniques like MAS and GWAS provide a deeper understanding of plant genotypes and phenotypes. Given the immense data volume, AI and machine learning (ML) are revolutionizing plant breeding by enabling rapid analysis of phenotypic, biochemical, and yield-related traits (Figure 16.2). This integration enhances breeding efficiency, allowing researchers to identify superior genotypes and accelerate crop improvement with greater precision (Rai, 2022).

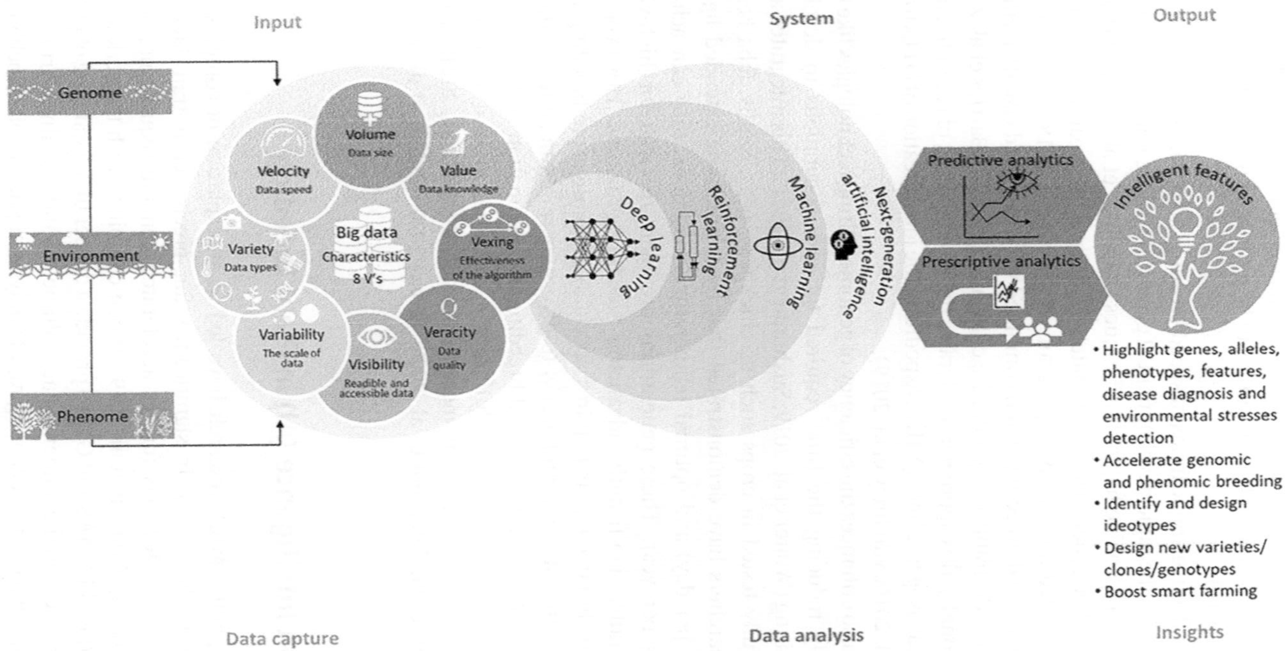

Figure 16.2 Next-generation (Next-Gen) artificial intelligence (AI) in plant breeding.

16.7 AI in Enhancing Plant Breeding Technologies

The integration of Next-Gen AI in plant breeding relies on advanced data mining techniques using intelligent algorithms and models (Harfouche et al. 2019). Researchers are continuously enhancing AI-driven image recognition to analyze complex datasets, making it a key tool for accelerating crop improvement (Harfouche et al. 2019; Godwin et al. 2019). Technologies like neural networks (NN) and deep learning (DL) are revolutionizing multi-omics data analysis by mimicking brain neurons for efficient classification (Parmley et al. 2019). Future AI models will not only assess breeding values but also adaptively refine predictions of disease resistance and agronomic traits under changing environmental conditions. Given that extensive hybridization and selection have altered crop phenotypic plasticity, modern breeding programs are now focusing on closing the genotype–phenotype gap to enhance crop resilience, particularly against abiotic stress (Parmley et al. 2019; Niazian and Niedbała 2020; van Dijk et al. 2021).

Researchers are leveraging genotypic, environmental, and phenotypic data to enhance abiotic stress breeding programs, aiming to identify superior genotypes with key agronomic traits (van Dijk et al. 2021). Given the complexity of these traits, precise monitoring is essential. To address this, scientists have developed an AI-driven physiological gravimetric system within the soil–plant-atmosphere continuum (SPAC), enabling real-time detection of subtle plant responses (Harfouche et al. 2019; Godwin et al. 2019). Next-Gen AI plays a crucial role in continuously analyzing phenotypic data to identify stress-responsive QTLs and other agronomically significant traits (Parmley et al. 2019; Niazian and Niedbała 2020). Additionally, the field phenomics suite, equipped with ML, UAVs, and ground-based sensors, streamlines breeding by capturing high-resolution images and generating large datasets for AI-driven analysis (Niazian and Niedbała 2020; van Dijk et al. 2021). These advanced tools, combined with MAS and GWAS, allow breeders to dissect plants at the molecular level, identifying novel genes and QTLs (van Dijk et al. 2021; Esposito et al. 2020). However, a key challenge remains in linking AI-generated phenotypic data with genetic markers to enhance genetic gain. Overcoming this hurdle requires refining AI models to better correlate complex traits with environmental variables. Advancing Next-Gen AI is essential for bridging the genotype–phenotype gap, ultimately accelerating crop improvement and enabling the rapid development of superior cultivars (Figure 16.3).

16.8 Studying Biochemical Phenotype Through AI

Advancements in technology have significantly improved the precision of recording genotypic and phenotypic variations, enabling researchers to extract valuable insights from complex datasets (van Dijk et al. 2021; Esposito et al. 2020; Jung et al. 2021). AI is now being leveraged to analyze intricate biochemical pathways, deciphering real-time molecular changes under abiotic stress conditions (Esposito et al., 2020). Since biochemical and metabolic processes involve intricate shifts across genomics, proteomics, metabolomics, and epigenomics, sophisticated instruments like NGS, ChIP, and MALDI-TOF-MS generate vast datasets that are difficult to interpret manually. AI has emerged as a powerful tool to process and analyze these complex datasets efficiently, aiding researchers in understanding

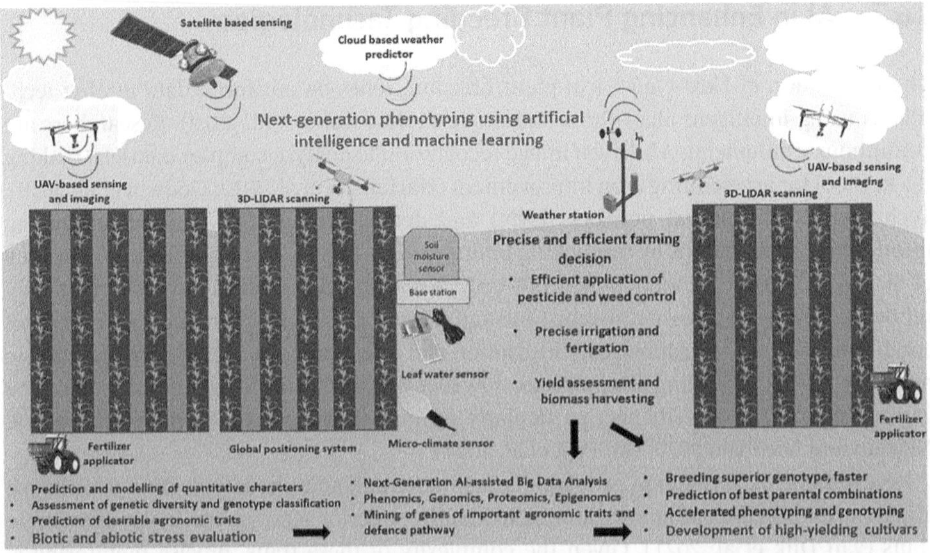

Figure 16.3 Artificial intelligence in augmenting plant breeding technology (Rai 2022).

plant stress biology (van Dijk et al. 2021). Studies have demonstrated AI's ability to predict genomic crossovers, classify genomic regions, and identify gene promoters in crops like maize and Arabidopsis (Demirci et al. 2021). AI also plays a crucial role in decoding metabolic regulatory networks, enhancing nitrogen use efficiency, starch biosynthesis, and secondary metabolite production in plants like rice and Arabidopsis (Li et al. 2020; Uygun et al., 2019; Varala et al. 2018).

Furthermore, AI applications extend beyond plant stress biology, contributing to biofuel production and bioenergy management (Meena et al. 2021). Its potential in single-cell RNA sequencing, DNA methylation, and post-translational modifications provides deep insights into key genomic regions influencing stress responses (Sartor et al. 2019). AI-driven breeding programs can predict complex traits such as photosynthesis, hormone regulation, and yield, facilitating the identification of QTLs and genetic markers through MAS and GWAS (Gupta et al. 2010). By integrating genomics, transcriptomics, proteomics, and metabolomics, AI is reshaping plant research, offering a holistic understanding of how crops respond to environmental stress and improving breeding strategies for sustainable agriculture.

16.9 Integrating Phenomics with Genomics for Smart Breeding

Classical breeding faces a major limitation in its inability to effectively correlate genotype with phenotype, restricting insights into plant genomic architecture (van Dijk et al. 2021). However, speed breeding combined with Next-Gen AI has revolutionized plant research by developing genomic selection models that predict trait expression through genetic markers

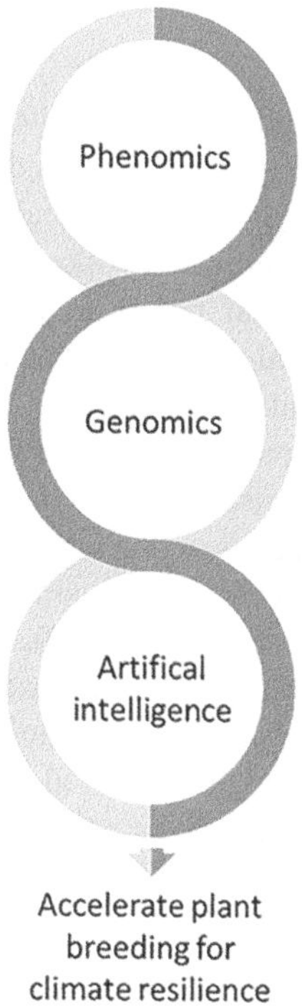

Figure 16.4 Integrating phenomics with genomics for smart breeding.

and extensive phenotyping (Niazian and Niedbała 2020). These models identify key genomic regions, such as SNPs and InDels, enabling breeders to forecast plant traits before maturity, which is crucial for developing climate-smart crops (Figure 16.4) (Parmley et al. 2019). AI further enhances this process by analyzing heritable components, shortening breeding cycles, and boosting crop yields (Nabwire et al. 2021). However, traditional genomic selection models often face accuracy issues due to complex genetic interactions (Nabwire et al. 2021; Shen et al. 2022). To overcome this, researchers have integrated neural networks, which simulate human-like decision-making to improve data interpretation (Parmley et al. 2019; Niazian and Niedbała 2020). Despite their promise, these AI models still struggle with complex datasets, revealing the need for human expertise to refine and align AI-driven insights with breeding objectives. Scientists are now exploring advanced AI-based algorithms, including transfer learning approaches that adapt existing datasets for more precise predictions (Uzal et al. 2018; Montesinos-López et al. 2019). A notable breakthrough is ARIGAN, a deep transfer learning model that synthesizes plant images and integrates multi-omics data, offering a predictive understanding of gene expression and transcriptional regulation (Valerio Giuffrida et al. 2017; Ubbens et al. 2018). However, AI's black-box nature still limits its full potential in plant breeding. Future advancements in Next-Gen AI systems capable of analyzing large multi-omics datasets will empower plant breeders to develop resilient, high-yield crops, paving the way for sustainable agriculture and global food security.

16.10 AI Technologies Benefiting Crop Breeding

AI replicates human decision-making and problem-solving abilities using advanced computing technologies (Jarrahi 2018). It encompasses ML, computer vision, and natural language processing (NLP), enabling systems to learn and function independently. AI-driven models, such as deep neural networks (DNN), artificial neural networks (ANN), and support vector machines (SVM), leverage vast computational power to identify patterns and make data-driven decisions faster than humans (Li et al. 2021; Kok et al. 2009; Harfouche et al. 2019; Baduge et al. 2022; Montesinos-López et al. 2018; Xu et al. 2021). A key aspect of AI is its integration with big data and ML, where AI processes massive datasets to uncover hidden relationships and automate complex decision-making (Cravero et al. 2021; Berente et al. 2021). For instance, in climate studies, AI analyzes meteorological data, refines predictive models, and issues automated alerts for extreme weather conditions (Huntingford et al. 2019). The growing use of GPUs has further enhanced AI's efficiency, making computations faster and more precise (Raschka et al. 2020). In agriculture and breeding, AI plays a transformative role by automating farm monitoring, streamlining data analysis, and enabling breeders to focus on high-value tasks (Talaviya et al. 2020). It reduces manual effort, enhances decision-making accuracy, and boosts efficiency in breeding programs. With continued advancements, AI is set to revolutionize digital breeding, offering unparalleled insights and precision in developing resilient, high-yield crops (Harfouche et al. 2019; Esposito et al. 2019; Reinoso-Peláez et al. 2022; Crossa et al. 2017).

AI is revolutionizing plant breeding by accelerating the development of new crop varieties through advanced genomics, phenomics, and precision agriculture. ML plays a crucial role in genomic prediction, selection, and marker-assisted breeding, helping breeders make data-driven decisions (Reinoso-Peláez et al. 2022; Crossa et al. 2017). Leading agricultural companies like Monsanto and John Deere have invested heavily in AI-driven technologies that analyze vast datasets including soil conditions, seed genetics, and weather patterns to optimize yields and reduce farming costs (Faulkner et al. 2014). AI-powered systems also integrate environmental sensors and big data analytics, as seen in precision-farming experiments by companies like NEC and Dacom (Stergiou and Psannis 2017; El Bilali and Allahyari 2018). However, extracting meaningful insights from these complex agricultural datasets remains a challenge, requiring advanced frameworks, algorithms, and AI-driven analytics (Priya and Ramesh 2020). Innovations in AI, including DL and predictive modeling, are enhancing our understanding of plant behavior under varying climate conditions (Jeong et al. 2016). The unveiling of Summit, a supercomputer with 27 000 GPUs, underscores AI's growing potential to transform agriculture, ensure food security, and drive an agricultural revolution in the near future (Streich et al. 2020).

16.11 Role of Artificial Intelligence in Addressing Phenomics Challenges

The rapid advancement of plant phenomics has paved the way for precision breeding, driven by cutting-edge technologies enabling high-throughput phenotyping (HTP) of complex plant traits. AI, particularly ML, DL, and computer vision, has revolutionized phenotyping by enhancing data collection and analysis through non-invasive imaging techniques. AI-powered tools facilitate data management, genotype–phenotype modeling, and open-source platforms, fostering collaboration and large-scale data sharing (Nabwire et al. 2021). However, a lab-to-field performance gap remains, as controlled environments fail to fully represent real-world conditions (Yang et al. 2020). While AI has been successfully integrated into indoor phenotyping, field-based HTP still lags behind, necessitating further investment in adaptable technologies. Smartphones, equipped with AI-driven imaging and data analysis tools, offer a promising, cost-effective solution for field-data collection, despite challenges like battery life and processing limitations (Nabwire et al. 2021). Additionally, citizen science initiatives can significantly expand data collection efforts (Kumar et al. 2022). The ultimate goal of integrating AI into plant phenomics is to streamline data analysis, improve breeding efficiency, and track plant development throughout the growing season, accelerating the creation of climate-resilient, high-yield crops (Song et al. 2021).

16.12 AI in Gene Function Analysis

The explosion of high-throughput technologies in biological sciences has led to the generation of vast amounts of "omics" data, which is far too complex for traditional analysis methods. This has fueled the adoption of AI and ML in plant genomics, where AI excels at processing massive datasets, uncovering hidden patterns, and refining predictions over time (Jung et al. 2021; Grapov et al. 2018) ML algorithms are now widely used in bioinformatics for tasks like gene identification, protein interactions, gene expression analysis, and genomic prediction (Mahood et al. 2020; Sharma et al., 2025a,b). Advanced models such as DeepBind and DeepSEA have been developed to analyze genetic features and predict gene functions. However, the sheer volume of genomic data presents storage and analysis challenges, which AI-driven bioinformatics helps overcome by enabling large-scale gene expression analysis across various conditions (Esposito et al. 2019; Alipanahi et al. 2015; Zhou and Troyanskaya 2015; Dasari et al. 2021; Deepika et al., 2024, 2025a,b). This empowers biologists to integrate data, validate hypotheses, and refine research outcomes, paving the way for more advanced genome exploration and precision-driven breakthroughs in plant science (Xia et al. 2019; Hu et al. 2018).

16.13 Artificial Intelligence in Genomics, Phenomics, and Envirotyping Data Accessibility

Climate change is reshaping agriculture, demanding new strategies for crop breeding and land management to ensure global food security (Xu 2016). A groundbreaking approach, envirotyping, has emerged alongside phenotyping and genotyping to decode environmental influences on crops, integrating factors like genotype-by-environment interactions (GEI), stress responses, and adaptive genes (Marsh et al. 2021). Researchers are now using genomic-environment associations (GEAs) to link climate data with evolutionary genomics, helping predict how crops will adapt to changing conditions

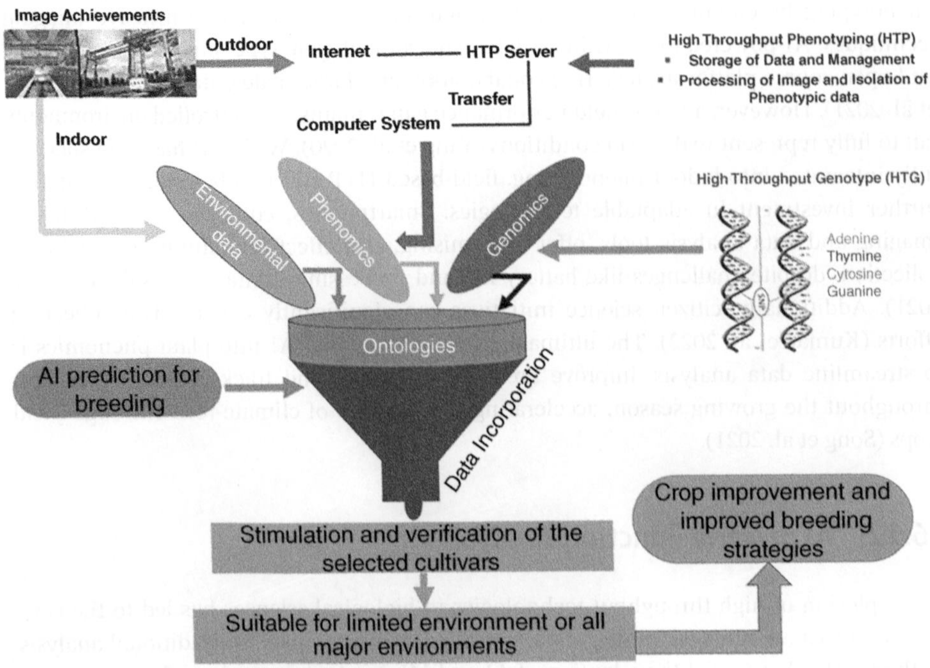

Figure 16.5 Integration and management of genomics, phenomics, and enviromics data by artificial intelligence for crop-breeding improvement. (Khan et al. 2022).

(Cortés et al. 2022). AI-driven technologies, including ML and DL, play a crucial role in processing vast datasets to identify climate-resilient traits (Shafiekhani et al. 2017). Breeding programs now leverage HTP and big data analytics to develop climate-smart crops and soils suited for evolving environmental challenges. The integration of genotypic, phenotypic, and envirotypic data through AI-powered selection models is revolutionizing crop breeding, enabling smarter, faster, and more adaptive agricultural solutions for the future (Figure 16.5).

16.14 Next-Generation (Next-Gen) Artificial Intelligence (AI) Augmented Farm

From Augmented Breeding to Smarter Farming Modern plant breeders have access to vast amounts of data, and Next-Gen AI is set to revolutionize how this data is processed to enhance sustainable agriculture. No single human can manually analyze all these complex data layers to optimize crop yield and resilience. However, breeders who possess decades of expertise and a deep understanding of genetics, climate, and biology can guide AI models by setting practical and economic constraints (Sun et al. 2015). The collaborative cycle between breeders and AI strengthens over time: as breeders integrate AI-driven insights, test suggested improvements, and feed new results back into the system, the AI refines its predictions with each growing season. This continuous learning loop makes plant breeding smarter, faster, and more precise. Similar advancements are already in use in cattle breeding, where genetic selection (GS) models evolve with real-time farmer-uploaded phenotypic data (Gao et al. 2018). With 5G networks, automation, and precision GPS technologies, plant breeding is poised to follow, bridging AI-driven insights with real-world agricultural practices for a future of intelligent and adaptive farming (Figure 16.6).

16.15 Future Prospects of AI Breeding

The rapid advancement of AI is transforming industries worldwide, and agriculture is no exception. To keep pace with the digital revolution, plant breeding must evolve, integrating AI-driven insights with real-world farming needs. AI enhances decision-making by analyzing complex datasets, identifying genetic traits, and optimizing breeding strategies. However, for AI to be truly effective, researchers and breeders must align its recommendations with farmers' practical challenges, ensuring that technological advancements lead to higher crop yields, resilience, and sustainability. While AI-driven automation and robotic technologies are increasing efficiency, a human-centered approach remains essential to ensure ethical, responsible AI applications that support rather than replace human expertise.

Despite its immense potential, AI alone cannot replace traditional breeding methods. While ML and DL models can analyze vast amounts of phenomics and genomics data, plant breeding remains a time-intensive and costly process that requires human intuition. Researchers are now shifting focus from algorithmic performance to sustainable farming models, aiming to revolutionize agriculture for both humanity and the environment. AI is expected to accelerate "omics" research and breeding management, rapidly screening genotypes and phenotypes to develop climate-resilient crops. By integrating speed-breeding techniques and fast-generation advancement, AI can shorten breeding cycles and improve genetic gains, paving the way for a smarter, more adaptive agricultural future.

16.16 Conclusion

Speed breeding is a promising solution for plant breeders to develop high-yielding cultivars that can withstand abiotic and biotic stresses. This method, which artificially controls light and temperature, accelerates plant growth while maintaining precision. When combined with molecular breeding techniques like MAS and GWAS, it can identify genes responsible for stress tolerance, nutrition, and yield improvements. The integration of Next-Gen AI enhances decision-making and data analysis. However, adoption in developing countries remains limited due to lack of trained breeders, infrastructure, and financial support. Public–private partnerships are needed to build capacity, enable technology transfer, and fund AI-driven crop improvement programs.

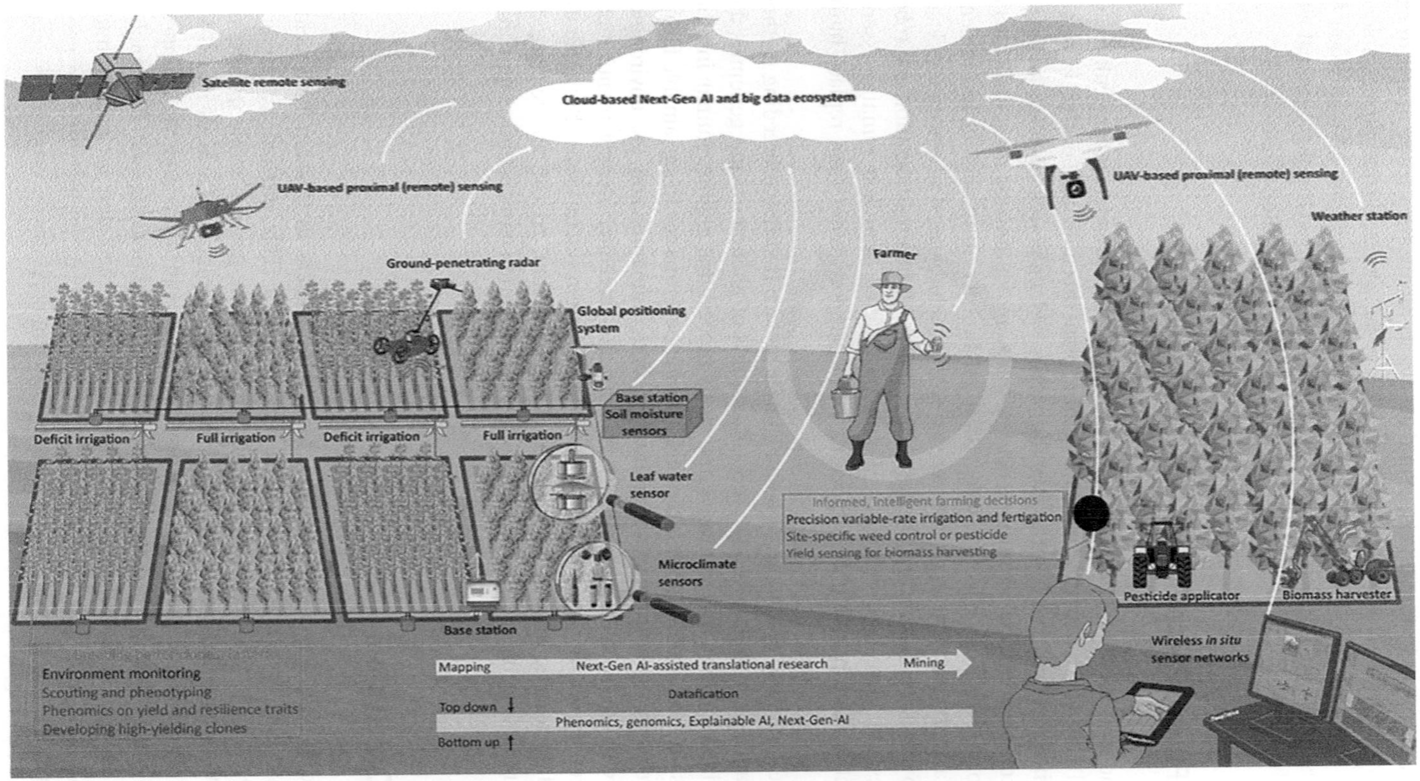

Figure 16.6 Next-generation (Next-Gen) artificial intelligence (AI)-augmented farm (Harfouche et al. 2019).

References

Ahmar, S., Gill, R.A., Jung, K.H. et al. (2020). Conventional and molecular techniques from simple breeding to speed breeding in crop plants: recent advances and future outlook. *International Journal of Molecular Sciences* 21 (7): 2590.

Alipanahi, B., Delong, A., Weirauch, M.T., and Frey, B.J. (2015). Predicting the sequence specificities of DNA-and RNA-binding proteins by deep learning. *Nature Biotechnology* 33 (8): 831–838.

Al-Khayri, J.M., Jain, S.M., and Johnson, D.V. (ed.) (2015). *Advances in Plant Breeding Strategies: Breeding, Biotechnology and Molecular Tools (Vol. 1)*. Cham: Springer International Publishing.

Ashkani, S., Rafii, M.Y., Shabanimofrad, M. et al. (2015). Molecular breeding strategy and challenges towards improvement of blast disease resistance in rice crop. *Frontiers in Plant Science* 6: 886.

Bado, S., Forster, B.P., Nielen, S. et al. (2015). Plant mutation breeding: current progress and future assessment. *Plant Breeding Reviews* 39: 23–88.

Baduge, S.K., Thilakarathna, S., Perera, J.S. et al. (2022). Artificial intelligence and smart vision for building and construction 4.0: machine and deep learning methods and applications. *Automation in Construction* 141: 1433–1450.

Berente, N., Gu, B., Recker, J., and Santhanam, R. (2021). Managing artificial intelligence. *MIS Quarterly* 45 (3).

Bhadauria, V., Ramsay, L., Bett, K.E., and Banniza, S. (2017). QTL mapping reveals genetic determinants of fungal disease resistance in the wild lentil species Lens ervoides. *Scientific Reports* 7 (1): 3231.

Bhat, J.A., Ali, S., Salgotra, R.K. et al. (2016). Genomic selection in the era of next generation sequencing for complex traits in plant breeding. *Frontiers in Genetics* 7: 221.

Bhatta, M., Morgounov, A., Belamkar, V., and Baenziger, P.S. (2018). Genome-wide association study reveals novel genomic regions for grain yield and yield-related traits in drought-stressed synthetic hexaploid wheat. *International Journal of Molecular Sciences* 19 (10): 3011.

Cazzola, F., Bermejo, C.J., Guindon, M.F., and Cointry, E. (2020). Speed breeding in pea (Pisum sativum L.), an efficient and simple system to accelerate breeding programs. *Euphytica* 216 (11): 178.

Chaudhary, J., Alisha, A., Bhatt, V. et al. (2019). Mutation breeding in tomato: advances, applicability and challenges. *Plants* 8 (5): 128.

Chiurugwi, T., Kemp, S., Powell, W., and Hickey, L.T. (2019). Speed breeding orphan crops. *Theoretical and Applied Genetics* 132: 607–616.

Cortés, A.J., López-Hernández, F., and Blair, M.W. (2022). Genome–environment associations, an innovative tool for studying heritable evolutionary adaptation in orphan crops and wild relatives. *Frontiers in Genetics* 13: 910386.

Cravero, A. and Sepúlveda, S. (2021). Use and adaptations of machine learning in big data—applications in real cases in agriculture. *Electronics* 10 (5): 552.

Crews, T.E. and Cattani, D.J. (2018). Strategies, advances, and challenges in breeding perennial grain crops. *Sustainability* 10 (7): 2192.

Crossa, J., Pérez-Rodríguez, P., Cuevas, J. et al. (2017). Genomic selection in plant breeding: methods, models, and perspectives. *Trends in Plant Science* 22 (11): 961–975.

Dasari, C.M., Amilpur, S., and Bhukya, R. (2021). Exploring variable-length features (motifs) for predicting binding sites through interpretable deep neural networks. *Engineering Applications of Artificial Intelligence* 106: 104485.

Deepika, Madhu and Upadhyay, S. K. (2024). Deciphering the features and functions of serine/arginine protein kinases in bread wheat. Plant Gene, 38, 100451. https://doi.org/10.1016/j.plgene.2024.100451.

Deepika, Madhu, Shekhawat, J. et al. (2025a). Pre-mRNA processing factor 4 kinases (PRP4Ks): Exploration of molecular features, interaction network and expression profiling in bread wheat. J Plant Growth Regul 44, 806–820. https://doi.org/10.1007/s00344-024-1148903.

Deepika, Shekhawat, J., Madhu et al. (2025b). Genome-Wide Analysis Unveiled the Characteristic Features and Functions of AFC Genes of Bread Wheat (Triticum aestivum L.). J Plant Growth Regul 44, 2466–2484. https://doi.org/10.1007/s00344-024-11559-6.

Delfina, B., Alessandro, T., Francesca, D. et al. (2016). Next generation breeding. *Plant Science* 242: 3–13.

Demirci, M., Gozde, H., and Taplamacioglu, M.C. (2021, June). Comparative dissolved gas analysis with machine learning and traditional methods. In: *In 2021 3rd International Congress on Human-Computer Interaction, Optimization and Robotic Applications (HORA)*, 1–6. IEEE.

Dormatey, R., Sun, C., Ali, K. et al. (2020). Gene pyramiding for sustainable crop improvement against biotic and abiotic stresses. *Agronomy* 10 (9): 1255. https://doi.org/10.3390/agronomy1009125.

El Bilali, H. and Allahyari, M.S. (2018). Transition towards sustainability in agriculture and food systems: role of information and communication technologies. *Information processing in agriculture* 5 (4): 456–464.

Esposito, S., Carputo, D., Cardi, T., and Tripodi, P. (2019). Applications and trends of machine learning in genomics and phenomics for next-generation breeding. *Plants* 9 (1): 34.

Faulkner, A., Cebul, K., and McHenry, G. (2014). Agriculture gets smart: the rise of data and robotics. *Cleantech Agriculture Report* 1.

Gao, X., Zhang, J., Wei, Z., and Hakonarson, H. (2018). DeepPolyA: a convolutional neural network approach for polyadenylation site prediction. *IEEE Access* 6: 24340–24349.

Ghosh, S., Watson, A., Gonzalez-Navarro, O.E. et al. (2018). Speed breeding in growth chambers and glasshouses for crop breeding and model plant research. *Nature Protocols* 13 (12): 2944–2963.

Godwin, I.D., Rutkoski, J., Varshney, R.K., and Hickey, L.T. (2019). Technological perspectives for plant breeding. *Theoretical and Applied Genetics* 132 (3): 555–557.

González-Barrios, P., Bhatta, M., Halley, M. et al. (2021). Speed breeding and early panicle harvest accelerates oat (Avena sativa L.) breeding cycles. *Crop Science* 61 (1): 320–330.

Grapov, D., Fahrmann, J., Wanichthanarak, K., and Khoomrung, S. (2018). Rise of deep learning for genomic, proteomic, and metabolomic data integration in precision medicine. *OMICS: A Journal of Integrative Biology* 22 (10): 630–636.

Gupta, P.K., Kumar, J., Mir, R.R., and Kumar, A. (2010). Marker-assisted selection as a component of conventional plant breeding. *Plant Breeding Reviews* 33 (4): 145–217. https://doi.org/10.1007/s11032-009-9359-7.

Harfouche, A.L., Jacobson, D.A., Kainer, D. et al. (2019). Accelerating climate resilient plant breeding by applying next-generation artificial intelligence. *Trends in Biotechnology* 37 (11): 1217–1235.

Hasegawa, T., Fujimori, S., Havlík, P. et al. (2018). Risk of increased food insecurity under stringent global climate change mitigation policy. *Nature Climate Change* 8 (8): 699–703.

Hickey, L.T., Hafeez, A., Robinson, H. et al. (2019). Breeding crops to feed 10 billion. *Nature Biotechnology* 37 (7): 744–754.

Hu, H., Scheben, A., and Edwards, D. (2018). Advances in integrating genomics and bioinformatics in the plant breeding pipeline. *Agriculture* 8 (6): 75.

Huntingford, C., Jeffers, E.S., Bonsall, M.B. et al. (2019). Machine learning and artificial intelligence to aid climate change research and preparedness. *Environmental Research Letters* 14 (12): 124007.

Idrissi, O. (2020). Application of extended photoperiod in lentil: towards accelerated genetic gain in breeding for rapid improved variety development. *Moroccan Journal of Agricultural Sciences* 1 (1): 30–3514-19.

Jähne, F., Hahn, V., Würschum, T., and Leiser, W.L. (2020). Speed breeding short-day crops by LED-controlled light schemes. *Theoretical and Applied Genetics* 133 (8): 2335–2342.

Jarrahi, M.H. (2018). Artificial intelligence and the future of work: human-AI symbiosis in organizational decision making. *Business Horizons* 61 (4): 577–586.

Jeong, J.H., Resop, J.P., Mueller, N.D. et al. (2016). Random forests for global and regional crop yield predictions. *PloS One* 11 (6): e0156571.

Jung, J., Maeda, M., Chang, A. et al. (2021). The potential of remote sensing and artificial intelligence as tools to improve the resilience of agriculture production systems. *Current Opinion in Biotechnology* 70: 15–22.

Khahani, B., Tavakol, E., Shariati, V., and Rossini, L. (2021). Meta-QTL and ortho-MQTL analyses identified genomic regions controlling rice yield, yield-related traits and root architecture under water deficit conditions. *Scientific Reports* 11 (1): 6942.

Khan, M.H.U., Wang, S., Wang, J. et al. (2022). Applications of artificial intelligence in climate-resilient smart-crop breeding. *International Journal of Molecular Sciences* 23 (19): 11156.

Kok, J.N., Boers, E.J., Kosters, W.A. et al. (2009). Artificial intelligence: definition, trends, techniques, and cases. *Artificial Intelligence* 1 (270-299): 51.

Kumar, R.S., Keerthana, G., Murali, L. et al. (2022). Enabling artificial intelligence and cyber security in smart manufacturing. *Artificial Intelligent Techniques for Wireless Communication and Networking* 269–286.

Kumar, A., Shumayla, and Upadhyay, S. K. (2023). CRISPR/Cas-Mediated Genome Editing in Plants: A Historical Perspective. Applications of Genome Engineering in Plants; Wiley; Portico. https://doi.org/10.1002/9781394183913.ch1.

Kushwah, A., Gupta, S., Bindra, S. et al. (2020). Gene pyramiding and multiple character breeding. In: *Chickpea: Crop Wild Relatives for Enhancing Genetic Gains*, 131–165. Academic Press.

Leal Filho, W., Wall, T., Mucova, S.A.R. et al. (2022). Deploying artificial intelligence for climate change adaptation. *Technological Forecasting and Social Change* 180: 121662.

Li, D., Wang, R., Xie, C. et al. (2020). A recognition method for rice plant diseases and pests video detection based on deep convolutional neural network. *Sensors* 20 (3): 578.

Li, Y., Zhou, X., and Li, G. (2021). *Bridging natural language and graphical user Interfaces*, 463–493. Artificial Intelligence for Human Computer Interaction: *A Modern Approach*.

Mahood, E.H., Kruse, L.H., and Moghe, G.D. (2020). Machine learning: a powerful tool for gene function prediction in plants. *Applications in Plant Sciences* 8 (7): e11376.

Marsh, J.I., Hu, H., Gill, M. et al. (2021). Crop breeding for a changing climate: integrating phenomics and genomics with bioinformatics. *Theoretical and Applied Genetics* 134: 1677–1690.

Meena, M., Shubham, S., Paritosh, K. et al. (2021). Production of biofuels from biomass: predicting the energy employing artificial intelligence modelling. *Bioresource Technology* 340: 125642.

Mobini, S., Khazaei, H., Warkentin, T.D., and Vandenberg, A. (2020). Shortening the generation cycle in faba bean (Vicia faba) by application of cytokinin and cold stress to assist speed breeding. *Plant Breeding* 139 (6): 1181–1189.

Montesinos-López, O.A., Martín-Vallejo, J., Crossa, J. et al. (2019). A benchmarking between deep learning, support vector machine and Bayesian threshold best linear unbiased prediction for predicting ordinal traits in plant breeding. *G3: Genes, Genomes Genetics* 9 (2): 601–618.

Montesinos-López, O.A., Montesinos-López, A., Crossa, J. et al. (2018). Multi-trait, multi-environment deep learning modeling for genomic-enabled prediction of plant traits. *G3: Genes, Genomes, Genetics* 8 (12): 3829–3840.

Nabwire, S., Suh, H.K., Kim, M.S. et al. (2021). Application of artificial intelligence in phenomics. *Sensors* 21 (13): 4363.

Niazian, M. and Niedbała, G. (2020). Machine learning for plant breeding and biotechnology. *Agriculture* 10 (10): 436.

Nzuki, I., Katari, M.S., Bredeson, J.V. et al. (2017). QTL mapping for pest and disease resistance in cassava and coincidence of some QTL with introgression regions derived from manihot glaziovii. *Frontiers in Plant Science* 8: 1168.

Parmley, K.A., Higgins, R.H., Ganapathysubramanian, B. et al. (2019). Machine learning approach for prescriptive plant breeding. *Scientific Reports* 9 (1): 17132.

Priya, R. and Ramesh, D. (2020). ML based sustainable precision agriculture: a future generation perspective. *Sustainable Computing: Informatics and Systems* 28: 100439.

Qaim, M. (2020). Role of new plant breeding technologies for food security and sustainable agricultural development. *Applied Economic Perspectives and Policy* 42 (2): 129–150.

Rai, K.K. (2022). Integrating speed breeding with artificial intelligence for developing climate-smart crops. *Molecular Biology Reports* 49 (12): 11385–11402.

Raina, A., Laskar, R.A., Khursheed, S. et al. (2016). Role of mutation breeding in crop improvement-past, present and future. *Asian Research Journal of Agriculture* 2 (2): 1–13.

Rajasundaram, D. and Selbig, J. (2016). More effort—more results: recent advances in integrative 'omics' data analysis. *Current Opinion in Plant Biology* 30: 57–61.

Raschka, S., Patterson, J., and Nolet, C. (2020). Machine learning in python: main developments and technology trends in data science, machine learning, and artificial intelligence. *Information* 11 (4): 193.

Ray, D.K., West, P.C., Clark, M. et al. (2019). Climate change has likely already affected global food production. *PloS one* 14 (5): e0217148.

Razzaq, A., Kaur, P., Akhter, N. et al. (2021). Next-generation breeding strategies for climate-ready crops. *Frontiers in Plant Science* 12: 620420.

Reinoso-Peláez, E.L., Gianola, D., and González-Recio, O. (2022). Genome-enabled prediction methods based on machine learning. In: *Genomic Prediction of Complex Traits: Methods and Protocols*, 189–218. New York, NY: Springer US.

Samineni, S., Sen, M., Sajja, S.B., and Gaur, P.M. (2020). Rapid generation advance (RGA) in chickpea to produce up to seven generations per year and enable speed breeding. *The Crop Journal* 8 (1): 164–169.

Sartor, R.C., Noshay, J., Springer, N.M., and Briggs, S.P. (2019). Identification of the expressome by machine learning on omics data. *Proceedings of the National Academy of Sciences* 116 (36): 18119–18125.

Saxena, K.B., Saxena, R.K., Hickey, L.T., and Varshney, R.K. (2019). Can a speed breeding approach accelerate genetic gain in pigeonpea? *Euphytica* 215: 1–7.

Shafiekhani, A., Kadam, S., Fritschi, F.B., and DeSouza, G.N. (2017). Vinobot and vinoculer: two robotic platforms for high-throughput field phenotyping. *Sensors* 17 (1): 214.

Sharma, Y., Dixit, S., Singh, K. et al. (2025a). Vascular plant one-zinc finger transcription factors: exploration of characteristic features, expression, coexpression and interaction suggested their diverse role in bread wheat (Triticum aestivum L.). Plant Growth Regul 105, 71–87. https://doi.org/10.1007/s10725-024-01264-5.

Sharma, Y., Deepika, Kaur, A., and Upadhyay, S. K. (2025b). Exploration of mitochondrial calcium uniporters suggested their role in abiotic stress response in bread wheat. Plant Physiology and Biochemistry, 227, 110180. https://doi.org/10.1016/j.plaphy.2025.110180.

Shen, Y., Zhou, G., Liang, C., and Tian, Z. (2022). Omics-based interdisciplinarity is accelerating plant breeding. *Current Opinion in Plant Biology* 66: 102167.

Singh, S., Mishra, V., Riyazuddin, R. et al. (2024). CRISPR/ Cas9-Mediated Genome Editing for Trait Improvement and Stress Tolerance in Leguminosae (Legume Family). Plant Breeding. Portico. https://doi.org/10.1111/pbr.13234.

Song, P., Wang, J., Guo, X. et al. (2021). High-throughput phenotyping: breaking through the bottleneck in future crop breeding. *The Crop Journal* 9 (3): 633–645.

Soriano, J.M., Colasuonno, P., Marcotuli, I., and Gadaleta, A. (2021). Meta-QTL analysis and identification of candidate genes for quality, abiotic and biotic stress in durum wheat. *Scientific Reports* 11 (1): 11877.

Stergiou, C. and Psannis, K.E. (2017). Recent advances delivered by mobile cloud computing and internet of things for big data applications: a survey. *International Journal of Network Management* 27 (3): e1930.

Streich, J., Romero, J., Gazolla, J.G.F.M. et al. (2020). Can exascale computing and explainable artificial intelligence applied to plant biology deliver on the United Nations sustainable development goals? *Current Opinion in Biotechnology* 61: 217–225.

Sun, L., Liu, H., Zhang, L., and Meng, J. (2015). lncRScan-SVM: a tool for predicting long non-coding RNAs using support vector machine. *PloS One* 10 (10): e0139654.

Supriya, M. and Deepa, A.J. (2020). Machine learning approach on healthcare big data: a review. *Big Data and Information Analytics* 5 (1): 58–75.

Sushmita, Kaur, G., Upadhyay, S. K., and Verma, P. C. (2021). An Overview of Genome-Engineering Methods. Genome Engineering for Crop Improvement; Wiley; Portico. https://doi.org/10.1002/9781119672425.ch1.

Talaviya, T., Shah, D., Patel, N. et al. (2020). Implementation of artificial intelligence in agriculture for optimisation of irrigation and application of pesticides and herbicides. *Artificial Intelligence in Agriculture* 4: 58–73.

Thudi, M., Palakurthi, R., Schnable, J.C. et al. (2021). Genomic resources in plant breeding for sustainable agriculture. *Journal of Plant Physiology* 257: 153351.

Ubbens, J., Cieslak, M., Prusinkiewicz, P., and Stavness, I. (2018). The use of plant models in deep learning: an application to leaf counting in rosette plants. *Plant Methods* 14: 1–10.

Ukhurebor, K.E., Singh, K.R., Nayak, V., and Gladys, U.E. (2021). Influence of the SARS-CoV-2 pandemic: a review from the climate change perspective. *Environmental Science: Processes & Impacts* 23 (8): 1060–1078.

Upadhyay, S. K. (2021). Genome Engineering for *Crop Improvement.* Wiley. https://doi.org/10.1002/9781119672425.

Upadhyay, S. K. (2024). *Applications of Genome Engineering in Plants.* Wiley. https://doi.org/10.1002/9781394183913.

Uygun, S., Azodi, C.B., and Shiu, S.H. (2019). Cis-regulatory code for predicting plant cell-type transcriptional response to high salinity. *Plant Physiology* 181 (4): 1739–1751.

Uzal, L.C., Grinblat, G.L., Namías, R. et al. (2018). Seed-per-pod estimation for plant breeding using deep learning. *Computers and Electronics in Agriculture* 150: 196–204.

Valerio Giuffrida, M., Scharr, H., and Tsaftaris, S. A. (2017). Arigan: synthetic arabidopsis plants using generative adversarial network. In Proceedings of the IEEE international conference on computer vision workshops (pp. 2064–2071).

van Dijk, A.D.J., Kootstra, G., Kruijer, W., and de Ridder, D. (2021). Machine learning in plant science and plant breeding. *Iscience* 24 (1): 2064–2071.

Varala, K., Marshall-Colón, A., Cirrone, J. et al. (2018). Temporal transcriptional logic of dynamic regulatory networks underlying nitrogen signaling and use in plants. *Proceedings of the National Academy of Sciences* 115 (25): 6494–6499.

Varshney, R.K., Bohra, A., Roorkiwal, M. et al. (2021a). Fast-forward breeding for a food-secure world. *Trends in Genetics* 37 (12): 1124–1136.

Varshney, R.K., Bohra, A., Yu, J. et al. (2021b). Designing future crops: genomics-assisted breeding comes of age. *Trends in Plant Science* 26 (6): 631–649.

Varshney, R.K., Pandey, M.K., Bohra, A. et al. (2019). Toward the sequence-based breeding in legumes in the post-genome sequencing era. *Theoretical and Applied Genetics* 132 (3): 797–816.

Wang, Y., Xu, J., Deng, D. et al. (2016). A comprehensive meta-analysis of plant morphology, yield, stay-green, and virus disease resistance QTL in maize (Zea mays L.). *Planta* 243: 459–471.

Watson, A., Ghosh, S., Williams, M.J. et al. (2018). Speed breeding is a powerful tool to accelerate crop research and breeding. *Nature Plants* 4 (1): 23–29.

Wolter, F., Schindele, P., and Puchta, H. (2019). Plant breeding at the speed of light: the power of CRISPR/Cas to generate directed genetic diversity at multiple sites. *BMC Plant Biology* 19 (1): 176.

Xia, E.H., Li, F.D., Tong, W. et al. (2019). Tea plant information archive: a comprehensive genomics and bioinformatics platform for tea plant. *Plant Biotechnology Journal* 17 (10): 1938–1953.

Xu, Y. (2016). Envirotyping for deciphering environmental impacts on crop plants. *Theoretical and Applied Genetics* 129: 653–673.

Xu, Y., Liu, X., Cao, X. et al. (2021). Artificial intelligence: a powerful paradigm for scientific research. *The Innovation* 2 (4).

Yang, W., Feng, H., Zhang, X. et al. (2020). Crop phenomics and high-throughput phenotyping: past decades, current challenges, and future perspectives. *Molecular Plant* 13 (2): 187–214.

Zhou, J. and Troyanskaya, O.G. (2015). Predicting effects of noncoding variants with deep learning–based sequence model. *Nature Methods* 12 (10): 931–934.

17

Artificial Intelligence (AI)-Based Strategies for Plant Health and Crop Safety

Muhammad Nauman[1,2], Nadeem Iqbal[1,2], Dil Khurram[2,3], Hafiz Muhammad Ansab Jamil[4], Moniba Zahid Mahmood[5], Kalpita Singh[6], and Riyazuddin Riyazuddin[7]

[1]*Chengdu University of Technology, State Key Laboratory of Geohazard Prevention and Geoenvironment Protection, Dongsanlu, Erxianqiao, Chengdu, 610059, Sichuan, P.R. China*
[2]*Chengdu University of Technology, College of Ecology and Environment, Chengdu, 610059, Sichuan, P.R. China*
[3]*Chengdu University of Technology, State Environmental Protection Key Laboratory of Synergetic Control and Joint Remediation & Water Pollution, College of Ecology and Environment, Chengdu, 610059, P.R. China*
[4]*COMSATS University Islamabad, Department of Biotechnology, Abbottabad Campus, 22060, Abbottabad, Pakistan*
[5]*COMSATS University Islamabad, Department of Environmental Sciences, Abbottabad Campus, 22060, Abbottabad, Pakistan*
[6]*HUN-REN Centre for Agricultural Research, Agricultural Institute, Martonvásár, 2462, Hungary*
[7]*Universidad Miguel Hernández, Instituto de Bioingeniería, 03202 Elche, Spain*

17.1 Introduction to AI in Agriculture

17.1.1 The Value of Pest and Disease Control in Agriculture

Agriculture serves as the foundation of human survival, playing a crucial role in supplying primary sources of food, feed, and raw materials for diverse industries. This critical sector, despite its significance, encounters numerous challenges primarily due to pests and diseases, which threaten the global food supply. The Food and Agriculture Organization (FAO, 2021) reports an annual loss of global crop yield ranging from 20% to 40% attributable to pests and diseases. These losses result in increased financial burdens and intensify food insecurity, especially in areas with limited resources. These figures indicate the necessity for interventions. These include emerging nations' economic stability and means of subsistence, the majority of which depend on agriculture. A further exacerbating aspect is the limited access to advanced technology in pest management. According to Pretty et al. (2023), outbreaks of locusts and fall armyworms, increasingly associated with climate change and altered ecosystems, result in smallholder farmers in sub-Saharan Africa losing up to 50% of their harvests. In addition to economic losses, a significant worry is the environmental deterioration resulting from ineffective pest control methods. Chemical pesticides are employed indiscriminately, resulting in the mortality of non-target species, including essential pollinators, and degrading soil and aquatic ecosystems (Adve et al. 2024). These solutions mitigate environmental degradation while preserving agricultural output by combining mechanical, biological, and cultural controls with the prudent application of chemicals (Savary et al. 2019; Pretty et al. 2023). Advanced technologies, including biotechnology and precision agriculture, have demonstrated considerable promise.

Precision agriculture employs data-driven technology, such as satellite images and Internet of Things sensors, to monitor crop health and improve pest control. Likewise, pest-resistant GM crops have been embraced in nations like the United States and India. There has been a significant reduction in both the number of pesticides used and the amount of yield losses (Pathirana and Carimi, 2024). The sole method to mitigate the effects of pests and illnesses on agriculture is through technical innovation, sustainable practices, and supportive legislation. Major investments are required to safeguard global food systems by fostering environmental sustainability and ensuring the stability of millions of people's livelihoods dependent on agriculture. These expenditures should prioritize research into agricultural education and the supply of resources for farmers in underdeveloped nations. Limitations of Traditional Methods Used for Control of Pest and Disease Problems Chemical insecticides and manual scouting were the two mainstays of traditional pest and disease management techniques (Hayat et al. 2019). Despite extensive adoption of them even now, these methods face harsh restrictions in modern agricultural fields. The excessive use of chemical pesticides has resulted in the development of pesticide-resistant insect populations, which gradually reduces the efficacy of the medications utilized (Mukhtar et al. 2022). The application of this practice has raised significant environmental issues, including soil erosion, water pollution, and a decline in biodiversity, particularly affecting crucial species like pollinators and predators (Hladik et al. 2018; Silva et al. 2023). For example, chronic conditions such as cancer and respiratory diseases pose significant health threats to individuals using this pesticide and the surrounding community (WHO, 2023). Manual scouting does not provide any advantages for monitoring the outbreak of pests and diseases. It is also prone to various issues due to its reliance on a greater number of human resources and its dependence on the availability and skill level of individuals.

Inaccurate estimates are always possible due to human error and limited diagnostic skills, especially when the disease is discovered early on or the pest infestation has already begun (Pinter et al. 2024). Delays in interventions, which typically occur during the initial stages, can lead to significant losses. Due to these constraints, the agricultural sector has increasingly turned to advanced technologies like AI to innovate pest and disease management strategies. The tools that utilize artificial intelligence (AI) depend on advanced machine learning (ML) algorithms, computer vision, and data analytics to accurately detect, diagnose, and predict outbreaks with unprecedented efficiency. AI-driven robotic systems and drones have the capability to survey extensive fields in real time, identifying subtle signs of infection or stress that might be overlooked by a human observer (Sahoo et al. 2024). Similarly, mobile applications powered by AI enable farmers to submit images of affected crops for prompt assessments and tailored management recommendations. AI plays a significant role in precision pest management beyond just detection. Predictive modeling tools enable farmers to strategically time their interventions by forecasting pest trends based on historical and real-time data inputs (Kaur et al. 2022).

Alongside decreasing reliance on chemical pesticides, it also minimizes labor, costs, and environmental effects (Jha et al. 2024). The implementation of AI in managing diseases and pests has demonstrated encouraging outcomes. According to ICRISAT (2023), cotton farmers in India managed to enhance their yields while simultaneously decreasing pesticide usage by 30% through the application of AI techniques. The spread of invasive species such as the spotted lanternfly has been notably curtailed in the United States thanks to AI-driven platforms, as reported by USDA (2024).

17.1.2 Rise of Artificial Intelligence Technologies in Agriculture

To create a future of productive, resilient, and sustainable farming, this is a revolutionary change in the way we handle disease and pests in agriculture. Extensive datasets drive these AI technologies to provide actionable insights that assist farmers in utilizing tools for early detection, real-time monitoring, and predictive analysis. This analytical approach is transforming conventional methods, substituting uncertainty with accuracy, and enabling farmers to make knowledgeable choices for improved productivity and sustainability. The application of AI significantly enhances the early detection and diagnosis of health issues in plants. Cutting-edge technologies powered by AI encompass drone imaging systems and mobile applications capable of detecting subtle indicators of stress, nutrient deficiencies, or pest infestations that are not easily visible to the human eye. An AI-based spectral analysis identifies variations in leaf color, indicating potential disease, thereby enabling farmers to prepare for its spread before it escalates. Recently, Sahoo et al. (2024) illustrated the effectiveness of AI technologies in decreasing crop losses by as much as 40% in pilot projects throughout Asia and Africa, leading to a notable increase in yields.

AI excels at real-time monitoring. Through Internet of Things (IoT) devices and sensor networks, AI systems can continually monitor environmental factors like soil moisture, temperature, and humidity that affect pest behavior and disease outbreaks. AI in agriculture enhances productivity, optimizes resource use, and supports sustainable farming practices (Figure 17.1). This delivers prompt notifications to farmers, thereby decreasing response times and mitigating harm. For example, utilizing AI to oversee pest populations in Brazilian soybean farms reduced pesticide usage by 25% without compromising productivity (Jha et al. 2024).

Predictive analytics amplifies the value of AI by furthering its usage in agriculture. Analyzing data from the past and through real-time, AI models can predict pest emergence trends and disease outbreaks with very high accuracy levels. Such insights will help implement targeted interventions on resource utilization and minimize environmental impacts. For instance, in India, predictive models have assisted wheat farmers in the preparation and management of rust outbreaks, a perennial threat that has been exacerbated by climate change (ICRISAT, 2023). The infusion of AI into agriculture also resonates with the overall sustainability objectives. Precision agriculture avoids the wasteful application of inputs such as water, fertilizers, and pesticides during agricultural activities, thus leading to the reduction of waste generation and ecological degradation. This saves the farmer expenses related to operations, and it becomes environmentally friendly. Europe has reduced greenhouse gas emissions from agriculture by 20% since the introduction of AI-based farming in 2020 (Pretty et al. 2023). This, therefore, implies that the adoption of AI in agriculture is a game-changer as it presents new innovative solutions to solving the biggest

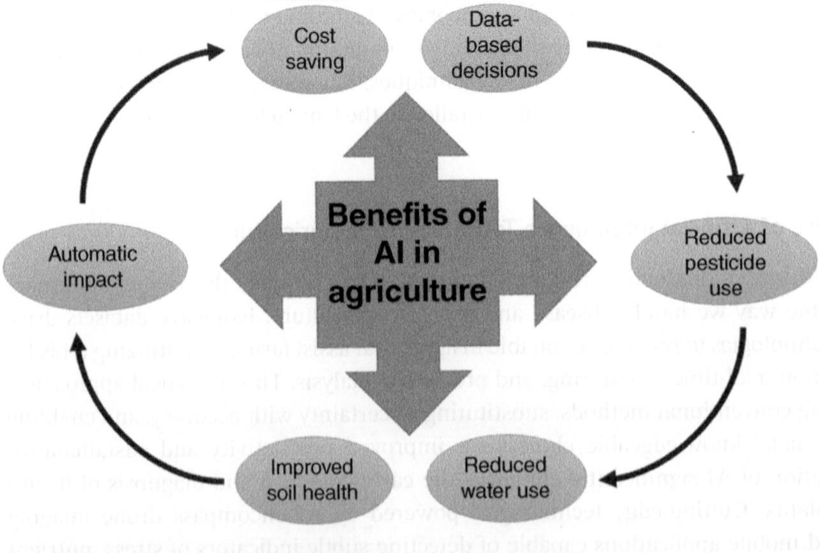

Figure 17.1 Benefits of AI in agriculture.

problems in food systems. It enables fostering precision, sustainability, and resilience so that farmers can manage these pressures in climate and population changes for food security across generations. The ongoing advancements in AI technology, coupled with supportive policies and infrastructure, will undoubtedly solidify its role as a cornerstone of modern agriculture (Singh and Kaur 2022).

17.1.3 Overview of Artificial Intelligence in Agriculture

AI denotes the emulation of human cognitive functions by machines, including learning, reasoning, and self-correction (Russell & Norvig, 2021). AI is a revolutionary influence on agriculture, altering the methods of food cultivation, management, and distribution. AI applications enhance agriculture by improving efficiency, sustainability, and resilience, from precision farming to supply chain optimization. Figure 17.2 illustrates a schematic overview of the benefits of AI in agriculture. AI, utilizing powerful algorithms to analyze extensive data sets, provides insights that enhance decision-making and operations, establishing it as a fundamental element of contemporary agribusiness (Wolfert et al. 2017).

17.1.4 Artificial Intelligence in Precision Agriculture

It is the most extensively utilized domain of AI, and by utilizing technology such as computer vision, drones, and Internet of Things sensors, precision agriculture assesses crop health, soil conditions, and weather patterns. It enables agriculture to utilize more precise inputs, including water, fertilizers, and pesticides, reduce waste, and optimize production. AI-driven irrigation systems can autonomously regulate water flow based on real-time soil moisture, thereby minimizing resource wastage and promoting healthier crops. Pretty et al. (2023) report that precision agriculture technology has decreased resource use by as much as 30% while enhancing production on European farms.

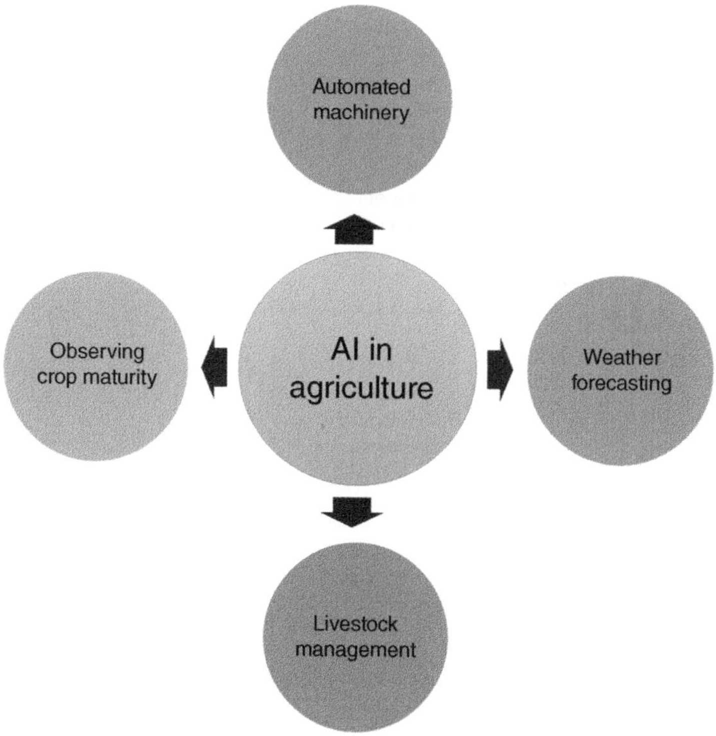

Figure 17.2 Reimbursements of AI in agriculture

17.1.5 Disease and Pest Management

Detect and manage the occurrence of diseases and pests among plants with AI. They can have images of their crops assessed for infections. These machines, using ML algorithms, detect infections quite early even before symptoms come into notice through human visual detection. Besides bad agricultural practices, growing soil salinity is also an important factor that causes low yield (Nauman et al. 2024; Iqbal et al. 2022). For instance, mobile applications with AI-powered diagnostic tools allow farmers to upload pictures of affected plants and receive immediate recommendations on treatment. In India, these tools have greatly reduced crop losses, especially in smallholder farms, where timely intervention is crucial (ICRISAT, 2023). AI facilitates predictive analytics, enabling farmers to forecast insect outbreaks using environmental and historical data. This has been crucial in combating invasive species like the fall armyworm, which has devastated maize fields throughout Africa. A 2024 study by Jha et al. demonstrated that AI-driven forecasting models decreased armyworm infestations by 40% in pilot areas.

17.1.6 Yield Optimization

AI systems scan through history and present data to improve planting dates, crop sequences, and harvesting dates that yield maximum crops. Some examples of AI platforms using satellite imagery and forecasts of weather patterns to make recommendations for the optimal

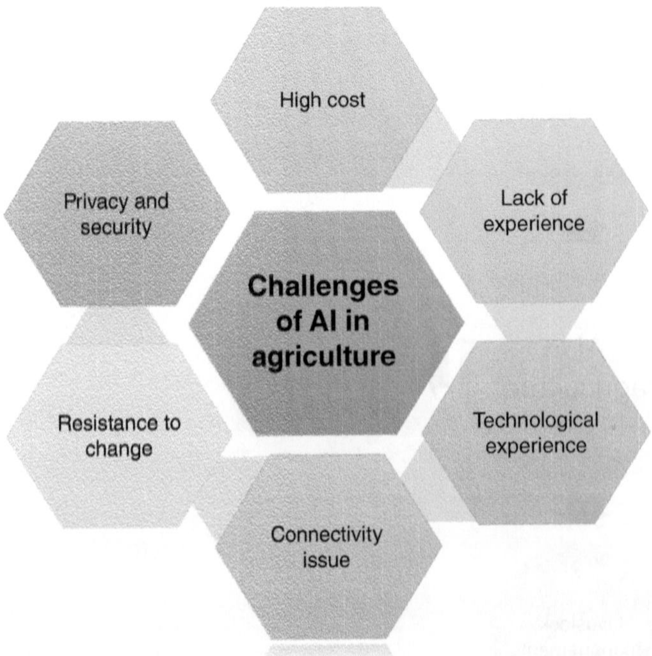

Figure 17.3 Challenges of AI in agriculture

times to plant will reduce climate variability risk for farmers. AI-assisted yield optimization tools increased corn production efficiency in the United States by 15% and reduced inputs (USDA, 2024).

17.1.7 Supply Chain Efficiency

AI improves supply chain efficiency by cutting down on food waste and simplifying logistics, which has applications beyond the farm. Algorithms forecast demand patterns, enhance transportation routes, and refine storage conditions, resulting in the expedited delivery of fresh products to consumers. A study by Kamilaris et al. in 2023 indicated that AI-driven supply chain solutions decreased post-harvest losses by 25% in Southeast Asia, where distribution network inefficiencies frequently result in significant waste. A schematic representation of the challenges faced by AI in agriculture is presented in Figure 17.3.

17.1.8 Policy and Global Implications

AI influences worldwide agricultural policies. The use of AI allows policymakers to tackle food security and sustainability issues by providing data-driven insights about crop production, trade patterns, and resource use. The FAO's AI program monitors global agricultural conditions, offering early warnings regarding potential food emergencies stemming from climate events or geopolitical tensions (FAO, 2023). Multiple fundamental technologies are collaboratively addressing difficulties and improving production in agriculture through the integration of AI into farming methods. These technologies transform traditional farming practices by enhancing precision, efficiency, and sustainability. Presented below are many significant innovations powered by AI that are advancing the agricultural revolution.

17.2 Machine Learning (ML)

ML constitutes the foundation of AI in agriculture, facilitating systems to assess historical and real-time data for trend prediction and decision-making. ML algorithms use meteorological trends, soil conditions, and pest activity to forecast pest and disease outbreaks. In India, ML algorithms have been utilized to forecast wheat rust outbreaks, enabling farmers to apply timely preventative measures, thereby mitigating up to 30% of production loss (Jha et al. 2024). The identical system in Brazil has been enhanced. ML constitutes the foundation of AI in agriculture, facilitating systems to assess historical and real-time data for trend prediction and decision-making. ML algorithms use meteorological trends, soil conditions, and pest activity to forecast pest and disease outbreaks. In India, ML algorithms have been utilized to forecast wheat rust outbreaks, enabling farmers to apply timely preventative measures, thereby mitigating up to 30% of production loss (Jha et al. 2024). The identical system in Brazil has enhanced soybean farming by identifying the region having pest attacks and the places where intervention should be concentrated (Silva et al. 2023).

17.2.1 Deep Learning (DL)

Deep learning (DL), an aspect of ML, excels in digesting intricate datasets and discerning complex patterns. In agriculture, DL models are utilized to differentiate symptoms induced by several pathogens that frequently exhibit similarities (Kamilaris and Prenafeta-Boldú 2018). In 2023, it was shown that DL algorithms may be utilized for the differential diagnosis of bacterial leaf blight and fungal leaf spot in rice, hence more accurately recommending treatments (Liakos et al. 2018). DL is utilized in cattle husbandry to monitor animal behavior, enabling the detection of early indicators of sickness to enhance herd management (Kowalska and Ashraf 2023).

17.2.2 Internet of Things

The Internet of Things is a network of interconnected sensors and gadgets that deliver real-time data regarding environmental and crop conditions. The Internet of Things facilitates the monitoring of soil moisture, temperature, humic content, and nutrient levels using sensors. All these elements are essential for input into AI to facilitate precise predictions. A smart irrigation system optimized for the Internet of Things would employ real-time soil data to dispense water accurately and at optimal times. This conserves up to 40% of water consumption in arid regions like California and Australia (Wolfert et al. 2017). The Internet of Things facilitates remote agricultural management, enabling farmers to oversee their operations from any location using mobile devices.

17.2.3 Natural Language Processing (NLP)

NLP is now being used more and more in agriculture, despite its lack of recognition, to give farmers access to AI capabilities. Farmers can interact with AI systems in their native languages via NLP-based chatbots and voice assistants. They can receive guidance on crop management, meteorological information, and pest control. In Kenya, an AI-driven chatbot facilitated the adoption of sustainable practices among farmers, leading to a 20% increase in yields over two years (FAO 2024).

17.2.4 Synergy of Technologies

These fundamental AI technologies frequently collaborate to deliver comprehensive solutions. A synergy between the Internet of Things and ML can enhance greenhouse conditions by forecasting temperature and humidity patterns and autonomously regulating ventilation and irrigation systems. Likewise, computer vision and DL improve pest and disease diagnostics by amalgamating image data with predictive analytics, guaranteeing accurate and prompt responses.

17.3 Role of Artificial Intelligence in Predictive Analytics and Real-Time Monitoring

Modern agriculture has experienced the rise of predictive analytics and real-time monitoring as two of the most revolutionary AI applications, providing farmers with novel ways to increase sustainability and productivity. These tools enable a farmer to make proactive, data-driven decisions, thereby optimizing resource utilization and executing timely remedial measures in anticipation of future dangers. Predictive analytics leverages historical data, meteorological patterns, and agricultural models to forecast disease outbreaks, insect invasions, and optimal planting periods. The role of AI in predictive analytics and real-time monitoring is shown in Figure 17.4. AI-driven predictive solutions provide farmers with actionable insights customized to their specific geographies and crops through the examination of extensive datasets. Research conducted by Jha et al. in 2024 revealed that predictive models significantly aided wheat farmers in South Asia,

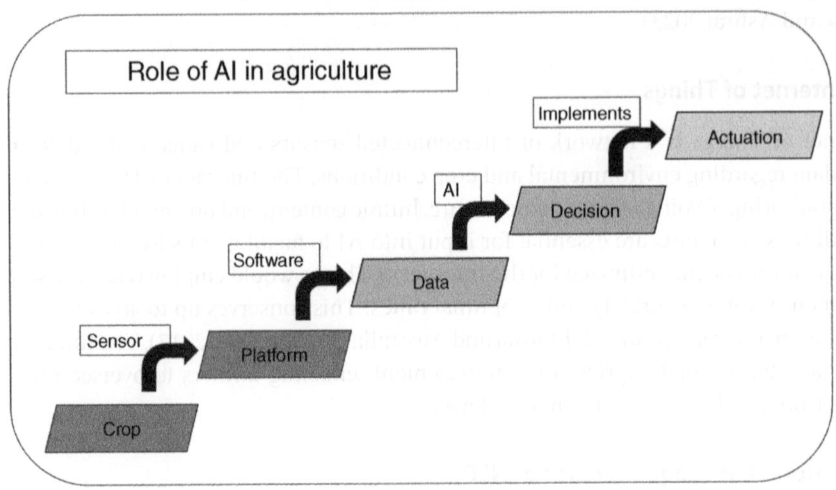

Figure 17.4 Role of AI in agriculture.

reducing losses from rust illnesses by 35% by enabling timely preventive actions before pathogen dissemination.

Another forecast from predictive analytics will anticipate the impacts of climatic variability, including extreme weather conditions, on crops. AI-based algorithms forecasted the effects of prolonged drought on almond production in California, prompting farmers to implement water-conserving irrigation methods, such as drip irrigation (Pretty et al. 2023). These technologies are an invaluable resource for pest control, as they can predict the life-cycle stages of the fall armyworm, allowing farmers to target its vulnerable phases and minimize pesticide usage.

17.3.1 Internet of Things and Drones for Real-Time Monitoring

Farmers can promptly identify and mitigate emergent threats through real-time monitoring, which is facilitated by drones and Internet of Things (IoT) devices. Internet of Things networks utilize sensors throughout agricultural fields to continuously monitor soil moisture, temperature, humidity, and nutrient levels (Demirel and Kumral 2021). These sensors transmit real-time data to AI platforms, which analyze the information and offer recommendations. In Israel, Internet of Things-based irrigation devices have decreased water consumption by 30% by providing accurate water quantities according to real-time soil conditions (Wolfert et al. 2017). It is now possible to efficiently monitor wide fields in real time using drones equipped with high-resolution cameras and multi-spectral photography. These drones detect subtle variations in color and canopy patterns in agricultural areas, signifying stress or disease in the crops. In Brazil, drone-based surveillance systems have enabled soybean farmers to detect fungal infections early, resulting in the preservation of over 20% of their total yield (Silva et al. 2023).

The integration of predictive analytics with real-time monitoring is highly effective. In the Netherlands, AI systems are amalgamating Internet of Things data regarding soil and crop conditions with meteorological forecasts to anticipate the probability of blight in potato farms. This integration enables farmers to administer fungicides with precision, reducing chemical usage by up to 40% without compromising production. Identical technology facilitates adaptable pest management tactics. In Africa, predictive models utilizing past migratory patterns of locusts and current weather conditions forecasted locust swarm movements. In conjunction with drones, it assisted authorities in executing targeted treatments, therefore alleviating the effects of locust infestations on food security (FAO 2024; Wijerathna-Yapa and Pathirana 2022).

17.4 Applications of Artificial Intelligence in Disease and Pest Management

17.4.1 Disease Detection

AI has transformed disease detection in agriculture. AI today provides precision and efficiency that were before unattainable. Advanced imaging and pattern recognition systems evaluate high-resolution images to identify the initial stages of fungal, bacterial, and viral diseases, enabling farmers to intervene promptly, therefore preserving crops and circumventing extensive chemical treatments (Boulent et al. 2019). A notable attribute of AI is its ability to differentiate between analogous symptoms arising from separate sources. Discoloration in leaves may come from bacterial infections, fungal diseases, or nutrient deficiencies. AI algorithms trained on diverse datasets will ascertain the problem, thereby enabling targeted treatment by farmers. A 2023 study revealed that AI-based diagnostic tools in Southeast Asian rice farms enhanced accuracy by 35% and significantly decreased excessive pesticide usage and related expenses. AI is essential for early detection, addressing the pervasive difficulties in agriculture. Cassava mosaic disease, a severe viral illness, has historically resulted in significant yield losses in sub-Saharan Africa. Mobile applications powered by AI, capable of diagnosing diseases from a mere snapshot, enable smallholder farmers to swiftly identify and isolate diseased plants. As reported by ICRISAT in 2023, this method has diminished cassava crop losses by up to 20% in experimental zones.

Artificially intelligent drones were employed to survey California vineyards for early indicators of powdery mildew. These machines monitor subtle texture variations and minute color alterations beyond human perception to identify and mitigate issues, thereby influencing intricate interventions. Employing these strategies conserves millions of dollars annually for the wine production sector by curtailing the spread of infections, in conjunction with fungicidal applications restricted to recognized problem regions (Jha et al. 2024).

17.4.2 Environmental Effects

AI also plays a role in the detection of diseases, which can contribute to environmental sustainability. Early and precise diagnoses reduce the use of excessive pesticides, thereby avoiding chemical runoff into the ecosystems surrounding the farms. In Brazil, soybean farmers who used AI diagnostic tools reported a 30% reduction in fungicide use without compromising their yields (Pretty et al. 2023). This saves production costs while promoting biodiversity and soil health.

17.4.3 The Future

Though the adoption of AI in disease detection is highly on the rise, access to technology and infrastructure in developing regions poses a challenge. However, efforts are underway to fill this gap. Governments, NGOs, and tech companies are working together to make AI tools more affordable and accessible to smallholder farmers, thereby enabling precision agriculture to reach those who need it most (FAO 2024).

17.4.4 Pest Identification and Monitoring

AI-driven pest identification tools are revolutionizing the way producers manage pests. They receive precise and prompt information on safeguarding their crops. These systems utilize picture classification techniques and algorithms to autonomously identify pest species, assess infestation severity, and suggest targeted actions (Fang 2024). AI integrates data-driven precision with accessible tools to empower farmers to manage pest issues more effectively and sustainably (Javaid et al. 2023). AI and DL algorithms are highly proficient in picture recognition, enabling the accurate identification of pest species. Common aphids, caterpillars, and whiteflies can be identified by training various AI models on diverse datasets, including multiple crop types such as wheat, maize, and cotton (Nauman et al. 2025). DL algorithms have successfully identified aphids in wheat fields with an accuracy over 95%, enabling farmers to take prompt action to prevent the spread of infestation (Chen et al. 2021).

Mobile applications and drone imaging systems are advancing this technology. Agriculturists can capture images of distressed vegetation using their smartphones, while drones equipped with high-resolution cameras can rapidly assess extensive fields. These technologies not only identify pests but also quantify the degree of damage, thus offering a detailed assessment that facilitates targeted treatment of specific regions. In India, AI-based solutions have diminished yield loss in maize from pest infestations by 20% and enhanced food availability in at-risk areas (ICRISAT 2023).

17.4.5 Pest Outbreak Prediction with Environmental Data

In AI, environmental variables such as temperature, humidity, and precipitation are crucial for monitoring pest populations. These variables exhibit a strong link with the behavior and patterns of pest life cycles. ML-driven predictive models utilize real-time weather data alongside historical records to forecast the likelihood of infections (Arsenijevic et al. 2022). In sub-Saharan Africa, AI systems have been employed to monitor and forecast the emergence of the fall armyworm, a significant pest affecting maize crops. They are engineered to amalgamate satellite data, Internet of Things sensors, and ML algorithms capable of identifying high-risk locations, enabling farmers to implement control measures prior to the escalation of infestations. This method has enabled farmers in trial areas to reduce pest management expenses by up to 30% while maintaining yields (FAO 2024).

17.4.6 Reducing Pesticide Use and Costs

AI reduces the application of broad-spectrum pesticides, which are typically expensive and ecologically destructive. Rather, farmers can only apply interventions such as applying biopesticides or introducing natural predators in areas that require it. According to Shamshiri et al. (2018), AI-enabled pest monitoring systems decreased pesticide use by 25% on average, thus lowering costs and reducing ecological harm significantly.

17.4.7 Case Study of Cotton Pest Management in the United States

AI is applied in the cotton farms in the United States to regulate the effects that pests such as bollworms and aphids have on cotton. The combination of drone surveillance with AI analytics allows these systems to identify hotspots for pest populations and predict their migrations. Such technology has reduced pesticide application among cotton farmers by 40%, thereby saving millions of dollars per year while preserving beneficial insects such as pollinators (USDA 2024). The applications in pest management are likely to grow as AI progresses. Real-time tracking of pests through Internet-of-Things networks and the use of AI-powered automated robots for pest removal are underway in high-tech farming processes. These developments will minimize the use of natural resources, increase sustainability even further, and make them more available to smallholder farmers with their operations in developing countries.

17.4.8 Decision Support Systems (DSS)

Decisions through AI-integrated decision support systems are changing agricultural decisions through the provision of a precision recommendation for disease and pest control that farmers can implement to achieve better yields. DSSs analyze numerous, diverse data sets such as those pertinent to soil conditions, crop health, and weather predictions; these insights have a significant impact at farm-specific levels, as these guide better decision-making processes that may lead to cost savings or higher yields with minimal application of chemicals or environmental disruption in farming (Botta et al. 2022).

17.4.9 What Artificial Intelligence-Powered DSS Work

AI-powered DSS uses a combination of ML algorithms, predictive analytics, and real-time data obtained from sources such as sensors of the Internet of Things, satellite imagery, and historical records. The system can then provide exact recommendations on critical farming issues, which are listed below.

17.4.10 Proper Use of Pesticides

DSS is able to suggest the best type, quantity, and timing of pesticide application with minimal chemical inputs and efficient effectiveness (Adetunji et al. 2023). For instance, in wheat cultivation, with DSS led by AI, pesticide usage has been decreased by 20% in predicting outbreaks of aphids based on localized weather conditions (Silva et al. 2023). Different agriculture management practices have been enlisted for reducing or withering in several crops (Iqbal et al. 2024).

17.4.11 Crop Rotation Plans

DSS proposes a rotation strategy to break up pest and disease cycles with the help of soil health and disease history, improving long-term productivity.

17.4.12 Biological Control Measures

AI-powered systems suggest the introduction of natural pest predators or biopesticides, thus providing environmentally friendly alternatives to chemical interventions.

17.5 Case Studies: DSS in Action

17.5.1 Integrated Pest Management (IPM) in Maize Farming

In Kenya, an AI-based DSS was implemented there to control fall armyworm infestations in the maize crop. The site-specific strategies for pest control were made through the integration of satellite data and observations coming from the farmers. Utilizing such a DSS, yield losses were 30%, and pesticide usage was lowered by 25%. That is how the AI would bring economic goals in alignment with environmental requirements (ICRISAT 2023).

17.5.2 Apple Orchards in Europe

Integrated AI with DSS has been employed in the fight against codling moths in Europe's apple orchards. The system analyzed pheromone trap data and weather patterns to predict moth emergence and suggested targeted interventions. This strategy reduced chemical usage by 40% while enhancing fruit quality, showcasing the capability of AI-powered DSS in high-value crops (Pretty et al. 2023; Dainelli et al. 2024).

Supporting Sustainability via Artificial Intelligence-Powered DSS

Among the many contributions AI-based DSS provides, perhaps its most significant contribution relates to sustainability. These systems have contributed significantly to the mitigation of soil degradation and water pollution by reducing dependence on chemical pesticides and fertilizers and have mitigated biodiversity loss. Jha et al. (2024) opined that adopting DSS for pest management has decreased the emission of greenhouse gases by 30% as compared to the traditional practice.

DSS systems are now available to smallholder farmers of developing regions through mobile applications. The apps often come with support for local languages and less complicated interfaces, bridging the technology gap and opening avenues for resource-limited farmers to benefit from advanced agricultural tools.

17.5.3 Challenges and Future Potential

There is much evidence that AI-based DSSs are of huge benefits, but the issues still remain: strong data infrastructure and farmer training. Governments, agricultural research institutions, and tech companies need to work together to remove such barriers. In the future, advancements in AI and ML are likely to further improve the capabilities of DSS. For example, next-generation systems might include advanced climate models that take into account the long-term effects of climate change on pest behavior and crop health, thus further strengthening the resilience of farming strategies.

17.6 Key Technologies in Artificial Intelligence-Driven Management

17.6.1 Computer Vision

Computer vision technologies are revolutionizing crop monitoring by enabling large-scale, high-precision surveillance through drones and smart cameras. Advanced systems capture high-resolution images of fields, which are then analyzed by AI algorithms for subtle anomalies such as discoloration, wilting, or pest damage. This reduces field inspection labor costs because of computer vision. Computer vision thus not only improves efficiency but also enables farmers to utilize actionable insights in protecting crops and maximizing yields (Zhang et al. 2020; Sharma and Shivandu 2024).

17.6.2 How Does Computer Vision Work in Agriculture

It basically lies in its ability to process and interpret visual data at a much higher scale and speed than human capacity. Using multispectral and thermal cameras, drones fly over the fields and capture images that allow analysis from different angles. It identifies early signs of stress, diseases, or pest infestation based on such images by applying AI models (Gkikas et al. 2023). For instance, computer vision will recognize chlorosis, a symptom that would represent nutrient deficiency or infection before it is seen to the naked human eye. Thermal imaging identifies alterations in transpiration in plants, which implies there might be water stress or damage to the roots. The vineyard utilizes computer vision to detect the health of grapes. Thus, vintners have identified the areas to focus their irrigation or pest control operations and produce high yields (Jha et al. 2024).

17.6.3 Precision Farming Applications

Computer vision technologies are key in precision farming, which allocates resources with high accuracy. In rice paddies, for example, AI systems in drones can detect patches of brown planthopper infestations and allow pesticides to be applied to only those infested areas. It helps protect yields but prevents environmental degradation by reducing chemical runoff into water systems (FAO 2024). Other uses are crop yield prediction. With the aid of computer vision models, plant density, height, and health enable it to predict the results of harvests, and therefore, enable farmers to anticipate the need for labor and storage well before harvest time. It has widely been used in soybean production in Brazil, increasing the accuracy of yield prediction by 20% (Pretty et al. 2023).

17.6.4 Environmental Impact

Computer vision has also helped sustain agriculture because it reduces the need for a lot of manual labor and maximizes resource utilization, which leads to the lowering of the carbon footprint of farming operations. Thus, accurate detection reduces high fertilizer and pesticide use, which helps save biodiversity and soil health (Getahun et al. 2024).

17.6.5 Future Scope

A more reliable farming system is expected as computer vision technology advances and integrates with other forms of AI, predictive analytics, and Internet of Things networks. Future applications may include autonomous drones that detect and repair the problem they have identified. This will be done by removing pests without human intervention through targeted spraying (Zhang et al. 2020).

17.6.6 Devices for the Internet of Things

The Internet of Things-powered sensors have revolutionized the agricultural industry by enabling real-time monitoring of vital environmental parameters such as temperature, humidity, soil moisture, and light intensity. Modern precision farming heavily relies on these sensors, which send massive data streams into AI models to generate insights that may be put to use. In doing this, technology allows the system to make better decisions concerning avoiding pest infestation, controlling plant diseases, and optimizing the use of resources (Javaid et al. 2022). All of these lead to healthier crops and greater yields.

17.6.7 Operating the Internet of Things-Powered Sensors

Fields are studded with Internet of Things sensors gathering soil and environmental data that are streamed continuously. AI models investigate the real-time data that is received by cloud-based platforms for trends and anomalies. A sudden spike in moisture levels in conjunction with mild temperature might suggest that everything is okay, and it would trigger alert systems for farmers to apply preventive measures before things turn worse.

Vineyard Internet of Things devices use AI algorithms for predicting outbreaks of powdery mildew, one of the most common fungal diseases. Such systems monitor humidity levels and leaf wetness duration and predict the likelihood of an outbreak with very high accuracy (Huang et al. 2024). Using a system like this in European vineyards reduced fungicide application by 30% because treatments were applied precisely where and when they were needed (Wolfert et al. 2017).

17.7 Advantages of Internet of Things-Based Monitoring

17.7.1 Real-Time Information for Preemptive Action

Internet of Things sensors are always updating in real-time, and farmers can thus act in time to combat potential threats. In wheat farming, for example, Internet of Things-based systems has been used to monitor rust disease conditions, saving up to 25% of crops through early action (Giller et al. 2021; Silva et al. 2023).

17.7.2 Improved Use of Resources

Internet of Things sensors help with water and fertilizer application optimization by offering detailed information on soil moisture and nutrient levels. Smart irrigation systems, for example, use the data above to precisely supply water where it is needed, lowering water use in arid regions by almost 40% in places like Australia and California (Pretty et al. 2023).

17.7.3 Cost-Effectiveness

Farmers can save money by using targeted interventions based on Internet of Things data instead of applying fertilizer and insecticides all at once. According to a case study on maize farming in sub-Saharan Africa, Internet of Things-based monitoring technologies made sustainable methods more affordable for smallholder farmers by lowering input prices by 20% (FAO, 2024).

17.7.4 Scalability Across Crop Types

Because of their versatility, Internet of Things systems can be tailored to monitor crops in a variety of settings. For instance, in paddies, it measures the amount of water present under ideal growing conditions for that crop; in an orchard, it measures temperature fluctuations over a period of time as though they would have an impact on fruit development. The Internet of Things' true potential lies in fusing those technologies with advancements in AI. AI models analyze data from the Internet of Things to forecast disease and pest outbreaks and offer interventions. For example, Internet of Things sensors and AI algorithms are used in Brazil's soybean farms to forecast the presence of caterpillar infestations (Jha et al. 2024). Drones, in fact, increasingly enabled by the Internet of Things technology are used to supplement ground sensors. This provides aerial views of crop health, and thus enables comprehensive monitoring of large farms without missing any area.

17.7.5 Challenges and Future Potential

While the Internet of Things-enabled agriculture has several benefits, its adoption is going to be limited, at least in poorer nations, by problems such as a high cost of up-front investment and the need for a robust network infrastructure. Internet of Things devices are becoming increasingly accessible, however, due to the continuous improvement of sensor affordability and satellite-based internet connectivity, which remove these barriers (Ayaz et al. 2019). Autonomous sensor networks that can adapt to changing conditions and self-calibrate are examples of future advancements. Such technologies will ensure even higher efficiency and sustainability through enhanced accuracy in prediction and resistance against climate unpredictability.

17.8 Machine Learning Models

ML models are changing the face of agriculture by analyzing diverse datasets ranging from weather patterns and crop growth stages to historical pest and disease records to predict potential outbreaks with remarkable accuracy (Xu et al. 2018; Domingues et al. 2022). These models, in uncovering patterns and correlations invisible to the human eye, provide farmers with actionable insights, enabling them to take proactive measures to safeguard their crops and optimize resources.

17.8.1 How Machine Learning Predicts Outbreaks

Large datasets of factors such as temperature, humidity, rainfall, soil conditions, and history of pest or disease incidents are used to train ML algorithms. Once these models are trained, they can identify the conditions leading to specific outbreaks and can send alerts in a timely manner (Khodadadi et al. 2023). For example, an ML model that looks at season-wise temperature trends and an increase in humidity levels could predict a higher chance of rust outbreaks in wheat fields (Polwaththa et al. 2024). Similarly, swarm movements have been forecasted using models that have been trained on the historical locust migration data. ML systems may classify areas that are at risk of locust invasion based on the satellite data about vegetation indices and weather patterns. These projections can really help farmers and governments to reduce significantly the damage caused to crops by coordinating management efforts (Mittal et al. 2024).

17.8.2 Successful Applications of ML in Agriculture

17.8.2.1 Rust Breakouts in Wheat

This system will be applied in South Asia, where stripe rust and stem rust are recurrent wheat threats. ML models in monitoring environmental conditions that enhance rust development have been made. A study in 2023 concluded that models reduced yield losses by 40% through the appropriate utilization of resistant cultivars and selective fungicide application (Silva et al. 2023).

17.8.2.2 Locust Swarm Forecasting

In East Africa, ML-based forecasting systems have been integrated into locust management programs. These systems, which combined satellite data, weather forecasts, and ground reports, predicted the movement of locust swarms in 2022 and 2023 with accuracy, enabling farmers and policymakers to implement control measures that reduced widespread devastation (FAO 2024).

17.8.2.3 Fall Armyworm Monitoring in Maize

ML algorithms have been introduced in sub-Saharan Africa tracking the lifecycle and dispersal of the fall armyworm, a pest significantly detrimental to maize production. This model allows for forecasts in infestation peaks as environmental and crop data go together, and it gives ways to target pest hotspots using biopesticides while reducing chemical application by 25% (ICRISAT 2023).

17.8.2.4 Integration with the Internet of Things and Remote Sensing

For even higher precision, the ML models are also connected to remote sensing and Internet of Things sensors. While satellite imaging offers information on the condition of vegetation and pest-prone areas, Internet of Things sensors provide real-time data on soil moisture, temperature, and humidity. This technological combination produces a comprehensive monitoring system that pays attention to every little aspect. For example, ML models using Internet of Things data have improved insect control methods in Brazilian soybean farms, resulting in lower pesticide use and higher-quality yields. According to a 2024 study, these technologies reduced input costs while increasing production by 15% (Pretty et al. 2023).

17.8.2.5 Challenges and Future Potential

Even though ML has shown enormous promise, issues including data accessibility, model regional flexibility, and the requirement for farmer training still exist. These obstacles are being addressed, nevertheless, by continued research and partnerships between governments, IT firms, and agricultural organizations. Future advancements might include real-time decision-making tools connected with mobile apps for farmer accessibility and localized machine-learning models that take region-specific factors into account (Akintuyi 2024).

17.8.3 Robotics

Robotic solutions powered by AI are revolutionizing the pest management industry with accuracy, effectiveness, and sustainability (Balaska et al. 2023). The advanced devices can self-navigate around greenhouses, orchards, and fields while carrying out specific tasks like removing diseased plants or applying pesticides.

17.8.4 What Artificial Intelligence-Powered Robot Work

Robots powered by AI utilize Internet of Things sensors, computer vision, and ML to carry out extremely accurate pest control activities (Khodadadi and Towfek 2023). These robots look for signs of pests or diseases on crops because of their high-resolution cameras and sensors (Botta et al. 2022). Once detected, they take specific measures to ensure maximum efficacy and lowest waste, such as manually removing the affected plants or applying pesticides only where necessary (Srivastava et al. 2025). For instance, robots with computer vision can detect the changes in leaf color or pest activity in greenhouses to identify whitefly infestations. Natural predators, such as parasitic wasps, or the use of specific sprays released by these robots reduce the usage of chemical treatments for general purposes (Shamshiri et al. 2018).

17.8.5 Artificial Intelligence Robots Applications in Pest Control

Precise Spraying: The robotic equipment uses GPS and AI algorithms to navigate vast fields with centimeter accuracy while spraying pesticides only where it is needed. AI robots spray fungicides in hotspots of powdery mildew to improve grape quality, cutting chemical usage by 30% (Pretty et al. 2023; Ali et al. 2024).

Self-sustained removal of pests: There have been developments in robots that can identify and pick plants with bacterial wilt infections in tomato greenhouses. This enhances the overall production chances by avoiding infection and saving the healthy stands of plants (Jha et al. 2024; Wang et al. 2025).

Implementation of biological pest control: AI-powered drones and ground robots are being used to deliver biological controls, including ladybugs or predatory mites, directly into the targeted areas. The strategy has been highly effective in organic farming where the use of chemicals is limited. A study conducted on cucumber greenhouses reported that the use of robotic delivery of predatory mites resulted in a 40% reduction in spider mite infestation (FAO 2024).

17.8.6 Benefits of Robotic Pest Management

Productivity of Tasks: These technologies relieve farm laborers of the pressure of strenuous work so that they have time for other vital things. This is particularly valuable for smallholder farmers as they do not have several workers to deal with their large fields (Amjadi et al. 2025).

Environmental sustainability: By greatly minimizing the misuse of pesticides, precision pest control maintains soil health and minimizes chemical runoff into water bodies. According to studies, the use of robotic systems for pest control has minimized the application of pesticides by up to 50% (Silva et al. 2023).

Economicalness: The cost of installing robotics is relatively high, but in the long term, the advantages like low labor and chemical input costs, as well as low crop losses, outweigh the majority of the expenditures.

Challenges and opportunities: Although the prospect of AI-based robots in agricultural applications is tremendous, it also comes with a few problems like:

High installation costs: The cost of this system is still a big deterrent, especially for small farms in developing countries.

Infrastructure requirements: Power supply, communication systems, and storage facility infrastructure are required to deploy these machines correctly (Hladik et al. 2018).

Work efficiency: By automating labor-intensive chores, these technologies relieve farmers' physical strain and free up time for other important activities. This is especially helpful for smallholder farmers who have limited staff to manage larger areas.

Sustainability of the environment: By significantly reducing the misuse of pesticides, precision pest control helps to maintain soil health and reduce chemical runoff into water bodies. Studies have even shown that the use of robotic systems for pest control has reduced the use of pesticides by up to 50% (Silva et al. 2021). Robots can be very costly when installed, but in the long run, the good things such as lower labor input costs and chemical input alongside reduced crop losses outweigh most expenditures (Gammanpila et al. 2024).

17.9 Conclusions and Future Perspectives

An emerging force that is revolutionizing modern agriculture, especially in the areas of disease and pest control, is AI. AI is helpful for predictive analytics, early discovery, and real-time monitoring that give farmers practical insights, lessen their need on chemical pesticides, and encourage sustainable practices. There is no denying the advantages of integrating AI technologies like robots, ML, and the Internet of Things, which include higher crop yields, better resource utilization, and reduced environmental effects. However, several obstacles stand in its way, such as the complexity of agricultural ecosystems, expensive costs, and restricted data availability (Chen et al. 2022).

Funding AI-based agricultural technologies should be a priority for governments, academic institutions, and the commercial sector. The goal of research should be to provide scalable and reasonably priced solutions that work in many agricultural environments. Adoption can be accelerated by implementing training programs to instruct farmers and other agricultural stakeholders on the application of AI technologies. Accessibility can be improved via farmer-friendly interfaces and simplified AI tools. To generate synergies, and bolster collaborations between governments, agribusinesses, and technology suppliers. Working together can facilitate the sharing of data, the pooling of resources, and the implementation of solutions in underserved areas. To promote the adoption of AI while addressing issues like data privacy, environmental impact, and fair technology distribution, supportive laws and regulatory frameworks should be established. Improving the precision and efficacy of AI models requires the creation of extensive datasets and consistent data collection procedures. To assist small-scale farmers in overcoming financial and technological obstacles, governments and non-governmental organizations ought to offer subsidies, loans, and technical assistance.

References

Adetunji, C.O., Olaniyan, O.T., Anani, O.A. et al. (2023). Artificial intelligence and automation for precision pest management. In: *Sensing and Artificial Intelligence Solutions for Food Manufacturing*, 49–70. CRC Press.

Adve, V.S., Wedow, J.M., Ainsworth, E.A. et al. (2024). AIFARMS: Artificial intelligence for future agricultural resilience, management, and sustainability. *AI Magazine* 45 (1): 83–88.

Akintuyi, O.B. (2024). Adaptive AI in precision agriculture: a review: investigating the use of self-learning algorithms in optimizing farm operations based on real-time data. *Research Journal of Multidisciplinary Studies* 7 (02): 016–030.

Ali, F., Rehman, A., Hameed, A. et al. (2024). Climate change impact on plant pathogen emergence: Artificial Intelligence (AI) approach. In: *Plant Quarantine Challenges Under Climate Change Anxiety*, 281–303. Cham: Springer Nature.

Arsenijevic, N., DeWerff, R., Conley, S. et al. (2022). Growth and development of multiple waterhemp (Amaranthus tuberculatus) cohorts in corn and soybeans. *Frontiers in Agronomy* 4: 1037483.

Ayaz, M., Ammad-Uddin, M., Sharif, Z. et al. (2019). Internet-of-Things (IoT)-based smart agriculture: toward making the fields talk. *IEEE Access* 7: 129551–129583.

Balaska, V., Adamidou, Z., Vryzas, Z., and Gasteratos, A. (2023). Sustainable crop protection via robotics and artificial intelligence solutions. *Machines* 11 (8): 774.

Botta, A., Cavallone, P., Baglieri, L. et al. (2022). A review of robots, perception, and tasks in precision agriculture. *Applied Mechanics* 3 (3): 830–854.

Boulent, J., Beaulieu, M., St-Charles, P.L. et al. (2019). Deep learning for in-field image-based grapevine downy mildew identification. In: *Precision Agriculture'19*, 8689. Wageningen Academic Publishers.

Chen, J.W., Lin, W.J., Cheng, H.J. et al. (2021). A smartphone-based application for scale pest detection using multiple-object detection methods. *Electronics* 10 (4): 372.

Chen, C.J., Li, Y.S., Tai, C.Y. et al. (2022). Pest incidence forecasting based on internet of things and long short-term memory network. *Applied Soft Computing* 124: 108895.

Dainelli, R., Bruno, A., Martinelli, M. et al. (2024). GranoScan: an AI-powered mobile app for in-field identification of biotic threats of wheat. *Frontiers in Plant Science* 15: 1298791.

Delfani, P., Thuraga, V., Banerjee, B., and Chawade, A. (2024). Integrative approaches in modern agriculture: IoT, ML and AI for disease forecasting amidst climate change. *Precision Agriculture* 1–25.

Demirel, M. and Kumral, N.A. (2021). Artificial intelligence in integrated pest management. In: *Artificial Intelligence and IoT-based Technologies for Sustainable Farming and Smart Agriculture*, 289–313. IGI Global.

Domingues, T., Brandão, T., and Ferreira, J.C. (2022). Machine learning for detection and prediction of crop diseases and pests: a comprehensive survey. *Agriculture* 12 (9): 1350.

Fang, J. (2024). Breeding 5.0: AI-driven revolution in designed plant breeding. *Molecular Plant Breeding* 15 (1): 27–33.

FAO (2024). *Precision Agriculture and Environmental Sustainability*. Food and Agriculture, Washington DC: Organization of the United Nations.

Gammanpila, H.W., Sashika, M.N., and Priyadarshani, S.V.G.N. (2024). Advancing horticultural crop loss reduction through robotic and AI technologies: innovations, applications, and practical implications. *Advances in Agriculture* 2024 (1): 2472111.

Getahun, S., Kefale, H., and Gelaye, Y. (2024). Application of precision agriculture technologies for sustainable crop production and environmental sustainability: a systematic review. *The Scientific World Journal* 2024 (1): 2126734.

Getahun, S., Kefale, H., and Gelaye, Y. (2024). Application of precision agriculture technologies for sustainable crop production and environmental sustainability: a systematic review. *The Scientific World Journal* 2024 (1): 2126734.

Giller, K.E., Delaune, T., Silva, J.V. et al. (2021). The future of farming: who will produce our food? *Food Security* 13 (5): 1073–1099.

Gkikas, D.C., Theodoridis, P.K., and Gkikas, M.C. (2023). Artificial Intelligence (AI) use for e-governance in agriculture: exploring the bioeconomy landscape. In: *Recent Advances in Data and Algorithms for e-Government*, 141–172. Cham: Springer International Publishing.

Hayat, M.T., Nauman, M., Nazir, N. et al. (2019). Environmental hazards of cadmium: past, present, and future. In: *Cadmium Toxicity and Tolerance in Plants*, 163–183. Academic Press.

Hladik, M.L., Main, A.R., and Goulson, D. (2018). Environmental risks and challenges associated with neonicotinoid insecticides. *Environmental Science & Technology* 52 (6): 3329–3335.

Huang, Y., Dong, Y., Huang, W. et al. (2024). Predicting the global potential suitable distribution of fall armyworm and its host plants based on machine learning models. *Remote Sensing* 16 (12): 2060.

ICRISAT (2023). *Artificial Intelligence-Driven Solutions for Fall Armyworm Management in Maize*. International Crops Research: Institute for the Semi-Arid Tropics.

Iqbal, N., Nazir, N., Numan, M. et al. (2022). Effects of polychlorinated biphenyls on plant growth. In: *Sustainable Plant Nutrition under Contaminated Environments*, 187–208. Cham: Springer International Publishing.

Iqbal, N., Riyazuddin, R., Nauman, M. et al. (2024). Role of plant defence system in crop protection against fusarium pathogens. In: *Fusarium-Recent Studies*. London: IntechOpen.

Javaid, M., Haleem, A., Singh, R.P., and Suman, R. (2022). Enhancing smart farming through the applications of agriculture 4.0 technologies. *International Journal of Intelligent Networks* 3: 150–164.

Javaid, M., Haleem, A., Khan, I.H., and Suman, R. (2023). Understanding the potential applications of artificial intelligence in agriculture sector. *Advanced Agrochem* 2 (1): 15–30.

Jha, P.K., Zhang, N., Rijal, J.P. et al. (2024). Climate change impacts on insect pests for high value specialty crops in California. *Science of the Total Environment* 906: 167605.

Kamilaris, A. and Prenafeta-Boldú, F.X. (2018). Deep learning in agriculture: a survey. *Computers and Electronics in Agriculture* 147: 70–90.

Kaur, I., Sandhu, A.K., and Kumar, Y. (2022). Artificial intelligence techniques for predictive modeling of vector-borne diseases and its pathogens: a systematic review. *Archives of Computational Methods in Engineering* 29 (6): 3741–3771.

Khodadadi, E. and Towfek, S.K. (2023). Internet of Things enabled disease outbreak detection: a predictive modeling system. *Journal of Intelligent Systems & Internet of Things* 10 (1): 84.

Liakos, K.G., Busato, P., Moshou, D. et al. (2018). Machine learning in agriculture: a review. *Sensors* 18 (8): 2674.

Mittal, M., Gupta, V., Aamash, M., and Upadhyay, T. (2024). Machine learning for pest detection and infestation prediction: a comprehensive review. *Wiley Interdisciplinary Reviews: Data Mining and Knowledge Discovery* 14 (5): e1551.

Mukhtar, S., Yasar, A., Mahfooz, Y. et al. (2022). Value addition and risk assessment of dairy digestate as biofertilizer on crop yield and soil fertility. *Arabian Journal of Geosciences* 15 (3): 245.

Nauman, M., Alghamdi, A.A., Ahmed, J. et al. Enhanced Resilience to Salt Stress: An Integrated Approach Addressing Physiochemical Attributes of Wheat (*Triticum aestivum* L.). *Polish Journal of Environmental Studies* 2323–2336.

Pathirana, R. and Carimi, F. (2024). Plant biotechnology—an indispensable tool for crop improvement. *Plants* 13 (8): 1133.

Pinter, P., Morichetta, A., and Dustdar, S. (2024, July). Distributed model serving for real-time opinion detection. In: *2024 IEEE International Conference on Service-Oriented System Engineering (SOSE)*, 64–73. IEEE.

Polwaththa, K.P.G.D.M., Amarasinghe, S.T.C., Amarasinghe, A.A.Y.D., and Amarasinghe, A.A.Y. (2024). Exploring artificial intelligence and machine learning in precision agriculture: a pathway to improved efficiency and economic outcomes in crop production. *American Journal of Agricultural Science, Engineering, and Technology* 8 (3): 50–59.

Pretty, J. (2023). Regenerative agriculture and redesign for sustainability. In: *Biological Approaches to Regenerative Soil Systems*, 13–24. CRC Press.

Russell, S.J. and Norvig, P. (2021). *Artificial Intelligence: A Modern Approach*. Pearson.

Sahoo, L., Mohapatra, D., Raghuvanshi, H.R. et al. (2024). Transforming agriculture through artificial intelligence: advancements in plant disease detection, applications, and challenges. *Journal of Advances in Biology & Biotechnology* 27 (5): 381–388.

Shamshiri, R., Kalantari, F., Ting, K.C. et al. (2018). Advances in greenhouse automation and controlled environment agriculture: a transition to plant factories and urban agriculture. *International Journal of Agricultural and Biological Engineering* 11 (1): 1–22.

Sharma, K. and Shivandu, S.K. (2024). Integrating artificial intelligence and Internet of Things (IoT) for enhanced crop monitoring and management in precision agriculture. *Sensors International* 100292.

Silva, V., Alaoui, A., Schlünssen, V. et al. (2021). Collection of human and environmental data on pesticide use in Europe and Argentina: field study protocol for the SPRINT project. *Plos One* 16 (11): e0259748.

Silva, J.C., Silva, M.C., Luz, E.J. et al. (2023). Using mobile edge AI to detect and map diseases in citrus orchards. *Sensors* 23 (4): 2165.

Singh, P. and Kaur, A. (2022). A systematic review of artificial intelligence in agriculture. *Deep learning for sustainable agriculture* 57–80.

Srivastava, V., Kumar, R., Wani, M.Y. et al. (2025). Role of artificial intelligence in early diagnosis and treatment of infectious diseases. *Infectious Diseases* 57 (1): 1–26.

Wang, J., Chen, Y., Huang, J. et al. (2024). Leveraging machine learning for advancing insect pest control: a bibliometric analysis. *Journal of Applied Entomology*.

Wijerathna-Yapa, A. and Pathirana, R. (2022). Sustainable agro-food systems for addressing climate change and food security. *Agriculture* 12 (10): 1554.

Wolfert, S., Ge, L., Verdouw, C., and Bogaardt, M.J. (2017). Big data in smart farming–a review. *Agricultural Systems* 153: 69–80.

World Health Organization (2023). Report of the 15th FAO/WHO joint meeting on pesticide management: Rome, Italy and online, 15–18 November 2022. *Food & Agriculture Organization*.

Xu, W., Wang, Q., and Chen, R. (2018). Spatio-temporal prediction of crop disease severity for agricultural emergency management based on recurrent neural networks. *GeoInformatica* 22: 363–381.

Zhang, X., Cao, Z., and Dong, W. (2020). Overview of edge computing in the agricultural internet of things: Key technologies, applications, challenges. *IEEE Access* 8: 141748–114176.

18

Challenges and Uncertainties Associated with AI in Agriculture

Vipin Bihari Mishra and Kanti Kiran

Gujarat Biotechnology University, GIFT City, 382355, Gandhinagar, Gujarat

18.1 Introduction

By 2050, the world's population will have reached 10 billion individuals, which will result in a 50% increase in agricultural demand over 2013. Current crop production occupies 37.7% of the world's land. Agriculture has a major role in both established and developing economies, providing jobs and national income. In India, 50% of the workforce is employed in agriculture, which drives rural development and transformation and accounts for 18% of the country's GDP (Singh, 2016).

The 2018 World Government Summit paper, "Agriculture 4.0: The Future of Farming Technology," emphasizes how artificial intelligence (AI), big data, precision agriculture, and the Internet of Things have the potential to completely transform the farming industry. Due to factors including population shifts, resource scarcity, climate change, and food waste, agriculture is at a turning point (Javaid et al., 2022c). The goal of AI technologies is to reduce their negative effects on the environment while improving crop yields, real-time monitoring, and decision-making. Robots and drones, among other AI-based devices, increase productivity in areas like automated irrigation, smarter seed and fertilizer delivery, soil and health monitoring, pest and disease surveillance, weed detection, soil sensing, and effective farm machinery management (Javaid et al., 2022c). This revolutionary strategy imagines an approach where individuals would be "producing more with fewer inputs."

Agriculture is changing quickly due to digital technologies, especially AI, which maximize resource utilization, decrease environmental effect, and increase efficiency. AI is a vital tool for agribusiness and is becoming more widely available because of developments in cloud computing, research, and processing capacity. Significant AI investment is anticipated in the agriculture sector, which is predicted to rise by 25.5% a year between 2020 and 2026 (Figure 18.1). AI helps tackle societal challenges like labor shortages and the need to increase food production while reducing environmental harm (Mohammed et al., 2024).

It builds on existing precision agriculture technologies, such as GPS guidance and yield mapping, by enhancing decision-making and management through data. Smart farming, which integrates AI, takes precision agriculture to the next level, helping farmers make better decisions and improve productivity.

Figure 18.1 Technologies used in agriculture. Together, these symbols highlight how AI may be utilized in numerous aspects of agriculture, from planting and planning to harvesting and monitoring. Farmers may enhance their operations, boost productivity and improve sustainability by utilizing AI's analytical and predictive powers. The picture highlights a data-driven farming strategy in which technology is essential for enhancing output and ensuring food security. This figure has been created with CANVA version (2.298).

18.2 Applicability of AI to Agricultural Practices

AI-powered agricultural robots are designed to increase the yield and biomass of the final product. These applications are used for issues like (Figure 18.2)

Internet of Things (IoT)-Based Solution:
These tools are linked to sensors and devices, providing real-time monitoring and encouraging data-driven decision-making for a safer and more resource-efficient agricultural practice.

Precision Water Management (PWM):
PWM optimizes water consumption in agriculture by providing water at the right time, location, and crop growth stage. It offers limited and precise uses of water. PMW seeks to improve overall farming efficiency, encourage sustainability, and preserve water resources.

Geographical Data Findings:
AI and big data technologies analyze geographical data to provide useful information on crop water requirements, weather patterns, and soil moisture levels.

Locating Inefficient Areas:
AI-based sensors examine field water usage and identify trends and insufficiencies. This aids Farmer in locating areas with wasteful or inefficient water consumption.

Water Waste Reduction:
AI optimizes irrigation techniques to reduce water waste. AI helps farmers optimize crop yields while reducing water use by accurately regulating water application. This

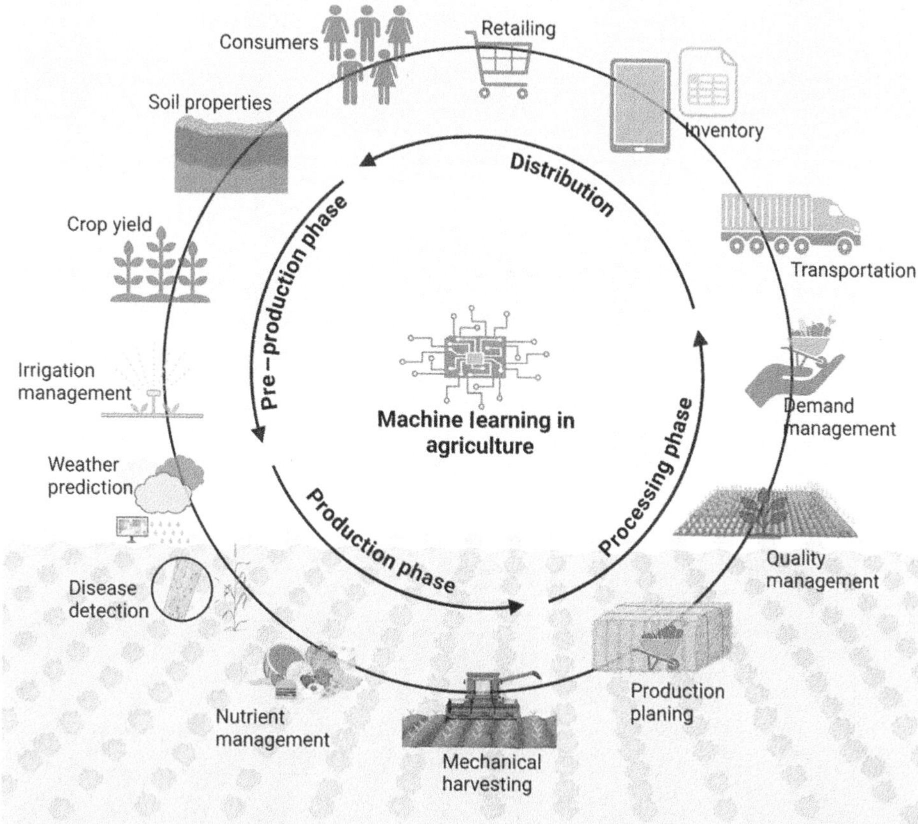

Figure 18.2 Over every step of the agricultural process, from pre-production to consumption, this picture represents the function of machine learning. To inform important decisions at every stage, such as irrigation, nutrient management, disease detection, harvesting, production planning, quality control, demand management, transportation, inventory, and, finally, retailing to consumers, it demonstrates how machine learning integrates data from multiple sources (soil, weather, crop yields, etc.). In essence, it shows a machine learning-powered, data-driven, and optimized agricultural system. This figure has been created with https://BioRender.com.

is especially important in areas with scarce water supplies or erratic weather patterns (Talaviya et al. 2020, Singh 2016, Ryan et al. 2023).

18.3 Scope of Improvement of AI Strategies in the Agriculture Sector

There are AI-based solutions that serve to increase efficiency in the agricultural sector, including crop yield, irrigation, soil content sensing, crop monitoring, wedding, and crop setup. The research indicates that the size of the worldwide AI market was approximately USD 1.4 billion in 2022 and is expected to increase at a compound annual growth rate (CAGR) of approximately 22.5% between 2023 and 2030, reaching USD 7.1 billion. https://www.zionmarketresearch.com/report/agriculture-ai-market. AI has the ability to provide

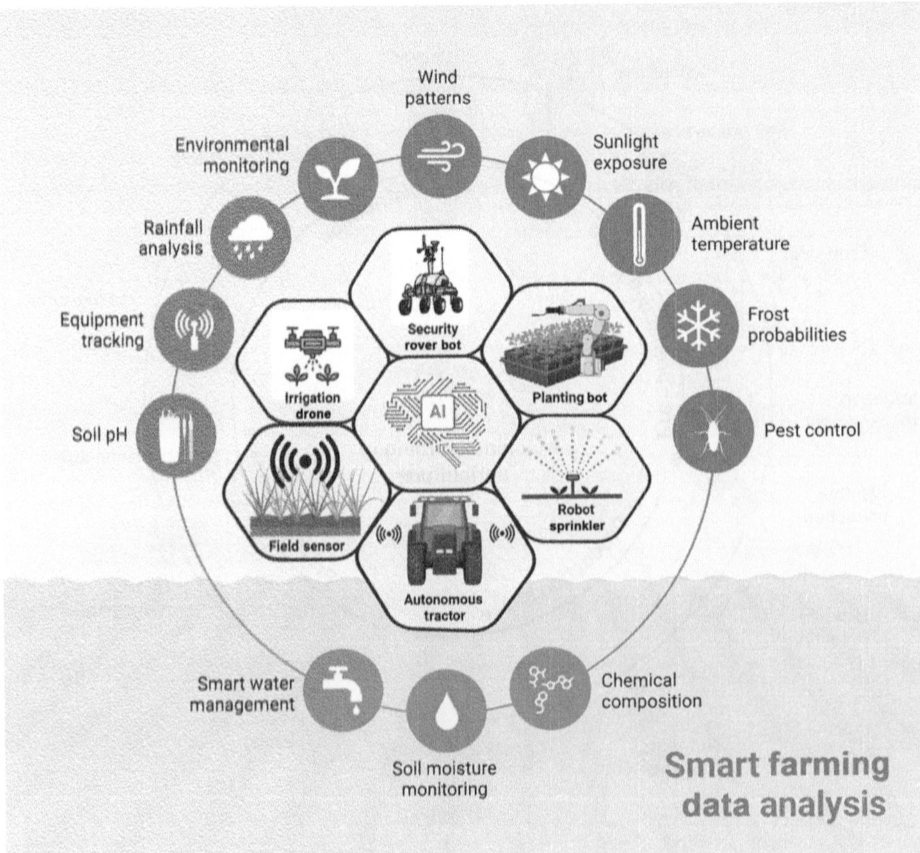

Figure 18.3 The picture illustrates smart agriculture, which optimizes agricultural practices by utilizing AI and data. Sensors gather information on operational elements (equipment tracking, pest activity) and environmental factors (rainfall, temperature, soil, etc.). AI uses this data to guide automated solutions such as autonomous tractors, drones for irrigation, and robotic systems for security, sprinkling, and planting. The goal of this data-driven strategy is to increase agricultural output, sustainability, and efficiency. This figure has been created with https://BioRender.com.

a much-needed answer in an era where the agriculture sector is experiencing an economic downturn as a result of the climate condition. By optimizing different farming aspects like water and energy management, precision agriculture, and smart farming techniques, AI has emerged as an effective tool for promoting sustainable agriculture (Figure 18.3). These AI-based technologies have allowed the farm to produce more with less input. Farmers can learn how to effectively control irrigation and save water supplies by using AI algorithms. Additionally, AI makes it possible to automate labor-intensive agricultural chores like crop harvesting, trimming, and plowing with the help of AI-guided autonomous tractors and harvesters, which lessens the need for significant human participation (Bakri et al., 2024,

Victoire et al., 2023). The different applications of AI under developing technologies for sustainable agriculture as follows:

- Monitoring the health of the crops and soil
- Managing weeds
- Using intelligent spraying
- Monitoring animal health
- Detecting pest and plant illnesses,
- Harvesting, plowing, and pruning,
- Grading and sorting production,
- Energy security

18.4 Recent Advancements in AI for Agriculture – Deep Learning in Agriculture

By examining enormous datasets and spotting complex patterns, deep learning, a subfield of AI, is transforming the agricultural industry. Drones can now take high-resolution pictures of fields via Convolutional Neural Networks (CNNs), a subset of deep learning models that are excellent at image recognition. After that, these photos are examined to determine soil conditions, identify weeds, and find crop illnesses. This reduces waste and the impact on the environment by enabling targeted actions, like the application of pesticides or fertilizer. Deep learning algorithms can also automate labor-intensive jobs, optimize irrigation schedules, and forecast crop yields, which will improve agricultural sustainability, efficiency, and cost-effectiveness (Sherkhane and Ratnaparkhi, 2024).

18.4.1 Precision Farming and Crop Monitoring

By evaluating data from satellites, drones, and Internet of Things sensors, precision agriculture uses AI to maximize crop yield. Real-time insights on weather patterns, crop health, and soil conditions are provided by this data. The best methods for fertilization, irrigation, and pest control are then suggested by AI algorithms using this data. Additionally, yields, soil requirements, and upcoming problems like disease outbreaks can all be predicted by machine learning algorithms. Companies like IBM and Microsoft provide AI-powered systems that forecast weather to decrease the effects of climate fluctuation and evaluate satellite data to advise farmers on the most favorable times to grow, allocate resources, and harvest (Singh 2016, Gangwani, 2024).

18.4.2 Identification of Pests and Illnesses

In agriculture, AI is essential for identifying and controlling pests and illnesses. AI-powered tools, such as machine learning algorithms and image recognition, examine crop photos to spot early infestation indicators, enabling farmers to take preventative action before serious crop loss happens. AI models are used by smartphone apps such as Plantix and Plant Village Nuru to identify plant illnesses from photos that farmers input, giving them prompt treatment recommendations. Furthermore, enormous agricultural areas can be easily surveyed by AI-enhanced drones fitted with high-resolution cameras. These drones can take pictures, which machine learning models can then use to identify indications of crop distress. Compared to traditional manual methods, this technology allows for faster and better initial detection and treatment of pest and disease outbreaks (Srikanth et al., 2020, Ahirwar et al., 2019).

18.4.3 Automated Farm Equipment and Robotics

The emergence of AI-driven autonomous equipment is transforming agricultural operations by decreasing the need for human labor and boosting operational efficiency. AI-enabled autonomous tractors, harvesters, and other farm machinery may work with little human oversight. They use GPS and AI algorithms to precisely navigate fields, sow seeds, apply fertilizer, and harvest crops. Machine learning is used by robotic systems, such as those created by Blue River Technology, to precisely execute tasks like weeding and applying pesticides. In order to minimize chemical consumption and lessen its impact on the environment, Blue River's "See & Spray" technology uses computer vision to identify and target weeds with herbicides. AI-powered robots are also being utilized in the harvesting of fruits and vegetables, ensuring high-quality produce while minimizing labor costs, traditionally a labor-intensive sector due to the delicate nature of the crops (Ahirwar et al., 2019).

18.4.4 AI in Livestock Management

AI is essential to livestock management, which includes productivity improvement, breeding optimization, and animal health monitoring. Wearable technology and sensors driven by AI monitor the vital signs, activity, and behavior of animals, facilitating prompt interventions to enhance their well-being and the early diagnosis of diseases or stress (Talaviya et al. 2020). AI is also used in precision breeding to evaluate genetic data, find beneficial qualities (growth rates, illness resistance, milk yield), and choose animals with the best attributes to increase the health and productivity of the herd. Furthermore, AI systems optimize nutrient intake and control feeding schedules in real-time, modifying feed composition according to the needs and growth stages of individual animals. This improves resource efficiency and boosts yields in the production of dairy, poultry, and meat (Patel et al., 2022).

18.4.5 AI in Supply Chain Optimization

By improving post-harvest management, forecasting, and transparency, AI is essential to agricultural supply chain optimization. In order to forecast demand for agricultural commodities, AI algorithms examine past data and market patterns. This allows farmers to adjust their production and harvesting schedules to meet market demands, reducing waste and guaranteeing that consumers receive fresh products on time (Pal et al. 2023). In order to improve processing efficiency and guarantee consistent product quality, AI-powered post-harvest management systems use computer vision to effectively sort and grade produce according to size, color, and quality. Additionally, by tracking items from farm to consumer, AI combined with block-chain technology improves supply chain transparency while guaranteeing food sustainability and safety. AI is used by platforms such as IBM Food Trust to enhance supply chain management, trace product origins, and mitigate the risks of contamination or fraud (Rane et al., 2024).

18.4.6 AI in Climate-Smart Agriculture

Agriculture faces major obstacles because of climate change. AI is an effective tool for farmers to help them adjust. AI models forecast future circumstances by evaluating climate data, which helps farmers select resilient crops, strategically schedule plantings, and modify farming practices to reduce the danger of extreme weather. By finding characteristics that increase a crop's resistance to climate stress, AI also aids in the development of superior crops. In order to help farmers make wise decisions and prosper in a changing environment, tools like Climate AI offer them useful information on climate risks and offer suggestions for how to modify their activities (Das et al. 2024).

18.4.7 AI in Food Security and Policy Planning

AI in food security and policy planning: By giving academics and policymakers vital data-driven insights, AI has the potential to greatly improve global food security. By simulating the effects of different agricultural techniques, legislative modifications, and technical breakthroughs on food production, AI models allow for more efficient strategic planning for future food demands. AI can detect areas at risk of food shortages and suggest ways to reduce such risks by evaluating global agricultural data (Talaviya et al. 2020). Additionally, in poor nations where trustworthy agricultural data is frequently hard to come by AI can enhance data gathering and interpretation. AI-powered solutions like Farm Logs provide farmers and policymakers with useful information about crop conditions, weather patterns, and market trends, allowing for better decision-making and resource management. This enables interested parties to make wise decisions and enhance agricultural outcomes (Rane, et al., 2024, Talaviya et al. 2020).

18.4.8 AI in Greenhouse Automation

By automating and streamlining processes, AI is transforming greenhouse agriculture. To provide optimal growing circumstances, AI-driven systems control important greenhouse environmental parameters like temperature, humidity, light intensity, and CO_2 levels. These devices track crop development and health in real time using computer vision and smart sensors, modifying conditions as necessary. In Atapattu et al. (2024), AI is used by businesses like Priva and iUNU to continuously track plant growth and modify greenhouse conditions in real time, improving agricultural production and quality. In vertical farming, where crops are produced indoors in stacked layers, AI is especially useful.

In highly regulated settings, AI optimizes environmental factors like light and nutrient distribution. AI systems improve resource usage by evaluating data on plant development and environmental circumstances; they enhance resource utilization, leading to increased yields while minimizing water and energy consumption (Rane et al. 2024).

18.4.9 AI-Driven Crop Breeding

AI is revolutionizing agricultural breeding initiatives by accelerating the development of novel, high-yielding, and resilient crop varieties. Traditional breeding methods are time-consuming and labor-intensive, but AI can significantly expedite the identification of desirable traits and the creation of new cultivars. Machine learning algorithms analyze vast datasets of genetic information and environmental variables to predict which plant varieties are most likely to thrive under specific conditions (Talaviya et al. 2020).

18.4.10 AI in Pollination Management

In areas where natural pollinator populations, such as bees, are declining, AI-driven robotic pollinators, such as drones outfitted with machine learning algorithms, are helping with agricultural pollination. These robots recognize flowers that are ready for pollination and precisely dispense pollen by using computer vision. This technology is especially useful in controlled settings where natural pollinators may not have easy access, such as greenhouses or indoor farms (Rane et al., 2024).

18.4.11 Weed Control with AI

Weeds compete with crops for vital resources like sunshine, water, and nutrients, which eventually results in lower yields. For this reason, weed control is critical in agriculture. Conventional weed control techniques, such as applying chemical herbicides, can be costly and environmentally damaging. However, by making more accurate and efficient methods possible, AI is completely changing the way weeds are managed (Blasco et al., 2002). Up to 90% less chemicals are used overall thanks to these robots' ability to spray herbicides only where weeds are present. In addition to reducing expenses, this focused strategy lessens its negative effects on the environment and helps stop the emergence of weed species that are resistant to herbicides. Furthermore, in order to assist farmers in implementing more sustainable farming practices, AI-driven weed management systems might suggest substitute, non-chemical control techniques like crop rotation or mechanical weeding (Ryan et al. 2023).

18.5 New Opportunities for AI-Driven Initiatives in Agriculture

Considering its significant effects on economies, labor markets, and industries, technology has long been important to social and economic studies. In recent years, digitalization – particularly through AI – has drawn more attention in the agricultural sector, as seen by the sharp increase in the number of businesses implementing AI technology (Davenport et al., 2020). The need to maximize output while avoiding overuse of pesticides, fertilizers, and medicines is one of the many intrinsic problems facing agriculture. Other problems include seasonality, erratic pests, and illnesses. By increasing output, resource efficiency, and food quality, AI is thought to be a viable way to address these issues (Eli-Chukwu, 2019).

18.5.1 AI for Agricultural Economic Performance

AI in agriculture has many economic advantages, such as increased analytics, cost savings, innovation, and better market strategies. By optimizing the use of resources (land, water, and pesticides), enhancing the control of weeds and pests, and providing accurate actions with fewer errors, AI technologies assist lower production costs (Eli-Chukwu, 2019). For example, compared to conventional sprayers, AI-powered smart aerial vehicles (drones) have been demonstrated to save up to 15% on costs. However, there are financial obstacles to AI adoption, including high initial expenses, costly servicing, and the requirement for specialized infrastructure (such as Wi-Fi and cloud storage). Additionally, the advantages of AI are sometimes greater for larger farms, which could leave smallholder farmers behind, particularly in emerging economies (Dharmaraj and Vijayanand, 2018).

By accurately forecasting crop conditions, pest attacks, and harvest results, AI also improves risk management in agriculture by lowering uncertainty and human error. With up to 95% accuracy in disease identification demonstrated by AI technologies like computer vision, they are more effective and economical than conventional techniques. But there are still worries about the dangers of relying too much on AI and the possibility of false signals, which emphasizes the necessity of human supervision throughout the process (Mishra et al., 2024).

AI-powered solutions are becoming more common in both developed and developing nations, assisting farmers who struggle with issues like inaccurate weather, pest, and market data. These technologies offer crucial information and guidance on risk management, increasing productivity, cutting expenses, and scheduling planting and harvesting according to market trends (Dharmaraj and Vijayanand, 2018). Farmers who suffer with restricted access to market information in emerging markets will find this very useful. By examining consumer demands, crop pricing, and market trends, AI assists them in making better selections. For instance, farmers can now buy mobile AI apps that gather data on crops, livestock, and food safety, which enhances decision-making and lessens market failures (Mishra et al., 2024).

AI is also anticipated to replace many manual farming operations, such as physical labor and pesticide application. It won't, however, take the role of human judgment in intricate circumstances. Rather, AI will require farmers to acquire new skills in data analytics and interpreting AI outcomes, as well as develop new jobs like digital farm consultants. Low-skilled workers may be displaced by AI, even as technology can benefit high-income nations with their labor shortages. Furthermore, because AI robots are not flawless, they could not be able to recognize items that are out of their field of vision, which could result in mishaps. Despite its benefit, the effects on employees and safety need to be carefully addressed (Dharmaraj and Vijayanand, 2018, Talaviya et al. 2020, Mishra et al., 2024).

18.6 Aspects of AI in Agriculture That are Ethical, Socioeconomic, and Environmental

Improving decision-making, adaptability, and offering customized solutions are the main goals of integrating Smart Information Systems (SIS) in agriculture. There are obstacles, too, such as the present machine learning algorithms' inability to efficiently handle big datasets. Additionally, possible errors brought on by local variances, limited data, and environmental conditions might lead to problems with data quality (Ryan et al., 2023). These restrictions may result in false conclusions, poor advice, and eventually detrimental effects on farmers, such as financial losses and harm to the environment. Because AI chatbots lack domain-specific knowledge and real-world experience, their ability to understand complex agricultural scenarios may be limited. AI has the potential to alter our conventional understanding of agriculture and allow farmers to modify their way of life. AI offers solutions depending on a variety of elements, including the type and volume of data gathered, modeling, design, and algorithm type, and it is heavily reliant on how it is all applied (Dara et al., 2022).

18.6.1 Trust Issues

It's difficult to fully trust agricultural and food processing equipment. Even if technology is transforming agriculture, we still have a way to go before we can completely depend on machines to do everything. While intelligent systems certainly help in risk management and performance optimization, they can give rise to security problems. Because of the possibility of hacking and cyberattacks, agriculture may be susceptible to disturbances or even violence (Dara et al., 2022). Furthermore, there may be moral conundrums if decision-making is completely left to data-driven algorithms. These systems might not have the "human touch" and function in a ruthless, rational manner. It is obvious that the future of agriculture depends on the establishment of precise ethical standards and laws for these cutting-edge technologies (Rane et al., 2024).

18.6.2 Artificial Intelligence's Effects on the Labor Market

The future of human labor is seriously threatened by agriculture's growing reliance on AI. Although AI can boost productivity and automate processes, it also poses a danger to human labor, which could result in job losses and a reduction in the role of humans in farming (Talaviya et al. 2020). Farmers may feel inadequate and frustrated because of this change, which can weaken their sense of autonomy and purpose. Additionally, the use of AI in the workplace creates new risks for workers, including greater mental strain, a blurring of the lines between work and personal life, and the possibility of data exploitation. These issues call for a cautious and deliberate approach to the application of AI in agriculture, putting worker welfare first and guaranteeing a seamless transition for the agricultural workforce (Ryan et al., 2024).

18.6.3 The Data-Driven Algorithms Question

Smart agriculture's future depends on the effective gathering and analysis of thorough and accurate data, which is now fraught with difficulties. Research, decision-making, and the improvement of agriculture worldwide depend on the creation of an open-source database that includes satellite imagery, data from IoT sensors, and information about soil, crops, water, and weather (Ryan et al., 2024). When combined with cloud-based technologies and subjected to machine learning models, this data can yield important insights about agricultural conditions both now and in the future. Promoting global data availability, accessibility, and usability still presents difficulties, nevertheless (Javaid et al., 2022). To improve the precision and realism of these prediction systems, researchers are actively integrating stochastic algorithms with deterministic models. Furthermore, using machine learning in agriculture has ethical and legal ramifications that call for thorough examination as well as the creation of suitable guidelines and standards (Ryan et al., 2024, Dara et al., 2022).

18.6.4 Logistical Problems

AI-based platforms can give farmers on-demand access to a variety of machinery services, such as tractors, increasing the affordability and accessibility of agriculture, especially for smallholders. To adjust to the uncertainties of climate change, this is essential. AI-driven decision-making systems can also improve efficiency, decrease corruption, and increase accountability in areas like public procurement and input programs by streamlining agricultural administration and logistics. Extensive study is required to collect information on current logistical services, their capabilities, and their limitations to efficiently utilize these technologies. Additionally, this study should investigate how these services may be scaled up and how the government might help put them into place. Additionally, this study should investigate how these services may be scaled up and how the government might help put them into place (Kamble et al., 2019). The study must consider the legal framework that governs the application of cutting-edge technology in agriculture, taking safety, security, and ethical issues into account. Enhancing efficiency, reducing corruption, and developing more inclusive and equitable agricultural systems that benefit farmers and customers can all be achieved by incorporating AI-driven decision-making systems (ADMs) into agricultural logistics. Farmers may increase their profitability and make better decisions by using the useful market information that ADMs can give them. ADMs can also be utilized to pinpoint farmers who are experiencing food insecurity, enabling more equitable and sustainable farming practices (Saiz-Rubio and Rovira-Más, 2020).

18.6.5 Climate Change

Climate change: One of the main causes of climate change is the widespread use of digital technology, which has led to an increase in energy consumption and greenhouse gas emissions every day. This trend is expected to continue since digital technology alone is expected to generate 14% of greenhouse gas emissions by 2040 and consume 20% of the world's electricity by 2025. Massive amounts of data storage require a lot of power, which is a major factor in climate change, whose consequences are already felt disproportionately in developing countries. Developing nations must take a multi-pronged strategy to combat climate change. Prioritizing awareness-building and the promotion of sustainable practices is crucial in addition to technology solutions. This can be accomplished by creating data and AI initiatives that are suited to the unique requirements and conditions of these countries. AI algorithms, for example, can help farmers maximize crop yields and minimize water use, while also assisting local communities in better managing natural resources such as rivers and forests. To lessen the negative effects of digital technology on the environment, it is also crucial to switch to renewable energy sources like solar, wind, and hydropower. One of the main causes of greenhouse gas emissions is the extensive use of fossil fuels to power computer facilities and data centers. Digital technology's carbon footprint can be significantly decreased by using renewable energy sources (Adegbite et al., 2024, Talaviya et al. 2020).

18.7 AI in Agriculture and Cross-Disciplinary Relationships

In an interdisciplinary study, specialists from several fields concentrate on a single topic or problem. Because to produce a solution that addresses a variety of challenges, it necessitates the understanding and synthesis of concepts from several disciplines. A complex array of issues encompassing technological, social, ethical, economic, and environmental aspects surrounds the application of AI in agriculture. Since these elements are intricately linked, a complete strategy is necessary. For instance, a digital divide is created when AI technology is too expensive for most farmers, which raises questions about accessibility and fairness. These discrepancies could result in employment losses in the agriculture industry and exacerbate socioeconomic inequality, especially for smallholder farmers. To address these issues, a comprehensive, multidisciplinary strategy that involves the broad societal and environmental impacts of AI in agriculture is needed.

To optimize its potential and prevent unforeseen negative effects, the integration of AI in agriculture requires an interdisciplinary approach. To guarantee that AI systems benefit all stakeholders, including farmers, the environment, and society at large, cooperation and communication across multiple disciplines are necessary. A concentration on technology without a sound business plan might lead to discontinuation, while a focus only on financial gains can result in privacy violations, crop destruction, and injury to farmers. On the other hand, concentrating solely on moral issues may impede the development of new technologies. Therefore, implementing AI in a way that is both socially and financially responsible requires a balanced strategy that integrates technological, economic, and ethical factors (Ryan et al., 2024, Aggarwal et al., 2025, Dara et al., 2022).

To develop sustainable, efficient, and morally sound AI systems that promote more just and fruitful farming methods, a variety of disciplines must actively collaborate in the future of AI in agriculture.

18.8 Acceptance and Challenges of AI in Agriculture

One of the biggest problems is a lack of high-quality agricultural data. Due to the inherent variability and complexity of agricultural systems, gathering sizable, reliable, and representative datasets can be especially difficult. Data gathering is challenging and resource-intensive since these systems are impacted by a wide range of unpredictable elements, including weather, soil conditions, and farming practices. It is very impossible to create AI models that can provide trustworthy, accurate predictions and recommendations without high-quality data (Linaza et al., 2021, Tantalaki et al., 2019).

One of the main obstacles to the application of AI in agriculture is the scarcity of high-quality data. A significant obstacle to the application of AI in agriculture is the scarcity of high-quality data. It is difficult to get reliable and consistent data throughout large farms and during several seasons because of things like weather fluctuations, soil variances, and different agricultural methods. Even when data are gathered, they are frequently inconsistent or lacking, which hinders AI algorithms' ability to learn. Concerns about privacy and competition may make farmers hesitant to share their data, which would further impede the creation of reliable AI models. For AI to be successfully applied in agriculture, these data issues must be resolved (Tantalaki et al., 2019).

One major obstacle to AI's widespread adoption in agriculture is its high implementation costs. Both the acquisition and upkeep of specialized gear, such as robots and drones, as well as the massive volumes of data needed to train AI models are costly (Eli-Chukwu, 2019). Costs are rising because of the increasing demand for skilled personnel for creating and maintaining these systems. Furthermore, as big businesses buy up startups and monopolize the market, intellectual property rights frequently prevent access to cutting-edge AI technologies, which drives up prices for farmers. These financial limitations prevent AI from being widely used in the agriculture industry and restrict its availability to smaller farms (Ritambara et al., 2024).

For many farmers, especially in developing nations, the high cost of AI technology poses a significant barrier to entry, which can exacerbate existing inequalities and hamper progress toward more equitable and sustainable agricultural systems. To address these issues, efforts should concentrate on making AI technologies more accessible, which includes expanding internet connectivity through programs like community mesh networks and utilizing low-cost devices like smartphones and tablets, as well as lowering the cost of adopting AI for farmers through the development of affordable sensors and drones, open-source software, and subsidized access to cloud computing. By giving accessibility and democratizing access to AI, we can make sure that the advantages of these technologies are distributed fairly across all communities and regions (Ritambara et al., 2024).

18.9 Data Bias and Its Implications in AI Agriculture

Precision agriculture AI systems frequently use large datasets that may be biased, which affects the fairness and accuracy of their judgments. Farmers in underrepresented areas may suffer from biased results if, for example, models built exclusively on data from regions with distinct soil and climate characteristics perform poorly in other contexts. Additionally, because small-scale and marginalized farmers, especially in developing nations, could not have access to high-quality data or the resources necessary to put AI-driven solutions into practice, algorithmic bias could reflect social and economic inequalities. By favoring farms with more resources and advanced technology, this digital divide runs the risk of increasing already-existing disparities in the agriculture industry. It is essential to use inclusive data collection techniques and create flexible algorithms to guarantee justice and equity (Ryan et al., 2024).

18.10 Future Advancements and Plans

AI-based technologies are providing farmers throughout the world with accurate and useful answers to their biggest agricultural problems in the future. AI could supply solutions for practically everything (Javaid et al., 2022). From helping with farm work to weed control and weather forecasting by utilizing AI in the future. The use of data to optimize yields for individual plant rows made farming much easier and helped farmers increase output. Numerous businesses are creating AI-powered sensors and robots that can be used for multitasking in the agriculture industry. Advanced agricultural robots that can complete jobs faster and more accurately than humans are being developed thanks to AI. These robots are programmed to check for weeds and crop quality while harvesting and packing crops.

These robots may reduce the obstacles that agricultural workers must overcome. Farmers can receive real-time warnings on their phones when AI systems detect and prevent insect infestations by analyzing satellite imagery and historical data. This proactive approach to pest management can prevent the use of hazardous pesticides and drastically cut crop losses (Ritambara et al., 2024). Even though AI has enormous potential in agriculture, it is imperative to guarantee that these technologies are accessible to everybody. Since smallholder farmers in poor nations frequently lack the infrastructure and resources to implement cutting-edge technology, the emphasis must be on creating and implementing AI solutions that are affordable for them. We may ensure that all farmers benefit from this revolutionary technology by closing the digital divide and democratizing access to AI.

18.11 Conclusions

Among the many issues facing the agriculture industry are inadequate irrigation, weed management, and crop monitoring challenges brought on by things like plant height and harsh weather. But technological developments, especially those involving AI, present encouraging paths to getting over these challenges. For example, remote sensors can precisely measure the amount of moisture in the soil, allowing an automated irrigation system to maximize water use. By efficiently weeding crops, autonomous robots can minimize the use of hazardous chemicals and eliminate the need for physical labor. While sophisticated image analysis techniques allow for effective crop monitoring, drones can be used for precision pesticide and herbicide application.

The process of describing plant features, known as phenotyping, can be made much more efficient using AI. Manual labor-intensive processes are frequently used in traditional ways of evaluating plant height, soil texture, and other qualities. However, high-throughput phenotyping data may be quickly and non-destructively acquired by AI-powered systems, offering important insights for crop optimization and development.

Agriculture could undergo a revolution with the integration of AI, IoTs, and SIS technologies. Increased yields and lower costs can result from these technologies' ability to optimize resource management, boost disease prediction, and improve crop monitoring. Large volumes of data gathered from sensors and drones may be analyzed by AI-powered systems, which can then spot trends and insights that people might overlook. For example, they can use photos to diagnose plant illnesses or anticipate crop performance. Farmers can benefit greatly from AI-based monitoring, which can improve communication within the agricultural community and offer tailored agricultural advice. However, several obstacles, such as exorbitant prices, restricted infrastructure accessibility, and the requirement for technical know-how, prevent these technologies from being widely adopted. Since trustworthy AI models need high-quality data, data availability and quality are also crucial considerations. Furthermore, ethical issues including data ownership, privacy, and the possible effects of automation on jobs in rural areas must be carefully considered.

A multidimensional strategy is needed to overcome these obstacles, including creating reliable data-sharing frameworks, increasing internet connectivity in remote places, and creating AI solutions that are both accessible and reasonably priced. We can use AI to make agriculture more efficient, inclusive, and sustainable in the future by resolving these problems and guaranteeing fair access to these technologies. Through AI-powered solutions that boost productivity and revenue without depleting valuable natural resources, farmers can meet the world's demand for more profitable and abundant food. AI benefits farmers by reducing the usage of herbicides, improving crop quality, increasing revenues, and realizing significant cost savings. AI solutions could help farmers with issues like climate change and plant and insect infestations that reduce crops.

References

Adegbite, N.A.O., Barrie, N.I., Osholake, N.S.F. et al. (2024). Artificial intelligence in climate change mitigation: a review of predictive modeling and data-driven solutions for reducing greenhouse gas emissions. *World Journal of Advanced Research and Reviews* 24 (1): 408–414. https://doi.org/10.30574/wjarr.2024.24.1.3043.

Aggarwal, S., Mehta R.A., Kumar P., et al. (2025). AI-powered innovations in agriculture: A systematic review on plant disease detection and classification. *Indian Journal of Agricultural Research*, 59(9): 1321–1330. https://doi.org/10.18805/IJARe.A-6371.

Ahirwar, S., Swarnkar, R., Bhukya, S., and Namwade, G. (2019). Application of drone in agriculture. *International Journal of Current Microbiology and Applied Sciences* 8 (01): 2500–2505. https://doi.org/10.20546/ijcmas.2019.801.264.

Atapattu, A.J., Perera, L.K., Nuwarapaksha, T.D. et al. (2024). Challenges in Achieving Artificial Intelligence in Agriculture. https://doi.org/10.1007/978-981-97-5878-4_2.

Bakri, K.Z.S.A., Flaiti, M.K.S.A., Al-Balushi, S.A.S., and Poorngalingam, J. (2024). Revolutionizing agriculture: harnessing the power of artificial intelligence for sustainable farming practices. *International Journal of Advanced IT Research and Development (IJAITRD)*, 01(01). https://doi.org/10.69942/1920184/20240101/01.

Blasco, J., Aleixos, N., Roger, J. et al. (2002). AE – automation and emerging technologies. *Biosystems Engineering*, 83(2), 149–157. https://doi.org/10.1006/bioe.2002.0109.

Bronson, K., Rotz, S., and D'Alessandro, A. (2021). The human impact of data bias and the digital agricultural revolution. In: *Edward Elgar Publishing eBooks*. https://doi.org/10.4337/9781839101748.00017.

Cham, Switzerland: Springer.Bronson, K., and Knezevic, I. (2024). Ethical data governance and digital agricultural transformation. Journal of Responsible Innovation, 11(2), 300–318.

Dara, R., Fard, S.M.H., and Kaur, J. (2022). Recommendations for ethical and responsible use of artificial intelligence in digital agriculture. *Frontiers in Artificial Intelligence* 5: https://doi.org/10.3389/frai.2022.884192.

Das, S., Kaur, M., Chhabra, V. et al. (2024). A systematic review of artificial intelligence: a future guide to sustainable agriculture. *International Journal of Environment and Climate Change*, 14(4), 562–573. https://doi.org/10.9734/ijecc/2024/v14i44139.

Davenport, T. H., Guha, A., Grewal, D., and Bressgott, T. (2020). How artificial intelligence will change the future of business. Journal of the Academy of Marketing Science, 48(1): 24–42. https://doi.org/10.1007/s11747-019-00696-0.

Dharmaraj, V., and Vijayanand, C. (2018). Artificial intelligence (AI) in agriculture. International Journal of Current Microbiology and Applied Sciences, 7(12): 2122–2128. https://doi.org/10.20546/ijcmas.2018.712.241.

Elbasi, E., Mostafa, N., Zaki, C. et al. (2024). Optimizing agricultural data analysis techniques through AI-powered decision-making processes. *Applied Sciences*, 14(17), 8018. https://doi.org/10.3390/app14178018.

Eli-Chukwu, N.C. (2019). Applications of artificial intelligence in agriculture: a review. *Engineering Technology & Applied Science Research* 9 (4): 4377–4383. https://doi.org/10.48084/etasr.2756.

Gangwani, N. (2024). AI-driven precision agriculture: optimizing crop yield and resource efficiency. *International Journal for Multidisciplinary Research* 6 (6): https://doi.org/10.36948/ijfmr.2024.v06i06.29913.

At a compound annual growth rate (CAGR) of approximately 22.5% between 2023 and 2030, the global agriculture AI industry is expected to rise from its expected USD 1.4 billion in 2022 to USD 7.1 billion. https://www.zionmarketresearch.com/report/agriculture-ai-market.

Javaid, M., Haleem, A., Khan, I.H., and Suman, R. (2022). Understanding the potential applications of artificial intelligence in agriculture sector. *Advanced Agrochem* 2 (1): 15–30. https://doi.org/10.1016/j.aac.2022.10.001.

Kamble, S. S., Gunasekaran, A., and Gawankar, S. A. (2019). Achieving sustainable performance in a data-driven agriculture supply chain: a review for research and applications. *International Journal of Production Economics*, 219, 179–194. https://doi.org/10.1016/j.ijpe.2019.05.022.

Linaza, M. T., Posada, J., Bund, J. et al. (2021). Data-driven artificial intelligence applications for sustainable precision agriculture. *Agronomy*, 11(6), 1227. https://doi.org/10.3390/agronomy11061227.

Mana, A., Allouhi, A., Hamrani, A. et al. (2024). Sustainable AI-based production agriculture: exploring AI applications and implications in agricultural practices. *Smart Agricultural Technology*, 7, 100416. https://doi.org/10.1016/j.atech.2024.100416.

Mishra, N. A. C., Das, N. J., and Awtar, N. R. (2024). An emerging era of research in agriculture using AI. *Journal of Scientific Research and Technology*, 1–7. https://doi.org/10.61808/jsrt93.

Ntoutsi, Y., and Nakano, J. (2022). International comparison of cross-disciplinary integration in industry 4.0: a co-authorship analysis using academic literature databases. *PLoS ONE*, 17(10), e0275306. https://doi.org/10.1371/journal.pone.0275306.

Mohammed, M.A., Jameel, S.H., Hussein, A.J. et al. (2024). The transformation of agriculture by artificial intelligence in smart farming. In: *36th Conference of Open Innovations Association (FRUCT), 2024*, 515–526. https://doi.org/10.23919/fruct64283.2024.10749901.

Ntoutsi, E., Fafalios, P., Gadiraju, U. et al. (2020). Bias in data-driven artificial intelligence systems – an introductory survey. *Wiley Interdisciplinary Reviews Data Mining and Knowledge Discovery*, 10(3). https://doi.org/10.1002/widm.1356.

Pal, S. (2023). Integrating AI in sustainable supply chain management: a new paradigm for enhanced transparency and sustainability. *International Journal for Research in Applied Science and Engineering Technology*, 11(6), 2979–2984. https://doi.org/10.22214/ijraset.2023.54139.

Patel, H., Samad, A., Hamza, M. et al. (2022). Role of artificial intelligence in livestock and poultry farming. *Sinker* 7 (4): 2425–2429. 10.33395/sinkron.v7i4.11837.

Rai, D. (2022). Artificial intelligence in agriculture: A data-driven approach to sustainable farming. International Journal of Science, Engineering and Technology, 13(3), 45–52.

Rai, K. K. (2022). Integrating speed breeding with artificial intelligence for developing climate-smart crops. *Molecular Biology Reports*, 49(12), 11385–11402. https://doi.org/10.1007/s11033-022-07769-4.

Rane, J., Kaya, Ö., Mallick, S.K., and Rane, N.L. (2024). Generative artificial intelligence in agriculture, education, and business. *Generative Artificial Intelligence in Agriculture, Education, and Business.* https://doi.org/10.70593/978-81-981271-7-4.

Ritambara, N., Kaushal, S., and Shubham, N. (2024). Frontiers of artificial intelligence in agricultural sector: trends and transformations. *Journal of Scientific Research and Reports*, 30(10), 970–980. https://doi.org/10.9734/jsrr/2024/v30i102518.

Ryan, M. (2022). The social and ethical impacts of artificial intelligence in agriculture: Mapping the agricultural AI literature. AI & Society. Advance online publication. https://doi.org/10.1007/s00146-022-01520-8.

Ryan, M., Isakhanyan, G., and Tekinerdogan, B. (2023). An interdisciplinary approach to artificial intelligence in agriculture. *NJAS Impact in Agricultural and Life Sciences*, 95(1). https://doi.org/10.1080/27685241.2023.2168568.

Saiz-Rubio, V. and Rovira-Más, F. (2020). From smart farming towards agriculture 5.0: a review on crop data management. *Agronomy*, 10(2), 207. https://doi.org/10.3390/agronomy10020207.

Sherkhane, P.D. and Ratnaparkhi, N.S. (2024). Deep learning techniques used in agriculture: a review. *International Journal for Research in Applied Science and Engineering Technology* 12 (3): 1–5. https://doi.org/10.22214/ijraset.2024.58567.

Singh, S. (2016). Progress and performance of agriculture in India. *Journal of Agroecology and Natural Resource Management* 2016: http://eprints.icrisat.ac.in/14557/.

Sivarethinamohan, R., Yuvaraj, D., Priya, S.S., and Sujatha, S. (2021). Captivating profitable applications of artificial intelligence in agriculture management. *Advances in intelligent systems and computing* 848–861. https://doi.org/10.1007/978-3-030-68154-8_73.

Srikanth, R., Rekha, K. S., Kiran, D. R., and Raju, P. V. (2020). Use of artificial intelligence in ipm#. *Indian Journal of Entomology*, 82(4), 609–616. https://doi.org/10.5958/0974-8172.2020.00183.2.

Talaviya, T., Shah, D., Patel, N. et al. (2020). Implementation of artificial intelligence in agriculture for optimization of irrigation and application of pesticides and herbicides. *Artificial Intelligence in Agriculture* 4: 58–73. https://doi.org/10.1016/j.aiia.2020.04.002.

Tantalaki, N., Souravlas, S., and Roumeliotis, M. (2019). Data-driven decision making in precision agriculture: The rise of big data in agricultural systems. Journal of Agricultural Informatics, 10(1): 1–12.

Victoire, T.A., Karunamurthy, A., Sandhiya, S., and Yuvaraj, S. (2023). Leveraging artificial intelligence for enhancing agricultural productivity and sustainability. *Quing International Journal of Innovative Research in Science and Engineering* 2 (2): 141–156. https://doi.org/10.54368/qijirse.2.2.0016.

19

Ethical and Regulatory Considerations in AI-Driven Crop Improvement

Praveen Kumar Maddheshiya[1], *Kapil Gupta*[1], *Shubhra Gupta*[1], *Kiran Gupta*[2], *and Ravindra Pratap Singh*[3]

[1]*Siddharth University, Department of Biotechnology, Siddharth Nagar, Kapilvastu 272205, UP, India*
[2]*Siddharth University, Department of Botany, Siddharth Nagar, Kapilvastu 272205, UP, India*
[3]*Rajarshi Tandon Open University, School of Sciences (Biochemistry), Prayagraj 211021, UP, India*

19.1 Introduction

The agriculture industry is searching for numerous solutions for such difficulties, one of which is data analytics. Big Data analytics is seen as the fourth technology revolution in agriculture, with the goal of providing a solution to our increasing food need (Kumari et al. 2016; Morota et al. 2018; O'Grady and O'Hare 2017). According to Zhang et al. (2015) and Carolan (2017), Big Data analytics are going to have an essential part in the future of agriculture. Agricultural Big Data, data analytics, and machine learning algorithms are projected to serve as catalysts for achieving the world's agricultural goals.

Agricultural Big Data analytics is the examination of massive datasets from a variety of sources, frequently utilizing artificial intelligence (AI) approaches. The agricultural sector's successful expansion is predicted to rely heavily on the integration of Big Data and AI (Smart Information Systems – SIS). Agribusinesses are changing their focus to data-driven agriculture solutions. While these advancements are viewed as viable approaches to attain the difficult goals ahead, they also create a number of major social and ethical issues. AI is one of the most quickly evolving digital technologies in agriculture (Ryan et al. 2023).

A promising connection between AI and the farming industry has emerged, with the potential to transform how resources are handled and food is produced around the world (Slimani et al. 2024). As we try to produce larger quantities of food using fewer resources, AI has the potential to spark a revolution in agriculture (Vijayakumar et al. 2022). The basic objective of AI is to create machinery that functions exactly like the human brain. With the help of AI, it is possible to collect massive amounts of data via public and open-access websites and to analyze it, and provide farmers with solutions to a variety of complex issues, as well as an additional intelligent method to handle inputs by taking into account all field variations, resulting in a higher yield.

19.1.1 The Use of SIS in Agriculture

Before exploring the social and ethical implications of employing SIS technology in agriculture, it is critical to recognize the reasons why these technologies are developed on the first place. To fully comprehend the societal and ethical implications of employing such technologies, it is important to elaborate on the different types of data being gathered, where they are retrieved from, and how they are used. To begin, agricultural SIS integration consists of numerous components, including data retrieval, analysis, and prescription. The different types of information retrieved vary from animal behavior and grazing patterns, moisture in the soil and nutrient levels, irrigation, rainfall and climate, land photography, crop development patterns, and market price (Bennett 2015; Kamilaris et al. 2017).

19.2 The Potential of AI for Agriculture

In recent years, AI has been widely applied in agriculture. AI technology is used in agriculture to improve numerous administrative chores associated with the food supply chain, as well as to grow better crops, control pests, monitor soil and cultivation circumstances, and analyze data for farmers (Javaid et al. 2023; Upadhyay 2026). It makes it harder for farmers to determine the best season for seed germination. Farmers can utilize AI to identify the optimal seed for a certain meteorological circumstance (Javaid et al. 2023). It also delivers weather predictions.

AI-powered technology will allow farms to produce more from fewer resources. Improving crop quality and reducing product time to market (Javaid et al. 2023). AI improves understanding of soil qualities. AI helps farmers by guiding them on which nutrients to add for enhancement of the soil. Farmers can utilize AI to help them choose the best time

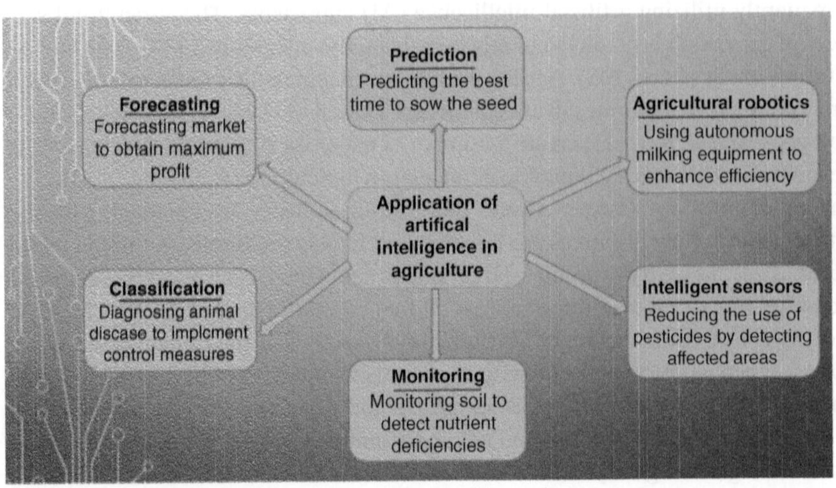

Figure 19.1 Application of AI in the agriculture field.

to sow their seeds. Intelligent gear can compute the optimum planting depth as well as the seed spacing. An AI-powered monitoring system can provide farmers with information about their crops' overall status as well as the nutrients they need to boost yield quality and volume (Javaid et al. 2023).

As discussed below (Figure 19.1), there are some sectors of agriculture where AI has great potential.

19.2.1 Data Used in Agriculture Field

Data are obtained from a variety of sources, including weather stations, surveys, static historical datasets, geospatial data, satellite imaging, farm sensors, farm equipment sensors, radiation sensors, temperature sensors, and GPS-based field maps (Tzounis et al. 2017; Ribarics 2016). The data are used in several contexts, including climate and weather, land, animal studies, crops, soil, weeds, availability of food and security, biodiversity, and farmer security and financing (Kamilaris et al. 2017). The data used in the agriculture sector are explained in Figure 19.2, Table 19.1.

Considering the potential value of using agricultural Big Data in any of these scenarios, the amount of data now being studied remains minimal (Hirafuji 2014). This is expected to increase significantly due to the numerous benefits offered by the proper application of agricultural Big Data. These include improving water and air quality, soil health, food quality and security, biodiversity protection, quality of life, increasing output, lowering costs, crop forecasting, and improving choice-making and effectiveness (Castle et al. 2015; Mintert et al. 2016; O'Grady and O'Hare 2017, Schönfeld et al. 2018, and Sonka and Cheng 2015).

Figure 19.2 Data used in the agriculture field.

Table 19.1 Data used in the agriculture field.

Soil Data	Data of soil composition, pH, moisture, and nutrient level in soil	Helpful for decision making to farmer, soil's health monitor and fertilizer applications	Karthika 2024
Weather Data	Rainfall, humidity, wind patterns, and temperature	Used for irrigation, crop planning, and pest control	Adamo et al. 2021
Satellite & Remote Sensing Data	Spectral data from satellites and drones which gives high-resolution images	Which enhance crop monitoring, land use mapping, and prediction of disaster	Sishodia et al. 2020
Genomic Data	Genetic details of crops and livestock	Helps in genetic modification, supports plant breeding and resistance of disease	Priyatham et al. 2025
Market Data	Demand trends, prices and supply chain insights	Helps the farmer for making decision on crop selection and sales	De Camillis and McAllister 2024
Pest & Disease Data	Resistance patterns, disease prevalence, and information on pest outbreaks	Early disease detection aids in integrated pest management (IMP)	Abdullah et al. 2023
Farm Management Data	Yield records, resource allocation and machinery usages	Helps in cost reduction, form optimization, and sustainability	Mohamed et al. 2021

19.3 Ethical Issues of Using SIS in Agriculture

19.3.1 Accuracy of Data and Recommendations

The primary objective of integrating SIS technology in agriculture is to enable improved decisions, adaption to situations, and recommendations (Talavera et al. 2017). However, some argue that machine learning is unfit for purpose because the algorithms used are only appropriate for limited datasets (Zhang et al. 2015). Because of their failure "to strike a compromise between timeliness and precision of processing," these algorithms cannot efficiently analyze Big Data or huge datasets (Zhang et al. 2015). This is a technical restriction that must be overcome in the agriculture sector, and, more broadly, the SIS industry as a whole. Limited data may sometimes result in incorrect conclusions. Agricultural Trade and Policy Section (ATPs) may make prescriptive recommendations that have negative consequences because they are based on incorrect or false information (Taylor and Broeder 2015).

It is also possible that data retrieval is false or incorrect due to environmental factors. Animals are a basic example that can cause problems with SIS technologies by influencing the radio signals used to communicate, being too close to sensors, or interfering with the equipment. Sensors can be protected against damage; however, there are also worries about conditions that cause incorrect readings, such as temperature extremes and humidity (Tzounis et al. 2017). Possible interferences must be evaluated in order to reduce erroneous readings, distorted analytics, and misleading prescriptions.

19.3.2 Data Ownership and Intellectual Property

There are issues regarding the sharing of farm data with third parties (Rosenheim and Gratton 2017). Farmers feel worried that their data would end up in the wrong hands and used against themselves (Ferris 2017). Some farmers are concerned that disclosing their data may leave them in a vulnerable situation in future (Coble et al. 2018). People are concerned about how their data will be collected and shared with regulatory organizations, agencies, and government officials (Sykuta 2016). Their data may be used against individuals in many kinds of conditions, including regulatory enforcement, the imposition of penalties, charges, and restrictions.

There is also a risk that their data would be used by traders in commodities on the stock exchange (Ferri 2017). This might be used against farmers by obtaining specific knowledge about the farm, allowing traders to purchase it for a lower price, or as an instrument of bargaining against the farmer. Many ATPs require that farmers control their data, although the ATPs may include a free from royalty license that allows the ATP to utilize the data independent of proprietorship (Darr 2014). If farmers control their data and wish to switch to a different ATP, they might be in breach of contract.

Monsanto, an American company is a perfect example of data ownership and their strict regulatory controls over their intellectual property and data analytics, and if a farmer violates their contract, they may face penalties and/or legal action. Furthermore, if a farmer is looking for a different ATP, it may be challenging or even impossible to identify another ATP because of data ownership issues: ATPs may have concerns about getting data from farmers that the farmer herself does not own, providing rise to potential breaches of intellectual property or licensing restrictions (Sykuta 2016).

19.3.3 Economic Issues and the Digital Divide

Globally, there are countless more small farms than large farms. In the United States alone, 66% of farms have annual sales of less than $1 million (USDA NASS, 2012). In LMICs (low-to-middle-income countries), most agriculture takes place on tiny farms with limited technology. However, most agricultural data analytics are conducted on vast monoculture industrial farms (Carbonell 2016). This may result in disproportionate growth of larger farms and the breakup of smaller farms. SIS technologies are somewhat expensive, which may deter poorer farmers from using them (Kosior 2017; Schönfeld et al. 2018).

This creates a "digital divide" between those who can and cannot apply SIS (Kamilaris et al. 2017). Farmers in rural and isolated areas may also face data transmission difficulties, preventing them from adopting these technologies. As a result, SIS technologies have the potential to generate or exacerbate disparities between people who can and cannot use them (Poppe et al. 2014).

19.4 Scenario of Digital Divide in Agriculture

The digital gap is a big concern in India's agricultural sector. There are several hurdles to agricultural expansion in India. Digitalization in agriculture is critical to reducing agricultural challenges and accelerating agricultural success in India. According to research by the National Remote Sensing Centre (NRSC), India has 182 million hectors of planted fields. As a result, digitalization helps to reduce the number of resources required for farming, which saves water, reduces costs, and saves agricultural fields. According to the Indian Council of Agriculture Research's 2021 experiment report, pests and diseases harm 20–30% of crops (Javaid et al. 2022; Paul et al. 2016; Warren 2002).

As a result, real-time data analysis is essential to prevent pest and disease destruction of crops. According to the Internet and Mobile Association of India (IAMAI), 60% of Indians have access to the Internet, with rural areas accounting for only 29% of the total. As a result, a substantial number of farmers do not have access to the internet. According to the World Bank survey, 23.4% of people in rural areas lack the necessary skills and digital literacy. As a result, all of these issues have an impact on the development of digital agriculture in India. 80% of Indian farmers are from poor and marginal families, and they lack the necessary capital to apply digitalization in agricultural regions in India.

19.5 Ethical and Regulatory Considerations in AI-Driven Crop Improvement

AI-based crop enhancement is revolutionizing agriculture by improving precision breeding, boosting resource utilization, and increasing food security (Ryan, 2023). Nonetheless, AI applications in agriculture face significant ethical challenges, including data privacy, algorithmic bias, and the potential exclusion of small-scale farmers from innovation (Dara et al. 2022) (Table 19.2). Furthermore, regulation of AI use presents issues, particularly in

Table 19.2 Ethical and regulatory considerations in AI-driven crop improvement.

Category	Consideration	Description	References
Ethical Considerations	Data Privacy & Ownership	Protection of farmers' genetic, environmental, and personal data collected for AI training. Risks of data misuse by corporations.	Stilgoe 2020
	Fair Access to AI Technologies	Ensuring smallholder farmers have equitable access to AI tools, preventing monopolization by agritech corporations.	Lassoued et al. 2019
	Transparency & Explainability	AI models must provide clear, interpretable results so that farmers and regulators can trust their recommendations.	Dara et al. 2022
	Algorithmic Bias & Social Inequalities	AI models may inherit biases from training data, disproportionately benefiting certain crops, regions, or farming communities.	Dignum 2019
	Environmental Sustainability	AI-driven breeding or genetic modification should avoid harmful ecological impacts such as biodiversity loss.	Barnes et al. 2019
	Ethical Use of GMOs	AI-assisted genetic modification must align with ethical concerns about altering crop genomes and potential long-term impacts.	Qaim 2009
	Autonomy & Farmer Decision-Making	AI-driven automation should not undermine farmers' ability to make independent choices about cultivation practices.	(FAO 2021)
	Public Trust & Perception	AI-driven crop improvements may face resistance if the public is not adequately informed about safety and benefits.	(Juma, 2016)
	Cultural Sensitivities	Some communities have cultural or religious objections to AI-driven genetic modifications of crops.	(Sharma and Tripathi 2021)

Table 19.2 (Continued)

Category	Consideration	Description	References
Regulatory Considerations	Compliance with Biosafety Laws	AI-driven genetic modifications must meet national and international biosafety regulations.	(ISAAA, 2021)
	Intellectual Property & Patents	AI-generated plant varieties may be subject to patent laws, raising concerns over monopolization.	(Graff and Sherkow 2020)
	Regulatory Approvals for AI-assisted GMOs	Government agencies (e.g., USDA, EFSA) assess the safety of AI-assisted genetically modified crops before market release.	(USDA, 2022)
	Labeling & Consumer Rights	Regulations may require clear labeling of AI-developed crops to inform consumers.	(Veale and Zuiderveen Borgesius 2021)
	Ethical AI Governance in Agriculture	International guidelines on ethical AI use in food and agriculture, ensuring responsible innovation.	(FAO 2021)
	Cross-Border Trade & Export Regulations	AI-assisted crop varieties must comply with international trade regulations, including biosafety protocols.	(OECD 2022)
	Liability & Accountability	Clear legal frameworks are needed to define responsibility if AI-driven crop improvements cause unintended consequences.	Stilgoe 2020
	Data Sharing & Open Science	Balancing proprietary AI research with open-access data-sharing policies to support global agricultural innovation.	Williams 2016
	Climate-Resilient Crop Development	Regulatory frameworks must support AI applications that enhance climate resilience while ensuring food safety.	(FAO 2021)

terms of genetic modifications and intellectual property rights (Falola et al. 2024). To ensure fair benefits and long-term viability, governments must pass open legislation, promote ethical AI design, and support multi-stakeholder involvement (Wójcik-Gront et al. 2024).

19.6 Ethical Framework for Assessment of Artificial Intelligence-Based Solutions in Agriculture

This section describes a framework for ethical AI. This paradigm throws light on the ethical considerations and issues that may develop as a result of using AI-based farming solutions. Studying the ethical framework can provide us with a clear perspective on how to approach the creation and development of AI-based technologies for digital agriculture. Building ethical technology ensures ethical validity and responsibility while also gaining the trust of stakeholders, particularly farmers, in the agricultural industry. The ethical standards are explored in six categories (Figure 19.3).

Fairness is one of the fundamental principles of ethical AI. Fairness is defined as monitoring and minimizing bias in the AI model, collecting and using data fairly, and providing equitable access to digital assets (Jobin et al. 2019). Bias in an AI model refers to the tendency to learn a desired pattern of data rather than the actual data distribution when building the model. The AI model will be biased if it makes prejudiced assumptions throughout the algorithm building-phase or if the data contains preconceptions. Furthermore, if the obtained data are not an accurate depiction of the context/problem, the AI model created from it may be biased. AI bias can have an impact on agricultural applications.

Figure 19.3 Ethical artificial intelligence principal in digital agriculture.

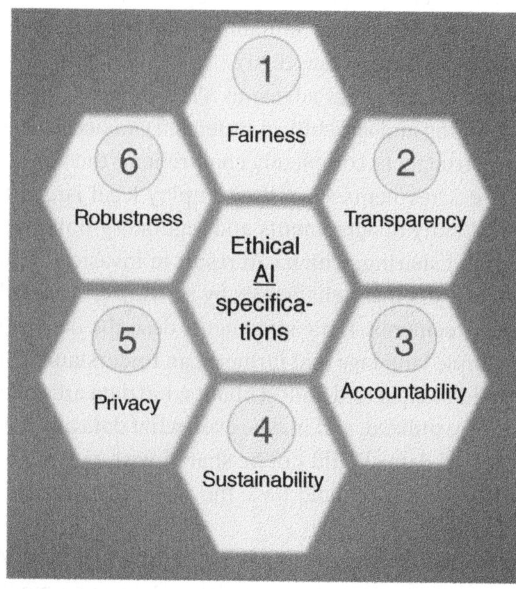

For example, suppose an autonomous apple-picking application is designed to detect ripe apples. If the photos acquired for the AI model are primarily of red apples, the final model will be biased toward these apples. In an orchard with green apples, this model may misidentify ripe green apples as raw apples and fail to pick them. Having a broad set of apple photos representing different types and phases of growth can help to build a more trustworthy and inclusive model. AI systems should not only be bias-free, but also adhere to society norms such as human rights, democratic principles, and diversity (Shukla 2024). Another unjust situation that may arise as a result of the employment of AI systems in agriculture is unequal power distribution caused by access to and control over farm data and AI technologies. When technology and service providers have access to data from an extensive number of farms and can operate AI systems, they can use this ability to manipulate the market, communicate data with third parties, and overrule farmers' rights to their own assets (Ryan 2020). With the right application of the fairness principle, all members of the community will benefit from AI tools in a fair and inclusive manner.

19.7 Recommendations for Agricultural Technology Providers

In the farming ecosystem, ATPs are companies that develop smart farming technologies and platforms, as well as install, integrate, and manage farm technologies (Dara et al. 2022). In certain ways, they might be viewed as custodians of agricultural data and technology. Technology stewardship in agriculture is the process of using, managing, or supervising technology on farms (Gow 2020). In many circumstances, ATPs acquire and oversee agricultural data; as technology stewards, they should be held accountable for data governance and management (Anidu and Dara 2021). ATPs can also access technology governance issues including security, sustainability, and privacy, and advise farmers on how to handle these ethical concerns (Gow 2020).

This section gives advice to ATPs on how to implement ethical and responsible AI use in farming and agriculture systems. Data license agreements ensure transparency. Farmers may struggle to completely comprehend the substance of most privacy policies and terms of use agreements since they employ legal jargon (Kaur et al. 2018). Another issue is the uncertainty of agreements and regulatory frameworks governing data collecting, processing, and sharing, which can result in invasions of farmers' privacy and data misuse. ATPs can address these challenges by providing clear, full, and unambiguous data and terms of use agreements. Data agreements describe data practices and approved use of technology in simple language that farmers can understand (Carbonell 2016). In the agreements, ATPs should be clear and honest about what data are collected from farmers and their farms, why data are utilized, and maybe even what data are used to construct AI models. Furthermore, acquired data should not be shared with outside parties unless farmers agree (Kaur et al. 2018). Farmers should have the option to give or withdraw authorization for the usage of their data.

Finally, these agreements should specify how ATPs would approach data ownership and farmer access to their data. Data agreements should also include specifics about data retention, farm data security, cross-border data integration and transfer (where applicable), and other privacy and confidentiality procedures (Tsesis 2019). Transparency can also be improved using processes and technologies. Data access, portability, and harmonization are some of the recommended techniques for increasing transparency. Farmers must be able to download, access, use, and transmit historical data to other farm management system suppliers in order to comply with these principles. To accomplish this, farmers' data must be available in a digitized, well-structured, and machine-readable format. For example, an aviculture farmer may desire to employ a newer technology given by another ATP to run their poultry farm or create tools to check their animals' health.

Farmers may also choose to share their agricultural data with cooperatives so that it can be sold, integrated, or used for other purposes. To accomplish this, as well as address data usability and sustainability, ATPs should provide data portability for agricultural data. Data portability refers to the capacity of users to move their data between other workstations, systems, and applications. ATPs should also address data interoperability and standards when creating their platform for digital systems for agriculture (Engel et al. 2006). Interoperability refers to a system's capacity to store, consume, and share data across a common platform. Interoperability allows you to collaborate with data from other domains and applications (Jaleel et al. 2020).

19.8 Conclusion

On the basis of AI, crop enhancement presents an unprecedented opportunity to enhance yields from agriculture, sustainability, and food security. AI enables precise farming, continuous observation, with minimum resource like utilization of machine learning, big data analysis, and automation. But along with its potential benefits, the utilization of AI in agricultural activities is presenting significant ethical and regulatory hurdles that need to be met so as to ensure responsible innovation. One of the biggest ethical concerns is confidentiality and ownership. AI technology relies on vast databases, like farm histories, genomic information, and climate data, posing questions regarding ownership and benefit of this information. In the absence of specific legislation, small farmers could be taken advantage of, and data monopolization by big agribusiness companies could further increase existing inequalities. From the regulatory point of view, AI-based genetic alteration, mechanized decision-making, and intellectual property protection are real challenges. An absence of harmonized global policies for AI use in crop improvement leads to discrepancies in safety evaluation, clearance procedures, and responsibility questions. Moreover, policy frameworks should take into account the environmental implications of AI-based agriculture, such as loss of biodiversity and soil contamination.

In order to overcome these obstacles, multi-stakeholder involvement is the way forward. Governments, scientists, agribusinesses, and farmers should join hands in forming open and inclusive regulatory frameworks. Ethical guidelines for AI usage, like fairness, accountability, and explainability, need to be used for deploying AI in agriculture. Moreover, policies ought to support the equitable accessibility of AI tools such that technology benefits all farmers irrespective of scale.

References

Abdullah, H.M., Mohana, N.T., Khan, B.M. et al. (2023). Present and future scopes and challenges of plant pest and disease (P&D) monitoring: remote sensing, image processing, and artificial intelligence perspectives. *Remote Sensing Applications: Society and Environment* 32: 100996.

Adamo, N., Al-Ansari, N., and Sissakian, V. (2021). Review of climate change impacts on human environment: past, present and future projections. *Engineering* 13 (11): 605–630.

Anidu, A. and Dara, R. (2021, October). A review of data governance challenges in smart farming and potential solutions. In: *2021 IEEE International Symposium on Technology and Society (ISTAS)*, 1–8. IEEE.

Barnes, J.C. et al. (2019). Genetic engineering and society. In: *Science, Technology, and Society: New Perspectives and Directions* (ed. T.L. Pittinsky), 203–233. Cambridge: Cambridge University Press.

Bennett, J.M. (2015). Agricultural big data: utilisation to discover the unknown and instigate practice change. *Farm Policy Journal* 12 (1): 43–50.

Carbonell, I.M. (2016). The ethics of big data in big agriculture. *Internet Policy Review* 5 (1): 1–13.

Carolan, M. (2017). Publicising food: big data, precision agriculture, and co-experimental techniques of addition. *Sociologia Ruralis* 57 (2): 135–154.

Castle, M., Lubben, B.D., and Luck, J. (2015) 'Precision Agriculture Usage and Big Agriculture Data', *Cornhusker Economics*, University of Nebraska–Lincoln, 27 May, pp. 1–4. Available at: University of Nebraska Digital Commons.

Coble, K.H., Mishra, A.K., Ferrell, S., and Griffin, T. (2018). Big data in agriculture: a challenge for the future. *Applied Economic Perspectives and Policy* 40 (1): 79–96.

Dara, R., Hazrati Fard, S.M., and Kaur, J. (2022). Recommendations for ethical and responsible use of artificial intelligence in digital agriculture. *Frontiers in Artificial Intelligence* 5: 884192.

Darr, M. (2014) "Big Data" The catalyst for a transformation to digital agriculture," in Proceedings of the Integrated Crop Management Conference. Iowa State University, Digital Press. Available at: https://doi.org/10.31274/icm-180809-136.

De Camillis, C. and McAllister, T. (2024). Agrifood systems transformation for climate action and environmental improvements, with co-benefits for food security and nutrition. *The International Journal of Life Cycle Assessment* 1–4.

Dignum, V. (2019). *Responsible Artificial Intelligence: How to Develop and Use AI in a Responsible Way*, vol. 2156. Cham: Springer.

Engel, K., Pharowe, P., and Blobel, B.G.M.E. (2006). Semantic Interoperability. *Methods of Information in Medicine* 45 (04): 343–353.

Falola, P.B., Adeniyi, A.E., Madamidola, O.A. et al. (2024). Artificial intelligence in agriculture: the potential for efficiency and sustainability, with ethical considerations. In: *Exploring Ethical Dimensions of Environmental Sustainability and Use of AI*, 307–329. IGI Global Scientific Publishing Available at: https://doi.org/10.4018/979-8-3693-0892-9.ch015.

FAO (Food and Agriculture Organization of the United Nations) (2021). *AI in food and agriculture: ethical and governance challenges*, 1–52. Rome: FAO. https://www.fao.org.

Ferris, J.L. (2017). Data privacy and protection in the agriculture industry: is federal regulation necessary? *Minn. JL Sci. & Tech.* 18: 309.

Gow, G. (2020). *Introduction to Technology Stewardship for Agricultural Extension and Advisory Services*, 1–146. Edmonton: University of Alberta, Open Educational Resource Available at: eCampusOntario Open Library.

Graff, G.D. and Sherkow, J.S. (2020). Models of technology transfer for genome-editing technologies. *Annual Review of Genomics and Human Genetics* 21 (1): 509–534.

Hirafuji, M., 2014, April. A strategy to create agricultural big data. In 2014 Annual SRII Global Conference (pp. 249–250). IEEE.

Indian Council of Agricultural Research. (2021). *Annual Report 2021* (pp. 1–3). ICAR–National Bureau of Agricultural Insect Resources, Bengaluru, India.

ISAAA. (2021) Global Status of Commercialized Biotech/GM Crops: 2021. ISAAA Brief No. 56, pp. 1–60. https://www.isaaa.org/resources/publications/briefs/56/default.asp.

Jaleel, A., Mahmood, T., Hassan, M.A. et al. (2020). Towards medical data interoperability through collaboration of healthcare devices. *IEEE Access* 8: 132302–132319.

Javaid, M., Haleem, A., Singh, R.P., and Suman, R. (2022). Enhancing smart farming through the applications of Agriculture 4.0 technologies. *International Journal of Intelligent Networks* 3: 150–164.

Javaid, M., Haleem, A., Khan, I.H., and Suman, R. (2023). Understanding the potential applications of artificial intelligence in agriculture sector. *Advanced Agrochem* 2 (1): 15–30.

Jobin, A., Ienca, M., and Vayena, E. (2019). The global landscape of AI ethics guidelines. *Nature Machine Intelligence* 1 (9): 389–399.

Juma, C. 2016, *Innovation and Its Enemies: Why People Resist New Technologies*. Oxford University Press, pp. 1–432, New York, 2016, ISBN: 9780190467036.

Kamilaris, A., Kartakoullis, A., and Prenafeta-Boldú, F.X. (2017). A review on the practice of big data analysis in agriculture. *Computers and Electronics in Agriculture* 143: 23–37.

Karthika, R. (2024). Role of precision agriculture in soil fertility and its application to farmers. In: *Remote Sensing in Precision Agriculture*, 137–147. Academic Press.

Kaur, J., Dara, R.A., Obimbo, C. et al. (2018). A comprehensive keyword analysis of online privacy policies. *Information Security Journal: A Global Perspective* 27 (5–6): 260–275.

Kosior, K., 2017, July. Agricultural education and extension in the age of Big Data. *In European Seminar on Extension and Education.*

Kumari, S.V., Bargavi, P., and Subhashini, U. (2016). Role of big data analytics in agriculture. *Int. J. Comput. Sci. Math. Eng* 3: 110–113.

Lassoued, R., Macall, D.M., Hesseln, H. et al. (2019). Benefits of genome-edited crops: expert opinion. *Transgenic Research* 28 (2): 247–256.

Mintert, J.R., Widmar, D., Langemeier, M., Boehlje, M., and Erickson, B., 2016. The Challenges of Precision Agriculture: Is Big Data the Answer?. 2016 Annual Meeting, February 6–9, 2016, San Antonio, Texas 230057, Southern Agricultural Economics Association. https://doi.org/10.22004/ag.econ.230057.

Mohamed, E.S., Belal, A.A., Abd-Elmabod, S.K. et al. (2021). Smart farming for improving agricultural management. *The Egyptian Journal of Remote Sensing and Space Science* 24 (3): 971–981.

Morota, G., Ventura, R.V., Silva, F.F. et al. (2018). Big data analytics and precision animal agriculture symposium: machine learning and data mining advance predictive big data analysis in precision animal agriculture. *Journal of Animal Science* 96 (4): 1540–1550.

NASS, U., 2012. Census of Agriculture Highlights: Farm Economics, 2014.

OECD (Organization for Economic Co-operation and Development) (2022). *AI in agriculture: global trade and regulatory perspectives*. OECD Publishing.

O'Grady, M.J. and O'Hare, G.M. (2017). Modelling the smart farm. *Information Processing in Agriculture* 4 (3): 179–187.

Paul, P.K., Kumar, P.K., Sinha, R.R. et al. (2016). Cloud computing and virtualization in agricultural space: a knowledge survey. *Palgo Journal of Agriculture* 4 (2): 202–206.

Poppe, K., Wolfert, S., and Verdouw, C., 2014. How ICT is Changing the Nature of the Farm: a Research Agenda on the Economics of Big Data.

Priyatham, K., Chakraborty, S., Mouli Paul, S.S. et al. (2025). Genomics-assisted plant breeding for sustainable agriculture, environmental harmony, and global food security: a comprehensive review. *Plant Cell Biotechnology And Molecular Biology* 26 (1-2): 73–86. https://doi.org/10.56557/pcbmb/2025/v26i1-29102.

Qaim, M. (2009). The economics of genetically modified crops. *Annu. Rev. Resour. Econ.* 1 (1): 665–694.

Ribarics, P. (2016). Big Data and its impact on agriculture. *Ecocycles* 2 (1): 33–34.

Rosenheim, J.A. and Gratton, C. (2017). Ecoinformatics (big data) for agricultural entomology: pitfalls, progress, and promise. *Annual Review of E ntomology* 62 (1): 399–417.

Ryan, M. (2020). Agricultural big data analytics and the ethics of power. *Journal of Agricultural and Environmental Ethics* 33: 49–69.

Ryan, M. (2023). The social and ethical impacts of artificial intelligence in agriculture: mapping the agricultural AI literature. *AI & SOCIETY* 38 (6): 2473–2485.

Ryan, M., Isakhanyan, G., and Tekinerdogan, B. (2023). An interdisciplinary approach to artificial intelligence in agriculture. *NJAS: Impact in Agricultural and Life Sciences* 95 (1): 2168568.

Schönfeld, M.V., Heil, R., and Bittner, L. (2018). Big data on a farm—smart farming. *Big Data in Context* 1: 109–120. https://doi.org/10.1007/978-3-319-62461-7_12.

Sharma, A. and Tripathi, P. (2021). Artificial Intelligence in Agriculture: A Literature Survey. *International Journal of All Research Education and Scientific Methods* 9: 510–513.

Shukla, S. (2024). Principles governing ethical development and deployment of AI. *Journal of Artificial Intelligence Ethics* 7 (1): 45–60.

Sishodia, R.P., Ray, R.L., and Singh, S.K. (2020). Applications of remote sensing in precision agriculture: a review. *Remote Sensing* 12 (19): 3136.

Slimani, H., El Mhamdi, J., and Jilbab, A. (2024). "Assessing the advancement of artificial intelligence and drones' integration in agriculture through a bibliometric study". *International Journal of Electrical & Computer Engineering* 14 (1): 2088–8708.

Sonka, S. and Cheng, Y.T., 2015. Big Data in farming: Why Matters!. *farmdoc daily*, University of Illinois at Urbana-Champaign, Department of Agricultural and Consumer Economics, vol. 5, pp. 1–3, November. Available at: https://doi.org/10.22004/ag.econ.229523.

Stilgoe, J. (2020). *Who's Driving Innovation. New Technologies and the Collaborative State*, 1–74. Cham: Palgrave Macmillan https://doi.org/10.1007/978-3-030-32320-2.

Sykuta, M.E. (2016). Big data in agriculture: property rights, privacy and competition in ag data services. *International Food and Agribusiness Management Review* 19: 57–74.

Talavera, J.M., Tobón, L.E., Gómez, J.A. et al. (2017). Review of IoT applications in agro-industrial and environmental fields. *Computers and Electronics in Agriculture* 142: 283–297.

Taylor, L. and Broeders, D. (2015). In the name of development: power, profit and the datafication of the global south. *Geoforum* 64: 229–237.

Tsesis, A. (2019). Data subjects' privacy rights: regulation of personal data retention and erasure. *U. Colo. L. Rev.* 90: 593.

Tzounis, A., Katsoulas, N., Bartzanas, T., and Kittas, C. (2017). Internet of Things in agriculture, recent advances and future challenges. *Biosystems Engineering* 164: 31–48.

United States Department of Agriculture, National Agricultural Statistics Service. (2014). *Farms and land in farms 2014 summary* (pp. 2–3).

Upadhyay, S.K. (2026). *Designer Crops: Implications of Artificial Intelligence and Modern Biotechnological Methods*. In: Elsevier Science.

USDA (2022). *How GMOs Are Regulated in the United States*. U.S. Food and Drug Administration. https://www.fda.gov/food/agricultural-biotechnology/how-gmos-are-regulated-united-states.

Veale, M. and Zuiderveen Borgesius, F. (2021). Demystifying the draft EU artificial intelligence act—analysing the good, the bad, and the unclear elements of the proposed approach. *Computer Law Review International* 22 (4): 97–112.

Vijayakumar, S., Kumar, R.M., Choudhary, A.K. et al. (2022). Artificial intelligence (AI) and its application in agriculture. *Chronicle of Bioresource Management* 6: 025–031.

Warren, M. (2002). Digital divides and the adoption of information and communication technologies in the UK farm sector. *International Journal of Information Technology and Management* 1 (4): 385–405.

Williams, H.L. (2016). Intellectual property rights and innovation: Evidence from health care markets. *Innovation Policy and the Economy* 16 (1): 53–87.

Wójcik-Gront, E., Zieniuk, B., and Pawełkowicz, M. (2024). Harnessing AI-powered genomic research for sustainable crop Improvement. *Agriculture* 14 (12): 2299.

Zhang, H., Wei, X., Zou, T. et al. (2015). Agriculture big data: research status, challenges and countermeasures. In: *Computer and Computing Technologies in Agriculture VIII: 8th IFIP WG 5.14 International Conference, CCTA 2014, Beijing, China, September 16-19, 2014*, Revised Selected Papers 8, 137–143. Springer International Publishing.

20

Challenges and Opportunities Using AI Toward Crop Improvement

Akhouri Nishant Bhanu, Bangar Vaibhav, and Saurav Raj

Rajmata Vijayaraje Scindia Krishi Vishwa Vidyalaya, Department of Genetics and Plant Breeding, R.A.K. College of Agriculture, Ichhawar Road, Sehore, 466001, Madhya Pradesh, India

20.1 Introduction

Agriculture has long been a cornerstone of human civilization, sustaining populations and driving economic development. However, as the global population is expected to reach nearly 10 billion by 2050, ensuring food security while maintaining environmental sustainability presents a major challenge (FAO 2021). Traditional farming methods, while instrumental in past agricultural success, are increasingly inadequate in addressing modern challenges such as climate change, soil degradation, and pest infestations (Tilman et al. 2011). These evolving concerns necessitate the adoption of innovative solutions to optimize crop yields and resource efficiency.

Artificial intelligence (AI) has emerged as a powerful tool in modern agriculture, offering advanced capabilities in data analysis, decision-making, and automation (Figure 20.1). AI-driven applications, including machine learning (ML), deep learning (DL), and big data analytics, are transforming various aspects of crop improvement, from precision agriculture to breeding and pest management (Jha et al. 2019). By leveraging AI, farmers and researchers can enhance productivity, reduce reliance on chemical inputs, and develop climate-resilient crops, ultimately promoting sustainable agricultural practices (Sharma et al. 2022).

The image illustrates the AI-Driven Workflow for Crop Improvement, highlighting how AI processes agricultural data to enhance breeding and farming decisions. It consists of three stages: (1) Input Data, including weather, soil health, and genetic traits; (2) AI Processing, where data analysis, modeling, and recommendations are generated; and (3) Final Output, leading to optimized crop varieties, predictive insights, and informed farming decisions. This workflow demonstrates AI's role in precision agriculture, enabling climate-resilient crop development and sustainable farming practices.

One of the most significant applications of AI in agriculture is predictive analytics, which utilizes real-time and historical data to forecast weather patterns, pest outbreaks, and disease infestations. AI-based systems, such as convolutional neural networks, can analyze plant health through image recognition, enabling early detection of diseases and facilitating

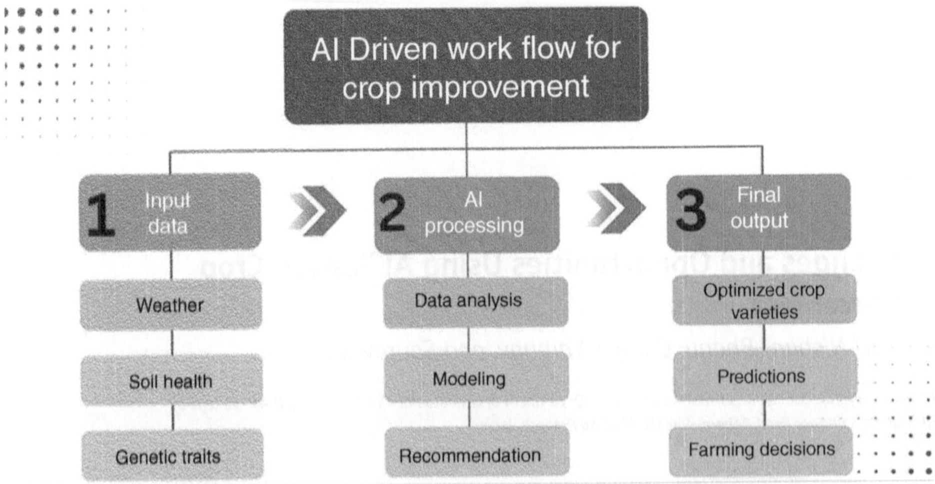

Figure 20.1 AI-driven workflow for crop improvement.

timely interventions (Mohanty et al. 2016). Furthermore, AI is accelerating genomic selection in plant breeding, allowing researchers to identify desirable traits with greater precision and efficiency, leading to the development of high-yield and stress-resistant cultivars (Crossa et al. 2017; Upadhyay 2026).

Another transformative aspect of AI in agriculture is its role in optimizing resource use. AI-powered precision farming systems integrate data from soil sensors, satellite imagery, and weather forecasts to guide irrigation and fertilization decisions. This not only enhances crop yields but also minimizes water and nutrient wastage, contributing to sustainable agricultural practices (Gebbers and Adamchuk 2010). Additionally, AI-driven automation, such as autonomous tractors and robotic weed control, is reducing labor dependency and improving overall efficiency in farm operations (Wang et al. 2024).

Despite its vast potential, the adoption of AI in agriculture is not without challenges. Data availability and quality remain significant barriers, as AI models require large, diverse datasets for effective implementation (Chen et al. 2023). High implementation costs and the need for advanced digital infrastructure further limit accessibility, particularly for smallholder farmers in developing regions (Gwagwa et al. 2021). Ethical concerns regarding data privacy and ownership also need to be addressed to ensure responsible AI deployment in agriculture (Ryan et al. 2023).

This chapter explores the role of AI in crop improvement, examining its applications in precision agriculture, pest and disease management, and climate resilience. It also discusses the challenges associated with AI adoption and highlights future directions for integrating AI into mainstream agricultural practices. By bridging the gap between technological advancements and practical farming applications, AI has the potential to revolutionize agriculture, making it more efficient, sustainable, and resilient to future challenges.

20.2 Opportunities of Using AI in Crop Improvement

In the rapidly evolving landscape of agriculture, AI has emerged as a transformative force, offering unparalleled opportunities to enhance productivity, sustainability, and resilience. By leveraging data-driven technologies, AI optimizes various aspects of crop improvement, from precision agriculture to advanced breeding techniques. This section explores the diverse applications of AI in agriculture (Table 20.1), highlighting its potential to revolutionize farming practices and address global food security challenges.

20.2.1 Precision Agriculture and Smart Farming

Precision agriculture, driven by AI, is redefining traditional farming methods by integrating advanced data analytics, remote sensing, and automation. By utilizing data from satellite imagery, unmanned aerial vehicles (UAVs), soil sensors, and weather forecasts, AI-powered systems provide farmers with real-time insights that enable precise decision-making (Wang et al. 2024). This leads to optimized irrigation, fertilization, and pest management, reducing resource wastage while maximizing yield potential.

ML models analyze multispectral and hyperspectral images to assess plant health, soil moisture, and nutrient deficiencies. AI-driven drones equipped with advanced computer vision technology continuously monitor field conditions, identifying potential stress factors such as water scarcity or pest infestations with remarkable accuracy (Yu 2024). Moreover, AI-powered robotic systems are being employed for targeted weed control and precision spraying, minimizing pesticide use and promoting sustainable farming practices (Gebbers and Adamchuk 2010).

Table 20.1 Applications of AI in crop improvement.

AI application	Description	Benefits to crop improvement	References
Precision agriculture	Uses AI to optimize sowing, irrigation, and fertilization schedules based on real-time data.	Improved resource efficiency, reduced waste, and higher yields.	Gangwani (2024)
Genomic analysis	AI tools to analyze plant genomes and identify key genetic traits for breeding.	Faster development of high-yield, pest-resistant, and climate-resilient crops.	Chen et al. (2023)
Pest prediction and control	AI algorithms predict pest outbreaks and suggest timely interventions.	Reduced crop losses and improved pest management strategies.	Zhang and Wang (2024)
Climate resilience development	AI models identify traits for climate-resilient crop varieties.	Enhanced ability to withstand environmental stresses, ensuring food security.	Wang et al. (2024)

Table 20.2 AI-driven approaches in crop breeding.

AI-driven approach	Description	Impact on crop improvement	References
Genomic selection	AI analyzes large datasets of genetic information to predict genotype performance under different conditions.	Accelerates breeding cycles and enhances accuracy in trait selection.	Zhang and Wang (2024)
Marker-assisted breeding	Uses AI to identify specific genetic markers linked to desirable traits like drought tolerance or pest resistance.	Speeds up the development of high-yield and resilient crop varieties.	Chen et al. (2023)
Trait discovery	Machine learning algorithms detect complex patterns to identify traits such as disease resistance or heat tolerance.	Enables targeted breeding strategies for stress-resilient crops.	Wang et al. (2024)
Simulation of crop responses	AI models simulate how various genotypes respond to environmental factors, including drought, heat, and flooding.	Helps breeders develop climate-adaptive crop varieties.	Sharma et al. (2022)

20.2.2 AI in Crop Breeding

AI is playing an instrumental role in accelerating crop breeding processes, making it possible to develop high-yielding, climate-resilient, and disease-resistant crop varieties in a fraction of the time required by conventional methods (Table 20.2). Through advanced computational models, AI enhances the selection of desirable genetic traits, ensuring the rapid development of superior cultivars.

20.2.2.1 Genomic Selection

Genomic selection leverages AI to analyze vast datasets encompassing genetic and phenotypic information. ML algorithms, such as deep neural networks, facilitate precise predictions of genotype performance under diverse environmental conditions, thereby optimizing breeding programs (Zhang and Wang 2024). By integrating AI with genome-wide association studies and quantitative trait locus mapping, researchers can identify key genes associated with traits such as drought tolerance, pest resistance, and improved nutritional content. This accelerates breeding cycles, ensuring food security in the face of climatic uncertainties (Crossa et al. 2017).

20.2.2.2 Marker-Assisted Breeding

Marker-assisted breeding benefits significantly from AI-powered genomic analysis, which identifies genetic markers linked to desirable traits. Computational models analyze large-scale genomic datasets to detect beneficial alleles, facilitating precision breeding strategies (Chen et al. 2023). AI-driven bioinformatics tools further enhance genome editing techniques, such as CRISPR-Cas9, enabling precise modifications for improved crop performance (Jaganathan et al. 2018).

20.2.3 Pest and Disease Management

AI has revolutionized pest and disease management by enabling early detection and intervention through image recognition and predictive analytics. By analyzing high-resolution satellite imagery and real-time field sensor data, AI-driven systems can identify disease symptoms and pest infestations with unparalleled accuracy (Wang et al. 2024).

Applications like Plantix and DeepMind utilize DL algorithms to diagnose plant diseases from smartphone images, providing farmers with instant recommendations for treatment (Minhans et al., 2025). Additionally, AI models predict pest outbreaks by analyzing climatic, ecological, and entomological data, empowering farmers with timely preventive measures to minimize crop damage (Mahlein et al. 2013).

20.2.4 Predictive Analytics for Yield Optimization

Predictive analytics, powered by AI, is revolutionizing yield forecasting by integrating historical and real-time data from various agricultural inputs. AI-driven models analyze factors such as soil health, climate patterns, genetic traits, and past yield records to provide highly accurate yield predictions (Chen et al. 2023).

By leveraging this predictive capability, farmers can optimize planting schedules, adjust resource allocations, and adopt precision-driven harvesting techniques. Additionally, AI-powered market analysis tools assess supply chain trends, helping farmers make informed decisions regarding crop selection and distribution strategies (Tilman et al. 2011).

20.2.5 AI in Climate-Resilient Crop Development

The development of climate-resilient crops is a pressing necessity in the wake of global climate change. AI-driven research integrates environmental modeling with genetic advancements to enhance crop adaptability to extreme conditions such as drought, heat stress, and flooding (Sharma et al. 2022).

AI models analyze vast datasets, including climate projections, soil health indicators, and genomic sequences, to identify genes associated with resilience. Through predictive breeding simulations, AI helps scientists select crop genotypes best suited for future climatic scenarios. Advanced phenotyping platforms utilizing infrared imaging and LiDAR sensors further aid in evaluating plant stress responses, facilitating the rapid identification of resilient crop varieties (Lobell et al. 2011).

20.2.6 Enhancing Resource Efficiency with AI

AI-driven technologies are enhancing resource efficiency by optimizing water and nutrient management strategies. Smart irrigation systems, powered by AI and IoT-enabled sensors, monitor soil moisture levels, weather forecasts, and crop water requirements in real time. This allows precise irrigation scheduling, significantly reducing water wastage (Wang et al. 2024).

Similarly, AI-driven nutrient management systems analyze soil composition and crop nutritional needs to develop customized fertilization plans. This reduces excessive fertilizer application, minimizing nutrient runoff and mitigating environmental impacts (Chen et al. 2023). Moreover, AI-powered precision agriculture techniques enable site-specific application of agricultural inputs, ensuring maximum efficiency and cost-effectiveness.

20.3 Challenges for Using AI in Crop Improvement

AI has immense potential to revolutionize crop improvement, particularly in developing countries like India. While developed nations are at the forefront of AI integration in agriculture, they too have not fully harnessed its capabilities. Despite the promising applications of AI, several key challenges hinder its widespread adoption and efficacy in crop improvement programs. These challenges range from data limitations and high costs to ethical concerns and infrastructural constraints, all of which must be addressed to realize AI's full potential in agricultural research and breeding.

20.3.1 Data Limitations and Quality Issues

The availability and quality of agricultural data play a crucial role in leveraging AI for crop improvement. While advances in genomics and phenomics have resulted in the generation of vast datasets, several challenges still limit their usability in AI applications. One of the primary issues is the limited availability of data for certain crops and traits. While major crops such as wheat, rice, and maize have extensive datasets, many understudied crops lack sufficient data for effective AI model training. Additionally, traits that are highly environment-dependent, such as drought resistance and pest tolerance, often have insufficiently detailed data, making it challenging to develop precise AI models (Tong and Nikoloski 2021).

Another significant challenge lies in the heterogeneity of data formats and standards. Data generated from diverse sources, including field trials, genomic sequencing, remote sensing, and high-throughput phenotyping, often vary in format, quality, and experimental protocols. This lack of uniformity complicates data integration and necessitates extensive preprocessing and standardization before AI models can effectively utilize the information. Furthermore, the process of collecting high-quality, labeled data for AI training—particularly for complex phenotyping tasks—is both expensive and time-consuming. Advanced phenotyping methods, such as hyperspectral imaging and automated field sensors, require significant investment and expertise (Wäldchen and Mäder 2018). To overcome these issues, there is a pressing need for collaborative open-data initiatives that promote the generation, curation, and sharing of high-quality datasets, alongside investments in standardized data collection protocols and digital infrastructure.

20.3.2 High Implementation Costs and Accessibility

The financial implications of AI adoption in crop improvement present a considerable barrier. AI-driven breeding platforms require substantial investments in computational infrastructure, software development, data acquisition, and skilled human resources. One of the major cost drivers is the expensive data acquisition methods required for AI applications. High-resolution satellite imagery, UAV-based (unmanned aerial vehicle) remote sensing, and precision phenotyping technologies significantly increase operational costs, making AI-driven breeding less accessible to small-scale research institutions and agricultural enterprises (Chen et al. 2023).

Moreover, the high initial and recurring costs associated with AI model development, maintenance, and periodic upgrades present an ongoing financial challenge. AI systems require continuous investment in hardware, cloud computing services, and skilled personnel for algorithm training and data processing (Gwagwa et al. 2021). Another concern is the uncertainty regarding the return on investment. Although AI can improve breeding efficiency and accelerate genetic gains, its benefits, such as increased crop yields and reduced breeding cycles, often take several years to materialize. Predicting the financial impact of AI-driven applications, such as yield forecasting and harvest risk management, adds further complexity to securing investment (Ryan et al. 2023). To mitigate these cost barriers and improve accessibility, alternative funding models, including public-private partnerships, government subsidies, and AI-as-a-service subscription-based platforms, should be explored (Ahmad et al. 2024).

20.3.3 Ethical Concerns and Privacy Issues in Data Collection

The rapid expansion of AI in agriculture raises several ethical and privacy concerns, particularly regarding data ownership, algorithmic transparency, and regulatory oversight. One of the primary issues is regulatory uncertainty. Current agricultural policies and regulations primarily focus on genetically modified organisms and conventional breeding practices, leaving AI-driven breeding techniques in a legal grey area. This lack of clear governance frameworks creates ambiguity and slows down AI adoption in plant breeding programs.

Algorithmic transparency and accountability are also crucial challenges. Many AI models, particularly DL algorithms, operate as "black boxes," making it difficult to interpret their decision-making processes. This opacity raises concerns regarding bias, reliability, and the unintended consequences of AI-driven breeding decisions (Kamilaris and Prenafeta-Boldú 2018). Additionally, the proprietary nature of AI algorithms and datasets used in crop improvement creates intellectual property concerns. Breeding companies and research institutions must balance data protection with collaboration and innovation to ensure fair and ethical AI development.

International regulatory disparities further complicate AI adoption in agriculture. Different countries have varying levels of oversight and governance concerning AI applications, creating barriers to global collaboration and data sharing. Establishing harmonized international regulatory frameworks is essential for enabling seamless AI-driven agricultural research and trade. Given the evolving nature of AI, adaptive policies that support responsible innovation while addressing public concerns about data privacy, equity, and sustainability must be developed.

20.3.4 Integration Challenges with Traditional Practices

Cultural and behavioral factors play a significant role in the adoption of AI in agriculture. Many farming communities have deep-rooted traditional practices and may resist transitioning to AI-driven decision-making systems. One major challenge is the resistance to change among farmers who have relied on conventional breeding methods and personal experience for generations. AI-generated recommendations may be met with skepticism, especially if farmers do not fully understand how the models work or if they perceive AI as undermining their expertise.

Lack of trust in AI technologies is another barrier to widespread adoption. Many farmers prefer relying on empirical knowledge and field observations rather than data-driven insights. To overcome this skepticism, it is essential to demonstrate AI's reliability, accuracy, and tangible benefits through pilot projects, farmer training programs, and participatory research approaches. Additionally, AI integration must align with local values and priorities. In some communities, concerns about environmental sustainability, biodiversity conservation, and social equity may influence attitudes toward AI adoption. Engaging with stakeholders through culturally sensitive outreach and participatory decision-making is key to ensuring successful AI implementation in crop improvement programs.

20.3.5 Dependency on Advanced Infrastructure and Connectivity

The deployment of AI in crop improvement is heavily reliant on robust digital infrastructure, which remains a significant limitation in many agricultural regions. One of the primary challenges is the limited internet connectivity in rural and developing areas. AI-powered tools, such as precision agriculture platforms and real-time data analytics systems, require stable internet access for seamless operation. However, poor network coverage and bandwidth limitations hinder the effective use of AI-driven technologies in many farming communities (Basso and Antle 2020).

Computational power constraints present another challenge. AI model training requires high-performance computing resources, which may not be available to smaller research institutions and breeding programs. Additionally, AI applications generate vast amounts of data that require efficient storage, processing, and management systems. Many agricultural research centers lack the necessary infrastructure to handle large-scale AI data processing, further limiting adoption.

Addressing these infrastructure gaps requires targeted investments in expanding internet access, enhancing computing capabilities, and developing affordable cloud-based AI solutions. Governments and private-sector stakeholders must collaborate to bridge the digital divide and ensure that AI-driven crop improvement technologies are accessible to all agricultural researchers and farmers.

20.4 SWOT Analysis of AI Technologies in Crop Improvement

AI has emerged as a transformative tool in crop improvement, offering innovative solutions to enhance breeding efficiency, trait selection, and stress resilience. However, its implementation presents various strengths, weaknesses, opportunities, and threats. A SWOT analysis (Table 20.3) provides a comprehensive evaluation of AI's role in crop improvement, enabling stakeholders to harness its potential while addressing associated challenges.

Table 20.3 SWOT analysis of AI technologies in crop improvement.

Strengths	**Weaknesses**
AI enables rapid analysis of large genomic and phenotypic datasets, facilitating data-driven breeding decisions and accelerating crop improvement.	The effectiveness of AI in crop improvement depends on high-quality and diverse datasets, which may be limited in certain crops or regions.
Machine learning algorithms enhance genomic selection and marker-assisted breeding by identifying key traits for improved yield, resilience, and disease resistance.	Implementing AI in breeding programs requires specialized expertise, posing challenges for researchers and institutions with limited technical capacity.
AI-driven predictive models improve breeding efficiency by forecasting plant performance under different environmental conditions.	High computational costs and infrastructure requirements may limit accessibility, particularly for smaller breeding programs and resource-constrained institutions.
AI automates phenotyping through image-based analysis, reducing the time and labor required for trait evaluation.	Ethical concerns regarding data privacy, ownership of genetic data, and potential biases in AI-driven breeding models require careful regulation.
AI enables precision gene editing by identifying optimal gene targets for CRISPR-based modifications, expediting the development of climate-resilient and high-yielding crop varieties.	Adoption of AI-driven breeding methods may be slow due to regulatory challenges, especially for gene-edited crops.
Opportunities	Threats
AI can accelerate the development of climate-resilient crops by integrating climate modeling with genomic insights, ensuring food security amid changing environmental conditions.	AI-driven breeding programs rely on large-scale genetic data, increasing concerns about data security, potential misuse, and unauthorized access.
AI enhances collaboration between researchers, breeders, and farmers by providing real-time insights into trait performance and agronomic adaptation.	Algorithmic biases in AI models can lead to unintended consequences, such as overlooking locally adapted crop varieties in favor of widely studied ones.
AI-powered digital platforms facilitate open-source breeding, allowing wider access to innovative genetic resources and decision-support tools.	Overreliance on AI without human validation may lead to errors in trait selection and genetic predictions, affecting breeding outcomes.
AI enables early detection of crop diseases and stress through automated image recognition and sensor-based monitoring, improving crop health and productivity.	The high cost of AI-based breeding technologies may create disparities, making them inaccessible to smallholder farmers and public breeding programs.
Increasing recognition of AI's potential in crop improvement is driving investments in research, fostering innovation in smart breeding techniques.	Resistance to AI adoption due to traditional breeding preferences, lack of awareness, and regulatory constraints may hinder widespread implementation.

20.5 Future Directions for AI in Crop Improvement

The future of AI in crop improvement hinges on overcoming existing challenges while fostering innovation and inclusivity. To fully harness AI's potential, researchers, policymakers, and agribusinesses must focus on developing scalable, efficient, and ethically responsible AI applications that address both large-scale agricultural operations and smallholder farming systems.

A critical area of advancement lies in standardized and inclusive data ecosystems. Current AI models for crop improvement often struggle due to fragmented and biased datasets. The development of open-access, high-quality datasets, integrating multi-environment trials and diverse germplasm collections, is essential for creating robust AI models (Kassem, 2025). Future AI frameworks must incorporate automated data harmonization techniques using blockchain and federated learning to ensure data integrity, security, and accessibility across research institutions and farming communities (Zhao 2019).

AI-driven predictive breeding and trait discovery will see significant advancements with the integration of generative AI and multi-omics data analytics. Instead of solely relying on conventional genomic selection, future models will predict complex trait interactions by integrating transcriptomics, proteomics, and metabolomics data, enabling precise crop trait enhancements (Shankar et al. 2024). Moreover, the rise of AI-enhanced synthetic biology may lead to the development of customized plant varieties with improved photosynthetic efficiency, stress resilience, and nutrient-use efficiency (Zhang et al. 2024).

For AI to have a transformative impact on smallholder farmers, cost-effective and easy-to-use AI solutions must be prioritized. Governments and agritech firms should focus on mobile-based AI advisory systems that function even in low-connectivity regions. AI-driven voice assistants in regional languages could provide real-time agronomic recommendations, enabling farmers to make informed decisions on planting, irrigation, and pest control (Chukwuma et al. 2024). Additionally, edge computing-based AI models, which run locally on low-power devices instead of relying on cloud infrastructure, can reduce dependency on stable internet connections, making AI-driven decision support systems more accessible (Padhiary et al. 2025).

Advancements in AI-powered climate-smart breeding will be essential to mitigate the impacts of climate change on crop productivity. Future AI models will integrate real-time climate data with phenotypic and genomic information to predict plant responses to changing weather conditions, aiding in the rapid development of climate-resilient cultivars (Boros et al. 2024). AI-driven simulations may enable researchers to virtually test new crop varieties under different environmental conditions before field trials, significantly reducing breeding timelines and costs.

The ethical and regulatory landscape surrounding AI in crop improvement must evolve to ensure responsible and equitable deployment. Clear policies on data ownership, privacy, and intellectual property rights will be crucial in preventing monopolization of AI-driven agricultural innovations (Ryan et al. 2023). Future regulations should promote AI democratization, ensuring that farmers, particularly in developing regions, can freely access AI-powered tools without being restricted by proprietary technologies (Zhao 2019).

Finally, AI's role in sustainable agricultural intensification will expand through circular AI-driven agroecosystems. AI models will optimize on-farm resource recycling, predicting the best ways to utilize agricultural waste for biofertilizer production, water conservation, and carbon sequestration. AI-powered precision carbon farming models will help farmers participate in global carbon credit markets, incentivizing sustainable practices while ensuring food security (Zhang et al. 2024).

By integrating these innovative AI applications with traditional breeding expertise and participatory farmer engagement, the future of AI in crop improvement holds immense potential to drive productivity, sustainability, and resilience in global agriculture.

20.6 Conclusion

The integration of AI in crop improvement marks a paradigm shift in modern agriculture, offering unprecedented opportunities for precision farming, genetic advancements, and resource optimization. As highlighted in this chapter, AI-driven innovations such as genomic selection, marker-assisted breeding, and predictive analytics are revolutionizing the way crops are developed and managed, ensuring higher yields, enhanced disease resistance, and resilience to climate change (Sharma 2024). Moreover, AI's ability to process vast datasets, analyze environmental variables, and provide real-time recommendations has empowered farmers and researchers to make more informed decisions, paving the way for sustainable agricultural practices (Padhiary et al. 2025).

However, the road to fully harnessing AI's potential is fraught with challenges. Issues such as data accessibility, high infrastructure costs, ethical concerns, and technological biases must be addressed to ensure equitable adoption, particularly among smallholder farmers (Devasia et al., 2024). The future of AI in agriculture depends on a collaborative ecosystem where governments, research institutions, private enterprises, and farmers work together to create scalable, inclusive, and transparent AI-driven solutions.

Looking ahead, the focus should shift toward developing AI models that are adaptable to diverse agro-climatic conditions, ensuring their applicability across different regions and farming scales. Moreover, investments in open-source AI tools, farmer-centric training programs, and ethical AI frameworks will be critical in democratizing access to these technologies. By bridging the digital divide and fostering AI literacy, policymakers and scientists can create a future where AI serves as a catalyst for global food security and climate resilience (Sharma 2024).

In conclusion, the true success of AI in crop improvement will not only be measured by yield gains but also by its ability to create a more resilient, inclusive, and sustainable agricultural landscape. The next frontier lies in harnessing AI's power while ensuring it remains a tool for empowerment rather than exclusion. If stakeholders embrace a forward-thinking approach—one that integrates technological innovation with ethical considerations and farmer inclusivity—AI will not only transform crop improvement but also redefine the future of global agriculture.

References

Ahmad, A., Liew, A.X., Venturini, F. et al. (2024). AI can empower agriculture for global food security: challenges and prospects in developing nations. *Frontiers in Artificial Intelligence* 7: 1328530.

Basso, B. and Antle, J. (2020). Digital agriculture to design sustainable agricultural systems. *Nature Sustainability* 3 (4): 254–256.

Boros, A., Szólik, E., Desalegn, G., and Tőzsér, D. (2024). A systematic review of opportunities and limitations of innovative practices in sustainable agriculture. *Agronomy* 15 (1): 76.

Chen, T., Lv, L., Wang, D. et al. (2023). Empowering agrifood system with artificial intelligence: a survey of the progress, challenges and opportunities. *arXiv preprint* arXiv:2305.01899.

Chukwuma, U., Gebremedhin, K.G., and Uyeh, D.D. (2024). Imagining AI-driven decision making for managing farming in developing and emerging economies. *Computers and Electronics in Agriculture* 221: 108946.

Crossa, J., Pérez-Rodríguez, P., Cuevas, J. et al. (2017). Genomic selection in plant breeding: methods, models, and perspectives. *Trends in Plant Science* 22 (11): 961–975.

Devasia, J.V., Deepanramkumar, P., Helensharmila, A., and Yenduri, G. (2024). Ethical Considerations and Social Implications. In: *Sustainable Farming through Machine Learning*, 267–279. CRC Press.

FAO (2021). *The State of Food and Agriculture 2021: Making agrifood systems more resilient to shocks and stresses*. Rome: FAO. https://doi.org/10.4060/cb4476en.

Gangwani, N. (2024). AI-driven precision agriculture: Optimizing crop yield and resource efficiency. *International Journal for Multidisciplinary Research* 6 (6): 1–10. https://www.ijfmr.com/papers/2024/6/29913.pd.

Gebbers, R. and Adamchuk, V.I. (2010). Precision agriculture and food security. *Science* 327 (5967): 828–831.

Gwagwa, A., Kazim, E., Kachidza, P. et al. (2021). Road map for research on responsible artificial intelligence for development (AI4D) in African countries: the case study of agriculture. *Patterns* 2 (12).

Jaganathan, D., Ramasamy, K., Sellamuthu, G. et al. (2018). CRISPR for crop improvement: an update review. *Frontiers in Plant Science* 9: 985.

Jha, K., Doshi, A., Patel, P., and Shah, M. (2019). A comprehensive review on automation in agriculture using artificial intelligence. *Artificial Intelligence in Agriculture* 2: 1–12.

Kamilaris, A. and Prenafeta-Boldú, F.X. (2018). Deep learning in agriculture: a survey. *Computers and Electronics in Agriculture* 147: 70–90.

Kassem, M.A. (2025). Harnessing Artificial Intelligence and Machine Learning for Identifying Quantitative Trait Loci (QTL) Associated with Seed Quality Traits in Crops. *Plants* 14 (11): 1727.

Lobell, D.B., Schlenker, W., and Costa-Roberts, J. (2011). Climate trends and global crop production since 1980. *Science* 333 (6042): 616–620.

Mahlein, A.K., Rumpf, T., Welke, P. et al. (2013). Development of spectral indices for detecting and identifying plant diseases. *Remote Sensing of Environment* 128: 21–30.

Minhans, K., Sharma, S., Sheikh, I. et al. (2025). Artificial intelligence and plant disease management: an agro-innovative approach. *Journal of Phytopathology* 173 (3): e70084.

Mohanty, S.P., Hughes, D.P., and Salathé, M. (2016). Using deep learning for image-based plant disease detection. *Frontiers in Plant Science* 7: 215232.

Padhiary, M., Hoque, A., Prasad, G. et al. (2025). Precision agriculture and AI-driven resource optimization for sustainable land and resource management. In: *Smart Water Technology for Sustainable Management in Modern Cities* (ed. J. Ruiz-Vanoye and O. Diaz-Parra), 197–232. IGI Global Scientific Publishing.

Ryan, M., Isakhanyan, G., and Tekinerdogan, B. (2023). An interdisciplinary approach to artificial intelligence in agriculture. *NJAS: Impact in Agricultural and Life Sciences* 95 (1): 2168568.

Shankar, J., Verma, P., and Shah, M.P. (2024). *Microbial Approaches for Sustainable Green Technologies*. CRC Press.

Sharma, R. (2024). Revolutionizing agriculture: AI-driven smart farming analytics platform. *International Journal of Research in Computer Applications and Information Technology (IJRCAIT)* 7 (2): 1056–1068.

Sharma, S., Partap, A., Balaguer, M. A. D. L. et al. (2022). DeepG2P: fusing multi-modal data to improve crop production. *arXiv preprint* arXiv:2211.05986.

Tilman, D., Balzer, C., Hill, J., and Befort, B.L. (2011). Global food demand and the sustainable intensification of agriculture. *Proceedings of the National Academy of Sciences* 108 (50): 20260–20264.

Tong, H. and Nikoloski, Z. (2021). Machine learning approaches for crop improvement: leveraging phenotypic and genotypic big data. *Journal of Plant Physiology* 257: 153354.

Upadhyay, S. K. (2026). *Designer Crops: Implications of Artificial Intelligence and Modern Biotechnological Methods*. In: Elsevier Science.

Wäldchen, J. and Mäder, P. (2018). Machine learning for image-based species identification. *Methods in Ecology and Evolution* 9 (11): 2216–2225.

Wang, G., Xuan, J., Wang, P. et al. (2024). Lstm autoencoder-based deep neural networks for barley genotype-to-phenotype prediction. In *Australasian Joint Conference on Artificial Intelligence* (pp. 342–353). Singapore: Springer Nature Singapore.

Zhang, D., Xu, F., Wang, F. et al. (2024). Synthetic biology and artificial intelligence in crop improvement. *Plant Communications* 6 (2).

Zhao, W. (2019). Blockchain technology: development and prospects. *National Science Review* 6 (2): 369–373.

Index

C

d

e

h

i

Printed and bound by CPI Group (UK) Ltd, Croydon, CR0 4YY
17/09/2026
14964300-0004